VIBRATIONAL SPECTRA OF ORGANOMETALLIC COMPOUNDS

VIBRATIONAL SPECTRA OF ORGANOMETALLIC COMPOUNDS

EDWARD MASLOWSKY, Jr.
Loras College
Dubuque, Iowa

A Wiley-Interscience Publication

JOHN WILEY & SONS

New York • London • Sydney • Toronto

Library of Congress Cataloging in Publication Data:

Maslowsky, Edward.
Vibrational spectra of organometallic compounds.

"A Wiley-Interscience publication."
Includes index.
1. Organometallic compounds—Spectra. 2. Vibrational spectra. 3. Spectrum, Infra-red. 4. Raman spectroscopy. I. Title.

QC463.07M35 547'.05 76-18694
ISBN 0-471-58026-0

Printed in the United States of America

10 9 8 7 6 5 4 3 2 1

TO MY PARENTS

Preface

Several books are available that deal with the theory of infrared and Raman spectroscopy, and the application of these techniques to the study of organic, inorganic, or coordination compounds. No comprehensive treatment has been made, however, of the infrared and Raman spectra of organometallic compounds. The Specialists Reports of the Chemical Society of London have somewhat filled this void, but these are annual reviews and do not give a complete general survey of the reported literature. The reviews that have appeared of the spectra of organometallic compounds in general or of specific types of compounds have been by their very nature limited in treatment.

The present book is intended as a comprehensive review of the literature dealing with the infrared and Raman spectra of organometallic compounds. Included are all compounds that contain a direct metal-carbon interaction. Excluded are those compounds that contain carbonyl or cyano ligands, or both, unless other types of metal-carbon interactions are also found in these compounds. The definition used of a metal is rather broad in that it includes the elements boron, silicon, germanium, phosphorus, arsenic, antimony, selenium, and tellurium in addition to those elements more generally defined as metals. The literature reviewed includes that which was available as of the end of April, 1976.

The book has been divided into three chapters. Chapter I deals with those compounds where the organic ligand is a saturated alkyl group. Chapter II is limited to compounds where the organic ligand is unsaturated and noncyclic. Chapter III includes organic ligands that are unsaturated and cyclic.

The completion of this book has been aided by Dr. R. G. Goel and Dr. C. V. Senoff of the University of Guelph and Dr. T. G. Spiro of Princeton University, who provided an atmosphere that allowed me to begin this work while a postdoctoral fellow under their direction. The availability of excellent library facilities at the University of Guelph, Princeton University, the University of Rochester, and the John Crerar Library were also very much appreciated. A special note of thanks is also due the James C. Pe family of Quezon City and the Gregorio E. Uymatiao family of Dumaguete City, who provided a leisurely setting in the Philippines this summer that allowed me to revise the first draft, as well as to Loras College, which has provided me with the freedom to complete this work. I also want to thank Professor K. Nakamoto of Marquette University for reading the entire text and providing suggestions for improvements, and my wife Gerry for aiding in the typing as well as for many nights of extra duty babysitting our son J. P. while I searched for typing errors.

Edward Maslowsky, Jr.

Dubuque, Iowa
May, 1976

Contents

VIBRATIONAL SPECTRA OF ORGANOMETALLIC COMPOUNDS

I

Alkyl Organometallic Derivatives

A. METHYL COMPOUNDS

1. Monomeric $(CH_3)_nM$ Complexes

One of the earliest vibrational studies for an organometallic compound was that of Venkateswaren in 1930 (1). This work included a report of the Raman spectrum of $(CH_3)_2Zn$. Since that study, vibrational data have been reported for neutral, covalent, methyl compounds of nearly every main group metal and metalloid.

For CH_3M, the vibrational modes illustrated in Figure 1.1 are expected. Table 1.1 summarizes the frequency ranges found for these modes in methyl and deuterated methyl derivatives, including those complexes with other ligands bonded to the metal atom. The frequency range for the metal-carbon stretching mode, $\nu(MC)$, is not included in Table 1.1 since it will be discussed in conjunction with the other metal-carbon skeletal modes.

Since in most organometallic methyl compounds, more than one methyl group is bonded to the metal atom, more than one peak can often be observed in the regions expected for the modes illustrated for CH_3M. This arises from the possible in-phase and out-of-phase combinations of the modes expected for one methyl group. In the solid state, additional bands might also be observed because of distortion of the molecule from the

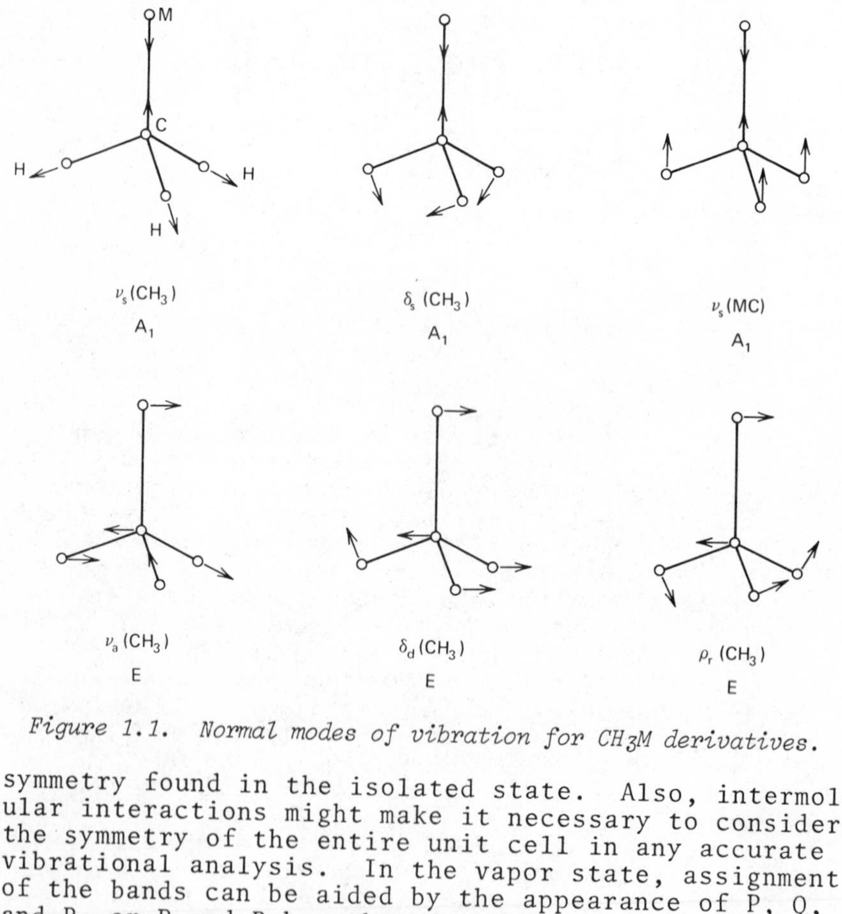

Figure 1.1. Normal modes of vibration for CH_3M derivatives.

symmetry found in the isolated state. Also, intermolecular interactions might make it necessary to consider the symmetry of the entire unit cell in any accurate vibrational analysis. In the vapor state, assignment of the bands can be aided by the appearance of P, Q, and R, or P and R branches for the various peaks; different profiles are expected for modes of different symmetry (2).

The band intensity for each mode varies with the number of methyl groups bonded to a metal atom, the phase in which the sample is being studied, and the metal being studied. Therefore, no absolute rule can be

TABLE 1.1
Frequency Ranges (cm^{-1}) for Terminal Methyl and Deuterated Methyl Modes

Mode	CH_3	CD_3
ν_a	3050-2810	2275-2035
ν_s	2950-2750	2175-2000
δ_d	1475-1300	1125-960
δ_s	1350-1100	1000-870
ρ_r	975-620	750-475

used to predict band intensity. In general, however, the rocking modes show relatively strong infrared intensity, weak Raman intensity, and percentage variation over the largest frequency range. Although the degenerate bending mode shows very strong intensity in the Raman spectra of $(CH_3)_2Se$ and $(CH_3)_2Te$ (3), the intensity is usually medium to weak and has not in fact been observed in either the infrared or Raman spectra of several compounds (4-10). The frequency variation of each mode within a given family of elements is illustrated in Figure 1.2. The $\nu(CH_3)$ modes are relatively stationary, showing little frequency variation from one metal to another. A detailed study has been made of the $\nu(CH_3)$ and $\nu(CD_3)$ modes for $(CH_3)_n(CD_3)_{4-n}M$ (n=0 to 4; M=C,Si,Ge,Sn,Pb) (25). The position of the $\delta_d(CH_3)$ modes is also relatively constant. The $\delta_s(CH_3)$ and $\rho_r(CH_3)$ modes, however, appear at progressively lower frequencies as the mass of the metal atom increases in a homologous series, the modes of $(CH_3)_2Hg$ being the only exceptions.

In addition to the C-H modes, $\nu(MC)$ and carbon-metal-carbon bending modes, $\delta(CMC)$, are expected. The number and activity of the skeletal modes depend on the number of methyl groups bonded to the metal atom. Table 1.2 summarizes the activity of the metal-carbon skeletal modes for methyl derivatives with various stoichiometries.

The advantage of using both infrared and Raman data for determining the symmetry about the central metal atom can be observed in Table 1.2. For example, the methyl groups (assumed to be single atoms) and metal atom for $(CH_3)_3M$ can be either coplanar or pyramidal. For a coplanar skeleton, the two $\nu(MC)$ modes and one $\delta(CMC)$ mode are Raman active, and one $\nu(MC)$

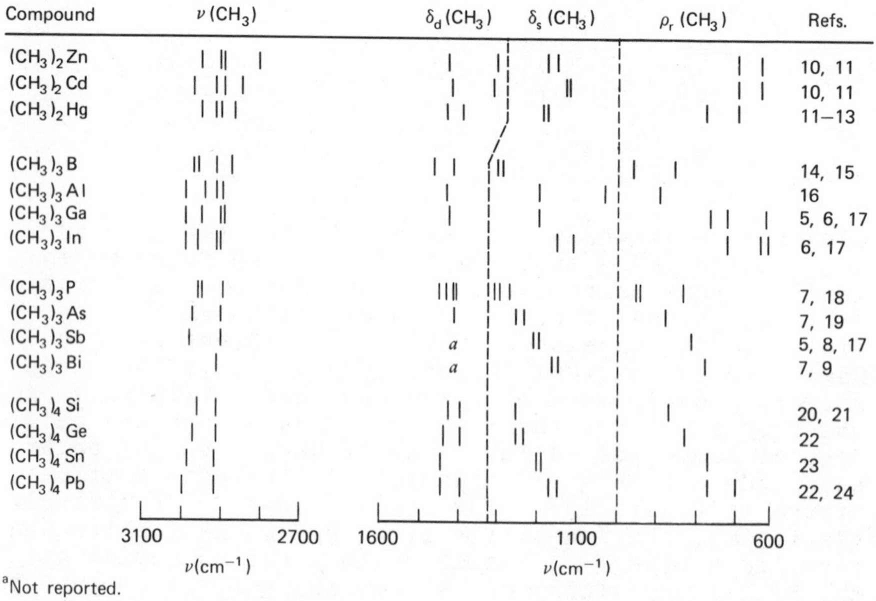

Figure 1.2. Frequency variation of the methyl modes for different groups of the periodic table.

mode and the two δ(CMC) modes are infrared active. For a pyramidal skeleton, the two ν(MC) and two δ(CMC) modes are both infrared and Raman active.

It is dangerous, however, to base structural conclusions on the absence of a band, for two reasons. The first is that the band of interest may accidentally overlap another band. For both $(CH_3)_3Sb$ and $(CH_3)_3Bi$, only one ν(MC) band is observed in the infrared spectrum (26). If only infrared data were available, one might be tempted to conclude that the skeletons of these compounds are planar, although they are not. The Raman spectra of each of these compounds also show one ν(MC) band. It has therefore been concluded that both ν(MC) modes are accidentally degenerate because of the relatively heavy mass of the metal atoms, which insulates the methyl groups from interacting with one

TABLE 1.2
Activity of the Metal-Carbon Skeletal Modes for $(CH_3)_nM$ Derivatives of Various Symmetries

Compound	Skeletal Structure	Stretching Mode Sym.	Stretching Mode Activity	Bending Mode Sym.	Bending Mode Activity
CH_3M		A_1	IR,R		
$(CH_3)_2M$	Linear	A_1	R	E'	IR
		A_2''	IR		
	Bent	A_1	IR,R	A_1	IR
$(CH_3)_3M$	Planar	A_1'	R	A_2''	IR
		E'	IR,R	E'	IR,R
	Pyramidal	A_1	IR,R	A_1	IR,R
		E	IR,R	E	IR,R
$(CH_3)_4M$	Tetrahedral	A_1	R	E	R
		T_2	IR,R	T_2	IR,R
$(CH_3)_5M$	Trig. bipyram.	$2A_1'$	R	A_2''	IR
		A_2''	IR	2E'	IR,R
		E'	IR,R	E''	R
	Square pyram.	$2A_1$	IR,R	A_1	IR,R
		B_1	R	B_1	R
		E	IR,R	B_2	R
				2E	IR,R
$(CH_3)_6M$	Octahedral	A_{1g}	R	T_{2g}	R
		E_g	R	T_{2g}	ia.
		T_{1u}	IR	T_{1u}	IR

another. As a rule, as the mass of the metal atom increases, the frequency separation of the expected ν(MC) modes and of the expected δ(CMC) modes decreases.

The other possible reason for the absence of an expected band is that the intensity may be too low to be detected. For example, the ν_s(MC) mode of a pyramidal $(CH_3)_3M$ molecule might not be observed in the infrared spectrum for this reason. To describe such an occurrence, in which the vibrational spectrum of a compound with a nonplanar skeleton appears to be consistent with planar selection rules, the compound is said to be "pseudoplanar." It is difficult to predict the angle required before the symmetric mode becomes observable in the infrared spectrum. This phenomenon, in fact, depends not only on the geometry of the compound but also on the nature of the atoms involved (26). In Table 1.3, the metal-carbon skeletal mode assignments are summarized for $(CH_3)_nM$ and $(CD_3)_nM$ compounds.

Some normal coordinate analyses have been reported for methyl derivatives. These calculations have used either the entire molecule or models in which the methyl groups are assumed to be single atoms of mass 15 or in which the effective mass is calculated for the methyl group (37). The calculations show the extent of mixing between the modes and provide values for the force constants of a molecule. The metal-carbon stretching force constant, K(MC), is especially useful since it has been empirically related to the strength of the metal-carbon bond. Within a given family of the main group elements, both the strength of the metal-carbon bond and K(MC) decrease as the mass of the metal atom increases. It is of more value to relate the bond strength to K(MC) rather than to the frequency of the ν(MC) mode since the latter depends not only on the strength of the metal-carbon bond but also on the mass of the metal atom and the extent to which this mode might be mixed with other modes.

Although solid methyllithium exists as a tetramer (38), the infrared spectrum of the monomer has been determined using an argon matrix (28). The $\nu(CH_3)$ modes of the monomer appear at among the lowest frequencies (2820 and 2780 cm^{-1}) reported for methyl derivatives. Dimethylberyllium normally exists in the associated form. In the unsaturated vapor, however, the monomer is mainly present. The infrared spectrum of the monomer and its deuterated analog have been reported (4). The ν_a(BeC) mode of the monomer is observed at a relatively high frequency relative to similar frequencies for other beryllium alkyls; it also increases in frequency on deuteration. The former data indicate an unusually strong Be-C bond in $(CH_3)_2Be$. This has been attributed to hyperconjugation in which the methyl groups release electrons to vacant beryllium 2p orbitals (1.1). Hyper-

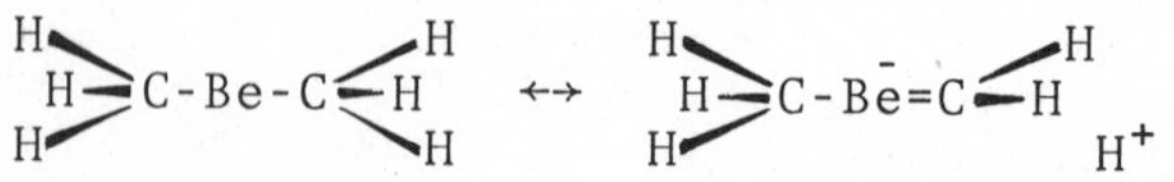

(1.1)

conjugation has also been used to explain the stability of monomeric $(CH_3)_3B$ (39). The shift of the ν_a(BeC) mode for $(CH_3)_2Be$ on deuteration to higher rather than lower frequencies has also been observed for the ν_a(BC) mode of $(CH_3)_3B$ (14). This effect in both compounds has been attributed to vibronic interactions and resonance effects. The $\delta_s(CH_3)$ (1222 cm^{-1}) and ν_a(BeC) (1081 cm^{-1}) modes of $(CH_3)_2Be$ might normally be expected to shift to approximately 924 and 1050 cm^{-1}, respectively, on deuteration. Both modes, however, are of A_{2u} symmetry, and such shifts would violate the noncrossing rule (40). On deuteration, therefore, the character of the modes is switched with a band at 1150 cm^{-1} assumed to have mainly ν_a(BeC) character and a band at 994 cm^{-1} assumed to have mainly $\delta(CD_3)$ character. The shapes and intensity ratios of the relevant $(CH_3)_2Be$ and $(CD_3)_2Be$ bands also seem to support the above conclusions (4). Since two natural isotopes of boron exist in relatively high abundance and boron has a relatively high mass, bands attributed to both isotopes are usually observed for modes involving the motion of the boron atom in boron derivatives. This has been observed for the ν_a(BC) mode of $(CH_3)_3B$ (15,30). Only one band, however, is observed for the ν_s(BC) mode since the boron atom and carbon atoms are coplanar and the boron atom does not move during this vibration. Trimethylaluminum, like dimethylberyllium, is found as both a monomer and a dimer in the vapor state. The vapor-phase Raman spectrum of monomeric trimethylaluminum has been reported and assigned (16). A complete normal coordinate analysis has been reported for dimethylmercury (41).

In addition to the ν(MC) and δ(CMC) skeletal modes, torsional modes are expected for methyl derivatives. Most vibrational studies have been in solution or in the liquid or vapor phases in which free rotation of the methyl groups is found. The E and A_2 torsional modes have been assigned at 224 (vw) and 210 (vw,sh) cm^{-1}, respectively, for $(CH_3)_3P$, and 164 (vw) and 152 (vw,sh) cm^{-1}, respectively, for $(CD_3)_3P$ (18). In low-frequency infrared studies of solid $(CH_3)_2Se$ and $(CH_3)_2Te$ at -190°C, torsional modes have also been assigned. The A_2 mode has only been observed for $(CH_3)_2Se$ (175 (w) cm^{-1}) while the B_2 mode has been observed for both $(CH_3)_2Se$ (207 (m) cm^{-1}) and $(CH_3)_2Te$ (185 (w) cm^{-1}) (42).

TABLE 1.3
Metal-Carbon Skeletal Mode Assignments (cm^{-1}) for $(CH_3)_nM$ and $(CD_3)_nM$ Derivatives

Compound	Skeletal Structure	ν_a(MC)	ν_s(MC)	δ(CMC)	Refs.
CH_3Cu			645		27
$CH_3{}^6Li$			558		28
$CD_3{}^6Li$			536		28
$CH_3{}^7Li$			530[a]		28
$CD_3{}^7Li$			510		28
$(CH_3)_2Be$	Linear	1081	b	b	4
$(CD_3)_2Be$	Linear	1150	b	b	4
$(CH_3)_2Zn$	Linear	613	503	134	10,11
$(CD_3)_2Zn$	Linear	554	458	96	11
$(CH_3)_2Cd$	Linear	534	459	120	10,11
$(CD_3)_2Cd$	Linear	492	419	109	11
$(CH_3)_2Hg$	Linear	540	515	161	10,11
$(CD_3)_2Hg$	Linear	491	471	141	11,12
$(CH_3)_2Se$	Bent	604	589	233	3
$(CD_3)_2Se$	Bent	563	548	201	29
$(CH_3)_2Te$	Bent	528	528	198	3
$(CH_3)_3{}^{10}B$	Planar	1177	680	341 321	14,15,30
$(CH_3)_3{}^{11}B$	Planar	1149	680	341 321	14,15,30
$(CD_3)_3B$	Planar	1205	b	b b	14
$(CH_3)_3Al$	Planar	760	530	c 170	16
$(CH_3)_3Ga$	Planar	577	521.5	c 162.5	5,6,17
$(CH_3)_3In$	Planar	500	467	c 132	6,17
$(CH_3)_3P$	Pyramidal	711	657	315 288[d] 269[d]	18,31,32

$(CD_3)_3P$	Pyramidal	648	594	269	242[d]	18,32
		629			224[d]	
$(CH_3)_3As$	Pyramidal	583	568	238	223	7,31,32
$(CH_3)_3Sb$	Pyramidal	513	513	188[e]		7,8,17
$(CD_3)_3Sb$	Pyramidal	472	472	159[e]		33
$(CH_3)_3Bi$	Pyramidal	460	460	171[e]		7,9
$(CH_3)_4Si$	Tetrahedral	696	598	239	202	20,21
$(CH_3)_4Ge$	Tetrahedral	595	558	195	175	22
$(CH_3)_4Sn$	Tetrahedral	529	508	157[e]		23
$(CH_3)_4Pb$	Tetrahedral	476	459	120[e]		24
$(CH_3)_4Ti$	Tetrahedral	577	489	180[e]		34
$(CH_3)_5Sb$	Trig. bipyram.	456[f]	414[f]	104[f]		35
		516[g]	456[g]	213[g]	195[g]	
				239		
$(CH_3)_6W$	Octahedral	482[h]				36

[a]Estimated value.
[b]Not observed.
[c]$\delta_{\parallel}$, not observed.
[d]The δ(CPC) mode of E symmetry is split into two bands.
[e]The two expected δ(CMC) modes are accidentally degenerate.
[f]An axial mode.
[g]An equatorial mode.
[h]Since the band was observed in the infrared spectrum, it has T_{1u} symmetry.

Tetramethyltitanium is one of only two neutral monomeric methyl compounds of a transition element for which vibrational data have been reported. The low-temperature vibrational spectrum of $(CH_3)_4Ti$ has been obtained using diethyl ether as the solvent (34). The K(TiC) calculated for $(CH_3)_4Ti$ is about 20% lower than expected on the basis of a comparison with analogous force constants calculated for the tetramethyl derivatives of the group IVb elements. Limited infrared data have been reported for $(CH_3)_6W$ (36). Figure 1.3 illustrates the infrared spectrum of a typical methyl compound, tetramethylsilane (43).

2. Monomeric $(CH_3)_nM^{m+}$ and $(CH_3)_nM^{m-}$ Complexes

Cationic methyl compounds form in aqueous solution or are found as ionic solids. In solution, water may coordinate to the metal atom. For complexes in which metal-water bonds have appreciable covalent character, it is often possible to observe metal-water stretching modes. Such complexes are discussed in Sec. E.4 of this chapter.

Vibrational data have been reported for only one monomethyl organometallic cation, namely, CH_3Zn^+. The infrared and Raman spectra of solid CH_3ZnBH_4 and CH_3Zn-BD_4 indicate that these compounds are best viewed as coupled CH_3Zn^+ and BH_4^- or BD_4^- ions (44).

The Raman spectra of aqueous solutions of several $(CH_3)_2M^{n+}$ compounds have been reported. Although the skeletons of $(CH_3)_2M^{n+}$ complexes in solution can be either linear (1.2) or angular (1.3), only linear skeletons have been found for all $(CH_3)_2M^{n+}$ derivatives that have been studied. The first such compound studied was $(CH_3)_2Tl^+$ (45). While this Raman study involved aqueous solutions of the nitrate and perchlorate

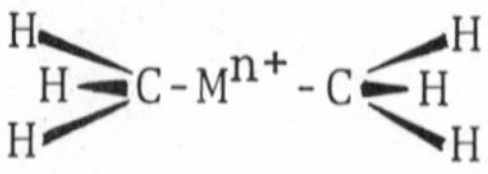

(1.2)

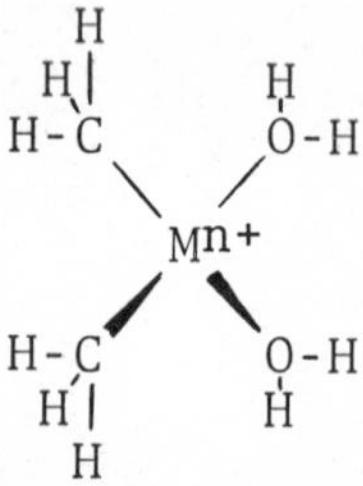

(1.3)

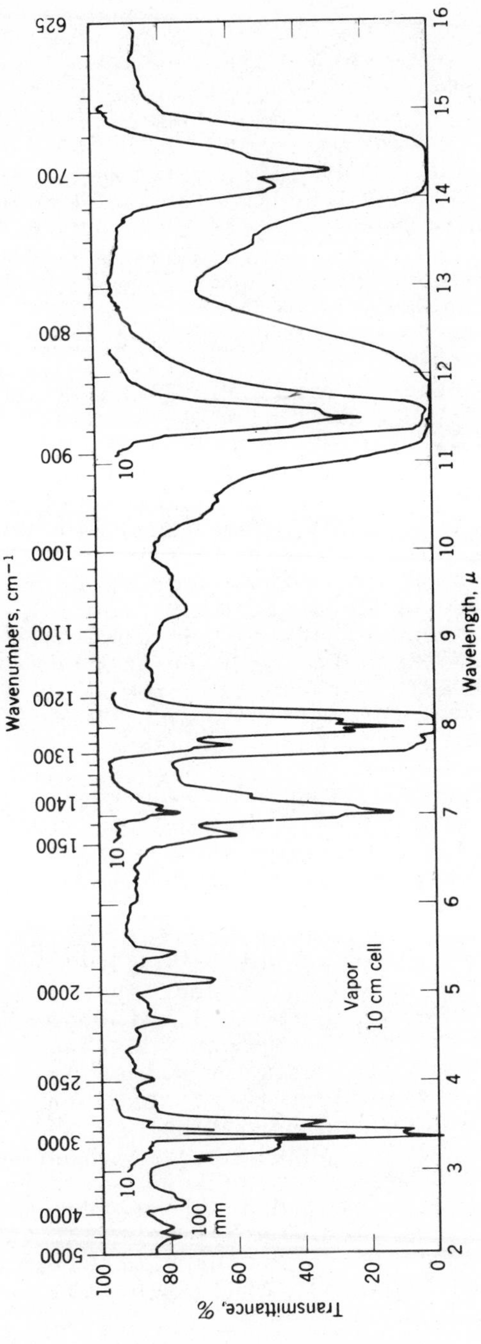

Figure 1.3. Infrared spectrum of $(CH_3)_4Si$ *(43).*

salts, a more recent study has been made of the Raman spectrum of an aqueous solution of the chloride salt and the solid-state infrared spectra of the nitrate and chloride salts (46). These two studies are in agreement except in the assignment of one band. Therefore, in the original study (45), a very weak intensity band at 569 cm^{-1} was assigned to the $\rho r(CH_3)$ mode. This is a relatively low frequency for the $\rho r(CH_3)$ mode. Based on recent infrared data, it is more reasonable to assign the 569 cm^{-1} band to the $\nu a(TlC)$ mode and a very strong intensity band at 803 cm^{-1} to the $\rho r(CH_3)$ mode. A comparison has been made of the Raman spectra of $(CH_3)_2Sn^{2+}$ and $(CH_3)_2Tl^+$ as single crystals and in aqueous solutions (47). The infrared and Raman spectra of aqueous solutions of $(CH_3)_2SnX_2$ ($X=ClO_4,NO_3$) (48,49), $(CH_3)_2InCl$ (50), and $(CH_3)_2InClO_4$ (50), and of solid $[(CH_3)_2In]InI_4$ (51) have all been interpreted in terms of linear metal-carbon skeletons. Raman data for aqueous solutions of $(CH_3)_2PbX_2$ ($X=ClO_4,NO_3$) show $(CH_3)_2Pb^{2+}$ to be isostructural with $(CH_3)_2Sn^{2+}$ with a weak intensity band at 425 cm^{-1} assigned to the $\nu(PbO)$ mode of coordinated water (52). No $\nu(MO)$ mode assignments could be made for aqueous solutions of any of the other $(CH_3)_2M^{n+}$ complexes mentioned above. Low-frequency vibrational data are now available for two isostructural $(CH_3)_2M^{n+}$ series, namely, $(CH_3)_2Cd$, $(CH_3)_2In^+$, and $(CH_3)_2Sn^{2+}$, and $(CH_3)_2Hg$, $(CH_3)_2Tl^+$, and $(CH_3)_2Pb^{2+}$. Complete normal coordinate calculations have been reported for all six of these compounds using a Urey-Bradley force field (49, 50). Figure 1.4 illustrates the variation of the K(MC) force constants for each series. While the expected increase is observed in the K(MC) force constants for the former series as the metal charge increases, the opposite trend is observed for the latter series. Solvation effects alone cannot be used to explain these data since they would be present in both series. It has been suggested that the stabilization of the metal-carbon bonds in these compounds depends on metal-carbon orbital overlap. The largest values for K(MC) are found for the bonds most inert to attack by acids and bases, namely Hg-C, Tl-C, and Sn-C (50).

Both planar and nonplanar skeletons have been found for $(CH_3)_3M^{n+}$ complexes. No bands attributable to ν(Sn-halogen) or $\nu(SnO)$ modes have been observed in aqueous solutions of trimethyltin halides (53). Similarly, no $\nu(SnO)$ modes have been observed in the spectra of 6M NaOH solutions of $(CH_3)_3SnOH$ (54). The spectra of

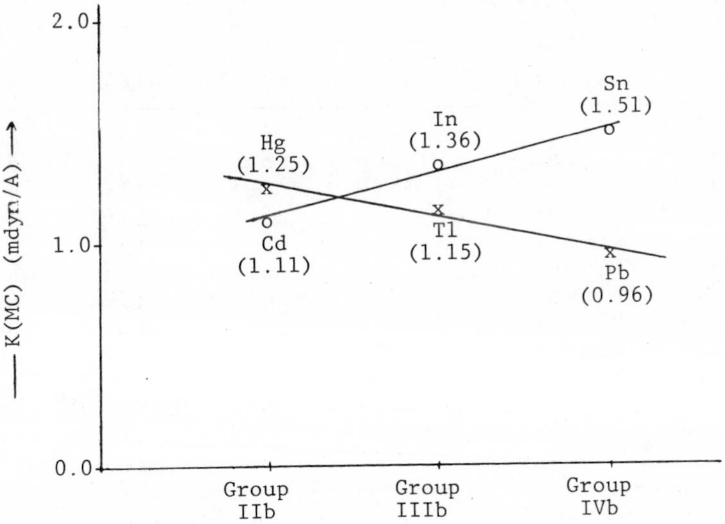

Figure 1.4. Variation of the Urey-Bradley stretching force constant for the metal-carbon bond [K(MC)] in the isoelectronic series $(CH_3)_2Cd$, $(CH_3)_2In^+$, and $(CH_3)_2Sn^{2+}$, and $(CH_3)_2Hg$, $(CH_3)_2Tl^+$, and $(CH_3)_2Pb^{2+}$.

these derivatives are consistent with the presence of the $(CH_3)_3Sn^+$ cations and a planar C_3Sn skeleton. The Raman spectra of aqueous solutions of the perchlorate and nitrate salts of $(CH_3)_3Pb^+$ (55) and $(CH_3)_3Sb^{2+}$ (56) have also been interpreted in terms of a planar skeleton (56). The expected increase in the value of K(MC) (mdynes/A) with increasing nuclear charge has been observed in the isoelectronic series:

Compd.	$(CH_3)_3In$ <	$(CH_3)_3Sn^+$ <	$(CH_3)_3Sb^{2+}$
K(MC)	1.93	2.39	2.55

A single crystal X-ray study of $(CH_3)_3SeI$ has shown the presence of $(CH_3)_3Se^+$ cations with a nonplanar skeleton (57). The infrared spectra of solid $[(CH_3)_3Se]^+Cl^-$ (58) and $[(CH_3)_3Te]^+Br^-$ (59) are also consistent with a nonplanar skeleton similar to that found for $(CH_3)_3SeI$.

Solid-state infrared and solution infrared and Raman data have appeared for several halide salts of

TABLE 1.4
Metal-Carbon Skeletal Mode Assignments (cm^{-1}) for $(CH_3)_nM^{m+}$ Derivatives

Compound	Skeletal Structure	$\nu_a(MC)$	$\nu_s(MC)$	$\delta(CMC)$	Refs.
CH_3Zn^+			557		44
$(CH_3)_2In^+$	Linear	566	502	a	50
$(CH_3)_2Tl^+$	Linear	559	498	114	45,46
$(CH_3)_2Sn^{2+}$	Linear	582	529	180	49
$(CH_3)_2Pb^{2+}$	Linear	a	480	150	49,52
$(CH_3)_3Sn^+$	Planar	557	521	152	53
$(CH_3)_3Pb^+$	Planar	504	473	127	55
$(CH_3)_3Sb^{2+}$	Planar	582	536	166	56
$(CH_3)_3Se^+$	Nonplanar	602	580	272	58
$(CH_3)_3Te^+$	Nonplanar	534	a	a	59
$(CH_3)_4P^+$	Tetrahedral	783	649	285 170	60,61
$(CH_3)_4As^+$	Tetrahedral	652	590	217	7
$(CH_3)_4Sb^+$	Tetrahedral	574	535	178	7

[a]Data not reported.

$(CH_3)_4M^+$ (M=P (60-63), As (7), Sb (7,64)). The $(CH_3)_4M^+$ complexes have tetrahedral skeletons in both the solid state and solution. Table 1.4 summarizes the skeletal mode assignments for several $(CH_3)_nM^{m+}$ cations. Figure 1.5 illustrates the infrared spectrum of solid $[(CH_3)_4Sb]^+I^-$ (64).

Limited vibrational data have been reported for anionic methyl compounds. The lithium salts of both $(CH_3)_2Au^-$ and $(CH_3)_4Au^-$ have been isolated in the solid state by complexing the lithium ion with N,N,N',N",N"-pentamethyldiethylenetriamine (65). Solvent separated ion pairs of Li^+ and $(CH_3)_2Au^-$ or $(CH_3)_4Au^-$ have also been synthesized in diethyl ether solution (66). Solid state and solution Raman data for $(CH_3)_2Au^-$ ($\nu(AuC)$= approximately 525 and 491 cm^{-1}) and $(CH_3)_4Au^-$ ($\nu(AuC)$= approximately 530, 521, and 487 cm^{-1}) indicate that change of state has little effect on the structure of these aurates and that the two anions are perturbed from the expected free ion symmetries of $D_{\infty h}$ and D_{4h}, respectively (65,66).

Infrared data for $M^+[(CH_3)_4In]^-$ (M=Li,Na,K,Rb,Cs) and X-ray crystallographic data for $M^+[(CH_3)_4In]^-$ (M= Li,Na) show the presence of regular tetrahedral $(CH_3)_4In^-$ anions that do not interact with the alkali

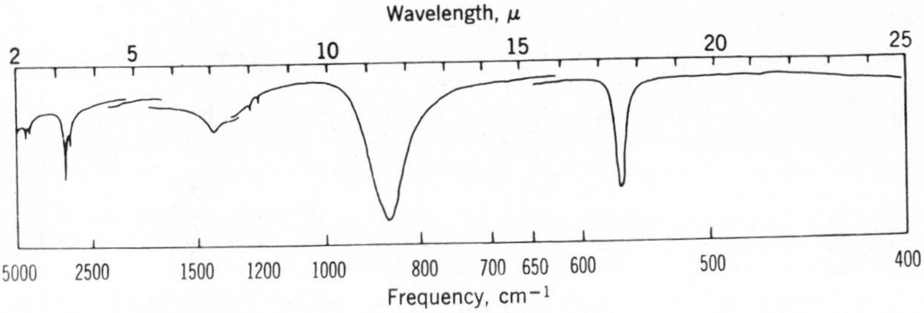

Figure 1.5. Infrared spectrum (5000-400 cm^{-1}) of [$(CH_3)_4Sb$]I (64).

metal cations (67). All of these derivatives show an infrared band at about 440 cm^{-1}, which has been assigned to the ν_a(InC) mode. Raman spectra of diethyl ether solutions prepared from [$(CH_3)_3PtI$]$_4$ and methyllithium are consistent with the presence of essentially octahedral $(CH_3)_6Pt^{2-}$ anions with an intense, polarized band at 508 cm^{-1} and a depolarized band at 505 cm^{-1} assigned to the ν(PtC) modes of A_{1g} and E_g symmetry, respectively (68). In some M_n[$(CH_3)_mM'$] derivatives, there is an interaction between the anions and cations through bridging methyl groups to give associated structures. These complexes are discussed in Sec. A.3 of this chapter.

3. Bridging CH_3 and CH_2 Groups

An X-ray crystallographic study of methyllithium has shown it to be a polymer of weakly associated tetramers in which each methyl group of the tetramer bridges a face of a Li_4 tetrahedron (38). Previous investigators have assigned the ν(LiC) modes of polymeric methyllithium at 1052 and 820 cm^{-1} (69,70) and at 880 cm^{-1} (71). More recently, these modes have been reassigned to bands below 520 cm^{-1} on the basis of the frequency shifts observed using both ^{6}Li and ^{7}Li, and deuterated methyllithium (72). These latter assignments have been confirmed in another recent infrared study (73). The low frequency of the $\nu(CH_3)$ modes of polymeric methyllithium (2840 and 2780 cm^{-1}) has been reported to be characteristic of methyl groups involved in electron-deficient bonding (72). The fact that the $\nu(CH_3)$ modes of monomeric methyllithium (2820 and 2780 cm^{-1}) are in the same region, however, implies little change in the

C-H bond strength on association of the monomer in the solid state. Halide-free solutions of methyllithium in diethyl ether give rise to a characteristic low-frequency $\nu(CH_3)$ mode (2766 cm^{-1}) in the Raman spectrum as well as low-frequency Raman bands at 486, 299, 209, and 171 cm^{-1} (65). It was concluded in this study that the tetrameric units found in solid methyllithium are retained in solution and that the low-frequency Raman bands arise from the cluster modes of these units.

A single crystal X-ray study of dimethylberyllium shows the presence of a long-chain polymeric structure (1.4) (74). A similar polymeric structure has been

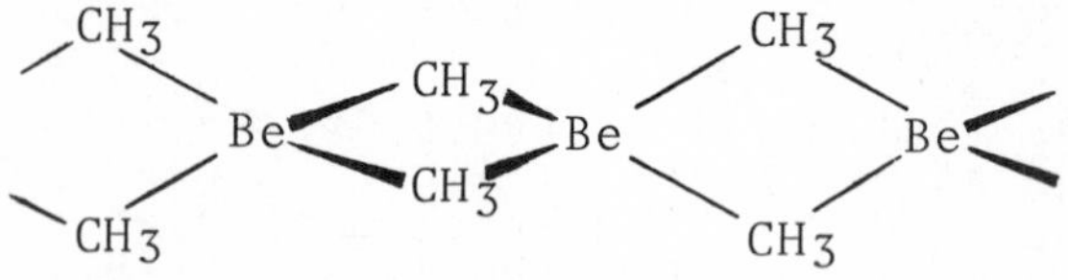

(1.4)

found in an X-ray study of dimethylmagnesium as a crystalline powder (75). Characteristic features of the infrared spectra of polymeric $(CH_3)_2Be$ (73,76,77) and $(CH_3)_2Mg$ (77-79) include relatively low frequencies for the $\nu(CH_3)$ modes and the absence of bands in the $\delta_d(CH_3)$ region from about 1500 to 1400 cm^{-1}. Vibrational data and assignments for tetrameric CH_3Li, and polymeric $(CH_3)_2Be$ and $(CH_3)_2Mg$ in the solid state are compared in Table 1.5. The infrared spectra (1300 to 300 cm^{-1}) have also been reported for the diethyl ether solutions of $LiBe(CH_3)_3$, $Li_2Be(CH_3)_4$, and $Li_3Be(CH_3)_5$ (81).

Solid trimethylaluminum is dimeric with bridging methyl groups (82). The most recent vibrational studies of this compound have included normal coordinate analyses in which the methyl groups were assumed to be single atoms (83), or in which the entire molecule was treated (84). Although an attempt has been made to differentiate between the skeletal and methyl modes by measuring deuterium isotopic shifts and substituting methyl groups with chlorine atoms (85), the assignment of the vibrational spectrum of dimeric $(CH_3)_3Al$ is complicated by extensive mixing of the skeletal stretching and the methyl bending modes (84). It was also observed to be difficult to distinguish between the bridging and terminal C-H modes of dimeric $(CH_3)_3Al$ (84).

TABLE 1.5
Vibrational Assignments (cm^{-1}) for Tetrameric Methyllithium and Polymeric Dimethylberyllium and Dimethylmagnesium

	$[CH_3Li]_4$[a]	$[CD_3Li]_4$[a]	$[(CH_3)_2Be]_n$		$[(CH_3)_2Mg]_n$[d]
Assignment	Infrared	Infrared	Infrared[b]	Raman[c]	Infrared
$\nu(CH_3)$	2840	2150	2912	2970	2850
	2780	2027	2885	2900	2780
$\delta_d(CH_3)$	1480	1100		1440	
	1427	1043			
$\delta_s(CH_3)$	1096	827	1255	1250	1200
	1061	815	1243		1186
$\rho_r(CH_3)$				800	712
				728	
				680	
Skeletal modes	517[e] 514[f]	455[e] 436[f]	835	918	575
	446[e] 417[f]	348[e] 336[f]	567	505	440
			535	455	400
			427	410	310
			403		
			292		

[a]Reference 72.
[b]References 73,76,77,80.
[c]Reference 80.
[d]References 77-79.
[e]Data obtained using ^{6}Li isotope.
[f]Data obtained using ^{7}Li isotope.

TABLE 1.6
Metal-Carbon Stretching Mode Assignments (cm^{-1}) for $[(CH_3)_3Al]_2$ and Compounds with $Li\cdots CH_3$—M Interactions

$[(CH_3)_3Al]_2$[a]	$Li[(CH_3)_4Al]$[b]	$Li_2[(CH_3)_4Zn]$[b]	$Li_3[(CH_3)_6Cr]\cdot$ 2Dioxane[c]
Terminal	ν(AlC)	ν(ZnC)	ν(CrC)
ν(AlC)			
	632	599	484
697	567	416	462
683	493		450
592		$\nu(^6LiC)$	
564	$\nu(^6LiC)$		
		321	
Bridging	348		
ν(AlC)	295	$\nu(^7LiC)$	
480	$\nu(^7LiC)$	289	
453			
390	320		
367	284		

[a]Reference 84.
[b]Reference 87.
[c]Reference 89.

The infrared spectra of $M[(CH_3)_4Al]$ (M=Li,Na) have been interpreted in terms of a model of C_{2v} symmetry (86). A more recent solid-state infrared and Raman study of $Li[(CH_3)_4Al]$ has been reported (87) using ^6Li and ^7Li isotopes and deuterated methyl groups. Contrary to the conclusions of the previous study (86), the data indicate that $Li[(CH_3)_4Al]$ forms linear polymeric chains (1.5) in which the local symmetry around both aluminum

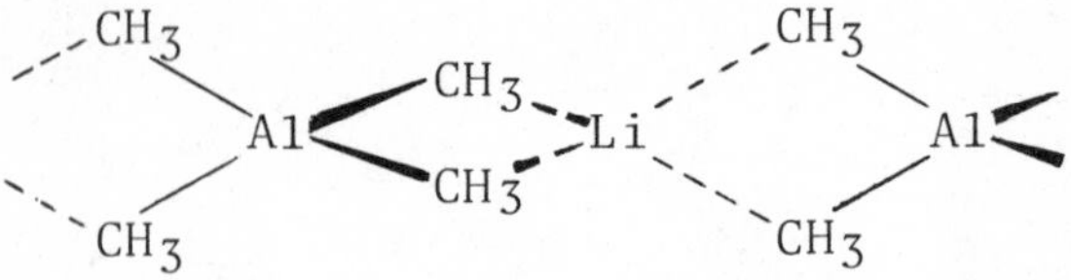

(1.5)

and lithium is D_{2d}. Vibrational data for $Li_2[(CH_3)_4Zn]$ and its isotopic analogs (87) indicate that the local symmetry around zinc and lithium is T_d although crystallographic data (88) show a deviation from T_d symmetry. It has also been concluded on the basis of the following observations that the Li-C bonds in both $Li[(CH_3)_4Al]$ and $Li_2[(CH_3)_4Zn]$ are highly ionic (87). First, the ν(LiC) modes are observed in the infrared but not the Raman spectra. Second, the $\nu(CH_3)$ modes appear at relatively low frequencies. Finally, the value of the ^{13}C-^{1}H coupling constant seems relatively low. The structure of $Li_3[(CH_3)_6Cr]\cdot$2Dioxane has been characterized using both X-ray crystallographic data and low-temperature (100°K) infrared spectra (89). The six methyl groups form an octahedral skeleton around each chromium atom. The octahedron is slightly distorted through interactions of the three lithium atoms with three of the edges. Each lithium atom is tetrahedrally surrounded by two oxygen atoms from two different dioxane molecules and two carbon atoms from the methyl groups. Table 1.6 compares the metal-carbon stretching mode assignments for dimeric $(CH_3)_3Al$ and compounds in which methyl groups bridge both a lithium and another metal atom.

The normal modes expected for compounds with a bridging methylene group, $-CH_2-$, are illustrated in Figure 1.6 while the frequency ranges observed for these modes are summarized in Table 1.7. In general, methylene modes appear at lower frequencies than related modes found for the methyl group. The frequencies observed for the $\nu(Si_2X)$ modes in $[(CH_3)_3Si]_2X$ (X=O,NH,CH_2) have been interpreted as implying that the SiXSi angle is relatively large (90). Using similar arguments based on the observed $\nu(Si_2C)$ and $\nu(CH_2)$ mode frequencies, the Si-C-Si bond angle of $(H_3Si)_2CH_2$ has been calculated to be a minimum of 116°, while the H-C-H angle is concluded to be smaller than the tetrahedral angle (91). Several studies have been made of metal complexes of the trimethylsilylmethyl group, $(CH_3)_3SiCH_2-$. Relatively complete infrared and Raman data and approximate assignments have been reported for $[(CH_3)_3SiCH_2]_2Hg$ (92), $[(CH_3)_3SiCH_2]HgX$ (X=Cl,Br,I) (92), $[(CH_3)_3SiCH_2]_4M$ (M=Sn,Pb,Cr,V) (93), $[(CH_3)_3SiCH_2]_3VO$ (93), and $[(CH_3)_3SiCH_2]_6M_2$ (M=Mo,W) (93). Less complete data have been reported for trimethylsilylmethyl derivatives of zinc (94), aluminum (95), thallium (96), titanium (97), zirconium (97), hafnium (97), and rhenium (98).

Replacement of alkyl groups with trimethylsilylmethyl groups sometimes gives considerable stability to organometallic compounds. It has been reported that

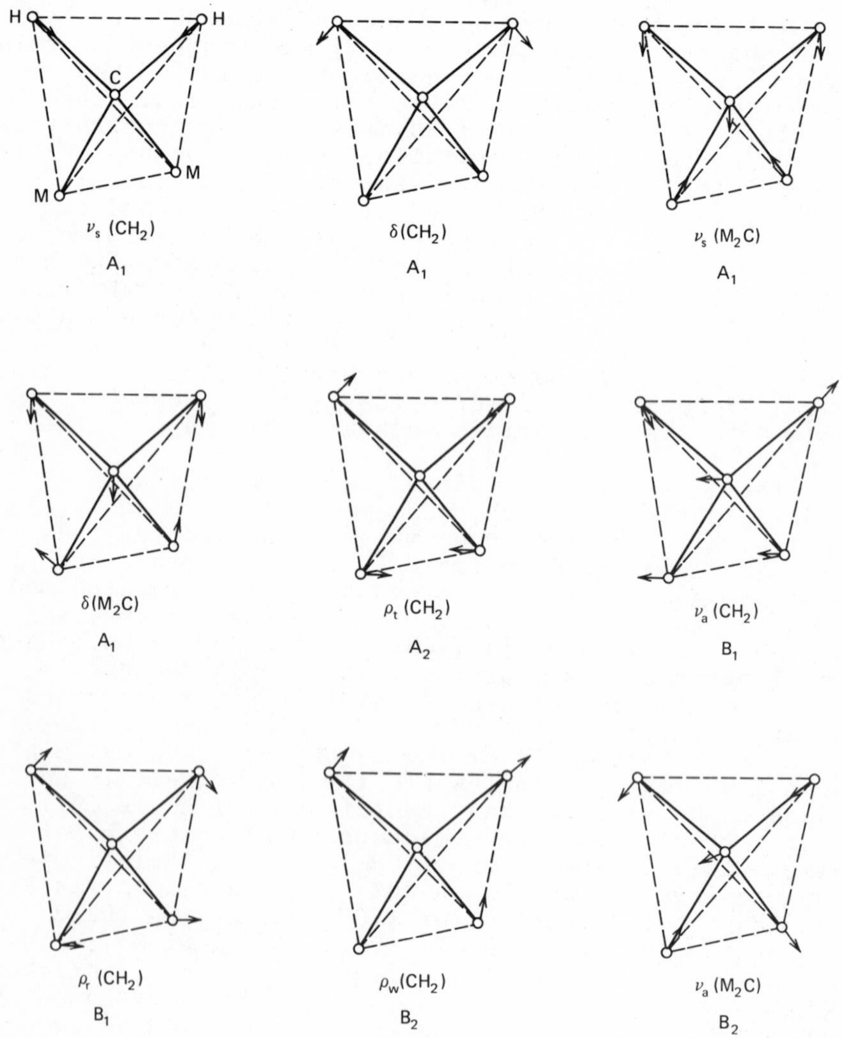

Figure 1.6. Normal modes of vibration for the bridging methylene group, M-CH_2-M.

TABLE 1.7
Frequency Ranges (cm^{-1}) for the Carbon-Hydrogen Modes of the Bridging Methylene Group

Mode	Frequency Range
$\nu(CH_2)$	3000-2800
$\delta(CH_2)$	1400-1250
$\rho_w(CH_2)$ } $\rho_t(CH_2)$	1150-875
$\rho_r(CH_2)$	850-600

this stabilizing effect is increased by the $(CH_3)_3P{=}CH_2$ group. Therefore, the metalated ylides $[(CH_3)_2P\text{-}(CH_2)_2]_2M_2$ (M=Cu (99),Ag (99),Au (100)) (1.6) have all

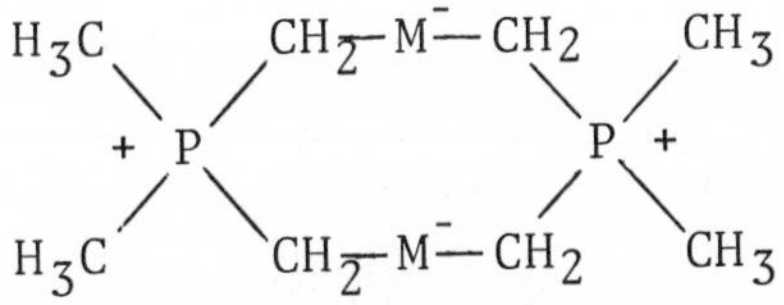

(1.6)

been isolated as stable products. Infrared bands at 557 and 509 cm^{-1}, 518 and 470 cm^{-1}, and 551 cm^{-1} were assigned to $\nu(MC_2)$ modes for M=Cu, Ag, and Au, respectively. The transformation of the square planar complexes $[((CH_3)_3PCH_2)_2AuX_2]Br$ (X=Br,I) from the *trans* to the *cis* isomer has resulted in an increase in a $\nu(AuC)$ mode frequency from 549 to 560 cm^{-1} for X=Br and from 543 to 556 cm^{-1} for X=I in the infrared spectra, while a band assigned to a $\nu(AuBr)$ mode for the bromide (256 cm^{-1}) was shifted below the experimental limit of 250 cm^{-1} (101). The infrared spectra have been assigned for $(I_2As)_2CH_2$ and $(X_2AsCH_2)_4C$ (X=CH_3,I) (102). Vibrational assignments have been made for several compounds with the $SbCH_2Sb$ unit (103-105). A linear relationship has been found for these compounds between the $\rho_r(CH_2)$ mode frequency and the average of the Sb-C stretching mode frequencies of the Sb_2C unit, ν_{av} ($\nu_{av} = 0.5[\nu_a(Sb_2C) + \nu_s(Sb_2C)]$) (105). This relationship has been explained in terms of the amount of antimony *s*-orbital character used to form the Sb-C bonds in these compounds. Data have been reported for several compounds containing the $SiCH_2Ge$ unit (106).

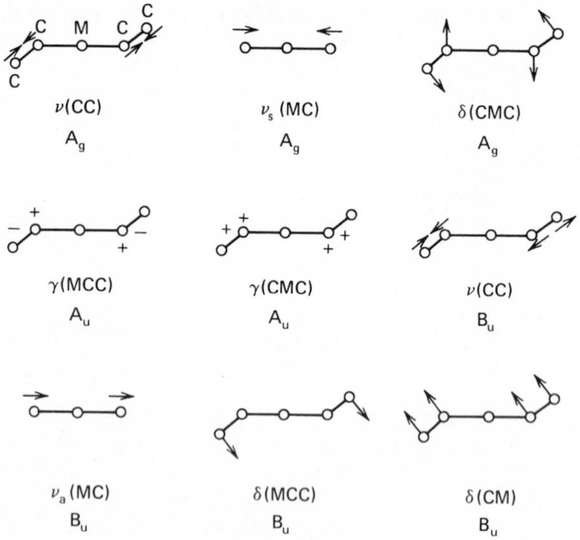

Figure 1.7. Normal modes of vibration for the C_2MC_2 skeleton of $(C_2H_5)_2M$ derivatives with C_{2h} symmetry.

The infrared and Raman spectra have been assigned for $(CH_3)_3P{=}CH_2$ with the ν(P=C) mode assigned to a Raman band at 998 cm^{-1} (107). A sharp infrared band at 1407 cm^{-1} has been reported as characteristic of a compound formulated as $(CH_3)_2Si{=}CH_2$ (108).

B. ETHYL COMPOUNDS

The vibrational modes arising for an ethyl group include those discussed for the C-H bonds of the methyl and methylene groups. In addition, a methyl torsional mode, C-C stretching mode, and carbon-carbon-metal bending modes are expected. In Figure 1.7 the skeletal modes are illustrated for $(C_2H_5)_2M$, assuming a local symmetry of D_{2h}.

The two ν(CC) modes are due to the in-phase and out-of-phase combinations of the ν(CC) mode expected for one ethyl group. Such combinations can produce a more complex spectrum than is expected for one ethyl group. The degree of coupling is most severe when the mass of the metal atom is relatively light. Table 1.8

TABLE 1.8
Frequency Ranges (cm^{-1}) for Terminal Ethyl Modes

Mode	Frequency Range
$\nu_a(CH_3)$, $\nu_a(CH_2)$	3000-2925
$\nu_s(CH_3)$, $\nu_s(CH_2)$	2975-2850
$\delta_d(CH_3)$	1480-1450
$\delta(CH_2)$	1430-1400
$\delta_s(CH_3)$	1390-1370
$\rho_w(CH_2)$, $\rho_t(CH_2)$	1300-1150
$\rho_r(CH_3)$	1200-900
$\nu(CC)$	1050-900
$\rho_r(CH_2)$	800-600
$\rho_\tau(CH_2)$	300-150

summarizes the frequency ranges found for the modes in organometallic ethyl compounds. In many cases, it has proved difficult to make complete vibrational assignments or establish frequency ranges for the ethyl modes, for one or more of the following reasons: (1) overlap of the related methyl and methylene modes, (2) interethyl-group coupling as illustrated above for the $\nu(CC)$ modes of $(C_2H_5)_2M$, and (3) intra-ethyl-group coupling between the modes in one ethyl group.

The possible intramode coupling makes some of the assignments given for the $\nu(CC)$ and $\rho_r(CH_3)$ modes somewhat meaningless. Recent normal coordinate analyses for $C_2H_5SiCl_nH_{3-n}$ (n=1 to 3) have shown the presence of severe vibrational mixing of the ethyl modes (109). For $C_2H_5SiCl_3$, a band at 1010 cm^{-1} has been shown to involve changes in the H-C-C, H-C-H, and H-C-Si angles (and to a lesser extent the C-C-Si angle) and the C-C bond, a band at 975 cm^{-1} changes in the H-C-C and H-C-Si angles and C-C bond, and a band at 965 cm^{-1} (that can only approximately be described as an out-of-plane $\rho_r(CH_3)$ mode) changes in the H-C-C, H-C-H, and H-C-Si angles. Three similar bands have also been observed in the spectra of $(C_2H_5)_4Si$ (1022,1007,974 cm^{-1}) (60), $(C_2H_5)_4P^+$ (1044,1003,975 cm^{-1}) (60), $(C_2H_5)_3P$ (1041,982,934 cm^{-1}) (110), and $(C_2H_5)_3Sb$ (1020,960, 935 cm^{-1}) (110). In recognition of the possible mixing of the $\nu(CC)$ and $\rho_r(CH_3)$ modes, rather than attempting to assign bands to pure modes, the three bands have been given the designations A, B, and C, respectively (60,109). It has been suggested that additional infra-

TABLE 1.9
Metal-Carbon Skeletal Mode Assignments (cm^{-1}) for Monomeric Organometallic Ethyl Derivatives

Compound	Skeletal Structure	ν_a(MC)	ν_s(MC)	δ(MCC)	δ(CMC)	Refs.
$(C_2H_5)_2Zn$	Linear	561	484			111
$(C_2H_5)_2Hg$	Linear	515	488	267 262	140 85	112,113
$(C_2D_5)_2Hg$	Linear	465				114
$(C_2H_5)_3{}^{10}B$	Planar	1135			287	14,115
$(C_2D_5)_3{}^{10}B$	Planar	1135				14
$(C_2H_5)_3{}^{11}B$	Planar	1120			287	14,115
$(C_2D_5)_3{}^{11}B$	Planar	1110				14
$(C_2H_5)_3Ga$	Planar	496				17,116
$(C_2H_5)_3In$	Planar	470				17
$(C_2H_5)_3P$	Pyramidal	690	619	410-249		110
$(C_2H_5)_3As$	Pyramidal	540	570			117
			563			
$(C_2H_5)_3Sb$	Pyramidal	505	505			17,110,118
$(C_2H_5)_3Bi$	Pyramidal	450	450	253 213	124	119
$(C_2H_5)_4Si$	Tetrahedral	731	549	392-233	170	60,120
$(C_2H_5)_4Ge$	Tetrahedral	572	532	332	152	121,122
$(C_2H_5)_4Sn$	Tetrahedral	508	490	272	132 86	123,124
$(C_2H_5)_4Pb$	Tetrahedral	461	443	240	122 107 86	119,123
$(C_2D_5)_4Pb$	Tetrahedral	416	401	217 196	95	119
$(C_2H_5)_2Tl^+$	Linear	502	437	263 or 225		46
$(C_2H_5)_4P^+$	Tetrahedral	787	590	390-290	182	60
$(C_2H_5)_4As^+$	Tetrahedral	613	548	349 327 312		122
$(C_2H_5)_4Al^-$	Tetrahedral	564	492	340 302		125

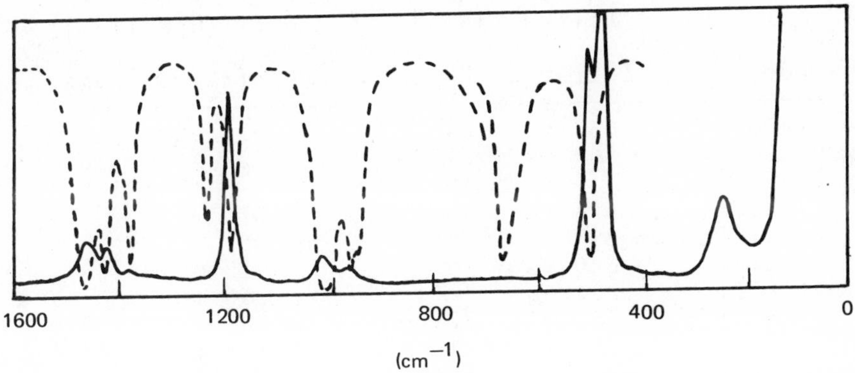

Figure 1.8. Infrared (---) and Raman (—) spectra of $(C_2H_5)_4Sn$ (128).

red bands observed in the same frequency region for $(C_2H_5)_3P$ may arise from the A, B, and C modes of more than one rotational isomer (109). The metal-carbon skeletal mode assignments are summarized in Table 1.9 for monomeric organometallic ethyl derivatives. The frequencies of the ν(MC) and δ(CMC) modes are lower for the ethyl derivatives than those found for the corresponding methyl derivatives, and they decrease within a homologous series as the mass of the metal atom increases. The ethyl-metal torsion mode has been assigned in the Raman spectrum of solid $(C_2H_5)_2Hg$ (65 cm^{-1}) (126).

Since normal coordinate calculations have only been reported for $(C_2H_5)_2Hg$ (126), no comparison can be made of the K(MC) force constants within an isoelectronic or homologous series. The infrared spectrum has been illustrated for $(C_2H_5)_5Sb$ (127). The infrared and Raman spectra below 1600 cm^{-1} are shown for $(C_2H_5)_4Sn$ in Figure 1.8 (128).

In the solid state, ethyllithium shows extended chains of interacting tetramers (129), while in hydrocarbon solvents, it is hexameric (130). The ν(LiC) modes have been assigned to infrared bands between 570 and 318 cm^{-1} for ethyllithium in various states (131).

Liquid diethylberyllium is an ethyl-bridged dimer (132,133). The infrared and Raman spectra of dimeric diethylberyllium have been interpreted in terms of a

TABLE 1.10
Low-Frequency Assignments (cm^{-1}) for $Li[(CH_3)_4Al]$ and $Li[(C_2H_5)_4Al]$

	$Li[(CH_3)_4Al]$[a]		$Li[(C_2H_5)_4Al]$[b]	
Mode	Infrared	Raman	Infrared	Raman
ν(AlC)	632 vs	646 m	616 sh	618 m
ν(AlC)	567 vs	570 s	527 s	526 m
ν(AlC)		494 vs		468 s
ν(^{6}LiC)	348 vs		326 m	
ν(^{6}LiC)	295 vs		262 m	
ν(^{7}LiC)	320 vs		312 m	
ν(^{7}LiC)	284 vs		256 m	
δ(CAlC)	252 w	271 m	283 m	279 m
δ(CAlC)	214 w	216 m	130 m	143 m
δ(CAlC)		186 s		109 s
δ(CAlC)		108 m		95 s

[a]Reference 87.
[b]Reference 125.

structure of $D_{2'h}$ symmetry (134). The vibrational spectrum of triethylaluminum, which has a dimeric ethyl-bridged structure, has been reported and assigned (128,135). The vibrational spectrum of $Na[(C_2H_5)_4Al]$ (86,125) indicates that $(C_2H_5)_4Al^-$ has a local symmetry of T_d, as has also been found in an X-ray study of $K[(C_2H_5)_4Al]$ (136). An X-ray crystallographic analysis of $Li[(C_2H_5)_4Al]$, however, shows a structure of infinite chains with a local symmetry of D_{2h} (137); this is also the conclusion of a vibrational study of $Li[(C_2H_5)_4Al]$ (125). The vibrational data also indicate that the Li-C bonds of $Li[(C_2H_5)_4Al]$ are highly ionic. The low-frequency assignments for $Li[(C_2H_5)_4Al]$ are compared in Table 1.10 with those of $Li[(CH_3)_4Al]$.

C. PROPYL, BUTYL, AND PENTYL COMPOUNDS

For ethyl organometallic compounds, there is the possibility of intra- and intergroup coupling. There is also the possibility of different torsional configurations about the metal-carbon bond arising from different possible arrangements of one alkyl group relative to another. An additional feature adds to the complexity of the spectra of organometallic compounds with *n*-propyl or larger alkyl groups. This is the possible presence of either a *trans* (1.7) or *gauche* (1.8) tor-

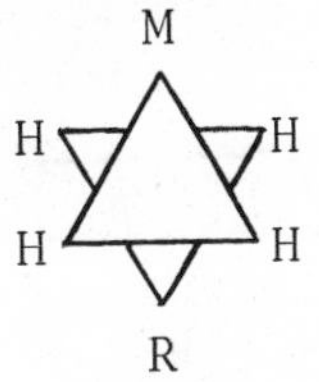

trans

(1.7)

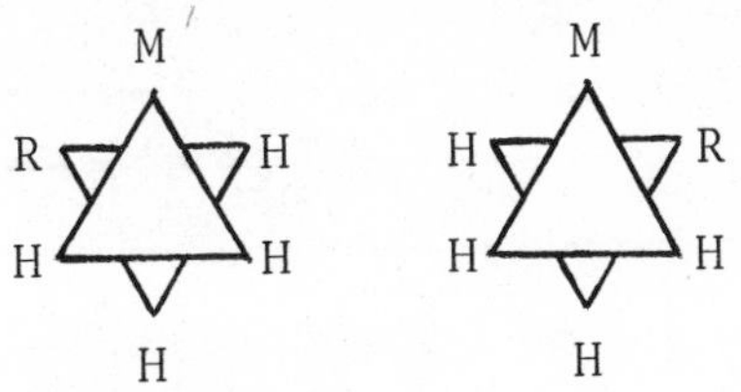

gauche

(1.8)

sional configuration with respect to orientation about the C_1C_2 bond. Unlike the torsional configuration found about the metal-carbon bond of ethyl derivatives, the isomerism found for larger alkyl groups depends only on the presence of one alkyl group. The results of this type of isomerism on the vibrational spectra have been discussed for *n*-propyl halides (138). The infrared spectra of *n*-propyl halides in the liquid state are more complex than those in the crystalline state. This has been attributed to the presence of both the *trans* and *gauche* isomers in the liquid state while only the *trans* isomer is present in the solid state. Two different carbon-halogen stretching mode frequencies were found for the two isomers. The carbon-halogen stretching mode frequency for the *gauche* isomers was not very different from that for the corresponding ethyl derivatives, while the carbon-halogen stretching mode frequency for the *trans* isomers was at an appreciably higher frequency. Similar results have been found for *n*-propyl and *n*-butyl organometallic derivatives. These are summarized in Table 1.11. The infrared spectra of solid $[(n\text{-}C_3H_7)_2Tl]Cl$ and $[(n\text{-}C_4H_9)_2Tl]Cl$ (46) show both to be the *trans* isomers. Liquid $(n\text{-}C_3H_7)_4Ge$ (139), $(n\text{-}C_4H_9)_4Ge$ (142), $(n\text{-}C_3H_7)_4Sn$ (123,140,141), and $(n\text{-}C_4H_9)_4Sn$ (140,141) consist of mixtures of both the *trans* and *gauche* isomers.

Complete infrared and Raman data have been reported for $(n\text{-}C_3H_7)_3B$ (143). The vibrational spectra of both $(n\text{-}C_3H_7)_3Al$ and $(n\text{-}C_4H_9)_3Al$ are consistent with a dimeric structure in solution (128). Similar data for $(i\text{-}C_3H_7)_3Al$, however, show it to be monomeric (128,144). Infrared data for R_3M (M=Ga (17,145),In (17),Sb (17);

TABLE 1.11
Comparison of ν(MC) Mode Assignments (cm^{-1}) for Ethyl Derivatives and the *Trans* and *Gauche* Torsional Configurations of *n*-Propyl and *n*-Butyl Derivatives

Compound	Torsional Configuration[a]	ν(MC)	Refs.
$(C_2H_5)_2Tl^+$		502 437	46
$(n\text{-}C_3H_7)_2Tl^+$	*Trans*	607 564	46
$(n\text{-}C_4H_9)_2Tl^+$	*Trans*	615 571	46
$(C_2H_5)_4Ge$		572 532	121,122
$(n\text{-}C_3H_7)_4Ge$	*Trans*	670-640	139
	Gauche	600-550	139
$(n\text{-}C_4H_9)_4Ge$	*Trans*	641	142
	Gauche	556	142
$(C_2H_5)_4Sn$		508 490	123,124
$(n\text{-}C_3H_7)_4Sn$	*Trans*	590	123,140,141
	Gauche	500	123,140,141
$(n\text{-}C_4H_9)_4Sn$	*Trans*	592	140,141
	Gauche	503	140,141

[a]The *trans* and *gauche* torsional configurations are illustrated in 1.7 and 1.8, respectively.

R=n-C_3H_7,n-C_4H_9,i-C_3H_7,i-C_4H_9,*sec*-C_4H_9) show all of these compounds to have a monomeric structure. Additional but limited vibrational data have also been reported for propyl and/or butyl derivatives of cadmium (146), mercury (147), gallium (148), silicon (149), lead (150), phosphorus (110), arsenic (151), antimony (117,118), and chromium (152).

In addition to the CH_3 modes, an analogous set of modes is expected for the C_3C skeleton in *tert*-butyl derivatives. In solution, *tert*-butyllithium is a tetramer (130,153) with the lithium and α-carbon atoms at the vertices of two concentric, interpenetrating tetrahedra (153). A normal coordinate analysis that neglected the hydrogen atoms has been reported for $(t\text{-}C_4H_9Li)_4$ (154). It has been suggested in this study that the extent of Li-Li interaction is small and amounts to less than 5% of the total bonding-electron density in the Li_4C_4 cage. Contrary to what was found for dimethyl- and diethylberyllium, the vibrational spectrum of $(t\text{-}C_4H_9)_2Be$ shows it to be monomeric with a linear C_2Be skeleton (155,156). In the vapor and liquid states, the vibrational spectra of $(t\text{-}C_4H_9)_2Be$ are consistent with a structure of D_{3d} symmetry (157). On crystallization, the spectra of $(t\text{-}C_4H_9)_2Be$ show increased complexity, indicating that the symmetry has become D_3, although a lower symmetry of C_{2v} could not be ruled out (157). In the original vibrational study of $(t\text{-}C_4H_9)_2Be$ (155), the ν_s(BeC) mode (545 cm^{-1}) was assigned at a higher frequency than the ν_a(BeC) mode (448 cm^{-1}). The same trend was reported for the ν_s(ZnC) (505 cm^{-1}) and ν_a(ZnC) (308 cm^{-1}) modes of $(t\text{-}C_4H_9)_2Zn$ (155). In the more recent vibrational study of $(t\text{-}C_4H_9)_2Be$, the ν_a(BeC) and ν_s(BeC) modes were assigned at 580 and 549 cm^{-1}, respectively, for the vapor and liquid states and 668 and 565 cm^{-1}, respectively, for the solid state (156). The infrared spectrum of $(t\text{-}C_4H_9)_4Cr$ in heptane shows medium intensity bands at 572, 500, 428, and 385 cm^{-1}, with the 385 cm^{-1} band assigned to a ν(CrC) mode (157).

Qualitative infrared and Raman assignments have been made for the neopentyl derivatives $[(CH_3)_3CCH_2]_4M$ (M=Ge,Sn,Ti,Zr,Hf) (158). Some low-frequency infrared data have been listed for $[(CH_3)_3CCH_2]_4Cr$ (157), while infrared data above 600 cm^{-1} have been reported for $[(C_6H_5)(CH_3)_2CCH_2]_2Hg$ (159).

D. MISCELLANEOUS ALKYL COMPOUNDS

1. Cycloalkyl Complexes

Vibrational data have been reported for several cycloalkyl, $c\text{-}C_nH_{2n-1}$, complexes. Solid cyclohexyllithium has been characterized by infrared and single crystal X-ray studies. A hexameric structure was found in which the core of lithium atoms forms a distorted octahedron. Also, the cyclohexyl rings are in the chair configuration with the α-carbon atom of each ring bonded to the six smaller of the eight triangular faces formed by the lithium atoms. The ν(CH) mode frequencies associated with the α-carbon atom (2800 and 2720 cm^{-1}) are lower than the ν(CH) mode assignments made for cyclohexane (2900 and 2800 cm^{-1}). Bands at 600 and 570 cm^{-1} were tentatively assigned to the ν(LiC) modes (160).

Infrared data have been listed for the complexes R_2Zn (R=$c\text{-}C_3H_5$,$c\text{-}C_4H_7$,$c\text{-}C_5H_9$,$c\text{-}C_6H_{11}$) (161). In a series of cycloalkyl chlorides and bromides, the C-H

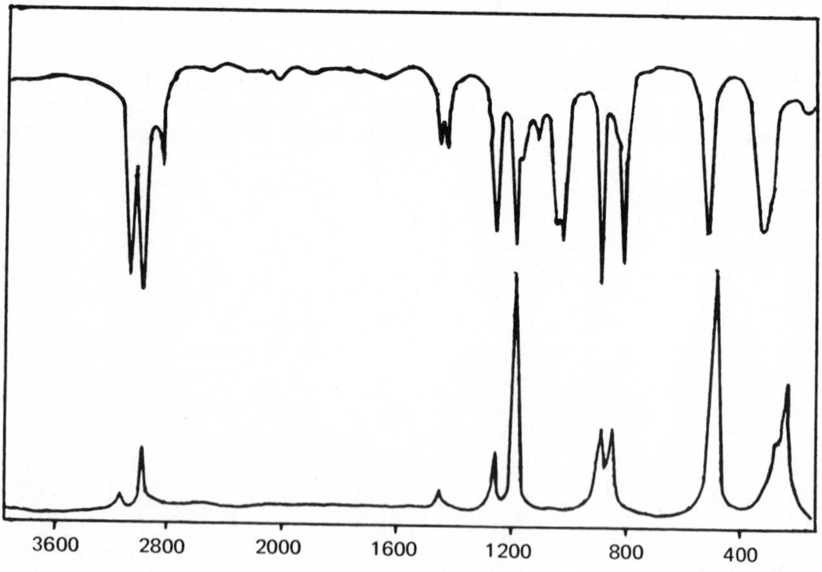

Figure 1.9. Infrared (top) and Raman (bottom) spectra of (c-C_3H_5)$_2$Hg (163).

stretching mode frequencies have been observed to increase as the size of the cycloalkyl ring decreases (162). A similar trend has been observed for (*c*-C_3H_5)$_2$Hg (ν(CH)=3058 cm^{-1},ν(CH_2)=2986 to 2864 cm^{-1}), (*c*-C_5H_9)$_2$Hg (ν(CH)=2936 cm^{-1},ν(CH_2)=2854 cm^{-1}), and (*c*-C_6H_{11})$_2$Hg (ν(CH)=2915 cm^{-1},ν(CH_2)=2860 to 2815 cm^{-1}) (163). The ν(CH) mode frequencies for this series are more sensitive than the ν(CH_2) mode frequencies to the ring size. Although the infrared and Raman spectra of (*c*-C_5H_9)$_2$Hg and (*c*-C_6H_{11})$_2$Hg show both to have a symmetry of C_s or C_1, a *trans* structure of C_{2h} symmetry (1.9) has been reported for (*c*-C_3H_5)$_2$Hg (163). The infrared and Raman spectra for (*c*-C_3H_5)$_2$Hg are illustrated in Figure 1.9 (163).

The principal infrared bands have been listed for (*c*-C_3H_5)$_3$B (164). Infrared (165,166) and Raman data support a dimeric structure for (*c*-C_3H_5)$_3$Al in the solid (165,166) and liquid (165) phases. The infrared

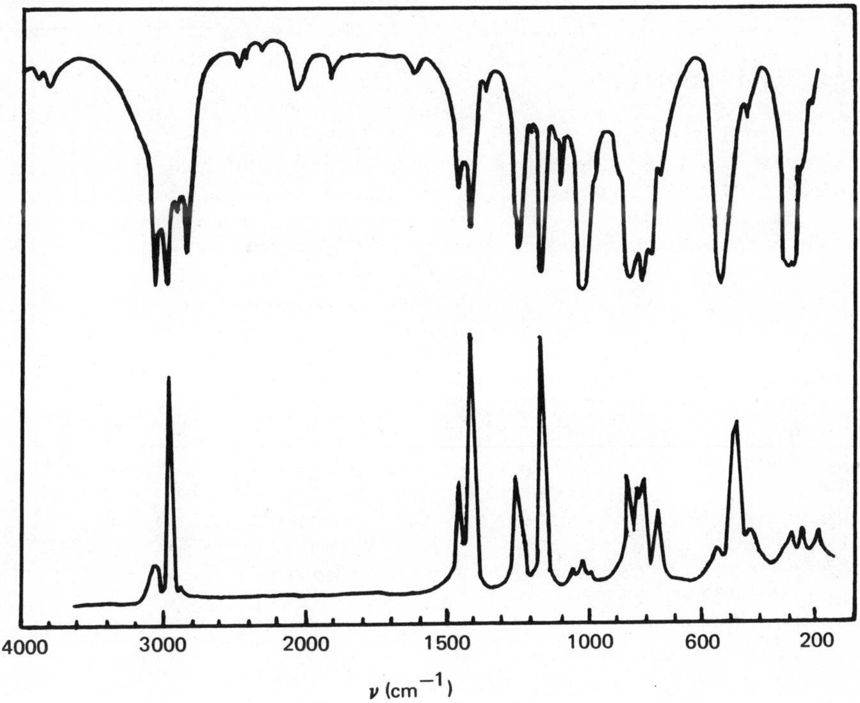

Figure 1.10. Infrared (top) and Raman (bottom) spectra of $(c\text{-}C_3H_5)_3Ga$ (167).

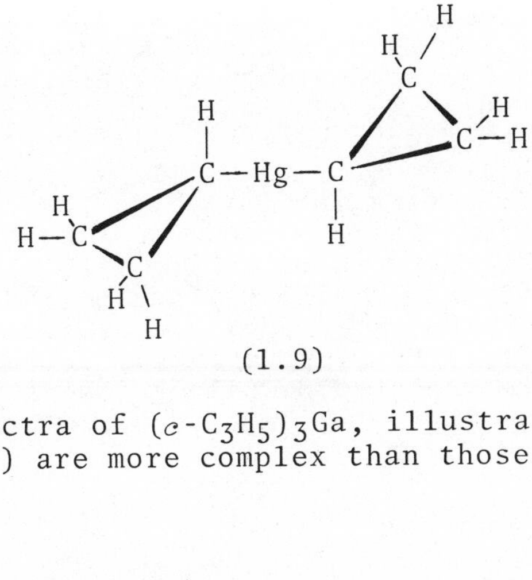

(1.9)

and Raman spectra of $(c\text{-}C_3H_5)_3Ga$, illustrated in Figure 1.10 (167) are more complex than those illustrated in Figure 1.9 for $(c\text{-}C_3H_5)_2Hg$ (163). This indicates that $(c\text{-}C_3H_5)_3Ga$ has a relatively low symmetry of C_1, because of the cyclopropyl rings (167).

The infrared and Raman spectra have been assigned for $(c\text{-}C_3H_5)_4M$ (M=Si,Ge,Sn) on the basis of D_{2d} symmetry, which, however, is strictly found only for the tin compound (168).

A complete study has been reported of the infrared and Raman spectra of liquid $(c\text{-}C_3H_5)_5Sb$ (169). An analysis of the low-frequency (600 to 200 cm^{-1}) spectral region is consistent with a square pyramidal skeleton rather than the trigonal bipyramidal skeleton found for most five-coordinate antimony(V) derivatives.

2. Heterocyclic Complexes

Infrared and microwave spectroscopy have proved very valuable in characterizing the structures of heterocyclic compounds (170). The degree of planarity found for small ring compounds is determined in part by the opposing forces of ring strain, which is decreased in planar four- and five-membered rings, and by the tendency of the methylene groups to adopt a staggered configuration with respect to one another, which favors the formation of nonplanar ring skeletons.

Complete vibrational assignments have been made for the highly strained heterocycle phosphiran, $(CH_2)_2PH$ (171,172), and its deuterated analogs, phosphiran-1-d_1 (171,172) and phosphiran-2,3-d_4 (171).

Complete vibrational spectra and normal coordinate analyses have been reported for the series $(CH_2)_3SiX_2$ (X=H,D,F,Cl) (173). Considerable mixing of the vibrational modes was observed in all four compounds (173). The far-infrared spectra and the assignment of the ring-puckering mode for this series have also been thoroughly investigated (174). The ring-puckering mode of four-membered rings provides valuable information about the structure of the ring skeleton. All four molecules are nonplanar with a symmetry of C_s (173). The dihedral angle of the C_3Si ring in $(CH_2)_3SiH_2$ was calculated to be 36° (174). The infrared and Raman spectra of $(CH_2)_3Se$ and its deuterated analog (175) show both to have C_s symmetry and a dihedral angle of $32.5^\circ \pm 2^\circ$, as compared to a value of $29.5^\circ \pm 1^\circ$ determined from microwave data (176). The ring potential determined from the far-infrared spectra of silacyclopent-3-ene, $C_4H_6SiH_2$, shows that the compound has a planar five-membered ring with no barrier to internal rotation in the CH_2SiCH_2 unit (177).

Although one ring-puckering mode is expected for four-membered rings and unsaturated five-membered rings, two out-of-plane ring modes (1.10) are expected for sat-

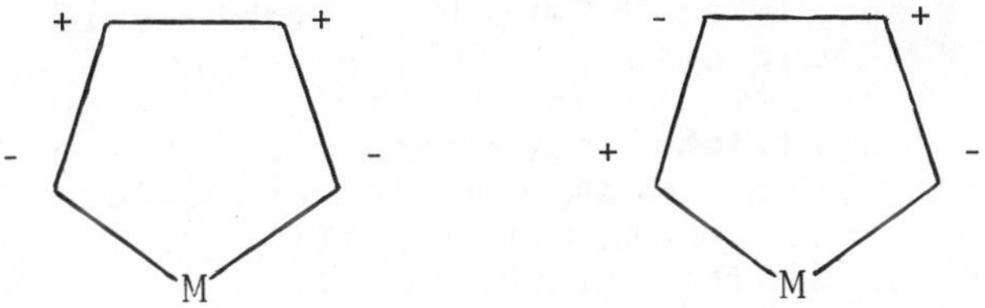

(1.10)

urated five-membered rings. In some compounds, coupling of these modes can produce a ring vibration in which the location of the ring puckering rotates about the ring. This vibrational motion is a type of pseudorotation. Far-infrared data for $(CH_2)_4SiH_2$ (178,179) and $(CH_2)_4SiD_2$ (178) are consistent with a nonplanar C_2 structure. Hindered pseudorotation is observed in both compounds with the relatively high barrier to such motion calculated to be 3.9 and 4.2 kcal/mole for the nondeuterated (178,179) and deuterated (178) compounds, respectively. Vibrational data for $(CH_2)_4GeH_2$ and $(CH_2)_4GeD_2$ show that the rings in these compounds also have nonplanar, twisted structures of C_2 symmetry (180,181). As with the corresponding silicon compounds, hindered pseudorotation is observed, with the barriers calculated to be 5.9±0.1 and 6.0±0.3 kcal/mole, respectively. Similar results have been reported for $(CH_2)_4Se$ (170), with a barrier to pseudorotation (5.1 kcal/mole) intermediate between those of $(CH_2)_4SiH_2$ and $(CH_2)_4GeH_2$. A complete infrared assignment has been reported for $(CH_2)_4BC_6H_5$ (182).

Empirical assignments have been made for the infrared spectra of 21 cyclopentamethylenedialkylsilanes (183). In a more detailed infrared and Raman study of the series $(CH_2)_5SiX_2$ (X=H,Cl,CH_3), the ν(SiC) mode frequencies are not very different from those of the corresponding silane derivatives, $(CH_3)_2SiX_2$ (X=H,Cl,CH_3) (184). This has been interpreted as indicating that very little strain is present in the six-membered rings. The upward shift in the frequency of the ν_s(SiC) mode of the exocyclic methyl groups for $(CH_2)_5Si(CH_3)_2$ (586 cm^{-1}), $(CH_2)_4Si(CH_3)_2$ (598 cm^{-1}), and $(CH_2)_3Si(CH_3)_2$

(612 cm^{-1}), however, shows increased ring strain with decreasing ring size (184). Infrared data have also been tabulated for $(CH_2)_4Pb(C_6H_5)_2$ and $(CH_2)_nPb(C_2H_5)_2$ (n=4,5) (185).

The infrared spectra (4000 to 200 cm^{-1}) of $(CH_2)_5Se$ in the vapor and liquid states and as amorphous and crystalline solids at -170°C have been recorded (186). In addition, Raman data, including semiqualitative polarization measurements and low-frequency (400 to 50 cm^{-1}) infrared data, have been recorded for the liquid. The vibrational assignments were based on infrared vapor contours, Raman polarization measurements, and normal coordinate calculations.

3. Benzyl Complexes

Because of the relatively heavy mass of the metal atoms, little vibrational coupling has been observed between the benzyl rings in the infrared and Raman spectra of $(C_6H_5CH_2)_2Hg$, $(C_6H_5CH_2)_4Ge$, and $(C_6H_5CH_2)_4Sn$ (187). Therefore, the models $C_6H_5CH_2M$ and C_nM (n=2,4) were used in analyzing the vibrational data. Using these models, however, more bands were observed in the ν(MC) region than are permitted by the selection rules. One possible explanation given was the presence of rotational isomers because of hindered rotation about the phenyl-carbon bond. Infrared data have been reported for $(C_6H_5CH_2)_4Ti$, and its structure has been discussed on the basis of these data (188). The crystal and molecular structures have been reported for $(C_6H_5CH_2)_4Ti$ (189,190) and the analogous tin and hafnium derivatives (190). The infrared spectrum has been assigned for $(C_6H_5CH_2)_4W$ (ν_a(WC)=556 (m),535 (m),523 (w) cm^{-1}; ν_s(WC)=485 (vw) cm^{-1}) (191) while the infrared spectra (3200 to 400 cm^{-1}) have been tabulated and some assignments made for $(C_6H_5CH_2)_4M$ (M=Zr,ν_a(ZrC)=536 (m),518 (s), 505 (m) cm^{-1},ν_s(ZrC)=475 (vw) cm^{-1};M=Th,ν_a(ThC)=539 (m), 516 (s) cm^{-1},ν_s(ThC)=463 (vw) cm^{-1}) (192).

4. Mixed Alkyl and Alkyl-Aryl Complexes

Although $(CH_3)_3Al$ is dimeric and $(i\text{-}C_4H_9)_3Al$ is monomeric, various mixed trialkylaluminum compounds with bridging methyl groups are produced on mixing these compounds. This exchange has been followed by molecular weight measurements and observation of changes in the infrared (Figure 1.11) and Raman (Figure 1.12)

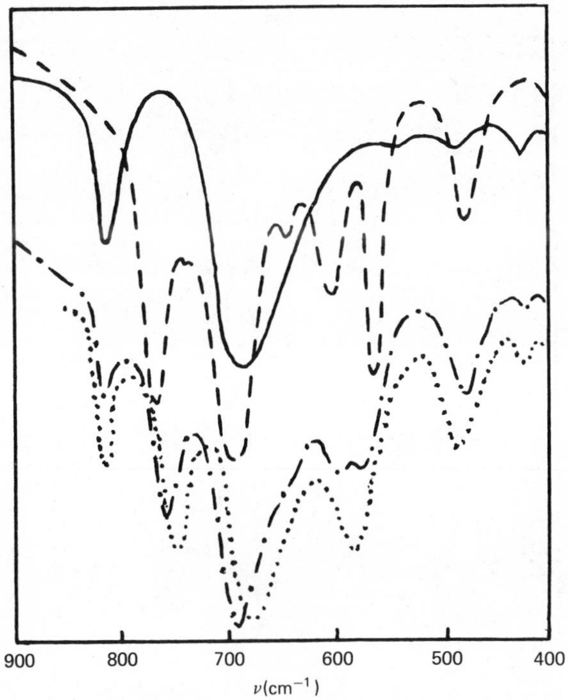

Figure 1.11. Infrared spectra of the cyclohexane diluted $[(CH_3)_3Al]_2$ *(---) and* $(i\text{-}C_4H_9)_3Al$ *(——), and their 2:1 and 1:2 mixtures which produce the methyl bridged dimers* $[(i\text{-}C_4H_9)_2(CH_3)Al]_2$ *(···) and* $[(i\text{-}C_4H_9)(CH_3)_2Al]_2$ *(-·-), respectively, through alkyl exchange (144).*

spectra of solutions containing different ratios of these alkyl derivatives (144). Vibrational data have also been reported for mixed alkyl and/or aryl derivatives of silicon (193-197), germanium (121,122,142,198), tin (124), lead (119), phosphorus (199) and arsenic (122,151,198,200).

E. ALKYL COMPOUNDS WITH OTHER FUNCTIONAL GROUPS

1. Hydrides

Vapor density measurements show CH_3BeBH_4 to be largely dimeric in the vapor phase (80). The similarity of the infrared spectra of the vapor and solid phases indicate

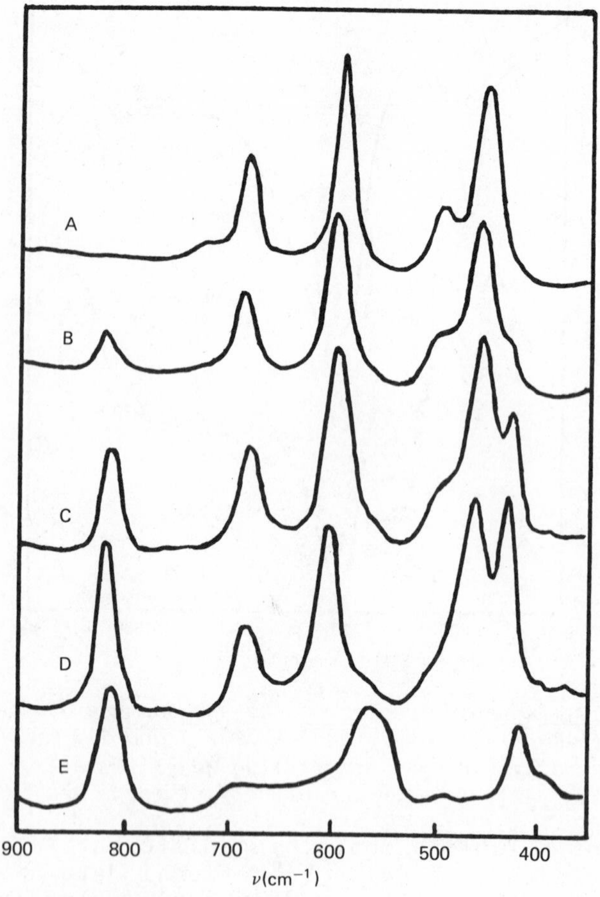

Figure 1.12. Raman spectra of undiluted mixtures of $[(CH_3)_3Al]_2$ *and* $(i\text{-}C_4H_9)_3Al$ *in which the* $(CH_3)_3Al/(i\text{-}C_4H_9)_3Al$ *ratios were (A) 6:0, (B) 5:1, (C) 4:2, (D) 3:3, and (E) 0:6 (144).*

no significant structural changes occur on condensation. The solid state infrared and Raman spectra of dimeric CH_3BeBH_4 and CH_3BeBD_4 also indicate the presence of bridging methyl groups between the beryllium atoms and double hydrogen bridges between the beryllium and boron atoms (80). The analysis of the vibrational data include assignments of the modes for the BeH_2BH_2 unit.

Vibrational data have been reported and assigned for most members of the series $R_nB_2X_{6-n}$ (n=1 to 4;X= H,D;R=CH_3,CD_3,C_2H_5,C_2D_5) (201-205). Raman data have also been reported for $CH_3B_2H_5$, 1,1-$(CH_3)_2B_2H_4$, and $(CH_3)_3B_2H_3$ (205-207). The boron atoms in these compounds are bridged by two hydrogen atoms. Four bridging B-H stretching modes, ν(BH'), (1.11 to 1.14) are ex-

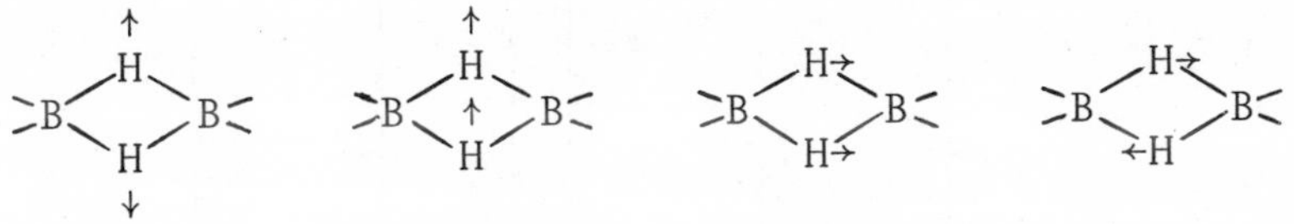

Symmetric in-phase	Symmetric out-of-phase	Asymmetric in-phase	Asymmetric out-of-phase
(1.11)	(1.12)	(1.13)	(1.14)

pected for these bridging hydrogen ligands. These have been assigned from approximately 2150 to 1525 cm^{-1} for the nondeuterated methyl- and ethyl-boron hydrides and from approximately 1550 to 1150 cm^{-1} for the corresponding deuterides, while the terminal ν(BH) and ν(BD) modes have been assigned from 2600 to 2500 cm^{-1} and 1965 to 1840 cm^{-1}, respectively. The Raman (206,207) spectra in the 2000 cm^{-1} region are complicated by the presence of overtone and combination bands, the intensity of which is enhanced through Fermi resonance at the expense of the ν(BH') modes occuring in the same frequency region. The total infrared assignments made for the methyl- and ethyl-boron hydrides are compared in Figure 1.13 (208).

For symmetric tetraalkyldiboranes, $(R_2BH)_2$, such as the methyl (204,208), ethyl (205,208), *n*-butyl (209), *iso*-butyl (209), siamyl (210), and cyclohexyl (210) derivatives, a characteristic in-phase ν(BH') mode (1.13) is observed from approximately 1605 to 1550 cm^{-1}. While the frequency of this mode is at the lower end of the 1605 to 1550 cm^{-1} range for 9-borabicyclo[3.3.1]nonane (1.15) (1560 cm^{-1}) (211), bisborolane (1.16) (1570 cm^{-1}) (212), and bisborinane (1.17) (1560 cm^{-1}) (213,214), it appears at the higher end of this range for the transanular B-H bridged compounds 1,6-diboracyclodecane (1.18) (1610 cm^{-1}) (212,215) and 2,7-dimethyl-1,6-diboracyclodecane (1.19) (1595 cm^{-1}) (214). Hydroboration of tetramethylene has resulted in the hydrogen-bridged

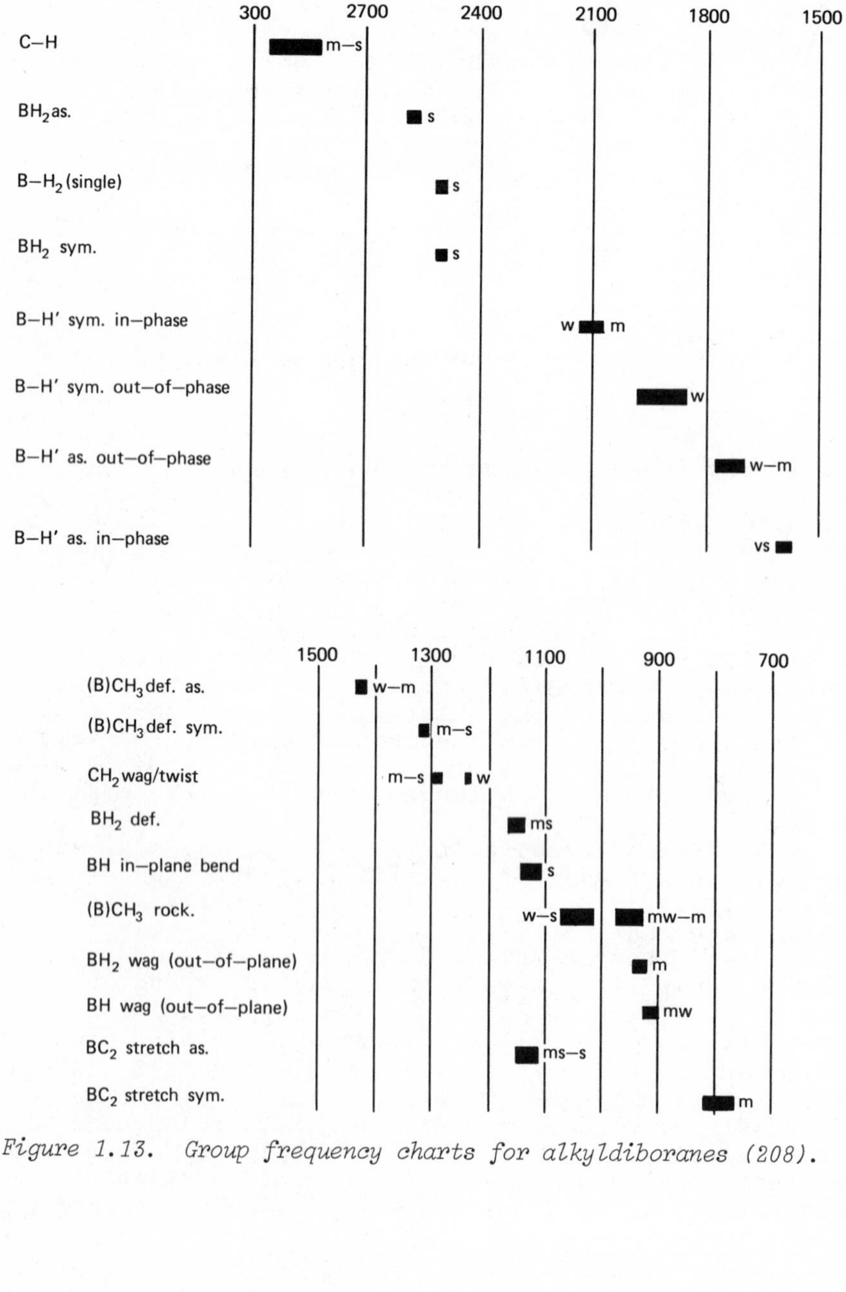

Figure 1.13. Group frequency charts for alkyldiboranes (208).

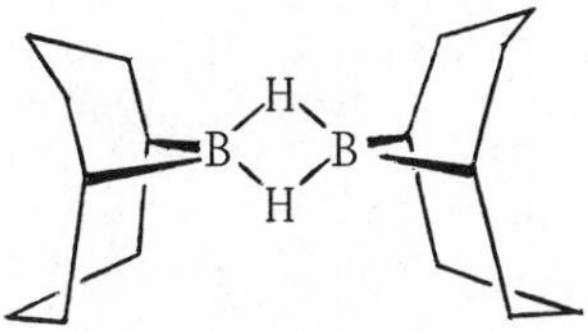

(1.15)

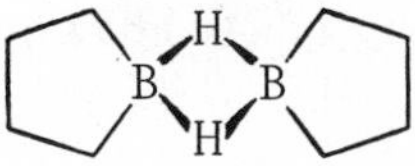

(1.16)

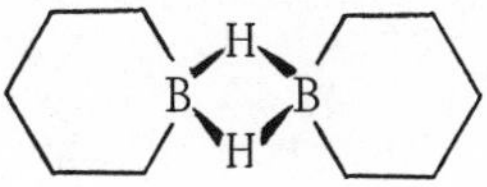

(1.17)

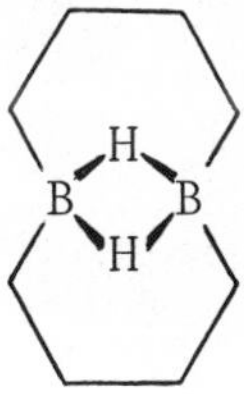

(1.18)

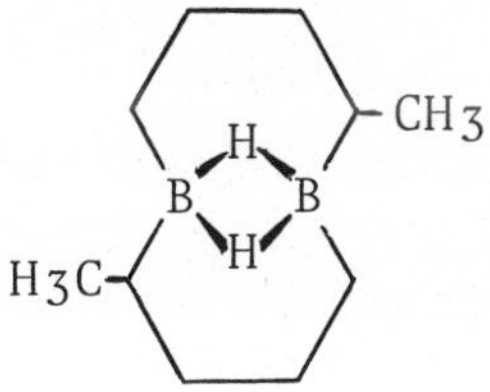

(1.19)

dimers 1,2-dithexyldiborane and 1,1,2-trithexyldiborane, both of which show characteristic terminal and bridging B-H stretching modes at 2540 and 1565 cm^{-1}, respectively (216,217). An attempt to replace the remaining terminal hydride ligand produced monomeric dithexylborane (1.20),

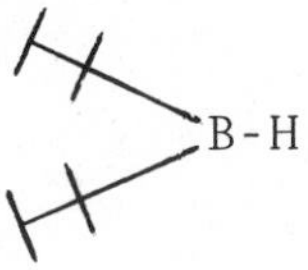

(1.20)

which exhibited a band at 2470 cm^{-1} but practically no bands in the 1600 to 1500 cm^{-1} region (216,217). Although dithexylborane is monomeric and exhibits an infrared band characteristic of only a terminal hydride ligand (2470 cm^{-1}), and all other dialkylboranes are dimeric and exhibit an infrared band characteristic of only a bridging hydride ligand (1600 to 1500 cm^{-1}), the infrared spectra of several thexylmonoalkylboranes exhibit infrared bands at both 2470 cm^{-1} and from 1600 to 1500 cm^{-1} (217). The intensity of the 2470 cm^{-1} band for these derivatives relative to that of the band in the 1600 to 1500 cm^{-1} region increases in the following order and has been observed to correlate with the steric requirements of the olefins: isobutylene < cyclopentene

< 2-butene < 1-methylcyclopentene < norbornene < 2-methyl-2-butene. These results have been interpreted as strongly suggesting that these thexylmonoalkylboranes exist as a monomer-dimer equilibrium and that the large steric requirements of the two alkyl groups are primarily responsible for the existence of such monomeric dialkylboranes.

Cryoscopic measurements have shown that the dialkylaluminum hydrides exist as trimers in both the pure liquid and solution states (218). A strong intensity infrared band at about 1800 cm^{-1} has been assigned to the ν(AlH) mode; this region is extended with increasing association of the R_2AlH molecules (218). The ν(AlH) modes have also been assigned from 1900 to 1700 cm^{-1} in donor-acceptor complexes formed between dialkylaluminum hydrides and various amines and ethers (219).

The gallium complexes $(CH_3)_nGaH_{3-n}[N(CH_3)_3]$ (n=1,2) show a strong intensity infrared band at about 1845 cm^{-1} because of the ν(GaH) modes; this band shifted to about 1300 cm^{-1} on deuteration (220).

The ν(MH) modes of alkyl-metal hydrides of the group IVb, Vb, and VIb elements give rise to relatively strong intensity infrared and Raman bands. This is illustrated by the infrared (Figure 1.14) and Raman (Figure 1.15) spectra of CH_3AsH_2 and its deuterated analogs (221). Generally, the metal-hydride stretching modes of MH_2 and MH_3 units are accidentally degenerate. The metal-hydride and -deuteride deformation modes are not as easily identified as the ν(MH) and ν(MD) modes because of vibrational coupling. Normal coordinate calculations have shown a strong interaction between the $\rho_r(CH_3)$, $\rho_w(PH_2)$, and $\rho_t(PH_2)$ modes of CH_3PH_2 (222) and the ν(AsC) and $\rho_w(AsD_2)$ modes of CH_3AsD_2 and CD_3AsD_2 (221). Similarly, no characteristic Si-H and Si-D deformation modes occur for $(CH_3)_2SiH_2$, $(CH_3)_3SiH$, and the analogous deuterides because of coupling between these modes, the $\rho_r(CH_3)$ modes, and, to a lesser extent, the ν(SiC) modes (223). Similar interactions have been found in normal coordinate calculations for $(CH_3)_nSiH_{4-n}$ (n=1 to 3) (225,226). Normal coordinate calculations have also shown mixing of the $\delta_d(GeH_3)$ and $\rho_r(GeH_3)$ modes, and the $\delta_d(SnH_3)$ and $\rho_r(SnH_3)$ modes of CH_3GeH_3 and CH_3SnH_3, respectively, with the $\rho_r(CH_3)$ modes of these compounds (227). The $\rho_r(CH_3)$ modes become more characteristic when the hydrogen atoms of the MH_3 units are replaced with the heavier deuterium atoms.

The ν(MH) mode frequencies decrease for a family of elements as the mass of the metal atom increases. The

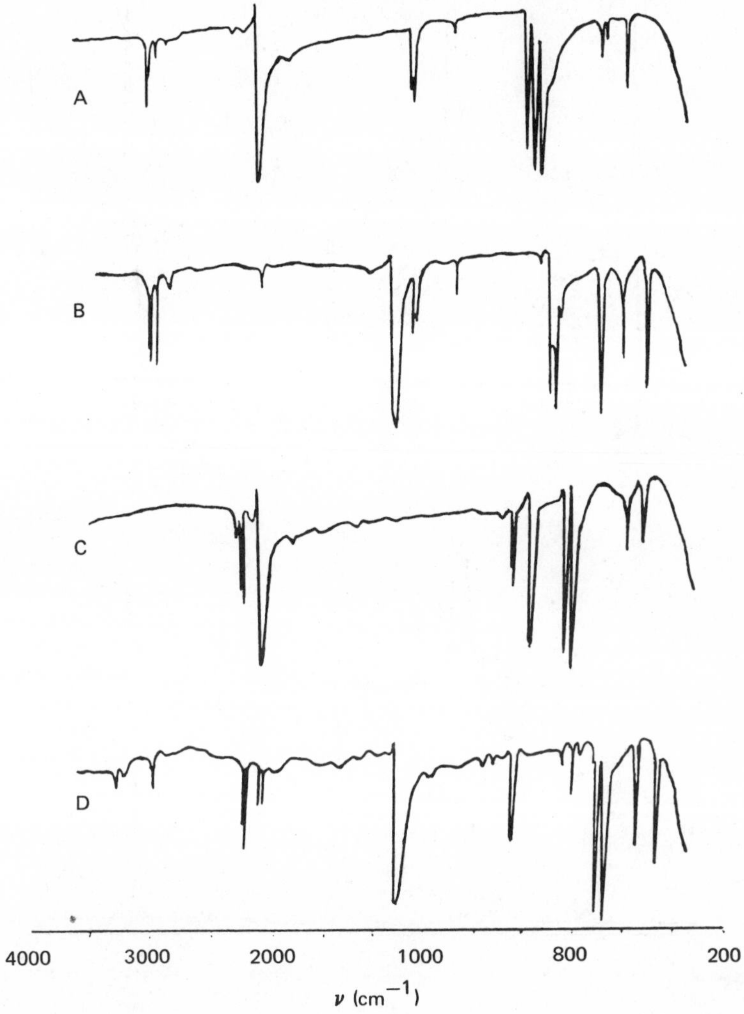

Figure 1.14. Infrared spectrum of solid films of (A) CH_3AsH_2, (B) CH_3AsD_2, (C) CD_3AsH_2, and (D) CD_3AsD_2 at -190°C (221).

ν(MH) mode frequency also decreases for a given metal as each hydrogen atom is progressively replaced with an alkyl group. Although phosphiran, $(CH_2)_2PH$, contains a strained three-membered ring, the position of the ν(PH)

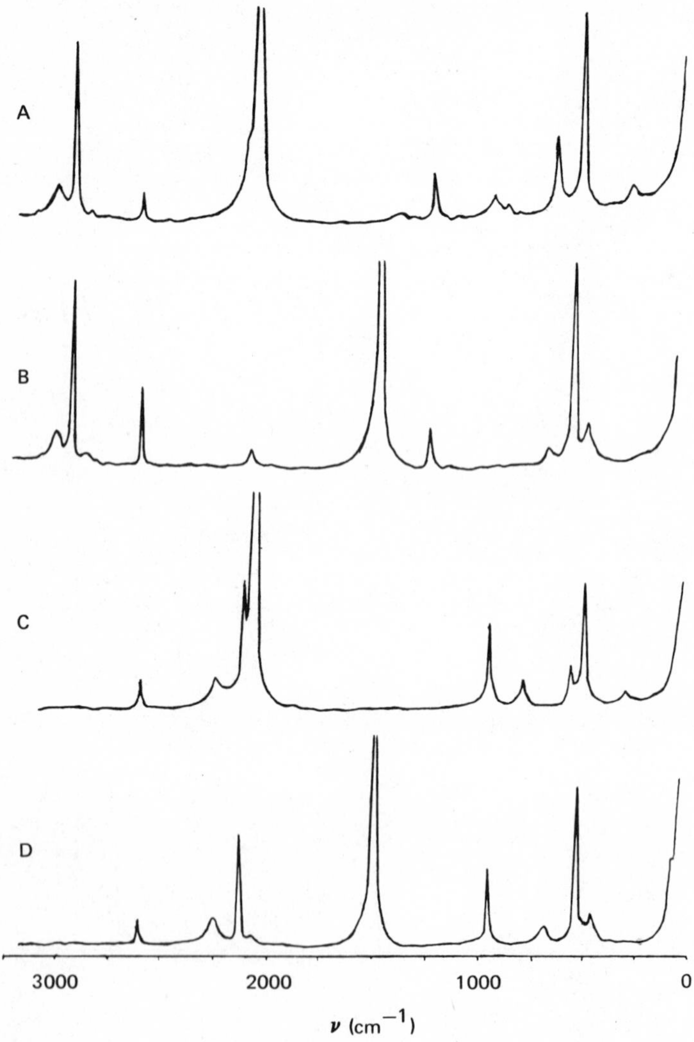

Figure 1.15. Raman spectrum of liquid (A) CH_3AsH_2, (B) CH_3AsD_2, (C) CD_3AsH_2, and (D) CD_3AsD_2 at room temperature (221).

mode (2291 cm^{-1}) does not appear to be influenced by this structure (171) and is not very different from the position of the ν(PH) mode for $(CH_3)_2PH$ (2290 cm^{-1}) (228). There is also little difference between the ν(MH) mode frequencies reported for other heterocyclic organometallic hydrides (173,178,180) and those of the corresponding dimethyl-metal hydrides.

The ν(MC) modes for alkyl-metal hydrides are found at higher frequencies than those of the corresponding fully alkylated compounds. The several Raman bands observed in the ν(PC) region of $(C_2H_5)_2PH$ (110,229) might indicate the presence of several rotational isomers. A similar assumption has been made in assigning the ν(SiC) modes for various alkylsilanes (230,231). The order of the K(MC) force constantes for monomethyl-group IVb hydrides,

Compound	CH_3SiH_3 >	CH_3GeH_3 >	CH_3SnH_3
K(MC) (mdynes/A)	2.97	2.70	2.25

is expected since the metal-carbon bond strength should increase with decreasing polarity of the metal-carbon bond (232).

In a study of the low-temperature, far-infrared and Raman spectra of $C_2H_5SiH_3$, $C_2H_5SiD_3$, and $C_2H_5GeH_3$, the methyl torsion modes have been assigned at 211, 210, and 189 cm^{-1}, respectively, and barriers to internal rotation calculated to be 2.82, 2.88, and 2.33 kcal/mole, respectively (233). The MH_3 and MD_3 torsional mode frequencies have also been assigned for $C_2H_5SiH_3$ (142 cm^{-1}) and $C_2H_5SiD_3$ (113 cm^{-1}) (233). Table 1.12 summarizes the ν(MC) and ν(MH) mode assignments for metal-hydrides and -deuterides, and the methyl derivatives of these compounds, while Table 1.13 includes ν(MH) mode assignments for various higher alkyl derivatives of the group IVb hydrides.

The ν(SiH) mode frequencies have been correlated with a series of empirical constants which indicate that the substituent effects at the silicon atom are additive (262), and also with the sum of the Taft inductive constants, $\Sigma\sigma^*$ (263,264). A significant deviation observed on correlating the ν(SiH) mode frequencies with $\Sigma\sigma^*$ indicates that the inductive influence of other substituents is not solely responsible for the observed variations (265). Similarly, while the ν(GeH) mode frequency for several triorganogermanes, R_3GeH, has been related to $\Sigma\sigma^*$ for the R groups when there is no possibility of $d\pi$-$p\pi$ bonding between the R groups and the germanium atom, this was not the case when such an interaction was possible (266). For these latter derivatives, the ν(GeH) mode frequency is determined by the $d\pi$-$p\pi$ interaction as well as the presence of inductive effects.

TABLE 1.12
Metal-Carbon, Metal-Hydride, and Metal-Deuteride Stretching Mode Assignments (cm^{-1}) for Metal-Hydrides and -Deuterides, and the Methyl Derivatives of These Compounds

Compound	ν(MH)	ν(MC)	Refs.	Compound	ν(MD)	ν(MC)	Refs.
SeH_2	2358		234	SeD_2	1698		234
	2345				1687		
CH_3SeH	2330	582	235,236	CH_3SeD	1680	599	235,236
CH_3TeH	2016	516	237	CH_3TeD	1449	525	237
PH_3	2328		238	PD_3	1698		238
	2323				1694		
CH_3PH_2	2309	676	222,239	CH_3PD_2	1681	686	222
	2305				1674		
$(CH_3)_2PH$	2290	709[a]	228				
		665					
$CH_3PH_3^+$	2505	753	240	$CH_3PD_3^+$	1815	776	240
	2453				1769		
AsH_3	2123		241	AsD_3	1529		241
	2116				1523		
CH_3AsH_2	2102	563	221	CH_3AsD_2	1517	585	221
SiH_4	2191		238	SiD_4	1597		238
	2187				1558		
CH_3SiH_3	2169	701	224,242	CH_3SiD_3	1577	b	224,242
	2166				1558		
$(CH_3)_2SiH_2$	2145	728	223	$(CH_3)_2SiD_2$	1570	700	223
	2142	659			1554	636	
$(CH_3)_3SiH$	2123	711	223	$(CH_3)_3SiD$	1548	704	223
		624				622	
GeH_4	2114		243	GeD_4	1522		243
	2106				1504		

CH_3GeH_3	2106	602	244,245	CH_3GeD_3	1500	b	244
	2082				1496		
$(CH_3)_2GeH_2$	2080	604	245,246	$(CH_3)_2GeD_2$	1480	b	246
	2062	590					
$(CH_3)_3GeH$	2049	601	245-247	$(CH_3)_3GeD$	1466	b	246
		573					
SnH_4	1901		248	SnD_4	1368		248
	b				b		
CH_3SnH_3	1875	527	227	CH_3SnD_3	1352	509	227
$(CH_3)_2SnH_2$	1869	536	249	$(CH_3)_2SnD_2$	1338	531	249
		514				515	
$(CH_3)_3SnH$	1837	521	249,250	$(CH_3)_3SnD$	1326	537	249
		516				520	
$(CH_3)_3PbH$	1709	b	251				

[a]Average value.
[b]Value not reported.

TABLE 1.13
Metal-Hydride Stretching Mode Assignments (cm^{-1}) for the Alkyl Derivatives of the Group IVB Hydrides

R=	C_2H_5	Refs.	*i*-C_3H_7	Refs.	*n*-C_3H_7	Refs.	*n*-C_4H_9	Refs.
$RSiH_3$	2164	252					2165	260
R_2SiH_2	2130	253	2110	257	2127	258	2135	260
R_3SiH	2105	254	2092	254	2105	253	2105	254
$RGeH_3$	2080	246,252, 253,255			2067	246	2062	261
R_2GeH_2	2032	246,247, 255					2032	247
R_3GeH	2010	246,247	1989	247	2003	247	2003	247
$RSnH_3$	1853	256	1853	256	1860	256	1862	256
R_2SnH_2	1822	256	1820	256	1833	256	1835	256
R_3SnH	1797	256	1794	256	1811	256	1813	256
R_2PbH_2							1680	259
R_3PbH					1675	259	1680	259

It has further been concluded in this study that the silicon atom is more capable of dπ-pπ interaction than the germanium atom. The ν(GeH) mode frequencies of several derivatives have also been correlated with the nuclear magnetic resonance (nmr) chemical shift (267-271), the ^{119}Sn-^{1}H spin-spin coupling constant (256,267, 270), and calculated bond polarities (272). Vibrational data have been used to calculate the thermodynamic properties of CH_3GeH_3 (232) and CH_3SnH_3 (273).

The ν(MH) mode has been assigned for $(C_2H_5)_2PH$ (2270 cm^{-1}) (110,229), $(C_2H_5)_2SbH$ (1835 cm^{-1}) (117), and $(n\text{-}C_4H_9)_2SbH$ (1855 cm^{-1}) (274). Additional extensive correlations are available for alkyl-silicon hydrides (275) and alkyl-phosphorus hydrides (276).

2. Halides

The intensity of an infrared active ν(M-halide) mode increases as the electronegativity of the halogen increases. The intensity of a Raman active ν(M-halide) mode, however, decreases as the electronegativity of the halogen increases. This is illustrated in Figures 1.16 and 1.17, in which the infrared (600 to 175 cm^{-1}) and Raman (600 to 100 cm^{-1}) spectra, respectively, are illustrated for $(CH_3)_3SbX_2$ (X=F,Cl,Br) (277). It is assumed that the relative intensity of the signals due to the ν(SbC) modes is relatively constant for all three compounds in either the infrared or Raman spectra. This being the case, the intensity difference between the ν(Sb-halide) modes is very evident, especially in the Raman spectra (Figure 1.17) in which the ν(SbF) modes are very weak, while the ν(SbBr) modes are very strong relative to the ν(SbC) modes.

Terminal and bridging halogen ligands can, in general, be distinguished by the fact that the bridging ν(M-halide) modes appear at lower frequencies than the corresponding terminal ν(M-halide) modes. The ν(M-halide) modes of some compounds vary by as much as 40 to 50 cm^{-1} depending on the phase in which the spectrum was recorded. The frequencies are the highest for the gas-phase spectra and lowest for the solid-phase spectra. This has been attributed to intermolecular association in the solid or liquid phases. The association most likely takes place through the halogen atoms and is thus reflected in the differences observed in the ν(M-halide) mode frequencies. The fact that the ν(MC) modes for these compounds remain relatively constant in different

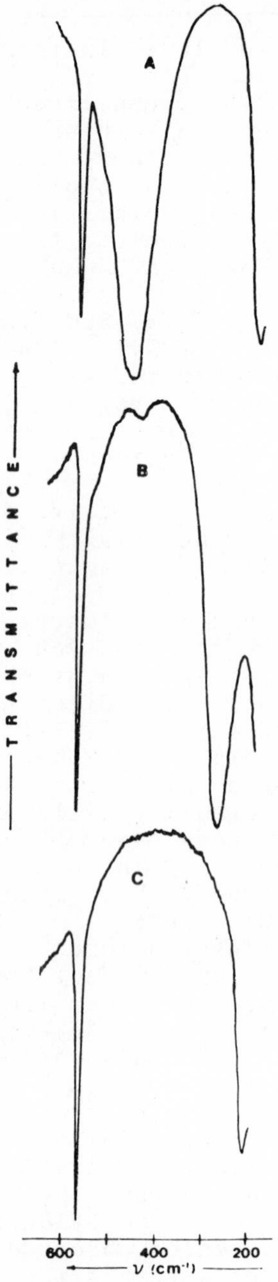

Figure 1.16. Infrared spectra (600-200 cm^{-1}) of (A) $(CH_3)_3SbF_2$, (B) $(CH_3)_3SbCl_2$, and (C) $(CH_3)_3SbBr_2$ (277).

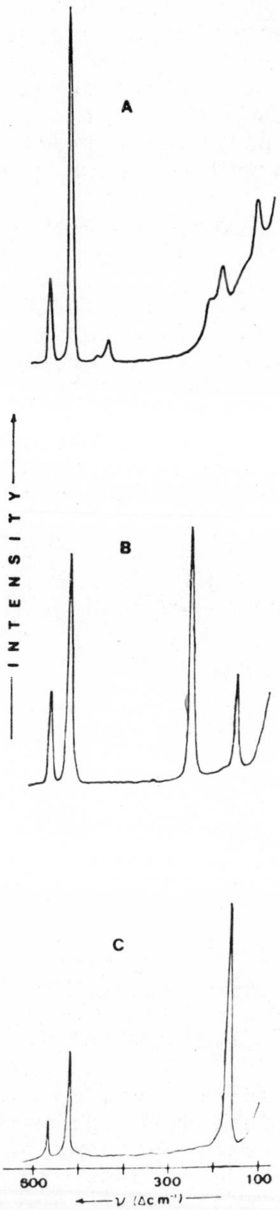

Figure 1.17. *Raman spectra (600-100 cm^{-1}) of (A) $(CH_3)_3SbF_2$, $(CH_3)_3SbCl_2$, and $(CH_3)_3SbBr_2$ (277).*

phases reflects the fact that the alkyl groups are not involved in the association. Several adducts of alkyl-metal halides have been reported. The metal-halide stretching modes for the alkyl-metal halides are at higher frequencies than those of the corresponding adducts. The vibrational data for alkyl-metal halides are discussed separately for each group of elements.

a. Group IA

The reaction products of mixtures of atomic lithium or sodium with CH_3Br, CH_3I, and CD_3I, or atomic potasium with CH_3I in a solid argon matrix gave infrared data that are consistent with the presence of methyl-alkali halides (278). The structure of these species has been pictured as consisting of a linear carbon-metal-halide skeleton in which the *s*-orbital of the metal atom interacts with a *p*-orbital of the halide and a *p*-orbital of a planar methyl group (1.21) through a

$$\begin{array}{c} H \\ | \\ C\text{-}M\text{-}X \\ H \diagdown \\ H \end{array}$$

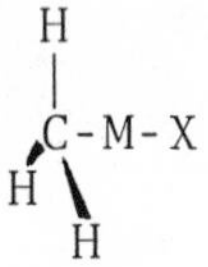

(1.21)

three-centered bond.

b. Group IIA

The Schlenk equilibrium (1.22 to 1.25) is useful

$$R\diagdown M \diagup X \diagdown M = R_2M\ +\ MX_2 = 2RMX = RM \diagup X \diagdown MR$$

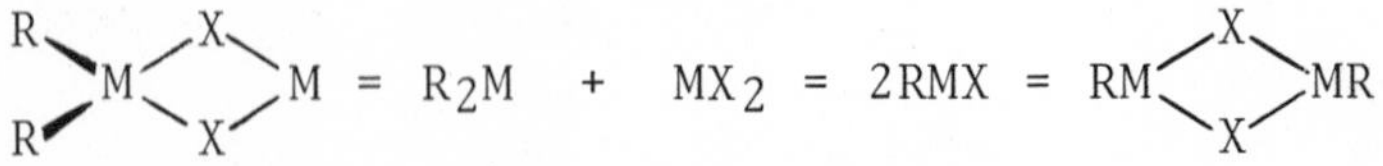

(1.22)   (1.23) (1.24) (1.25)

in describing the possible species in solutions of organomagnesium halides, with the nature of the individual compounds present dependent on the solvent and type of alkyl and halide groups. In Table 1.14, the infrared $\nu(MgC)$ mode assignments are compared for alkylmagnesium fluorides, chlorides, and bromides and dialkylmagnesium compounds in diethyl ether and tetrahydrofuran (279,280). The shoulder on the high-fre-

TABLE 1.14
Comparison of the Infrared ν(MgC) Mode Assignments (cm^{-1}) for $(Alkyl)_2Mg$ and AlkylMgX (X=F,Cl,Br) Compounds[a]

RMgX		ν(MgC)	
R	X	Diethyl Ether Solution	Tetrahydrofuran Solution
CH_3	F		535
	Cl		527
	Br	520	513
	CH_3	525 (593)[b]	535
C_2H_5	F		505 (560)
	Br	508	500
	C_2H_5	512	512
n-C_6H_{13}	F	501 (550)	500 (548)
	n-C_6H_{13}	501 (551)	500 (548)

[a]Data from Ref. 279 for all compounds except for those reported for the alkylmagnesium fluorides and dihexylmagnesium, which are from Ref. 280.
[b]Bands appearing as shoulders are given in parentheses.

quency side of the main absorption for $(CH_3)_2Mg$ and $(n\text{-}C_6H_{13})_2Mg$ is characteristic of bridging alkyl groups and is absent from the spectra of these compounds in tetrahydrofuran where monomeric structures are likely. Likewise, the absence of this shoulder in the infrared spectra of alkylmagnesium chlorides and bromides indicates the lack of such an interaction (280). Molecular weight and/or vibrational data show that while the chlorides and bromides are monomeric in tetrahydrofuran (280), in diethyl ether, monomeric species or dimeric species with halogen bridges may be predominant with the degree of association a function of the solution concentration (280-282). The presence of the shoulder for all of the alkylmagnesium fluorides, with the exception of CH_3MgF in tetrahydrofuran, is explained as possibly resulting from a mixed alkyl-fluoride-bridged species (1.26) (282).

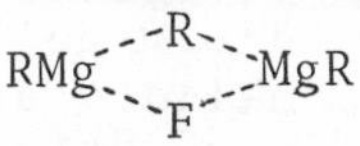

(1.26)

c. Group IIB

A Schlenk equilibrium (1.22 to 1.25) is possible in solutions consisting of equal molar quantities of dialkylzinc or -cadmium compounds and a dihalide of the corresponding metal. For a compound with structure 1.22, both ν(MC) modes are infrared active, while only one ν(MC) mode is infrared active for compounds with structures 1.24 or 1.25. Structures 1.24 and 1.25 are distinguishable since the metal-halogen stretching mode would be at a higher frequency for the former than for the latter.

Although polymeric in the solid state, molecular weight measurements indicate that C_2H_5ZnI is monomeric in an ethyl iodide solution (283). The infrared spectra of C_2H_5ZnI and mixtures of $(CH_3)_2Zn$ and ZnX_2 (X=Cl, Br,I) in tetrahydrofuran (284), and the infrared and Raman spectra of $(CH_3)_2Cd$ and CdX_2 (X=Cl,Br,I) in tetrahydrofuran (285) have been reported. In both studies, only one ν(MC) band was observed for the resulting products; it was concluded, therefore, that the Schlenk equilibrium lies predominantly on the side of the solvated RMX species. For the CH_3ZnX series, structure 1.25 could not be ruled out since no metal-halogen stretching modes were assigned. For the CH_3CdX series, however, the Raman assignments were in the regions expected for terminal cadmium-halogen stretching modes; this was given as evidence for the presence of structure 1.24 rather than the halogen-bridged structure 1.25. The reactions of *n*-butyl Grignard reagents with CdX_2 (X=Cl,Br,I) in diethyl ether using 1:1 and 1:2 molar ratios have been studied using infrared spectroscopy (146).

Organomercury halides are monomeric in all phases and have linear carbon-mercury-halogen skeletons. The K(MC) and K(MX) force constants calculated for CH_3HgX (X=Cl,Br,I,CH_3) are

Compound	CH_3HgCl	CH_3HgBr	CH_3HgI	$(CH_3)_2Hg$
K(MC) (mdynes/A)	2.69	2.60	2.50	2.45
K(MX) (mdynes/A)	2.01	1.80	1.55	

and show the Hg-C bond strength to increase as the electronegativity of X increases (286). Table 1.15 summarizes the ν(MC) and ν(M-halide) mode assignments made for the group IIb halides and alkyl-group IIb halides. Vapor-phase infrared data have also been reported for CH_3HgX (X=Cl,Br,I) (298).

d. Group IIIB

Several vibrational studies have appeared for R_nBX_{3-n} (n=1,2;X=F,Cl,Br;R=CH_3 (299-304),C_2H_5 (304)); normal coordinate calculations were included in many of these studies (301-303). The skeletal mode assignments for these compounds and RBXY (R=CH_3,X=F,Cl,Br,Y=F,Cl,Br;R=C_2H_5,X=F,Y=Br) (304) are given in Table 1.16. The methyl groups rotate freely (299,300), and all compounds are monomeric in the vapor and liquid states. The infrared spectra have also been assigned for some cycloalkyl-boron halides (305).

Although alkyl-boron halides are monomeric, various degrees of association are found for the alkyl derivatives of the other group IIIb halides. Both $(CH_3)_2AlF$ (306) and $(C_2H_5)_2AlF$ (307) are tetramers in benzene with a cyclic skeleton consisting of an alternating Al-F backbone. The original conclusion (306) that the infrared and Raman spectra of $(CH_3)_2AlF$ (illustrated in Figure 1.18) and $(C_2H_5)_2AlF$ satisfy the rule of mutual exclusion was not justified (308). A recent electron-diffraction study of $(CH_3)_2AlF$ vapor has shown that the tetrameric structure is retained in this phase and that the Al_4F_4 ring is puckered (309). A trimeric structure has been found for $(n\text{-}C_3H_7)_2AlF$ and $(n\text{-}C_4H_9)_2AlF$, with the infrared and Raman spectra of both having been reported (310). The infrared and Raman spectra of $(CH_3)_2GaF$ and $(C_2H_5)_2GaF$ also show the presence of trimeric, cyclic structures with bridging fluoride ligands (311,312). On storing $(CH_3)_2GaF$ at 5 to 15°C for 2 to 3 weeks, however, the tetramer was produced (313). The skeletal mode assignments made for dialkylaluminum fluorides and dialkylgallium fluorides are given in Table 1.17. Although the infrared spectrum of $(CH_3)_2InF$ (ν_a(InC)=548 cm^{-1}, ν_s(InC)=493 cm^{-1}) is similar to those of $(CH_3)_2InX$ (X=Cl,Br,I), the relatively high melting point of $(CH_3)_2InF$ and its low solubility in chloroform and diethyl ether reflect some structural differences between it and the other dimethylindium halides (314). The infrared spectrum of solid $(C_2H_5)_2InF$ shows a strong intensity ν_a(InC) mode (518 cm^{-1}) but a weak intensity ν_s(InC) mode (461 cm^{-1}). This, together with its high melting point and low solubility in methylene chloride and benzene, suggests a polymeric structure with a nearly linear C-In-C skeleton (315). The infrared and Raman spectra of the tetramethylammonium salt of $[(CH_3)_3Tl]_2F^-$ (ν_a(TlF)=250 cm^{-1}) are consistent with a structure of D_{3h} symmetry for this compound (315).

TABLE 1.15
Metal-Carbon and Metal-Halide Stretching Mode Assignments (cm^{-1}) for the Group IIB Halides and Alkyl-Group IIB Halides

MX_2 / RMX	Phase	$\nu(MC)$	$\nu_a(MX)$	$\nu(MX)$	$\nu_s(MX)$	Refs.
CH_3ZnCl	THF soln.	536 (IR)				284
CH_3ZnBr	THF soln.	531 (IR)				284
CH_3ZnI	THF soln.	530 (IR)				284
C_2H_5ZnI	THF soln.	506 (IR)				284
$CdCl_2$	Vapor		409 (IR)			287
CH_3CdCl	THF soln.	476 (R)		247 (R)		285
$CdBr_2$	THF soln.,vapor		315 (IR)		173 (R)	285,288
CH_3CdBr	THF soln.	475 (R)		206 (R)		285
CdI_2	THF soln.		265 (IR)		138 (R)	285,288
CH_3CdI	THF soln.	482 (R)		167 (R)		285
CH_3HgF	Solid	561 (IR)		482 (IR)		289
		573 (R)		414 (R)		289
$HgCl_2$	Vapor		413 (IR)		360 (R)	290,291
CH_3HgCl	Benzene soln.	556 (R)		334 (R)		286
	Nujol mull	539 (IR)		304 (IR)		286,292
CD_3HgCl	Solid	510 (R)		298 (R)		293
C_2H_5HgCl	Nujol mull			314 (R)		294
n-C_3H_7HgCl	Nujol mull			322 (IR)		294
$C_6H_5CH_2HgCl$	Nujol mull	552 (IR)		307 (IR)		295

$HgBr_2$	Vapor		293 (IR) 225 (R)	290,291
CH_3HgBr	Benzene soln.	546 (R)	228 (R)	286
	Solid or mull	538 (IR)	202 (R)	292,296
CD_3HgBr	Solid or mull	597 (IR)	201 (R)	292,296
C_2H_5HgBr	Nujol mull	513 (IR)	209 (IR)	112
n-C_3H_7HgBr	Nujol mull		214 (IR)	112
n-C_4H_9HgBr	Nujol mull		246 (IR)	294
$C_6H_5CH_2HgBr$	Nujol mull	550 (IR)	226 (IR)	295
HgI_2	Vapor		237 (IR) 156 (R)	290,291
CH_3HgI	Benzene soln.	535 (R)	181 (R)	286
	Solid or mull	526 (IR)	170 (R)	292,297
CD_3HgI	Solid or mull	483 (IR)	170 (R)	292,297
$C_6H_5CH_2HgI$	Solid or mull	573 (IR)	172 (R)	112

TABLE 1.16
Boron-Carbon and Boron-Halide Stretching Mode Assignments (cm^{-1}) for Methyl- and Ethyl-Boron Halides[a]

Compound	ν(BC)	ν(BX)	Deformations	Refs.
CD_3BF_2	718	1348 1193	470 305	299,304
CH_3BF_2	779	1350 1250	485 478	299,301,304
$(CH_3)_2BF$	1152 710	1261	405 310	301,304
CH_3BCl_2	1078	1018 540	276	302,304
$(CH_3)_2BCl$	1120 1080	590	310	302,304
CH_3BBr_2	1060	965 423	250 175	303,304
$(CH_3)_2BBr$	1130 1152	500	285	303,304
$C_2H_5BF_2$	752	1360 1220	480	304
$(C_2H_5)_2BF$	1100 680	1220	410 355	304
$C_2H_5BCl_2$	1065	890 550	265 190	304
$(C_2H_5)_2BCl$	1080 1010	551	345 210	304
$C_2H_5BBr_2$	1050	930 410	223 172	304
$(C_2H_5)_2BBr$	1075 1010	470	308 170	304
CH_3BFCl		618[b]		304
CH_3BFBr		540[c]		304
CH_3BClBr		990[b] 480[c]		304
C_2H_5BFCl		585[b]		304

[a]Assignments are for derivatives of the ^{11}B isotope.
[b]ν(BCl) mode.
[c]ν(BBr) mode.

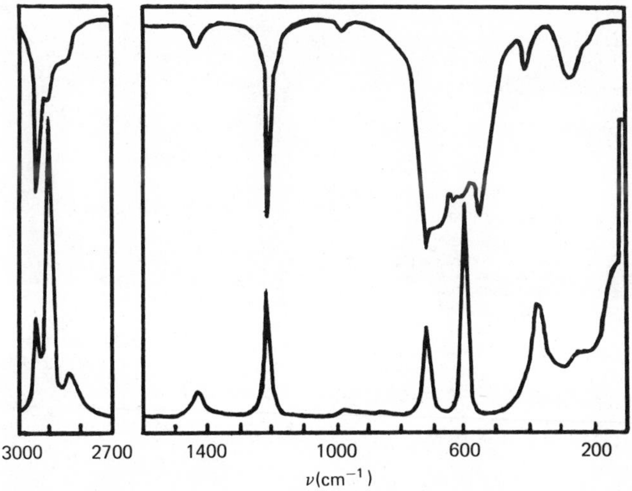

Figure 1.18. Infrared (top) and Raman (bottom) spectra of liquid $[(CH_3)_2AlF]_4$ *(306).*

A halogen-bridged dimeric structure has been found for those members of the $(alkyl)_nMX_{3-n}$ (n=1,2;M=Al,Ga; X=Cl,Br,I) series that have been characterized. The normal modes expected for bridged M_2X_6 molecules are illustrated in Figure 1.19 (317). More structural variety is found for the corresponding indium derivatives. Therefore, $(CH_3)_2InX$ (X=Cl,I) are dimeric in benzene, and $(CH_3)_2InX$ (X=Cl,Br,I) show both the ν_a(InC) and ν_s(InC) modes in the infrared spectra (314). Based on these data, it was suggested that all three compounds are halogen-bridged dimers with nonlinear C-In-C skeletons, although the possibility of polymeric solid-state structures was not ruled out (314). A single crystal X-ray study has now shown that solid $(CH_3)_2InCl$ is polymeric with each indium atom at the center of a distorted octahedron formed by two carbon atoms and four chlorine atoms with distinctly different In-Cl bond lengths and a C-In-C angle of 167° (318). In addition, while dimeric halogen-bridged structures seem likely for $RInX_2$ (R= n-C_3H_7,n-C_4H_9,X=Br;R=C_2H_5,n-C_3H_7,n-C_4H_9,X=I) (319,320), polymeric structures have been proposed for solid $RInBr_2$ (R=CH_3,C_2H_5) on the basis of vibrational data (320).

TABLE 1.17
Skeletal Mode Assignments (cm^{-1}) for Alkyl-Aluminum and -Gallium Fluorides

Compound		ν(MC)	ν(MF)	δ(CMC)	δ(MF)	Refs.
$[(CH_3)_2AlF]_4$	IR	726vs	638m	485vw	415w-m	306
		560w-m	614m	282w-m	245vw	
					217w-m	
	R	717w-m		370m	243w	
		599vs			145vw	
$[(C_2H_5)_2AlF]_4$	IR	680vs	872m	435w	412w	306
		552s	835vw	330w	275w	
	R	668m	875w-m	382m	262w	
		643w-m	822w		175w	
		565m				
$[(n\text{-}C_3H_7)_2AlF]_4$	IR	705vs	620vs	440m	390m	310
				285m		
	R	690m	620sh	423w	385vw	
		573s	600w-m	270w-m		
$[(i\text{-}C_4H_9)_2AlF]_4$	IR	710vs	650sh	440s	395w	310
			615vs	240vw		
	R	700w		405s	383vw	
		610m		280m		
				250m		
$[(CH_3)_2GaF]_4$	IR	618vs	511sh	308w-m	228w-m	312
		559w-m	498vs	297w-m	217w-m	
			439w-m	162w-m		
	R	611m	518w-m	389w-m	217w-m	
		550sh	501vw	293w-m		
			455w	165vvw		
				140m		
$[(C_2H_5)_2GaF]_4$	IR	589	490vs	305w-m	225vw	312
		528sh	429s	478w		
	R	580m	484w	310vvw	245 vw	
		525vvs	430vw	265w-m	200vvw	

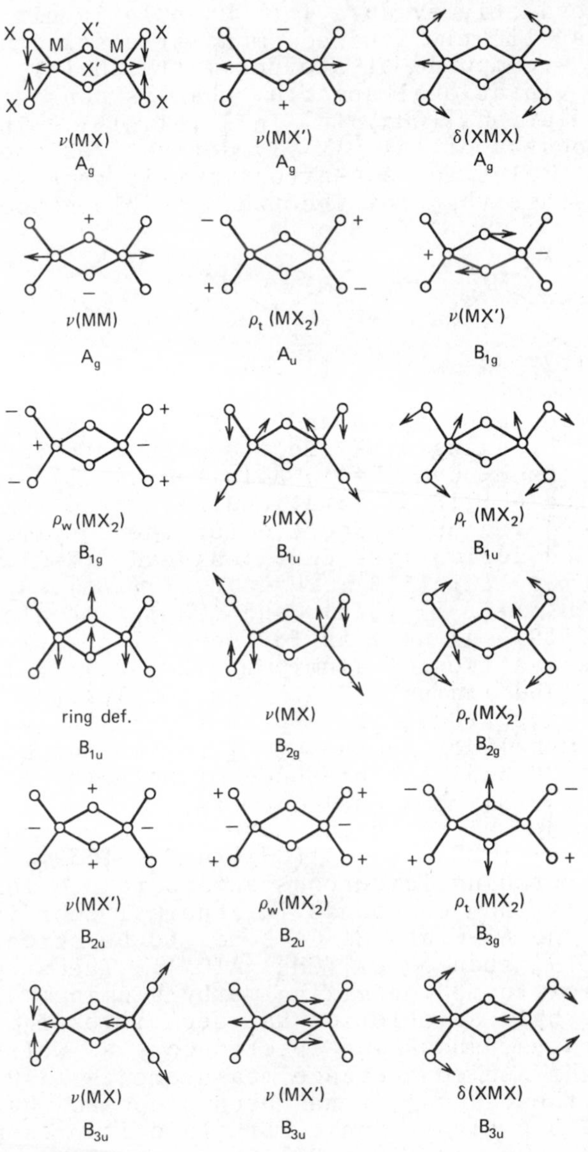

Figure 1.19. Normal modes of vibration of bridged M_2X_6 molecules.

Also, while solid CH_3InBr_2 is apparently polymeric (320), conductivity data for CH_3InCl_2 in nitromethane support a slightly ionized, monomeric structure (319, 321) and a compound of stoichiometry CH_3InI_2 has a solid-state vibrational spectrum that is consistent with the formulation $[(CH_3)_2In]^+InI_4^-$ (51,319). The vibrational spectra of $(alkylMX_2)_2$ derivatives obey the selection rules for a centrosymmetric *trans* structure (1.27) rather than for the two possible noncentrosym-

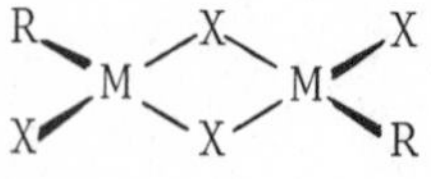
(1.27)

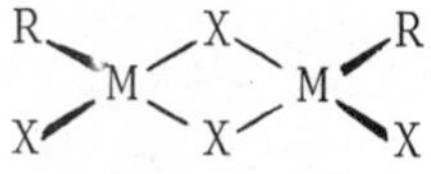
(1.28)

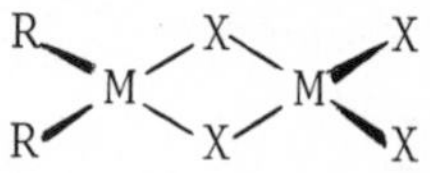
(1.29)

metric structures (1.28 and 1.29). The ν(MC) and terminal and bridging ν(MX) mode assignments for R_nMX_{3-n} ($n=0$ to 2;$R=CH_3,C_2H_5$;M=Al,Ga,In;X=Cl,Br,I) are summarized in Table 1.18. Vibrational data and some assignments have also been reported for the unsymmetrically substituted derivatives $CH_3(X)GaX_2GaX_2$ (X=Cl,Br (332)) (335), $CH_3Ga_2Cl_4I$ (335), $CH_3Ga_2Br_4I$ (335), $R_2GaX_2Ga(X)R$ ($R=CH_3,C_2H_5$;X=Br,I) (336), and $(CH_3)_n(I)_{2-n}InI_2In(I)CH_3$ ($n=0,2$) (319), as well as for several propyl and butyl halide derivatives of aluminum (128,324), gallium (148, 333,337), and indium (320). The complexity of the vibrational spectra of the *n*-propyl and *n*-butyl gallium and indium halides in the ν(MC) region has been attributed to (320,333) the presence of both the *trans* and *gauche* internal torsional isomers.

Tetrahedral skeletons have been proposed for $Na[CH_3AlCl_3]$ (328) and $Cs[(CH_3)_3AlCl]$ (339). The values of the stretching force constants obtained for both anionic aluminum compounds in a normal coordinate analysis show the Al-C and Al-Cl bonds to be stronger in $Na[CH_3AlCl_3]$ than in $Cs[(CH_3)_3AlCl]$. The structure of the complex formed on mixing methylaluminum dichloride and trimethyllead chloride has been investigated using nmr, infrared, and Raman spectroscopy as well as molecular weight and conductance measurements (340). On the basis of these data, it has been proposed that in the solid state a highly ionic structure is present in which $CH_3AlCl_3^-$ and $(CH_3)_3Pb^+$ ions are alternatively set in a polymer chain (1.30). Vibrational assignments have also been made for anionic alkyl halide derivatives of gallium (341) and indium (321). The metal-

TABLE 1.18

Skeletal Stretching Mode Assignments (cm^{-1}) for R_nMX_{3-n} (n=0 to 2; R=CH_3,C_2H_5; M=Al,Ga,In; X=Cl,Br,I) Derivatives

Compound: Empirical Formula	Phase	Spectra	ν(MC)	ν(MX): Terminal	ν(MX): Bridging	Refs.
$AlCl_3$	Vapor	IR		625 484	420 284	322
	Liquid	R		605 506	438 340	323
CH_3AlCl_2	Solid and solution	IR and R	704 688	495 485	380 345	324
$(CH_3)_2AlCl$	Vapor	IR	720 585		443	128,325,326
	Liquid	R	713 588		330	128,325,326
$C_2H_5AlCl_2$	Liquid	IR and R	663 657	502 486	396 346 323 274	324
$(C_2H_5)_2AlCl$	Liquid	R	670 556		339	135
$AlBr_3$	Mull	IR		502 375	345 198	327
	Melt	R		489 409	340 210	327,328
CH_3AlBr_2	Solid and liquid	IR and R	692 675	515 400	360 350 265 257	324
$C_2H_5AlBr_2$	Liquid	IR and R	638 635	426 391	348 320 269 221	324
$(C_2H_5)_2AlBr$	Liquid	R	664 547		197	135
AlI_3	Mull	IR		415 320	291 140	327
	Melt	R		408 339	348 147	327,328
$(C_2H_5)_2AlI$	Liquid	R	664 544		170	135
$GaCl_3$	Vapor	IR		477 400	310 280	329,330
	Melt	R		462 413	318 243	328-330
CH_3GaCl_2	Solution	IR and R	608 603	398 380	312 290 265 228	324

TABLE 1.18 (continued)

Compound: Empirical Formula	Compound: Phase	Spectra	$\nu(MC)$	$\nu(MX)$: Terminal	$\nu(MX)$: Bridging	Refs.
$(CH_3)_2GaCl$	Liquid	IR	604 545			331
$C_2H_5GaCl_2$	Liquid	IR and R	570 568	389 363	303 284 263 240	324
$(C_2H_5)_2GaCl$	Liquid	IR and R	580 578 520 516		298 290 250 235	324,330
$GaBr_3$	Mull	IR		347 268	232 188	327,330
	Melt	R		341 288	241 201	327,328,330
CH_3GaBr_2	Mull	IR	591			332
	Solid	R	595			332
GaI_3	Mull	IR		273 213	189 134	327
	Melt	R		273 228	203 143	327
CH_3GaI_2	Mull	IR	572			333
	Solid	R	582		231 169	333
$C_2H_5GaI_2$	Mull	IR	546			333
	Solid	R	550		211 164	333
$InCl_3$	Solid	IR		255	119	334
	Solid	R		279	127	334
CH_3InCl_2	Mull	IR	525	285		318
	Solid	R	530	280 269	179	318
$(CH_3)_2InCl$	Mull	IR	563 495			50,314,318
	Solid	R	560 500		197	50
$(C_2H_5)_2InCl$	Mull	IR	518 462			315,318
	Solid	R	519 468		237 205	318
$InBr_3$	Solid	IR		180	82	334
	Solid	R		171	84	328,334

CH_3InBr_2	Mull	IR	523 516			318,320
	Solid	R	524 517	196 148	97 85	318,320
$(CH_3)_2InBr$	Mull	IR	554 488			314
	Solid	R	555 496			318
$C_2H_5InBr_2$	Mull	IR	496 491			320
	Solid	R	502 494	213-187	135 91	320
$(C_2H_5)_2InBr$	Mull	IR	510 459			315,318
	Solid	R	512 463		110 98	318
InI_3	Mull	IR		226 179	157 125	327,334
	Solid	R		225 187	135	327 334
CH_3InI_2	Mull	IR	557			51,318,320
	Solid	R	484	185 142		51,318,320
$(CH_3)_2InI$	Mull	IR	548 480			314
$C_2H_5InI_2$	Mull	IR	489			320
	Solid	R	484	171 144	144	320
$(C_2H_5)_2InI$	Mull	IR	508 458			315,318
	Solid	R	504 462		87	318

TABLE 1.19
Metal-Carbon and Metal-Halide Stretching Mode Assignments (cm-1) for Anionic Alkyl Halide Derivatives of Aluminum, Gallium, and Indium

Compound	ν(MC)	ν(MX)	Refs.
$CH_3AlCl_3^-$ [a]	657	448 360	338
$(CH_3)_3AlCl^-$ [b]	617 571	394	339
$(CH_3)_3GaF^-$ [c]	543 523	380	341
$(C_2H_5)_3GaF^-$ [c]	524 493	361	341
$[(CH_3)_3Ga]_2F^-$ [d]	548 517 534	379	341
$[(C_2H_5)_3Ga]_2F^-$ [d]	524 498	429	341
$(CH_3)_3GaCl^-$ [d]	539 512 532	246	341
$[(C_2H_5)_3Ga]_2Cl^-$ [d]	527 491	288	341
$(CH_3)_3GaBr^-$ [d]	535 515	172	341
$[(C_2H_5)_3Ga]_2Br^-$ [d]	525 488	147	341
$(CH_3)_2InCl_2^-$ [e]	512 483 465		321

[a] Na^+ salt.
[b] Cs^+ salt.
[c] K^+ salt.
[d] $(CH_3)_4N^+$ salt.
[e] $(C_2H_5)_4N^+$ and $(C_6H_5)_4As^+$ salts.

carbon and metal-halide stretching mode assignments for anionic alkyl halide derivatives of aluminum, gallium, and indium are summarized in Table 1.19.

The structures of alkyl-thallium halides are quite different from those of the alkyl halide derivatives of the other group IIIb elements. The vibrational spectrum of $(CH_3)_2TlCl$ (46,342-344) is consistent with the linear C-Tl-C skeleton found for this compound (345). It is not known whether the chlorine atoms are ionic or whether there are four bridging chlorine atoms weakly bonded to the thallium atom (46,342); in any case, no ν(TlCl) mode assignments have been reported for $(CH_3)_2Tl$-Cl. The halogen atoms in $(C_2H_5)_2TlBr$ and $(n\text{-}C_3H_7)_2TlCl$ are also probably ionic or weakly bridging (46). The ν_a(TlC) and ν_s(TlC) modes for the above compounds as well as for $(CH_3)_2TlF$ (346), $(CH_3)(C_2H_5)TlCl$ (343), and $(C_2H_5)_2TlCl$ (343) have been assigned to bands in the 550 to 500 cm^{-1} and 500 to 435 cm^{-1} regions, respectively.

e. Group IVB

The alkyl halide derivatives of silicon and germanium are monomeric, covalent, and, in nearly all cases, neutral four-coordinate compounds with tetrahedral skeletons. Vibrational data have also been reported for the anionic derivatives $CH_3SiF_4^-$ (347) and $CH_3SiF_5^{2-}$ (348); trigonal bipyramidal and octahedral skeletons, respectively, have been found for these two compounds. In Table 1.20, frequency ranges are given for the skeletal modes of methyl and ethyl halide derivatives of silicon and germanium. The spectra of the ethyl derivatives in the skeletal mode region are more complex than those of the corresponding methyl derivatives because of the presence of more than one rotational isomer and/or vibrational coupling (371,375). This has also been observed for R_3GeF ($R=C_2H_5, n\text{-}C_3H_7, n\text{-}C_4H_9$) (371), $(n\text{-}C_3H_7)_nGeCl_{4-n}$ (n=1 to 3) (139), and several alkylsilicon trichloride compounds (376).

Vapor-phase spectra are preferred to liquid- or solid-phase spectra since intermolecular association decreases in going from the solid to the vapor phase. Frequency variations observed for the skeletal modes of several alkyl halide derivatives of silicon and germanium in different phases are shown in Table 1.21. The low-frequency infrared and Raman spectra have been reported for $(CH_3)_2SiH_nX_{2-n}$ (n=0,1;X=Cl,Br,I) (377), $(CH_3)_3MX$ (M=C,Si,Ge;X=Cl,Br) (378), CH_3GeI_3 (368), and CD_3GeI_3 (368) at low temperatures. Torsional modes and intermolecular vibrations were also assigned in these studies

TABLE 1.20
Skeletal Mode Frequency Ranges (cm^{-1}) for Methyl and Ethyl Halide Derivatives of Silicon and Germanium

M	R	ν(MC)	ν(MF)	ν(MCl)	ν(MBr)	ν(MI)	Defor.	Refs.
Si	CH_3	805-619	982-828	577-450	453-314	331	390-90	43,349-357
	C_2H_5	746-571	835	600-449	370	a	377-131	357-361
Ge	CH_3	663-572	744 662	430 387	309 263	251 195	277 67	357,362-373
	C_2H_5	613-533	648 620	422-362	350-280	255-245	330-100	357,371 374,375

[a]Data not reported.

TABLE 1.21
Frequency Variations (cm^{-1}) of the Metal-Carbon and Metal-Halide Stretching Modes in Different Phases for Alkyl Halide Derivatives of Silicon and Germanium

Compound	Phase	ν(MC)	ν(MX)	Refs.
$(CH_3)_3SiF$	Gas	712 619	912	349,357
	Liquid	704 635	898	
$(CH_3)_3SiCl$	Gas	700 640	487	350,357
	Liquid	704 635	472	
$(CH_3)_3SiBr$	Gas	702 633	386	350
	Liquid	704 632	373	
$(C_2H_5)_3GeBr$	Gas	594 [a]	648	357,371,374
	Liquid	597 551	625	
	Solid	597 550	581	
$(CH_3)_3GeF$	Gas	623 577	659	357,362,371
	Liquid	623 576	623	
	Solid	626 [a]	579	
$(CH_3)_3GeCl$	Gas	619 576	399	363
	Liquid	612 569	378	
$(CH_3)_2GeCl_2$	Gas	619 576	436 410	365
	Liquid	612 569	400 388	

[a]Data not reported.

While the intermolecular interactions of alkyl-silicon and -germanium halides in the liquid and solid phases are relatively weak, halogen bridged structures are found for alkyl-tin and -lead halides. Coordination expansion via fluorine bridges is a common feature of alkyl-tin fluorides. In Figure 1.20, the proposed structures are illustrated for some methyl-tin fluorides. These structures have been confirmed in single crystal X-ray studies of $(CH_3)_2SnF_2$ (379) and $(CH_3)_3SnF$ (380), while infrared, Raman, and ^{119}Sn Mossbauer spectroscopy have been used to characterize the structures of CH_3SnF_3, CH_3SnCl_2F, and $(CH_3)_2SnClF$ (381). Different investigators have concluded that the Sn-F bonds of $(CH_3)_2SnF_2$ are either predominantly ionic (379,382) or appreciably covalent (383,384). The absence of a Raman band that can be attributed to the ν_s(SnF) mode of $(CH_3)_2SnF_2$ has been given as evidence for an ionic formulation (382). It has more recently been observed (385), however, that for the structure determined for $(CH_3)_2SnF_2$ (379) (Figure 1.20), no ν_s(SnF) mode is possible since "*trans*"-fluorine ligands are primatively

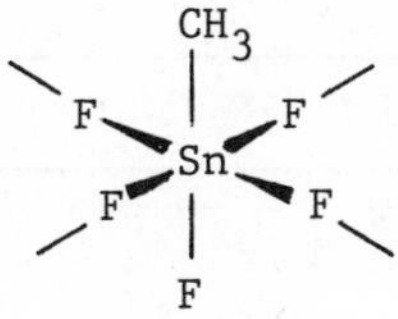

CH_3SnF_3

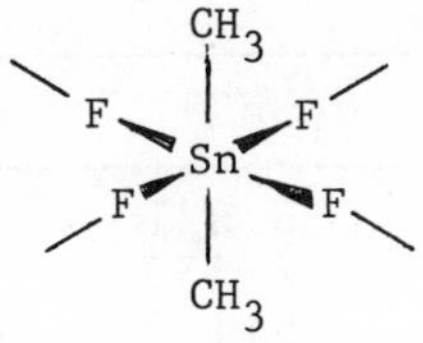

$(CH_3)_2SnF_2$

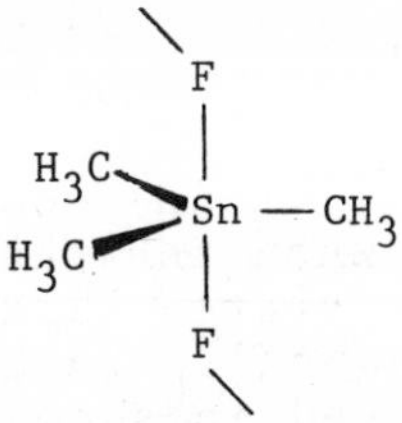

$(CH_3)_3SnF$

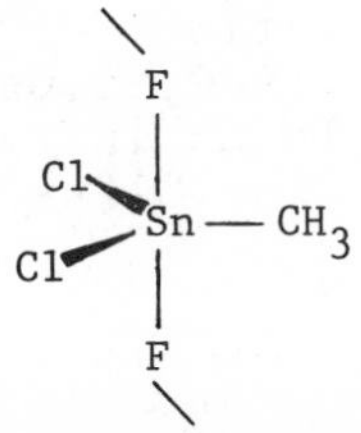

CH_3SnCl_2F

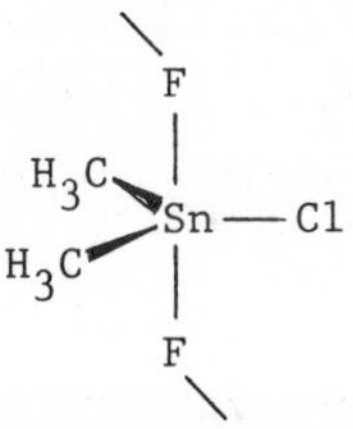

$(CH_3)_2SnClF$

Figure 1.20. Solid-state structures of CH_3SnF_3, $(CH_3)_2SnF_2$, $(CH_3)_3SnF$, CH_3SnCl_2F, and $(CH_3)_2SnClF$.

related and must therefore move identically. The skeletal mode assignments reported for methyl-tin fluorides are summarized in Table 1.22. Although polymeric in the solid state, the R_3SnF (R=CH_3,C_2H_5,n-C_3H_7,n-C_4H_9, C_6H_5) derivatives are monomers in the vapor state (386, 387). Infrared, Raman, Mossbauer, and mass spectral data were used to characterize these compounds. As

TABLE 1.22
Skeletal Mode Assignments (cm^{-1}) for Solid Methyl-Tin Fluorides[a]

Compound	ν(SnC)	ν(SnF)	ν(SnCl)	Deformations
$(CH_3)_3SnF$	555 521	335[b]		
$(CH_3)_2SnF_2$	597 533 544	365[b] 644[c] 425[b]		263 148 84 328 322 287 278
$(CH_3)_2SnClF$	582 536	365[b]		
CH_3SnCl_2F	555	398[b]	385 365	

[a]References 381 and 385.
[b]Bridging ν(SnF) mode.
[c]Terminal ν(SnF) mode.

seen in Table 1.23, the ν(SnF) mode frequency increases by more than 200 cm^{-1} when the phase is changed form the solid to the vapor state. Trends in the skeletal mode frequencies for R_nSnX_{4-n} ($R=CH_3,C_2H_5,n\text{-}C_4H_9;X=Cl,Br,I$) have been discussed in some recent studies (388-391). The average of both the ν(SnC) and ν(SnX) mode frequencies decreases with increasing alkylation of the tin (388). The frequency ranges for the skeletal stretching modes are summarized in Table 1.24 for neutral alkyl-tin halides. Correlations of the normal mode frequencies for $(CH_3)_3MCl$ (M=C,Si,Ge,Sn) are illustrated in Figure 1.21 (363). In an infrared and Raman study of $(p\text{-}XC_6H_4CH_2)_nSnCl_{4-n}$ (X=F,Cl;n=2,3) it has

TABLE 1.23
ν(SnF) Mode Assignments (cm^{-1}) for Alkyl-Tin Fluorides in the Solid and Vapor Phases[a]

	ν(SnF)	
Compound	Solid	Vapor
$(C_2H_5)_3SnF$	349 366	588
$(n\text{-}C_3H_7)_3SnF$	340 364	574
$(n\text{-}C_4H_9)_3SnF$	350 364	573

[a]References 357,386 and 389.

TABLE 1.24
Frequency Ranges (cm^{-1}) for Skeletal Stretching Modes of Neutral Alkyl-Tin Chlorides, Bromides, and Iodides

R	ν(SnC)	ν(SnCl)	ν(SnBr)	ν(SnI)	Defor.	Refs.
CH_3	506-508	382-331	264-235	207-174	157-95	388
C_2H_5	533-482	377-317	260-222	198-176	135-91	388,389, 391
n-C_4H_9	602-501	378-328	261-228	196-90	129-90	388,390
$C_6H_5CH_2$	587-558	587-558	352-295	254-230	149-70	187

been concluded that *para* substitution has little influence on the frequencies of the ν(SnCl) modes but causes a slight decrease in the ν(SnC) mode frequencies (392).

The complexation of $(CH_3)_2Sn^{2+}$ by chloride and bromide ions in aqueous solution has been studied using emf measurements and Raman and infrared spectroscopy (393). A study of the Raman active ν(SnCl) band as a function of chloride ion concentration indicated that both inner and outer sphere complexes were formed. An

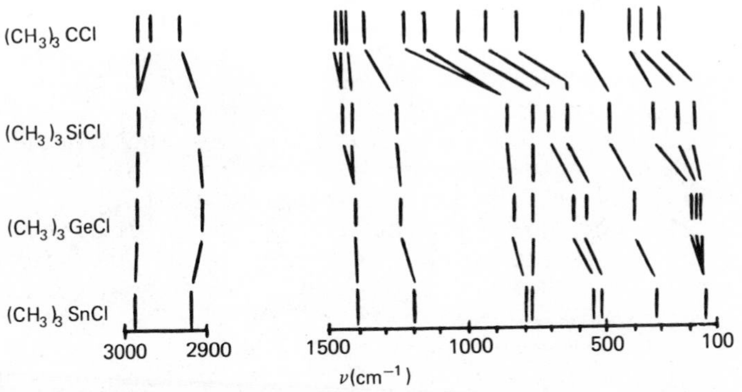

Figure 1.21. Correlation of normal modes for $(CH_3)_3MCl$ (M=C,Si, Ge,Sn). $(CH_3)_3SnCl$ was recorded in liquid except the 720 and 780 cm^{-1} bands which were recorded in a CS_2 solution (363).

appreciable concentration of the inner sphere complex was found in concentrated chloride ion solutions. The same was true for solutions of bromide ions. It is very difficult to displace the water molecules from the first coordination sphere of tin. Several alkyl-tin halide complexes have been characterized in which the coordination number of the tin(IV) atom is five or six. The five-coordinate complexes of mono-, di-, and trialkyltin(IV) have trigonal bipyramidal skeletons in which the alkyl groups preferentially occupy the equatorial positions (394,395). The six-coordinate complexes of mono- and dialkyltin(IV) have octahedral skeletons (382,394,395). When two alkyl groups are bonded to the tin atom in the six-coordinate complexes, they are *trans* to one another. In mixed, anionic methyltin pentahalide complexes of the type $CH_3SnX_3Y_2^{2-}$, the low symmetry makes it impossible to predict the structure about the tin atom (394); for $CH_3SnCl_4I^{2-}$, however, it has been proposed that the methyl and iodine ligands are *trans* to one another (394). Table 1.25 summarizes some assignments made for anionic five- and six-coordinate alkyl-tin halide complexes. In general, the ν(Sn-halide) modes for these derivatives appear at lower frequencies than the corresponding modes of neutral, four-coordinate alkyl-tin halides. Low-frequency infrared and/or Raman data have also been reported for the cesium salts of $(CD_3)_2SnX_4^{2-}$ (X=Cl,Br) and the tetraethylammonium salt of $(CD_3)_3SnCl_2^-$ (395). Normal coordinate calculations have been reported for $(CH_3)_2SnX_4^{2-}$ (X=F,Cl,Br) using both Urey-Bradley and general valence force fields (382). The K(SnC) force constants determined in these calculations decrease slightly from the fluoride to the bromide.

Several adducts have been reported between alkyl-tin halides (other than the fluorides) and Lewis bases (e.g., pyridine, 1,10-phenanthroline, and dimethylsulfoxide). The low-frequency assignments for these compounds have been thoroughly summarized (388,398,399). In general, the position of the ν(Sn-halide) modes is influenced very little by changes in coordination number, but is at lower frequencies for the adducts than for the uncomplexed derivatives, with the effect being greatest for the chlorides and smallest for the iodides (399). Also, the ease of adduct formation and the stability of the adducts decrease with increasing alkylation of the alkyl-tin halide; increasing alkylation decreases the effective charge of the metal, which decreases its ability to coordinate with electron-withdrawing ligands (388).

TABLE 1.25
Skeletal Stretching Mode Assignments (cm^{-1}) for Anionic Alkyl-Tin Halides

Compound	ν(SnC)	ν(SnX)	Refs.
$(CH_3)_3SnCl_2^-$	544 509	227	395,396
$(CH_3)_3SnBr_2^-$	555	140	396
$(CH_3)_3SnI_2^-$	553	134	394
$(CH_3)_2SnF_3^-$	553	592	396
$(CH_3)_2SnCl_3^-$	573 518	313 256 242	394,395
$(CH_3)_2SnBr_3^-$	567 512	228 218	397
$CH_3SnCl_4^-$	540	355 338 282 230	395
$(CH_3)_2SnF_4^{2-}$	582 539	347 397	382
$(CH_3)_2SnCl_4^{2-}$	580 508	235 207 197	382,394,395
$(CH_3)_2SnBr_4^{2-}$	571 498	220 183 174	382,394,395
$(CH_3)_2SnI_4^{2-}$	559	186	394
$CH_3SnCl_5^{2-}$	534	318 258 215	394,395
$CH_3SnCl_3Br_2^{2-}$	530	312[a] 294[a] 267[a] 210[b] 160[b]	394
$CH_3SnCl_2Br_3^{2-}$	528	306[a] 248[a] 209[b] 188[b] 160[b]	394
$CH_3SnBr_5^{2-}$	524	212 191 160	394
$CH_3SnCl_3I_2^{2-}$	530	328[a] 297[a] 267[a] 187[c] 163[c]	394
$CH_3SnCl_4I^{2-}$	534	307[a] 261[a] 163[c]	394

[a]ν(SnCl) mode.
[b]ν(SnBr) mode.
[c]ν(SnI) mode.

The alkyl-lead halides show a great tendency to polymerize; this is due in part to the preference of the relatively large lead atom for a coordination number greater than four. Infrared and Raman data for $(CH_3)_3PbX$ (X=F,Cl,Br,I) and $(CH_3)_2PbX_2$ (X=Cl,Br) and molecular weight measurements on $(CH_3)_3PbX$ (X=Br,I) indicate that while all of these compounds are monomeric in noncoordinating solvents, they are polymeric with bridging halogen ligands in the solid state (400). The skeletal stretching mode assignments for alkyl-lead halides are summarized in Table 1.26. The vibrational spectra of $(CH_3)_3PbX$ (X=Cl,Br,I) are consistent with structures analogous to that found for $(CH_3)_3SnF$ (Figure 1.20) (400). The structures proposed for $(CH_3)_2PbX_2$ (X=Cl,Br) (400), however, are similar to that found for $(C_6H_5)_2PbCl_2$ (401), with coplanar Cl_2PbCl_2 chains (1.31) rather than the sheet-like structure found for $(CH_3)_2SnF_2$.

TABLE 1.26
Skeletal Mode Assignments (cm^{-1}) for Alkyl-Lead Halides

Compound	Phase	$\nu_a(PbC)$	$\nu_s(PbC)$	$\nu(PbX)$	Deformations	Refs.
$(CH_3)_3PbF$	Solid	479	476	315	161 123 112	150,400
$(CH_3)_3PbCl$	Solid	494	464	191	125 111 102 81	150,400
	C_6H_6 soln.	489	464	281		400
$(CH_3)_3PbBr$	Solid	492	464	118	127 121 91	150,400
	C_6H_6 soln.	487	462	186		400
$(CH_3)_3PbI$	Solid	486	454	105	91	400
	C_6H_6 soln.	481	460	147	128	400
$(CH_3)_2PbCl_2$	Solid	534	460	202-159		400
$(CH_3)_2PbBr_2$	Solid	523	447	178-136		400
$(C_2H_5)_3PbCl$	Solid	467	442			150
$(n\text{-}C_3H_7)_3PbCl$	Solid	482	461			150

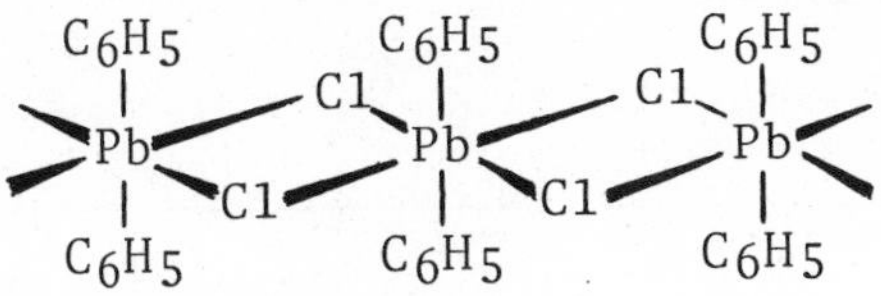

(1.31)

f. Group VB

With the exception of solid $(CH_3)_2PCl$, monomeric structures with pyramidal skeletons have been proposed for trivalent alkyl-group Vb halides. A complete study has been made of the infrared spectrum of $(CH_3)_2PCl$ in the vapor and solid states and of its Raman spectrum in the liquid and solid states (402). The infrared and Raman spectra below 1000 cm^{-1} for the solid differed from those of the liquid and vapor in that all of the intramolecular fundamental bands in this region doubled. The largest splitting was for the ν(PCl) mode. These results have been attributed to the presence of a noncentrosymmetric dimer in the solid state. Assignments for the skeletal modes of trivalent alkyl-group Vb halides are summarized in Table 1.27. Normal coordinate calculations have been reported for $(CH_3)_nAsX_{3-n}$ (n=0 to 3;X=Cl,Br) (416). The K(AsC) force constant increased slightly for a given compound when the chloride was replaced with the bromide group. The K(AsX) force constant increased from $(CH_3)_2AsX$ to AsX_3 (X=Cl,Br).

The pentavalent alkyl-group Vb halides are generally found with two types of structures: ionic with tetrahedral $(R_nMX_{4-n})X$ skeletons, or covalent with trigonal bipyramidal (rather than square pyramidal) R_nMX_{5-n} skeletons in which the R group(s) preferentially occupy equatorial positions. Some compounds are found with different structures in various phases. Therefore, on the basis of infrared and Raman data, CH_3PCl_4 has been reported to have an ionic structure in the solid state but a covalent structure with a trigonal bipyramidal skeleton in nonionizing solvents (417). Also, $(CH_3)_4PF$, which has an ionic solid-state structure, exhibited infrared bands that might be associated with the molecular compound when the vapor at 10^{-6} torr was condensed on a AgCl disk at -160°C (418). These bands disappeared on warming the sample. The structures proposed for the known pentavalent alkyl-group Vb halides, mainly on the basis of X-ray and vibrational data are

TABLE 1.27
Skeletal Mode Assignments (cm^{-1}) for Trivalent Alkyl-Group Vb Halides

Compound	ν(MC)	ν(MX)	Deformations	Refs.
CH_3PF_2	700	864 806	483 402 335	403
CH_3PFCl	701	797[a] 512[b]		404
CH_3PCl_2	692	487 473	286 237 194	405,406
CH_3PFBr	700	796[a] 413[c]		404
CH_3PBr_2	682			407
CH_3PI_2	675			407
$(CH_3)_2PCl$: vapor	700 675	485		402
liq.	708 674	462		
solid	776 717	542		
	736 679	420		
$C_2H_5PCl_2$		502 480		110,408
t-$C_4H_9PF_2$	615	820 814	477 363 288	409
$(t$-$C_4H_9)_2PF$	615 584	756	485 290 224	410
t-$C_4H_9PCl_2$	587	570 501	290 208 183	411
$(t$-$C_4H_9)_2PCl$	599 572	508	285 262 179	410
CH_3AsCl_2	583	388 360	226 205 155	412-414
CH_3AsBr_2	575	278 263	189 100	412,414
CH_3AsI_2	565	204 200	180 87	102,412,414
CD_3AsI_2	520	218 195	162 86	414
$(CH_3)_2AsCl$	580[d]	362	244 211 198	412,413
$(CH_3)_2AsBr$	582 573	267	236 188 165	412
$(CH_3)_2AsI$	576 567	230	226 169 106	102,412,415
$C_2H_5AsI_2$	546[e]			102
	518[e]			
$(CH_2)(SbCl_2)_2$		323 297		104
$(n$-$C_4H_9)_2SbCl$	505			118
$(n$-$C_4H_9)_2SbBr$	507			117

[a] ν(PF) mode.
[b] ν(PCl) mode.
[c] ν(PBr) mode.
[d] The ν_a(AsC) and ν_s(AsC) modes are accidentally degenerate.
[e] The appearance of two bands is attributed to the presence of both the *gauche* and *trans* conformational isomers.

summarized in Table 1.28. Although the data in Table 1.28 are far from complete, it appears that the ionic character of these compounds increases (1) as the halide is changed from fluoride to iodide, (2) as the central metal atom is changed from antimony to phosphorus, or (3) as the derivative becomes more highly alkylated.

TABLE 1.28

Proposed Skeletal Structures for Known Pentavalent Methyl-Group Vb Halide Derivatives

M	X	CH_3MX_4	$(CH_3)_2MX_3$	$(CH_3)_3MX_2$	$(CH_3)_4MX$
P	I			Tetrahedral	Tetrahedral
	Br			Tetrahedral	Tetrahedral
	Cl	Tetrahedral[a]	Tetrahedral	Tetrahedral	Tetrahedral
	F	Trigonal Bipyr.	Trigonal Bipyr.	Trigonal Bipyr.	Tetrahedral[b]
As	I			Tetrahedral	Tetrahedral
	Br			Tetrahedral	Tetrahedral
	Cl		Trigonal Bipyr.	Trigonal Bipyr.	Tetrahedral
	F			Trigonal Bipyr.	Tetrahedral
Sb	I			Trigonal Bipyr.	Tetrahedral
	Br			Trigonal Bipyr.	Tetrahedral
	Cl		Trigonal Bipyr.	Trigonal Bipyr.	Tetrahedral
	F			Trigonal Bipyr.	Trigonal Bipyr.

[a]Although an ionic structure with a tetrahedral skeleton is found in the solid state, a covalent structure with a trigonal bipyramidal skeleton is found when this compound is dissolved in nonionizing solvents (417).

[b]Although an ionic structure with a tetrahedral skeleton is found in the solid state, a covalent structure has been proposed as a possibility for the vapor state (418).

The vibrational assignments made for ionic, four-coordinate and covalent five-coordinate methyl derivatives of the pentavalent group Vb halides are summarized in Tables 1.29 and 1.30, respectively. Data for the ionic $[(CH_3)_4M]X$ derivatives (7,60-64) are not included in Table 1.29 since these compounds do not possess covalent metal-halogen bonds. Data for these compounds are discussed in Sec. A.2 of this chapter. The mass spectral, infrared, and Raman data for $(CH_3)_2SbCl_2X$ (X=N_3, NCO) show the presence of a dimeric structure in which the antimony atoms are bridged by the α-nitrogen atoms of the N_3 and NCO ligands, respectively (432). Low-frequency solid-state and nitromethane solution infrared and Raman data have been reported for the tetraethylammonium salt of $(CH_3)_2SbCl_4^-$ (395).

In addition to the vibrational data summarized above for methyl derivatives, additional data have been reported for $[(C_2H_5)_2PX_2]X$ (X=Cl,Br) (62), $(C_6H_5CH_2)_3MX_2$ (M=As,Sb;X=F,Cl) (433), $(C_2H_5)_2SbCl_3$ (427), $(C_2H_5)_3SbBr_2$ (430), $(i\text{-}C_3H_7)_3SbX_2$ (X=F,Cl,Br,I) (434) and $(n\text{-}C_4H_9)_3SbX_2$ (X=Cl (118),Br) (117). Both $t\text{-}C_4H_9PF_4$ (435) and $(t\text{-}C_4H_9)_2PF_3$ (436) were prepared with the expectation that steric factors would force the bulky *tert*-butyl group into an axial position. Complete infrared and Raman data for both compounds and vapor-phase dipole moment measurements for the former compound, however, show that the *tert*-butyl groups occupy equatorial positions. An attempt has been reported to prepare $t\text{-}C_4H_9PCl_4$ in which it was also hoped that the *tert*-butyl group would occupy an axial position (417); because of its instability in nonionizing solvents, however, the compound could not be characterized. The infrared and Raman spectra have been assigned for the tetramethylammonium salts of $[(CH_3)_2Cl_3Sb]_2X^-$ (X=F, CN) (437).

Normal coordinate calculations reported for $[(CH_3)_nMCl_{4-n}]Cl$ (n=0 to 4;M=P,Si) show that the K(MC) and K(MCl) force constants for M=P are larger than the corresponding force constants for M=Si (62). Normal coordinate calculations have also been reported for $(CH_3)_3AsCl_2$, $[(CH_3)_3AsBr]Br$, and their deuterated analogs (420), and for $(CH_3)_3SbF_2$ (277), $(CH_3)_3SbX_2$ (X=Cl, Br) (277,429), and their deuterated analogs (429).

Vibrational data have been reported for several organo phosphonic and organo thiophosphonic fluorides, chlorides, and bromides (408,409,438-442).

TABLE 1.29
Skeletal Mode Assignments (cm^{-1}) for Tetrahedral, Pentavalent Methyl-Group Vb Halides

Compound	ν(MX)	ν(MC)	Deformations	Refs.
$[CH_3PCl_3]Cl$	610 475	790	274 247	62,417
$[(CH_3)_2PCl_2]Cl$	580 496	768 743	364 281 168	62
$[(CH_3)_3PF]Cl$	668	780 708	476	62
$[(CH_3)_3PCl]Cl$	522	773 688	368 302 168	62,419
$[(CH_3)_3PBr]Br$	415	776 674	302	419
$[(CH_3)_3PI]I$	358	764 666	303	419
$[(CH_3)_3AsBr]Br$	299	646 585	281 185	420,421
$[(CD_3)_3AsBr]Br$	295	589 540	242 163	420,421
$[(CH_3)_3AsI]I$		629 576		421
$[(CH_3)_3PCl]SbCl_6$	522	767 684		422,423
$[(CH_3)_2PCl_2]SbCl_6$	593 506	769 735		423
$[CH_3PCl_3]SbCl_6$	627[a] 484 613[a]	786		423
$[(CH_3)_3AsCl]SbCl_6$		649 588		422
$[(CH_3)_3SbCl]SbCl_6$		669 524		422

[a]The presence of two bands is attributed to solid-state effects since the $AlCl_4^-$ salt of this compound shows only one band in this region at 618 cm^{-1}.

TABLE 1.30
Skeletal Mode Assignments (cm^{-1}) for Trigonal Bipyramidal, Pentavalent Methyl-Group Vb Halides

Compound	$\nu(MX)_{equat.}$	$\nu(MX)_{axial}$	$\nu(MC)$	Deformations	Refs.
CH_3PF_4	1009 932	843 725[a]	596[a]	538 514 467 412 397 179	424,425
CH_3PCl_4	567 441	382[b] 288 361[b]	743	278 271 251 178 141	417
$(CH_3)_2PF_3$	836		779 675	496 471 459 401 340 184	426
$(CH_3)_2AsCl_3$	413	315 278 302	581		413
$(CH_3)_2SbCl_3$	342		567 504		395,427
$(CH_3)_3PF_2$		670 501	778 647	422 392 367 190	426
$(CH_3)_3AsF_2$		525	662		421
$(CH_3)_3AsCl_2$		312 260	642 563		420
$(CD_3)_3AsCl_2$		291 258	599 527		420
$(CH_3)_3SbF_2$		484 465	591 546	245 215 210 146	277,428
$(CH_3)_3SbCl_2$		282 272	577 538	208 188 166 119	277,428,429
$(CD_3)_3SbCl_2$		280 266	526 479	189 174 155 112	429
$(CH_3)_3SbBr_2$		215 168	569 526	200 172 160 91	277,428,429
$(CD_3)_3SbBr_2$		207 167	521 473	165 135 88	429
$(CH_3)_3SbI_2$		144 122	559 508	196 173 163 80	428,430
$(CH_3)_4SbF$		385	568[c] 530[c] 508[d]		431

[a] The assignment of bands at 725 and 596 cm^{-1} to the $\nu(PF)_{axial}$ and $\nu(PC)$ modes respectively in Ref. 425 is a reversal of the original assignments given in Ref. 424. This reassignment seemed more consistent with the noncrossing rule. It was also concluded that these two modes are strongly coupled and that in the absence of coupling they would both be expected at approximately 660 cm^{-1} (425).

[b] The two bands are explained as being due to Fermi resonance.

[c] $\nu(SbC)_{equat.}$ mode.

[d] $\nu(SbC)_{axial}$ mode.

g. Group VIB

Several structures have been found for the alkylselenium or -tellurium halides $(alkyl)_nMX_{4-n}$ (n=1 to 3). A structure in which an unshared electron pair is presumed to occupy one of the three equatorial positions (1.32) has been found in a single crystal X-ray

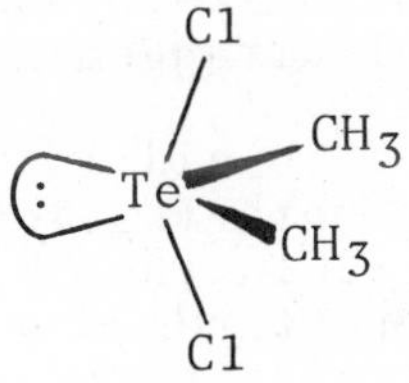

(1.32)

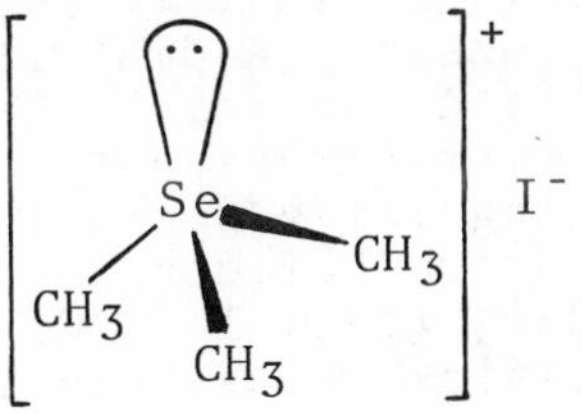

(1.33)

study of solid α-$(CH_3)_2TeCl_2$ (443). The bromide $(CH_3)_2TeBr_2$ was first reported to have an α and β form (444) with the β form postulated (445) to be a salt-like material, $[(CH_3)_3Te]^+[CH_3TeBr_4]^-$. A single crystal X-ray study (446) has shown solid β-$(CH_3)_2TeI_2$ to be best represented as $[(CH_3)_3Te]^+[CH_3TeI_4]^-$ with pyramidal $(CH_3)_3Te^+$ cations and square pyramidal $CH_3TeI_4^-$ anions. The ions are bridged by four weak Te-I interactions. A single crystal X-ray study of α-$(CH_3)_2TeI_2$ shows each tellurium atom to have a distorted octahedral environment with two iodine atoms in a *trans*-arrangement, two methyl groups *cis* to each other, and two contacts from iodine atoms attached to neighboring molecules (447). Finally, an ionic structure, in which one apex of the tetrahedron is presumed to be occupied by a lone electron pair, has been found in a single crystal X-ray study of $(CH_3)_3SeI$ (1.33) (57).

In a vibrational study of the products obtained on reacting $(CH_3)_2Se$ and $(CH_3)_2Te$ with molecular chlorine, bromine, and iodine, it has been concluded that the molecular compounds $(CH_3)_2MX_2$ were obtained in all instances except for the adduct formed between $(CH_3)_2Se$ and molecular iodine (448) for which a charge transfer complex (1.34) has been proposed. Reaction of the

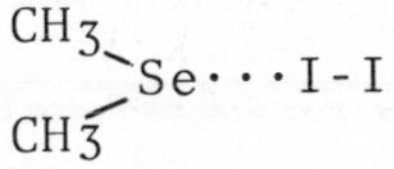

(1.34)

compound $(CH_3)_2TeI_2$ with molecular iodine in organic solvents produces $(CH_3)_2TeI_4$. The nearly identical nature of the infrared spectra of both $(CH_3)_2TeI_2$ and $(CH_3)_2TeI_4$, especially in the ν(TeI) mode region, indicates that the I_2 molecule in $(CH_3)_2TeI_4$ does not interact to any appreciable extent with the iodine atoms bonded to tellurium (449). It has therefore been proposed (449) that $(CH_3)_2TeI_4$ is a solid-state adduct with possibly a distorted octahedral arrangement similar to that found for α-$(CH_3)_2TeI_2$ (446).

Vibrational data have been used to aid in characterizing 1:1 adducts formed between methyl-selenium and -tellurium halides and $SbCl_5$ (450), BCl_3 (58,59,451), and BBr_3 (59,451). Of the three structures considered as possibilities, the ionic, pyramidal structure (1.35)

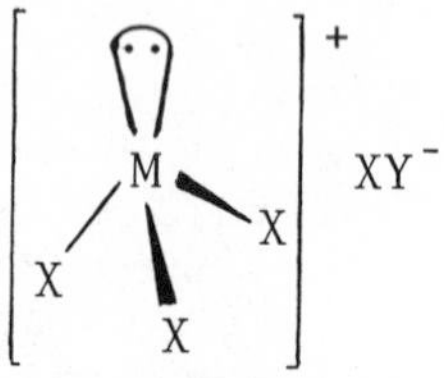

(1.35)

was the only one found. The ν(MX) mode frequencies of these adducts are higher than in the uncomplexed methyl compounds. For the adducts formed with $(CH_3)_2SeCl_2$, this has been attributed to an increase in *s*-character of selenium orbitals participating in the selenium-halogen bond when the halogen shifts from an axial position in the four-coordinate compound with a structure analogous to 1.32 to one of the positions in the three-coordinate, cationic, pyramidal complex (1.35).

Using molecular weight and infrared data, a dimeric, halogen-bridged structure has been proposed for CH_3SeCl_3 (58), CH_3TeX_3 (X=Cl,Br) (452), and $C_2H_5TeX_3$ (X=Cl,Br,I) (452). Vibrational assignments have been made for monomeric tetramethyl urea adducts of methyl- and ethyltellurium trihalides, $RTeX_3 \cdot SC[N(CH_3)_2]_2$, which have five-coordinate, square pyramidal skeletal structures with the R group in the apical position (452). In Table 1.31, the skeletal mode assignments are summarized for methyl-selenium and -tellurium halides. Although not included in Table 1.31, data for the ionic derivatives $[(CH_3)_3M]X$ (M=Se,Te;X=halide) are included in Sec. A.2 of this chapter.

TABLE 1.31
Skeletal Mode Assignments (cm^{-1}) for Methyl-Selenium and -Tellurium Halides

Compound	Phase	ν(MC)	ν(MX)	Deformations	Refs.
$(CH_3)_2SeF_2$	Liquid	622 605	506 486	295 248 210	452
$(CD_3)_2SeF_2$	Liquid	580 561	498 484	272 248 197	452
CH_3SeCl_3	Mull[a]	564	340 205		58
	CH_2Cl_2 soln.[a]	567	345 258 180[b]		58
$(CH_3)_2SeCl_2$	Mull	589 573	293 270		451
$(CH_3)_2SeBr_2$	Mull	588 572	265	244	448,451
$[CH_3SeCl_2]SbCl_6$	Mull	558	424 413	283	450
$[(CH_3)_2SeCl]BCl_4$	Mull	589 572	403		451
$[(CH_3)_2SeBr]BBr_4$	Mull	c	286		451
CH_3TeCl_3	Mull[a]		338 315 103[b]		452
$(CH_3)_2TeCl_2$	Solid	544[d]	281 248	213	448
CH_3TeBr_3	Mull[a]	470.5	234 208 143[b]	109 61	59,452
	C_6H_6 soln.[a]		220 200 120[b]	104	59
$(CH_3)_2TeBr_2$	Mull	540[d]	185 148	191 103	49,448
CH_3TeI_3	Mull	480	172 98[b]	139	452
$(CH_3)_2TeI_2$	Solid	527[d]	144 113	190	448,449
$[CH_3TeCl_2]SbCl_6$	Mull	526	375	283 265	450

[a]Associated through halogen bridges in this phase.
[b]Bridging metal-halogen stretching mode.
[c]Obscured by B-Br mode.
[d]The ν_a(TeC) and ν_s(TeC) modes are accidentally degenerate.

h. Transition Elements

Although several groups have reported vibrational data for CH_3TiCl_3 (454) and its deuterated analog (85, 455,456), and for $(CH_3)_2TiCl_2$ (454,456) and its deuterated analog (456), it has recently been concluded (456) that the assignments made for CH_3TiCl_3 and $(CH_3)_2TiCl_2$ are far from definite because of the possible impurities present in the sample. The ν(TiC) (490-440 cm^{-1}) and ν(TiCl) (approximately 370 cm^{-1}) mode assignments given for the six-coordinate titanium complexes formed between CH_3TiCl_3 and several bidentate ligands are at lower frequencies than the corresponding assignments for uncomplexed CH_3TiCl_3 (ν(TiC)=550 cm^{-1};ν(TiCl)=464, 391 cm^{-1}) (457,458).

Infrared and Raman assignments given for the halogen bridged complexes $[(CH_3)_2AuX]_2$ (X=Cl,Br,I) (459) and the monomeric complexes $Y^+[(CH_3)_2AuX_2]^-$ (X=Cl,Br; $Y^+=(C_6H_5)_4As^+$) (460) are summarized in Table 1.32. In all of these derivatives, the configuration about the gold atom is square planar with the methyl groups *cis* to one another. There is a very small frequency difference between the bridging and terminal ν(Au-halogen) modes of the corresponding halide although the bridging ν(AuCl) modes of $[(CH_3)_2AuCl]_2$ are about 30% lower than those of dimeric $AuCl_3$ (459). The ν_s(AuC) mode in *cis*-$(CH_3)_2AuX_2^-$ (X=Cl,Br) is assigned at a higher frequency than the ν_a(AuC) mode (460). Normal coordinate calculations for these two anions show the K(AuCl) and K(AuBr) force constants to be lower by 31% and 33%, respectively, than the corresponding force constants of $AuCl_4^-$ and $AuBr_4^-$ (459).

The vibrational spectra of solid trimethylplatinum chloride (461,462), bromide (461), and iodide (461,462) have been interpreted in terms of a tetrameric structure $[(CH_3)_3PtX]_4$; this structure has been found in partial X-ray crystallographic studies for X=Cl (463) and I (464). Polarization measurements on an oriented single crystal of the chloride have provided unambiguous determination of the symmetry species for the observed frequencies. The intensity of the totally symmetric Raman active modes of the chloride and iodide in the solid state and solution (CCl_4 and $CHCl_3$) has been measured. These data were used to calculate bond polarizability derivatives and to estimate bond orders. The results indicated weak and highly ionic bonds between the platinum atoms and the bridging halogen atoms. The intensity of the low-frequency Raman spectra provided

TABLE 1.32
Skeletal Mode Assignments (cm^{-1}) for Methyl-Gold(III) Halide Derivatives[a]

Compound		ν(AuC)	ν(AuX)	Deformations
$[(CH_3)_2AuCl]_2$	IR	550	256	
	R	571 561	273	289
cis-$(CH_3)_2AuCl_2^-$	IR	563	281 268	
	R	572		272
$[(CH_3)_2AuBr]_2$	IR	530		
	R	561 550	181	266
cis-$(CH_3)_2AuBr_2^-$	IR	560 552	197 179	
	R	558	193	
$[(CH_3)_2AuI]_2$	R	550 545	141 131	

[a]References 459 and 460.

evidence for a weak interaction between the platinum atoms (462). The infrared spectra of $(CH_3)_2PtX_2$ (X=Cl, Br,I) and $(CH_3)_2PtClBr$ indicate the presence of terminal and bridging halogen atoms (465). Although either a polymeric or tetrameric structure is possible, the data are more consistent with the tetrameric structure (465). The derivatives $[(CH_3)_3PtX\cdot(CH_3)_2PtX_2]_n$ (X=Br, I) may be made up of tetrameric units, each of which contains two $(CH_3)_3Pt(IV)$ moieties and two $(CH_3)_2Pt(IV)$ moieties per unit so that n=2. The infrared data contain features attributable to both moieties, but are not the simple superposition of the spectra of $[(CH_3)_3PtX]_4$ and $[(CH_3)_2PtX_2]_n$ (465). A tetrameric structure has been proposed for Tipper's compound $[C_3H_6PtCl_2]_4$ (1.36)

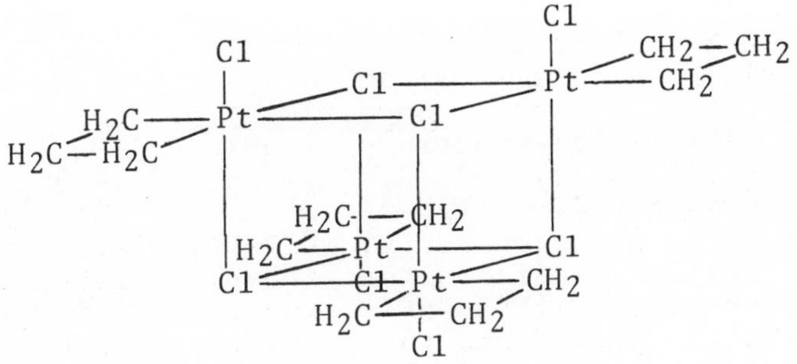

(1.36)

and its bromide analog using infrared and mass spectral data. Assignments were suggested for the terminal and bridging ν(PtX) modes for X=Cl (330 and 230 cm^{-1}, respectively) and X=Br (282 and 206 cm^{-1}, respectively).

Limited infrared data including ν(MC) mode assignments have been reported for CH_3MX_4 (M=Nb,X=Cl,Br;M=Ta, X=Cl) (467) and CH_3WCl_5 (468).

3. Pseudohalides

This section includes alkyl compounds containing terminal azido (-NNN), cyanato (-OCN), thiocyanato (-SCN), selenocyanato (-SeCN), fulminato (-CNO), and cyano (-CN) ligands, and the structurally isomeric isocyanato (-NCO), isothiocyanato (-NCS), isoselenocyanato (-NCSe), isofulminato (-ONC), and isocyano (-NC) ligands, as well as derivatives in which there is bridging through the same atom of the pseudohalide ($^{M}_{M}$>XYZ) or through the end atoms of the pseudohalide (M-XYZ-M). The metal atom may or may not be colinear with the atoms of the linear triatomic pseudohalide ligand. The normal modes associated with a triatomic pseudohalide ligand, XYZ, that is not colinear with the metal atom are illustrated in Figure 1.22. When the metal atom and the pseudohalide ligand are colinear, the MXY bending mode, ν_4, becomes doubly degenerate and the two XYZ bending modes, ν_5 and ν_6, become degenerate. The appearance of a strong intensity pseudohalide stretching band from about 2300 to 2000 cm^{-1} in the infrared spectrum (the corresponding Raman band may appear with either strong or weak intensity) is evidence for the presence of a pseudohalide ligand.

a. Azides

Alkylzinc azides (469) and alkylmercury azides (470,471) have covalent, monomeric structures.

The vibrational spectra of liquid R_2AlN_3 (R=CH_3 (472),C_2H_5 (473,474)) and $(C_2H_5)_2GaN_3$ (472,475) indicate the presence of a trimeric structure with six-membered metal-nitrogen rings, while similar data for solid $(C_2H_5)_2InN_3$ indicate the probable presence of a dimeric structure with a In_2N_2 ring (472). The infrared and Raman spectra and the proposed structures for trimeric $(CH_3)_2AlN_3$ and $(C_2H_5)_2GaN_3$ and dimeric $(C_2H_5)_2InN_3$ are illustrated in Figure 1.23. While the $\nu_a(N_3)$ modes give rise to the most intense band

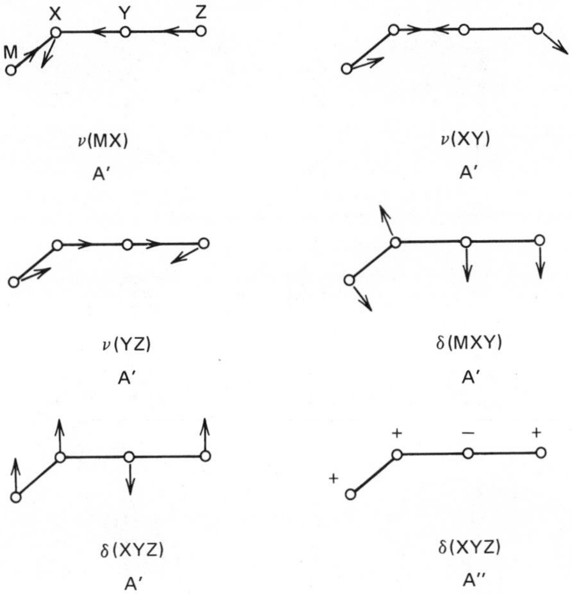

Figure 1.22. Normal modes of vibration for nonlinear M-XYZ pseudohalide derivatives.

in the infrared spectra of these three derivatives, this is not the case in the Raman spectra. An Al_3N_3 ring of C_{3v} symmetry is indicated from the infrared and Raman spectra of trimeric $C_2H_5Al(Cl)N_3$ (476); it has been proposed that the N_3 rather than the chloride ligand bridges the aluminum atoms. The infrared spectrum of solid $(C_2H_5)_2TlN_3$ is very similar to that of $(C_2H_5)_2Hg$ (472). This, together with the similarity of the frequencies of the N_3 modes found for $(C_2H_5)_2TlN_3$ ($\nu(N_3)$=2046,1338 cm^{-1},$\delta(N_3)$=640 cm^{-1}) to those of solid ionic KN_3 ($\nu(N_3)$=2041,1344 cm^{-1},$\delta(N_3)$=645 cm^{-1}) (477, 478), indicates that the thallium derivative has an ionic structure $[(C_2H_5)_2Tl]^+N_3^-$, as has been suggested for several alkyl-thallium halides (46,342,343,346). Using infrared and Raman data, structures have been proposed for the tetramethylammonium salts of $(R_3Ga)_nN_3^-$ (R=CH_3,C_2H_5;n=1,2) (479,480). For the complexes where

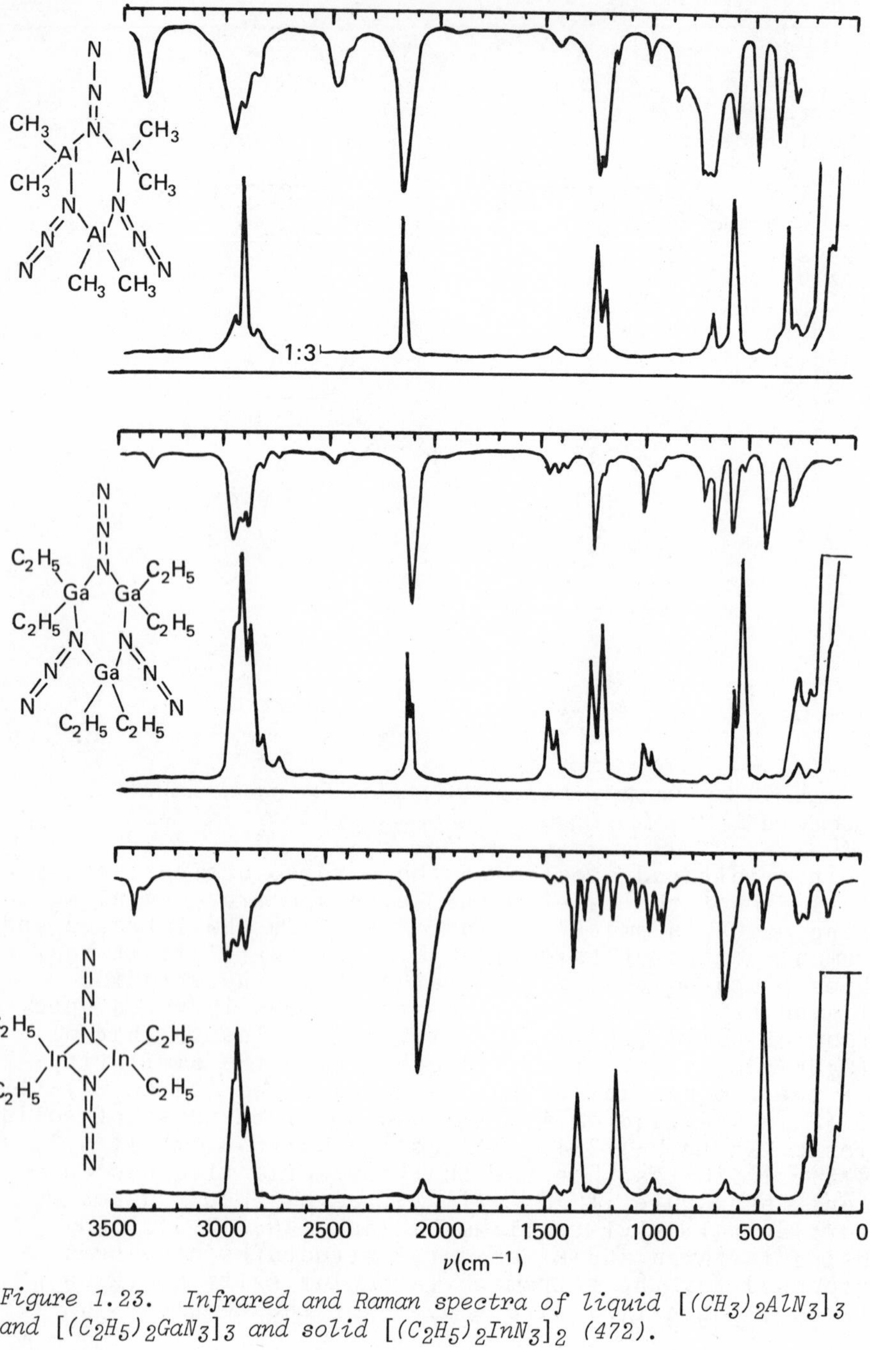

Figure 1.23. Infrared and Raman spectra of liquid $[(CH_3)_2AlN_3]_3$ and $[(C_2H_5)_2GaN_3]_3$ and solid $[(C_2H_5)_2InN_3]_2$ (472).

n=2, the gallium atoms are bridged through the α-nitrogen atom of the N_3 ligand. Similar conclusions have been reached in infrared and Raman studies of the tetramethylammonium salts of $[(CH_3)_3Al]_nN_3^-$ (n=1,2) (480,481). The frequency separation of the $\nu_a(N_3)$ and $\nu_s(N_3)$ modes for alkyl-aluminum and gallium azides decreases as electron density is withdrawn from the N_3 ligand either through changes in the electronegativity of the other ligands bonded to the central metal atom or through formation of an α-nitrogen bridge bond. The infrared and Raman spectra have been described for the tetramethylammonium salts of $(CH_3)_2M(N_3)_2^-$ (M=Al,Ga) and their addition products with trimethylaluminum, trimethylgallium, and dimethylmagnesium (482).

Although the metal-NNN skeleton is generally not colinear in most compounds with a terminal N_3 ligand, a nearly linear Si-NNN skeleton has been proposed for $(CH_3)_3SiN_3$ on the basis of a complete infrared and Raman study (483). The relatively high frequency of the $\nu_a(N_3)$ mode (2148 cm^{-1}) has been attributed to $p\pi\rightarrow d\pi$ back bonding between the α-nitrogen and silicon atoms (484), although the K(SiN) force constant found for $(CH_3)_3SiN_3$ (483) suggests that the Si-N bond order is not increased by such an interaction. Extensive intermolecular association has been suggested for $(CH_3)_2Sn(N_3)_2$ (485,486). The $\nu_a(N_3)$ and $\nu(SnN)$ modes of $(CH_3)_3SnN_3$ appear at 2045 and 464 cm^{-1}, respectively, in the solid state and 2088 and 401 cm^{-1}, respectively, in chloroform and methylene chloride solutions (484,487). Infrared data such as these are suggestive (485,486,488) of intermolecular N_3 bridging in solid $(CH_3)_3SnN_3$; this is also suggested by the physical properties of this compound (487). For $(CH_3)_3MN_3$ (M=Si,Ge,Sn,Pb), the frequencies of the $\nu(N_3)$, $\delta(N_3)$, and $\nu(MN)$ modes all decrease from silicon to lead (484). A *trans* octahedral structure has been assumed for the tetraphenylarsonium salt of $(CH_3)_2Sn(N_3)_4^{2-}$ on the basis of Mossbauer and vibrational data (489).

The infrared and Raman spectra of $(CH_3)_4SbN_3$ in the solid state and methylene chloride solution indicate the presence of slightly distorted, tetrahedral $(CH_3)_4Sb^+$ cations ($\nu_a(SbC)$=575 cm^{-1},$\nu_s(SbC)$=537 cm^{-1}) and N_3^- anions ($\nu_a(N_3)$=2024 cm^{-1},$\nu_s(N_3)$=1329 cm^{-1};$\delta(N_3)$=631 cm^{-1}) (490). The mass spectral, infrared, and Raman data for $(CH_3)_2SbCl_2N_3$ show the presence of a dimeric structure in which the antimony atoms are bridged by the α-nitrogen atom of the N_3 ligands (432).

A dimeric structure with an Au_2N_2 ring has been proposed for $(CH_3)_2AuN_3$ (491). In a study of trimethylplatinum azide using infrared, proton magnetic resonance (pmr), and mass spectral data, a cubane structure was proposed in which the α-nitrogen atom of each N_3 ligand links three platinum atoms in the tetrameric unit $[(CH_3)_3PtN_3]_4$ (492). This structure has been confirmed in a single crystal X-ray study (493). In Table 1.33, selected vibrational assignments are given for covalent alkyl-metal azides.

b. (Iso)Cyanates

To date, all of the organometallic NCO derivatives have the isocyanate structure with metal-nitrogen rather than metal-oxygen bonding. A limited amount of vibrational data are available for organometallic isocyanates. These data are summarized in Table 1.34. The potential of infrared spectroscopy in distinguishing between organometallic isocyanates and cyanates has been discussed (506) in terms of known organic (iso)cyanates. The pseudo ν_a(NCO) mode, also referred to as the ν(CN) mode, appears from 2300 to 2200 cm^{-1} in both normal and isocyanate derivatives. The pseudo ν_s(NCO) mode, also referred to as the ν(CO) mode, however, appears near 1400 cm^{-1} in organic isocyanates (507) and is expected below 1200 cm^{-1} in organic cyanates (508). As noted in Table 1.34, the ν_s(NCO) mode has been assigned above 1300 cm^{-1} for all compounds except CH_3HgNCO, for which two bands were observed in the ν_s(NCO) region (502) rather than the one expected. The two bands have been attributed to Fermi resonance between the ν_s(NCO) mode and twice the δ(NCO) mode (2x635 cm^{-1}) (502). The mass spectral, infrared, and Raman data for $(CH_3)_2SbCl_2NCO$ show the presence of a dimeric structure in which the antimony atoms are bridged by the α-nitrogen atom of the NCO ligands (432).

c. (Iso)Thiocyanates

Five types of bonding (1.37 to 1.41) have been used in characterizing the structures of organometallic NCS derivatives. The vibrational data associated with the NCS modes of alkyl organometallic derivatives of NCS, together with the structures and types of bonding found for these derivatives, are summarized in Table 1.35.

TABLE 1.33
Selected Vibrational Assignments (cm^{-1}) for Covalent Alkyl-Metal Azide Derivatives

Compound	$\nu_a(N_3)$	$\nu_s(N_3)$	$\delta(N_3)$	$\gamma(N_3)$	$\nu(MN)$	$\nu(MC)$	Refs.
CH_3ZnN_3	2106	1230	636	558			469
$C_2H_5ZnN_3$	2118	1280	669	621			469
	2105						
CH_3HgN_3	2020	1281	661	641	594	551	470
	2014						
$C_2H_5HgN_3$	2085	1281	657	642	597	532	470
	2037						
n-$C_3H_7HgN_3$	2023	1285	691	656	619	596	470
i-$C_3H_7HgN_3$	2037	1282		656	596	525	470
n-$C_4H_9HgN_3$	2075	1288	695	656	621	597	470
	2020						
c-$C_3H_5HgN_3$	2085	1290	673	622	398	518	471
	2058	1281	663	591	371	508	
c-$C_5H_9HgN_3$	2075	1279	655	595	393		471
	2047				372		
c-$C_6H_{11}HgN_3$	2060	1272	651	594	390	488	471
	2042				372		
$[(CH_3)_2AlN_3]_3$	2159	1246	723	580[a]		682 580[a]	472
	2146	1232	692				
			469				
$[(C_2H_5)_2AlN_3]_3$	2159	1250	736		557[a]	557[a]	473,474
	2145	1231	670				
			474				
			415				
$[C_2H_5Al(Cl)N_3]_3$	2179	1230	748		565	658 644	476
	2179	1215	465		495[b]486[b]		

TABLE 1.33 (continued)

Compound	$\nu_a(N_3)$	$\nu_s(N_3)$	$\delta(N_3)$	$\gamma(N_3)$	$\nu(MN)$	$\nu(MC)$	Refs.
$(CH_3)_3AlN_3^-$ [c]	2089	1303	665				480,481
$(CH_3)_2Al(N_3)_2^-$ [c]	2122	1367		608	457	675 575	482
	2093	1290			441		
$[(CH_3)_3Al]_2N_3^-$ [c]	2120	1270	660				480,481
$[(C_2H_5)_2GaN_3]_2$	2122	1252	717			573 520	472,485
	2108	1241	635				
			622				
			411				
$(CH_3)_3GaN_3^-$ [c]	2040	1344[d]	659	612	324	533 508	479
		1294[d]					
$(CH_3)_2Ga(N_3)_2^-$ [c]	2091	1346	662	614	381	589 544	482
	2062	1295			349		
$[(CH_3)_3Ga]_2N_3^-$ [c]	2079	1281	690	602	343	541 513	479,480
$(C_2H_5)_3GaN_3^-$ [c]	2010	1286	648			515	479
$[(C_2H_5)_3Ga]_2N_3^-$ [c]	2073	1280	686		341	521 489	479,480
	2052		641				
$[(C_2H_5)_2InN_3]_2$	2088	1358	655		288	518 468[e]	472
	2068	1302	468[e]				
$(CH_3)_3SiN_3$	2148	1313	685	530		702 623	483,484
$(CH_3)_2Si(N_3)_2$	2150						494
$(C_2H_5)_3SiN_3$	2136	1324	679				485
$(CH_3)_3GeN_3$	2100	1290	678		456		484,495
$(CH_3)_2Ge(N_3)_2$	2110	1282	680				495
$(CH_3)_3SnN_3$	2088	1286	674		401		484
$(CH_3)_2Sn(N_3)_2$	2062	1278	660				485
$(CH_3)_2Sn(N_3)_4^{2-}$ [f]	2040					555 497	489
$(CH_3)_3PbN_3$	2034	1279	655				484
$(CH_3)_3PN_3^+$ [g]	2170	1295	551		748	766 653	496

$(CH_3)_2P(N_3)_2^-$[g]	2190	1278	541		752		496
	2165	1258					
$CH_3P(N_3)_3^-$[g]	2195	1255	582	537	842	785	496
	2175	1235	572	525	775		
$(CH_3)_2AsN_3$	2081	1257	680		442	602 585	497,498
$(C_2H_5)_2AsN_3$	2082	1256	673		443	574 559	497,498
						541	
$(CH_3)_3As(N_3)_2$	2050	1268	665	585	345	640 594	499
$(CH_3)_3Sb(N_3)_2$	2080	1285	685		353	576 529	499-501
$[(CH_3)_2SbCl_2N_3]_2$	2102	1250	650	650	419	575 519	432
	2090	1243			363		
$[(CH_3)_2AuN_3]_2$	2087	1279	709			572 563	491
$[(CH_3)_3PtN_3]_4$	2089	1199	632	318	318	590 581	492

[a] ν_s(AlC) and ν(AlN) modes expected in the same region.
[b] ν(AlCl) mode.
[c] $(CH_3)_4N^+$ salt.
[d] The two bands are attributed to Fermi resonance.
[e] ν_s(InC), $\delta(N_3)$, and $\gamma(N_3)$ modes all expected in the same region.
[f] $(C_6H_5)_4As^+$ salt.
[g] $SbCl_6^-$ salt.

TABLE 1.34
Selected Vibrational Assignments (cm^{-1}) for Covalent Alkyl-Metal Isocyanates

Compound	ν_a(NCO)	ν_s(NCO)	δ(NCO)	ν(MN)	Refs.
CH_3HgNCO	2180	1285[a] 1195[a]	635	536	502
$(CH_3)_2BNCO$	2285	1505	580	1111	503
$(CH_3)_3GaNCO^-$[b]	2200	1337	621	318	479,480
$(C_2H_5)_3GaNCO^-$[b]	2198	1326	622	292	479,480
$CH_3Si(NCO)_3$	2280	1461	623	517	504
$(CH_3)_2Si(NCO)_2$	2265	1448	630	523	504
$(CH_3)_3SiNCO$	2290	1435	616	521	484,504
$(CH_3)_3GeNCO$	2240	1415	606	521	484
$(CH_3)_3SnNCO$	2243	1375	618	400	484
$(CH_3)_3PbNCO$	2190	1325	606		484
$(CH_3)_3Sb(NCO)_2$	2200	1365	620	325	500,501
$[(CH_3)_2SbCl_2NCO]_2$	2205	1374 1305	665	453 360	432
$[(CH_3)_2AuNCO]_2$	2192	1323	711 606	470 373	505

[a]The presence of two bands in the ν_s(NCO) mode region has been attributed to Fermi resonance between the ν_s(NCO) mode and twice the δ(NCO) mode (2x635 cm^{-1}).
[b]$(CH_3)_4N^+$ salt.

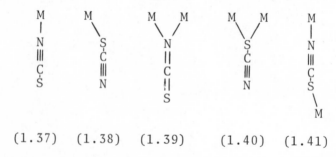

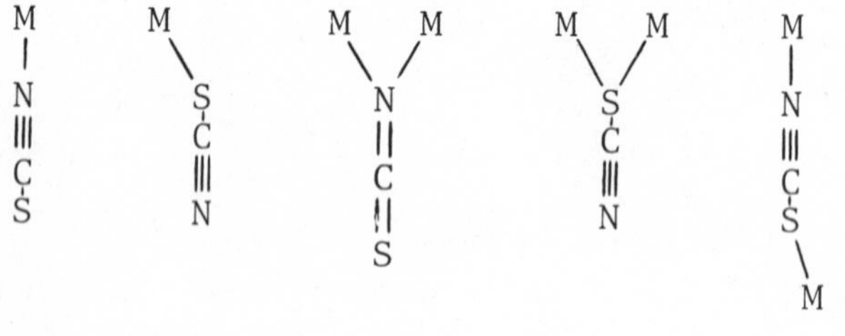

(1.37) (1.38) (1.39) (1.40) (1.41)

Vibrational data for RMSCN (R=CH_3,C_2H_5);M=Zn (509), Cd (510)) show the presence of polymeric solid-state structures. Similar data for CH_3HgSCN also indicate some intermolecular interactions, although a monomeric structure is found in methanol solution (511).

The presence of two bands in the ν_a(TlC) mode region (550 and 540 cm^{-1}) and the apparent absence of a band in the ν_s(TlC) mode region of the infrared spectrum of solid $(CH_3)_2TlNCS$ indicate the presence of

TABLE 1.35
Selected Vibrational Assignments (cm^{-1}) for Alkyl-Metal Isothiocyanates and Thiocyanates

Compound	Structure	Bonding	ν_a(NCS)	ν_s(NCS)	γ(NCS) or δ(NCS)	ν(MN) or ν(MS)	Refs.
$(C_2H_5Be)_2SCN^-$ [a]	Monomer	Be SCN Be	2100	740	415 355	865	134
CH_3ZnSCN	Polymer[b]	Zn SCN Zn	2190 2140	685	455		509
C_2H_5ZnSCN	Polymer[b]	Zn SCN Zn	2178 2130	618	447	280 228	509
CH_3CdSCN	Polymer[b]	Cd SCN Cd	2192 2145				510
C_2H_5CdSCN	Polymer[b]	Cd SCN Cd	2185 2142				510
CH_3HgSCN	Associated[b]		2181[c] 2043[c] 2131[d]	738[c]	440[c] 454[d]	238[d]	511
	Monomer[e]	Hg-SCN	2136[c] 2138[d]			283	511
$CH_3Hg(SCN)_3^{2-}$	Monomer[f]	Hg-SCN	2119			276	512
$(CH_3)_3AlSCN^-$ [a]	Monomer	Al-SCN	2097	847	485	335	480,513
$[(C_2H_5)_3Al]_2SCN^-$ [a]	Monomer	Al SCN Al	2140	800	479 466	318 178	480,513

TABLE 1.35 (continued)

Compound	Structure	Bonding	ν_a(NCS)	ν_s(NCS)	γ(NCS) or δ(NCS)	ν(MN) or ν(MS)	Refs.
$(CH_3)_2AlSCN$	Trimer	Al\SCN/Al	2075	627	501 438		514
$(CH_3)_3GaSCN^-$ [a]	Monomer	Ga-SCN	2089	808	476	236	479,480
$(C_2H_5)_3GaSCN^-$ [a]	Monomer	Ga-SCN	2065	804	477	238	479,480
$(C_2H_5)_2GaSCN$	Trimer	Ga\SCN/Ga	2105	665	476 454		514
$(C_2H_5)_2InSCN$	Trimer	In\SCN/In	2128	649	468 450		514
$(CH_3)_2TlNCS$	Polymer [g]	Tl-NCS	2050,sh 2035	750	470 465		515
$(CH_3)_2Tl(NCS)_2^-$ [h]	Polymer [g]	Tl-NCS	2050				515
$(CH_3)_3SiNCS$	Monomer	Si-NCS	2080	956	436	436	484,516
$(CH_3)_2Si(NCS)_2$	Monomer	Si-NCS	2074 2060	1015	465	433	516
$CH_3Si(NCS)_3$	Monomer	Si-NCS	2095	1050	483		517
$(CH_3)_3GeNCS$	Monomer	Ge-NCS	2075	892	476		484
$(CH_3)_3SnNCS$	Polymer [b]	Sn-NCS-Sn	2098 2079 2046	779	474 467		518
$(CH_3)_2Sn(NCS)_2$	Monomer [i]	Sn-NCS	2050	781	485 478		518
	Polymer [b]	Sn-NCS-Sn	2080 2062	846	485 459		518

$(C_2H_5)_2Sn(NCS)_2$	Polymer[b]	Sn-NCS-Sn	2079	835	483 465		518
$(n\text{-}C_3H_7)_2Sn(NCS)_2$	Polymer[b]	Sn-NCS-Sn	2075 2066	840	483 463		518
$(n\text{-}C_4H_9)_2Sn(NCS)_2$	Polymer[b]	Sn-NCS-Sn	2081	842	483 462		518,519
$(CH_3)_3Sn(NCS)_2^-$[a]	Monomer	Sn-NCS	2070		476		518
$CH_3Sn(NCS)_4^-$[a]	Monomer	Sn-NCS	2051	815	482		518
$[(C_2H_5)_2SnNCS]_2O$	Dimer	Sn-NCS	2037[b] 2030[j]				518
		Sn NCS Sn	1959[b] 1957[j]				
$(CH_3)_3PbNCS$	Polymer[b]	Pb-NCS-Pb	2090	760	463		520
	Monomer[k]	Pb-NCS	2045				520
$(CH_3)_3As(NCS)_2$	Monomer	As-NCS	2040	841	489 475		510
$(CH_3)_3Sb(NCS)_2$	Monomer	Sb-NCS	2045 2015	843	492 482	268	277,510
$(CH_3)_2BiSCN$	Monomer	Bi-SCN	2110	795	450 440		510
$(CH_3)_2AuNCS$	Dimer	Au-NCS-Au	2163	775	444 430	239[l]	459
$(CH_3)_3PtSCN$	Tetramer	Pt SCN-Pt Pt	2176	743	456[m] 446[m]	456[m] 446[m] 263[n] 238[n]	521,522

[a] $(CH_3)_4N^+$ salt.
[b] Solid sample.
[c] Infrared band.
[d] Raman band.
[e] Methanol solution.
[f] Aqueous solution.
[g] It has been suggested that $(CH_3)_2TlNCS$ might more correctly be formulated as $[(CH_3)_2Tl][(CH_3)_2Tl(NCS)_2]$ and that $(CH_3)_2TlNCS$ and $[(C_6H_5)_4As][(CH_3)_2Tl(NCS)_2]$

TABLE 1.35 (continued)

might therefore have similar structures in which polymeric $(CH_3)_2Tl(NCS)_2^-$ units and six-coordinate thallium atoms are present.

[h] $(C_6H_5)_4As^+$ salt.

[i] CS_2 or benzene solution.

[j] CCl_4 or benzene solution.

[k] Acetone solution.

[l] ν(AuN) mode.

[m] The 446 cm^{-1} band was assigned to the δ(NCS) mode in Ref. 521 while in Ref. 522 bands at 456 and 446 cm^{-1} were reported in the region expected for the δ(NCS) and ν(PtN) mode.

[n] In Ref. 521, bands at 234 and 224 cm^{-1} were reported in the ν(PtN) region while in Ref. 522 bands at 263 and 238 cm^{-1} were reported in the region expected for the δ(PtNCSPt) and ν(PtS) mode.

nonequivalent $(CH_3)_2Tl^+$ units and a linear C-Tl-C skeleton (515). Similarly, the tetraphenylarsonium salt of $(CH_3)_2Tl(NCS)_2^-$ shows only one infrared band in the ν_a(TlC) mode region (545 cm^{-1}) and none in the ν_s(TlC) mode region (515). It has therefore been suggested that $(CH_3)_2TlNCS$ might be more correctly formulated as $[(CH_3)_2Tl]^+[(CH_3)_2Tl(NCS)_2]^-$ and that both this compound and $[(C_6H_5)_4As]^+[(CH_3)_2Tl(NCS)_2]^-$ have similar structures in which there are polymeric $(CH_3)_2Tl(NCS)_2^-$ units and six-coordinate thallium atoms.

On the basis of the infrared spectrum of $(CH_3)_3Sn$-NCS, a polymeric solid-state structure and monomeric solution (CS_2, CCl_4, and C_6H_6) structure have been proposed (518). The proposed solid-state structure of $(CH_3)_3SnNCS$ has been confirmed in a single crystal X-ray study (524) that shows the presence of distorted tetrahedral dimethyltin diisothiocyanate molecules linked by Sn-S interactions to form infinite chains. A structure with both bridging and terminal isothiocyanate ligands (1.42) has been found in a single crystal

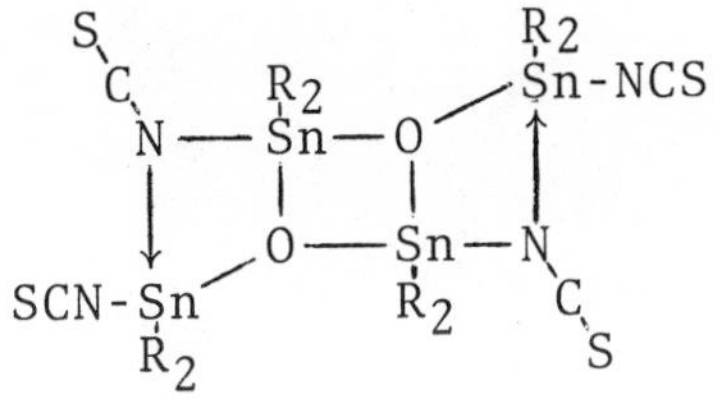

(1.42)

X-ray study of $[((CH_3)_2SnNCS)_2O]_2$ (525) and proposed for $[(R_2SnNCS)_2O]_2$ ($R=C_2H_5$, *i*-C_3H_7, *n*-C_4H_9) (518). Infrared and Raman data have been reported and assigned for $[((CH_3)_2SnNCS)_2O]_2$; this compound shows two bands in the ν_a(NCS) region at 2080 and 2015 cm^{-1} (526). The derivatives $[(R_2SnNCS)_2O]_2$ also show two infrared bands in the ν_a(NCS) region; the higher-frequency band (2037 cm^{-1}) was assigned to the terminal NCS ligand while the lower-frequency band (1959 cm^{-1}) was assigned to the bridging NCS ligand (518). The frequency difference between the ν_a(NCS) mode of $(CH_3)_3PbNCS$ in the solid state and acetone solution supports a structure similar to that found for $(CH_3)_3SnNCS$, namely a polymeric solid-state structure and a monomeric structure with terminal NCS ligands in nonpolar solvents (520).

An ionic structure has been proposed for $[(CH_3)_4Sb]^+$ NCS^- since the NCS mode frequencies (ν_a(NCS)=2064 cm^{-1},

ν_s(NCS)=748 cm^{-1},δ(NCS)=475 cm^{-1}) are similar to those found for KNCS (ν_a(NCS)=2053 cm^{-1},ν_s(NCS)=749^{-1},δ(NCS)= 484 and 470 cm^{-1}) (527).

While the α-nitrogen atoms bridge the gold atoms in dimeric $(CH_3)_2AuX$ (X=N_3,NCO) to form Au_2N_2 rings, an eight-membered ring of C_{2h} symmetry has been proposed for dimeric $(CH_3)_2AuNCS$ with the NCS ligand bridging through both the sulfur and nitrogen atoms (1.43). In-

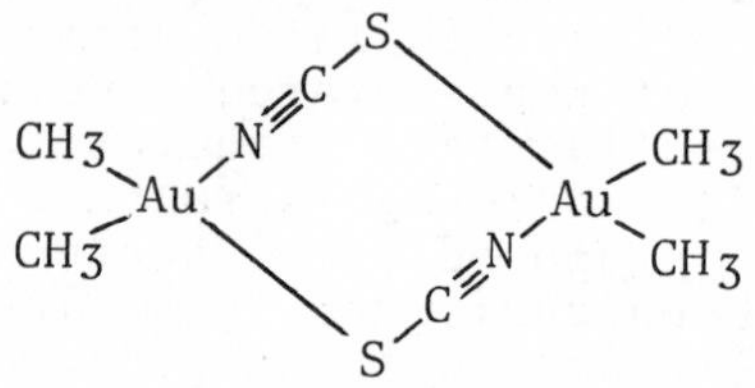

(1.43)

frared and nmr data for $(CH_3)_3PtSCN$ are consistent with a tetrameric structure in which the sulfur atom of each NCS ligand is bonded to two platinum atoms and the nitrogen atom is bonded to one platinum atom in each tetrameric unit (521,522). A dimeric structure similar to that found for $(CH_3)_2AuNCS$ (1.43) has also been proposed for $(CH_3)_3PtSCN\cdot$pyridine (521).

d. *(Iso)Selenocyanates*

Similar structures have been proposed for the alkyl-metal derivatives of NCS and their NCSe analogs with the exception of the trialkyllead(IV) derivatives. While an isothiocyanate structure with the nitrogen atom bonded to the lead atom has been proposed for $(CH_3)_3PbNCS$ in solution (520), a selenocyanate structure with the selenium atom bonded to the lead atom has been proposed (528) for $R_3PbSeCN$ (R=CH_3,C_2H_5) on the basis of the ν_a(NCSe) mode position (approximately 2100 cm^{-1}) in both compounds. The vibrational data for several alkyl-metal NCSe derivatives are summarized in Table 1.36.

e. *(Iso)Fulminates*

All of the reported alkyl-metal derivatives have an ionic structure or a covalent fulminate structure in which the metal atom is bonded to the carbon atom of the CNO ligand. No isofulminate structures have been

TABLE 1.36
Selected Vibrational Assignments (cm^{-1}) for Alkyl-Metal Isoselenocyanates and Selenocyanates

Compound	ν_a(NCSe)	ν_s(NCSe)	δ(NCSe)	ν(MN) or ν(MSe)	Refs.
KNCSe	2070	558	424 416		529
$(CH_3)_3SiNCSe$	2046	800	447	367	530,531
$(C_2H_5)_3SiNCSe$	2065	826			531
$(CH_3)_3GeNCSe$	2052	795	430		532
$(C_2H_5)_3GeNCSe$	2050		430		532
$(n\text{-}C_4H_9)_3GeNCSe$	2047		425		532
$(CH_3)_3PbSeCN$	2100	572			528
$(C_2H_5)_3PbSeCN$	2100	569			528
$CH_3HgSeCN$	2140	540[a] 536[a] 527[a]	395 374		533
$[(CH_3)_2Al]_2SeCN^-$ [b]	2132		425	272 154	480,513
$[(CH_3)_2AuSeCN]_2$	2169	620		199 172	505

[a]The ν(HgC) and ν_s(NCSe) modes are expected to overlap in this region.
[b]$(CH_3)_4N^+$ salt.

proposed for any CNO derivative. It has been noted (506) that the ν_a(CNO) mode is expected at higher frequencies in fulminate derivatives than in the fulminate ion. The similarities of the frequencies for the CNO modes in $(CH_3)_2TlCNO$ (ν_a(CNO)=2061 cm^{-1},ν_s(CNO)=1100, 1080 cm^{-1},δ(CNO)=461 cm^{-1}) to those in ionic CNO derivatives indicate that it has a highly ionic structure (346). Assignments have also been reported for $(n\text{-}C_3H_7)_3PbCNO$ (ν_a(CNO)=2135 cm^{-1},ν_s(CNO)=1156 cm^{-1}, δ(CNO)=477 cm^{-1}) (534), CH_3HgCNO (ν_a(CNO)=approximately 2200 cm^{-1},ν_s(CNO)=approximately 1200 cm^{-1}) (535), and $C_6H_5CH_2HgCNO$ (ν_a(CNO)=2122 cm^{-1},ν_s(CNO)=1228 cm^{-1}, δ(CNO)=493,432 cm^{-1}) (346,535). The only assignment for the ν(MC) mode of the metal-CNO bond is 313 cm^{-1} for $C_6H_5CH_2HgCNO$, while the ν(HgC) mode for the Hg-$CH_2C_6H_5$ unit has been assigned at 453 cm^{-1} (346).

f. (Iso)Cyanides

It has been noted (506) that although the ν(CN) mode in organic isocyanides generally appears at a lower frequency than in organic cyanides (nitriles) (535), the

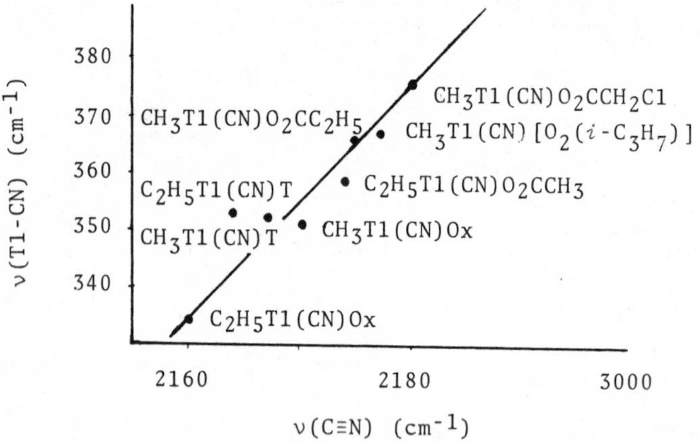

Figure 1.24. Infrared frequencies of the ν(C≡N) and ν(Tl-CN) modes in RTl(CN)Y (R = CH_3 or C_2H_5; Y = tropolonate (T), oxinate (Ox), or other carboxylate ligand) (541).

total frequency range is too narrow to permit the ν(CN) mode position to be used with certainty in determining the type of bonding in organometallic CN derivatives.

The infrared and Raman spectra have been assigned for the tetramethylammonium salt of $[(CH_3)_2MgCN]_4^{4-}$ (536). The vibrational spectrum of CH_3HgCN in the solid and molten states as well as in solution has been interpreted in terms of the carbon bonded cyanide structure and a molecular symmetry of C_{3v} (286,537). An X-ray and neutron diffraction study have confirmed this as the solid-state structure with a linear C-Hg-CN skeleton (538). Vibrational data for $[(CH_3Hg)_2CN]^+NO_3^-$ (539) are consistent with a structure in which the CN ligand bridges the mercury atoms through both the carbon and nitrogen atoms.

The infrared and Raman spectra have been assigned for tetrameric $(CH_3)_2MCN$ (M=Al,Ga,In) (536). Vibrational data for the tetramethylammonium salts of $(R_3M)_2CN^-$ (R=CH_3,M=Al (480,513),Ga (479,480,513);R=C_2H_5,M=Ga (479,480)) have been interpreted in terms of a structure in which the metal atoms are bridged by both the carbon and nitrogen atoms of the CN ligand. The presence of terminal carbon bonded CN ligands is consistent with vibrational data reported for $[(CH_3)_4N]^+[R_3GaCN]^-$ (R=CH_3,C_2H_5) (479,480), $K^+[(CH_3)_2Ga(CN)_2]^-$

(540), and $[(CH_3)_4N]^+[(CH_3)_3TlCN]^-$ (316). As illustrated in Figure 1.24, a linear relationship has been found between the ν(CN) and ν(Tl-CN) mode frequencies of several alkylthallium cyano carboxylates (541). This realtionahip is related to the strength of the thallium atom interaction with the noncyano ligands. The possibility of metal to CN π-back bonding, which is found in transition metal CN complexes, does not exist for thallium. Therefore, as the strength of the noncyano ligand-thallium interaction increases, there is a decrease in both the ν(CN) and ν(Tl-CN) mode frequencies.

Arguments have been advanced supporting the presence of pure cyanide or isocyanide isomers, or an equilibrium involving both of these linkage isomers for $(CH_3)_3SiCN$ (506,542,543). An equilibrium of both isomers was suggested because of the presence of two infrared bands, one of very weak intensity (2100 cm^{-1}) and one of relatively strong intensity (2190 cm^{-1}), in the ν(CN) mode region (544). The equilibrium theory has received strong support in a recent spectroscopic study (543). This study includes Raman data, variable-temperature infrared data on the normal and ^{13}C and $C^{15}N$-enriched samples of $(CH_3)_3SiCN$, variable-temperature pmr data on the normal and ^{13}CN-enriched samples, and $^1H(^{29}Si)$ double-resonance spectra of the enriched sample. The observed and calculated isotopic shifts of the ν(CN) mode frequencies of the $(CH_3)_3SiCN$ derivatives are summarized in Table 1.37. These data strongly support the presence of both the cyanide and isocyanide linkage isomers. In Figure 1.25, the variable-pressure, vapor-phase infrared spectra of $(CH_3)_3SiCN$ are illustrated (543). The similarity of spectra *a* and *d* illustrate the reversibility of the spectral changes; this would not be the case if the changes were due to decomposition. An analysis of the vapor-phase infrared spectra gives a value of 4.00±0.03 kcal/mole for ΔH^o for the reaction

$$(CH_3)_3SiCN \rightleftharpoons (CH_3)_3SiNC$$

The mole fraction of the isocyanide isomer in the liquid at 25°C was calculated to be 0.0015±0.0005. Similar equilibria have been found for the triethyl (544,545) and triisopropyl (545) derivatives. An analysis of the infrared spectrum of ^{13}CN-enriched samples of $(CH_3)_2Si(CN)_2$ at various temperatures shows the presence of an equilibrium between the two isomers $(CH_3)_2Si(CN)_2$ and $(CH_3)_2Si(CN)(NC)$ (546). Although no evidence could be

TABLE 1.37
Observed and Calculated Shifts (cm^{-1}) of the $\nu(^{13}CN)$ and $\nu(C^{15}N)$ Mode Frequencies Relative to the $\nu(CN)$ Mode Frequency at 2198 cm^{-1} and 2095 cm^{-1} for Vapor Samples of $(CH_3)_3SiCN$ and $(CH_3)_3SiNC$, Respectively[a]

	Band	
	2198 cm^{-1}	2095 cm^{-1}
Observed $\nu(^{13}CN)$ shift	-51	-40
Calc. $\nu(^{13}CN)$ shift assuming SiCN	-51	-49
Calc. $\nu(^{13}CN)$ shift assuming SiNC	-42	-40
Observed $\nu(C^{15}N)$ shift	-31	-35
Calc. $\nu(C^{15}N)$ shift assuming SiCN	-30	-28
Calc. $\nu(C^{15}N)$ shift assu ing SiNC	-38	-36

[a]Reference 543.

found for the existence of the isocyanide isomer in a microwave and vibrational study of $(CH_3)_2Si(CN)H$, it was concluded that these results do not necessarily exclude the presence of such an isomer in very low concentrations (547). Both a single crystal X-ray study of the solid (548) and a microwave study of the vapor (549) have shown $(CH_3)_3GeCN$ to have the cyanide structure with a Ge-CN bond and no bridging by the CN ligands, although the presence of a small amount of the isocyanide isomer could not be ruled our in either study. The presence of two infrared bands in the $\nu(CN)$ mode region of $(CH_3)_3GeCN$ in chloroform (2182 and 2090 cm^{-1}), together with chemical behavior similar to that of the analogous silicon compound, have been interpreted as indicating that $(CH_3)_3GeCN$ exists as an equilibrium mixture of the cyanide and isocyanide isomers (549,550). This comclusion is supported by infrared data for $(CH_3)_3GeCN$ and its ^{13}CN-substituted analog in the vapor state. The solid-state infrared spectra of several mono- and dialkyltin CN derivatives all show one $\nu(CN)$ band in the 2180 to 2145 cm^{-1} region (550,551). Infrared data for $(CH_3)_3SnCN$ have been interpreted as consistent with the cyanide structure and a highly polar Sn-CN bond (550). More recently, however, an X-ray crystallographic study of $(CH_3)_3SnCN$ has shown the presence of a polymeric solid-state structure consisting of infinite chains in which disordered CN ligands are aligned on either side of planar $(CH_3)_3Sn(IV)$ units, and the Sn-CN and Sn-NC bond distances are equal (552).

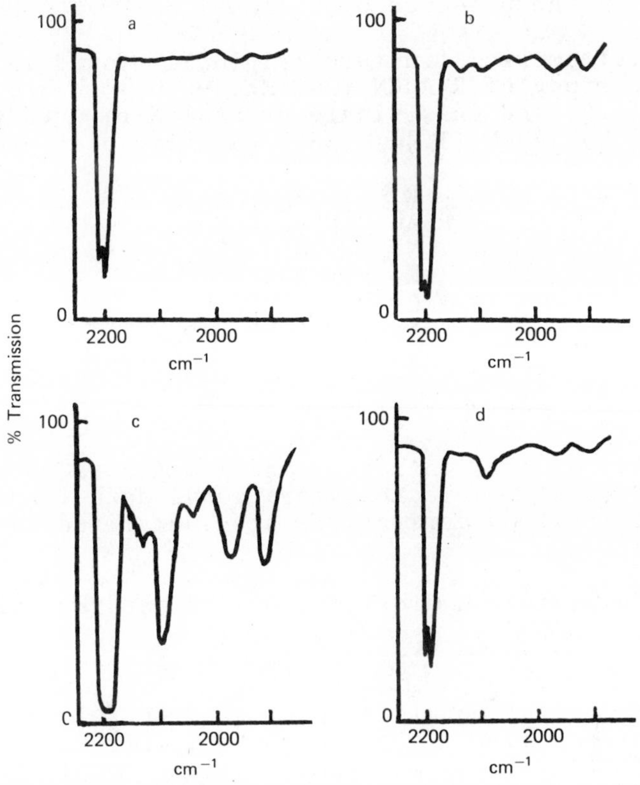

Figure 1.25. Infrared spectrum of $(CH_3)_3SiCN$ vapor. Path length = 1.25 m; spectral slit width = 1.3 cm^{-1}; successive vapor pressures = a, 2; b, 4; c, 12; and d, 2 mm Hg (543).

Similar polymeric structures have been found in single crystal X-ray studies of $(C_2H_5)_3SnCN$ and $(CH_3)_3PbCN$ (553).

The infrared spectra of $CH_3M(CN)_2$ (M=P,As) in the solid state, the Raman spectra of these compounds in the solid state, solution (methyl cyanide), and the vapor state, and the Raman spectrum of $CH_3P(CN)_2$ as a melt have been reported (554). The vibrational data for the condensed states indicate the presence of a weak M···N interaction which have lead to the assignment of the ν(MN) mode in addition to the assignment of the ν(M-CN)

mode. An ionic structure has been proposed for solid
$[(CH_3)_4Sb]^+CN^-$ ($\nu(CN)=2066$ cm^{-1} in the infrared and Raman spectrum) (512). The vibrational spectrum of the
tetramethylammonium dalt of $[(CH_3)_2Cl_3Sb]_2CN^-$ is consistent with the presence of a linear Sb-CN-Sb skeleton (437).

The tetrameric structure originally found in a
cryoscopic study of R_2AuCN ($R=C_2H_5,n-C_3H_7$) in bromoform
solution (555) and in a single crystal X-ray study of
solid $(n-C_3H_7)_2AuCN$ (1.44) (556) has also been proposed

$$
\begin{array}{ccc}
\overset{\displaystyle R}{|} & & \overset{\displaystyle R}{|} \\
R-Au-C\equiv N-Au-R \\
N & & C \\
\| \| \| & & \| \| \| \\
C & & N \\
R-Au-N\equiv C-Au-R \\
\underset{\displaystyle R}{|} & & \underset{\displaystyle R}{|}
\end{array}
$$

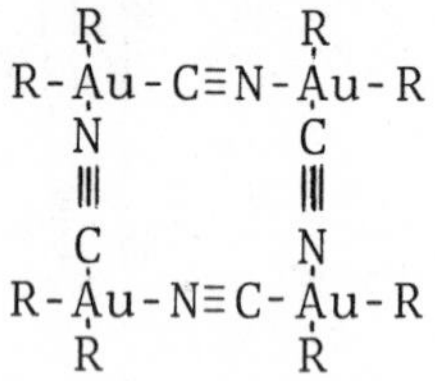

(1.44)

on the basis of nmr, infrared, and Raman data for
$(CH_3)_2AuCN$ (505). Selected vibrational data for covalent alkyl-metal CN derivatives are summarized in
Table 1.38.

4. Aquo Complexes, Hydroxides, Oxides, and Peroxides

a. Aquo Complexes

Aquated organometallic species often form on dissolving various salts (e.g., nitrates, sulfates, per-
chlorates, or halides) of organometallic complexes in
water. Infrared spectroscopy is of limited value in
the study of these complexes since solvent water gives
rise to very intense bands that obscure the modes a-
rising from coordinated water molecules. Raman spectroscopy has proved more useful in studying aquated
organometallic complexes (559) since the Raman scatter-
ing of water molecules is quite small. Even Raman
spectroscopy, however, has its limitations since the
O-H stretching and bending modes of coordinated water
overlap with those of the solvent water molecules.
Also, the $\nu(M-OH_2)$ modes expected for complexes involving coordinated water give rise to bands of relatively weak intensity, especially in those complexes
in which the metal-water interactions are highly polar.
Compounds with very weak or polar metal-water bonds are
treated in Sec. A.2 of this chapter.

TABLE 1.38
Selected Vibrational Assignments (cm^{-1}) for Covalent Alkyl-Metal Cyanides and Isocyanides

Compound	ν(CN)	ν(M-CN)	δ(MCN)	Refs.
$[(CH_3)_2MgCN]_4^{4-}$[a]	2145	371 290	230	536
CH_3HgCN	2180	386	306	286,537
$(CH_3Hg)_2CN^+$[b]	2207	565[c] 111[d]	330	539
$[(CH_3)_2AlCN]_4$	2223	510 386 370 290	250	536
$[(CH_3)_2GaCN]_4$	2207	429 365-316	198	536
$[(CH_3)_2InCN]_4$	2172	365 342 281 235	250	536
$[(CH_3)_3Al]_2CN^-$[a]	2185	459[c] 323[d]		480,513
$[(CH_3)_3Ga]_2CN^-$[a]	2158	351[c]		479,480
$[(C_2H_5)_3Ga]_2CN^-$[a]	2160	365[c]		479,480
$(CH_3)_3GaCN^-$[a]	2137	325		479
$(CH_3)_2Ga(CN)_2^-$[e]	2160 2138	370 350		540
$(C_2H_5)_3GaCN^-$[a]	2125	311		479,480
$(CH_3)_3TlCN^-$[a]	2100	225		316
$(CH_3)_3SiCN$	2198[f] 2095[g]	539	124	542,543
$(C_2H_5)_3SiCN$	2200[f] 2100[g]			544
$(CH_3)_3GeCN$	2182[f] 2090[g]			549,550
$(CH_3)_3Sn$	2163			548,549
$(CH_3)_2Sn(CN)_2$	2174			551
$(C_2H_5)_3SnCN$	2160			551
$(C_2H_5)_2Sn(CN)_2$	2146			551
$CH_3P(CN)_2$	2187	575 633[h]	476 316	554
$(CH_3)_2PCN$	2170	560	439 238	557
$CH_3As(CN)_2$	2181	463 411[h]	246 162	554

TABLE 1.38 (continued)

Compound	ν(CN)	ν(M-CN)	δ(MCN)	Refs.
$[(CH_3)_2Cl_3Sb]_2CN^-$ [a]	2205	380[i] 230-118[i]		437
CH_3SeCN	2152	521	401 365	558
C_2H_5SeCN	2153	565[j] 522[j]	401 363	558
i-C_3H_7SeCN	2152	540[j] 512[j]	419 363	558
n-C_3H_7SeCN	2152	565[j] 521[j]	413 363	558
n-C_4H_9SeCN	2152	565[j] 520[j]	400 362	558
$[(CH_3)_2AuCN]_4$	2202	473-354		505

[a] $(CH_3)_4N^+$ salt.
[b] NO_3^- salt.
[c] Assigned to ν(M-$\vec{C}$≡$\vec{N}$-M) mode.
[d] Assigned to ν(←M-C≡N-M→) mode.
[e] K^+ salt.
[f] Due to cyanide isomer.
[g] Due to isocyanide isomer.
[h] ν(M-NC) mode.
[i] Although the ν_a(Sb-CN-Sb) mode has been assigned at 380 cm^{-1}, the $\delta(SbCl_2)$, $\delta(SbC_2)$, and ν_s(Sb-CN-Sb) modes are all expected in the 230 to 118 cm^{-1} region and therefore no definite assignment could be made for the ν_s(Sb-CN-Sb) mode.
[j] Two bands attributed to the presence of more than one rotational isomer.

A limited number of solid compounds containing coordinated water have been characterized. The presence of one infrared band in the ν(SnC) mode region of solid $[(CH_3)_3Sn(H_2O)_2]^+[B(C_6H_5)_4]^-$ indicates that the C_3Sn skeleton is planar (560).

The solid-state and/or aqueous solution vibrational data for several organometallic derivatives containing coordinated water are summarized in Table 1.39. Also, infrared and Raman data for aquated $(CH_3)_2Pb^{2+}$ show the presence of a linear C-Pb-C skeleton with a weak Raman band at 425 cm^{-1}, possibly due to a $\nu(Pb\text{-}OH_2)$ mode (52).

b. Hydroxides and Oxides

Organometallic hydroxides are found with structures in which terminal (M-OH), oxygen-bridging ($^{M}_{M}\!\!>\!OH$), or hydrogen-bonded (M-OH···X) hydroxyl groups are present. Table 1.40 includes vibrational data for several alkylmetal hydroxide derivatives. The presence of hydrogen bonding is used to explain the relatively broad ν(OH) mode frequency ranges observed for silicon (568,570,571) and selenium (577) OH derivatives and $[(CH_3)_2AuOH]_4$ (578), and the complexity of the spectra for several phosphorus OH compounds (580,581). Structures with terminal and/or bridging oxygen atoms are found for organometallic oxides.

Using Raman data, it was originally concluded that the covalent structure CH_3HgOH found in the solid state was also present in a 4M aqueous solution of this compound (45). This interpretation of the solution Raman data has been questioned (295) since it has now been shown (582) that such a solution contains the compound $[(CH_3Hg)_3O]OH$; similarly, the product of the reaction of CH_3HgBr and silver oxide consists of a mixture of $[(CH_3Hg)_3O]OH$ and $(CH_3Hg)_2O$ (582). The infrared and Raman spectra of $[(CH_3Hg)_3O]ClO_4$ in the solid state and as a 1.5M solution in methyl cyanide seem consistent with the presence of a planar or nearly planar Hg_3O skeleton (583). It has been noted, however, that in the case of $[(CH_3Hg)_3O]OH$ it is possible to interprete the data in terms of a nonplanar Hg_3O skeleton, except for a band at 1050 cm^{-1} that cannot be explained in terms of either a planar or a nonplanar skeleton (295). Vibrational assignments have also been made for $(CH_3Hg)_2O$ (295).

Infrared data have been reported for $(R_2Al)_2O$ (R= CH_3,C_2H_5 (584),*i*-C_4H_9) (585) and $(C_2H_5AlCl)_2O$ (584). The cyclohexane solution infrared spectra of the former

TABLE 1.39
Vibrational Assignments (cm^{-1}) for Alkyl Organometallic Aquo Complexes

Compound	Phase	ν(OH)	δ(H_2O)	ν($M-OH_2$)	ν(MC)	ρ_r(H_2O)	Refs.
$(CH_3)_3Sn(H_2O)_2^+$	Mull of $(C_6H_5)_4B^-$ salt.	3450	1592		562	735	562
$CH_3HgH_2O^+$	4M soln. of ClO_4^- salt.	a	a	463	570	a	561
$(CH_3)_2Ga(H_2O)_2^+$	2.5M soln. of ClO_4^- or NO_3^- salt.	a	a		620 558	a	562
$(CH_3)_3Pt(H_2O)_3^+$	3M soln. of ClO_4^-, SO_4^{2-} or NO_3^- salt.	a	a	357	600	a	563,564
$(CH_3)_3Pt(D_2O)_3^+$	3M soln. of ClO_4^-, SO_4^{2-} or NO_3^- salt.	a	a	345	599	a	563,564
$(CH_3)_2Au(H_2O)_2^+$	1.35M or 2.25M NO_3^- or 2.6M ClO_4^- salt solns.	a	a	408	591	a	562,565
$(CH_3)_3SnNO_3H_2O$	Mull	3280	1645		540 510	606	566

[a]Absorptions obscured by solvent water.

TABLE 1.40
Selected Vibrational Assignments (cm^{-1}) for Alkyl Organometallic Hydroxides

Compound	Structure	Phase	ν(OH)	δ(M-OH)	ν(M-OH)	Refs.
$(CH_3)_2BOH$	Monomer	Solid	3630	950	1275	567
$CH_3B(OH)_2$	Monomer	Solid	3560-3000	1190 1117	820 1377	567
$(CH_3)_2Ga(OH)_2^-$	Monomer	2.4M soln. of Na^+ salt.	a	a		562
$(CH_3)_2GaOH$	Tetramer	Solid	3598	1033 1013	396 167	562
	Tetramer	CCl_4 soln.	3644	974		562
$(CH_3)_2GaOD$	Tetramer	Solid	2656	730	387 360	562
$(CH_3)_3SiOH$	Monomer	Vapor	3737	1074	915	568,569
$(CH_3)_3SiOD$	Monomer	Vapor	2754	779	903	568,569
$(CH_3)_2Si(OH)_2$	Monomer	Mull	3125			570
$(C_2H_5)_3SiOH$	Monomer	Solution	3676	805	856	568
$(C_2H_5)_3SiOD$	Monomer	Solution	2954		835	568
$(C_2H_5)_2Si(OH)_2$	Monomer	Mull	3220			571
$(CH_3)_2Ge(OH)_2$	Monomer	4.5M soln. of $(CH_3)_2GeO$	a	a	673	572
$(CH_3)_3SnOH$	Polymer	Mull	3620	920	580-500	573,574
	Dimer	1 to 6% CCl_4 soln.	3658		370	573
$(C_2H_5)_3SnOH$	Polymer	Mull	3630	890		574
$(CH_3)_2SnNO_3OH$	Polymer	Mull	3390	1000	403	566
$(C_2H_5)_2SnNO_3OH$	Polymer	Mull	3344	1006	413 398	566
$(CH_3)_3AsOH^+$	Monomer	Mull of NO_3^-, HSO_4^-, or ClO_4^- salts.	2540-2300	1392	770-750 595-580	421

TABLE 1.40 (continued)

Compound	Structure	Phase	ν(OH)	δ(M-OH)	ν(M-OH)	Refs.
$(CH_3)_3As(OH)Cl$	Monomer	Mull	2410	1220	760 692	421
$(CH_3)_3As(OH)Br$	Unknown	Mull	2340	1168	752 612	421
$(CH_3)_3As(OD)Br$	Unknown	Mull	1730	882	745 445	421
$(C_2H_5)_2As(OH)_2^+$	Monomer	Mull of Cl^- salt.	2325		772 355 or 309	421
$(CH_3)_4SbOH$	Monomer	Mull	3480 3060	995	482	431
$CH_3Se(OH)_2^+$	Monomer	Solid as Cl^- salt.	3050-2800 2990	1229	722 714	577
$C_2H_5Se(OH)_2^+$	Monomer	Solid as Cl^- salt.	2920-2840 2420	1230 1210	725 666	577
$(CH_3)_2Au(OH)_n^{1-n}$ [b]	Monomer	0.7M H_2O and 1.26M NaOH soln. of Na^+ salt.	a	a	441	565,578
$(CH_3)_2AuOH$	Tetramer	Mull	3400-3100		468 317	578
	Tetramer	CCl_4 soln.	3600		466 410	578
$(CH_3)_3PtOH$	Tetramer	Mull	3598	722	360	462,563, 579
$(CH_3)_3PtOD$	Tetramer	Mull	2654	559	358 340	563,579

[a] Absorption obscured by solvent water.
[b] $n \geq 3$.

three compounds all exhibit two bands in the ν(AlOAl) region at 815 to 780 cm^{-1} and 710 to 700 cm^{-1} (585). Since each of these compounds contains two electron-deficient aluminum atoms and one electron-donating oxygen atom per molecule, depending on the solution concentration, varying degrees of association are likely between the molecules. Therefore, the higher-frequency band, the position of which is dependent on the nature of the alkyl group, has been attributed to the isolated Al-O-Al unit, while the lower-frequency band, the position of which is independent of the nature of the alkyl group, has been attributed to the presence of an associated grouping (1.45) (585). The infrared spectra in

$$R_2Al-O{\rightarrow}\overset{|}{\underset{|}{Al}}- \qquad R_2Al-\underset{M}{\overset{|}{O}}{\rightarrow}AlR_3$$
$$\phantom{R_2Al-O{\rightarrow}}R_2Al$$

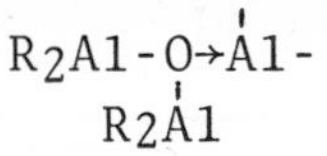

(1.45)   (1.46)

benzene or toluene solutions have been examined for $(R_2AlOM)AlR_3$ ($R=C_2H_5$,M=Li,Na,K,Cs;R=i-C_4H_9,M=Li) with a structure in which the R_2AlOM group acts as a Lewis base and R_3Al acts as a Lewis acid (1.46) having been proposed (586). Bands in the 900 to 800 cm^{-1} and 775 to 690 cm^{-1} regions have been assigned to the ν_s(AlOM) and ν_s(AlOAl) modes, respectively, while bands from 1110 to 1010 cm^{-1} were assigned to the coupled ν_a(AlOM) and ν_a(AlOAl) modes. A single crystal X-ray study of $(CH_3)_2$GaOH shows the presence of a tetrameric structure in which two oxygen and two carbon atoms are arranged tetrahedrally about each gallium atom to give a Ga_4O_4 ring (587). Vibrational data indicate that the tetrameric structure of solid $(CH_3)_2$GaOH is retained on dissolving in organic solvents (562). The vibrational spectrum of a concentrated aqueous solution of $(CH_3)_2$TlOH is similar to that of $(CH_3)_2$Tl$^+$ (45). The slight displacement of the frequencies in the spectrum of $(CH_3)_2$Tl-OH from those observed for $(CH_3)_2$Tl$^+$ indicates that in solution $(CH_3)_2$TlOH is found as electrostatically bound ion pairs (45).

Several compounds have been prepared that contain the $(CH_3)_3$Si or related units. For $[(CH_3)_n(C_2H_5)_{3-n}Si]_2O$ (n=1 to 3) and $(CH_3)_3SiOSi(C_2H_5)_3$, the ν_a(SiO) and ν_s(SiO) modes have been assigned to infrared bands in the 1072 to 1055 cm^{-1} and 553 to 495 cm^{-1} regions, respectively (90,588,589). Similar assignments have been reported for $[(C_2H_5)_2SiH]_2O$ (ν_a(SiO)=1068 cm^{-1},ν_s(SiO)=

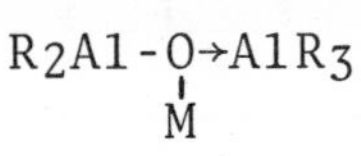

553 cm^{-1}) (589), and [(CH$_3$)$_2$SiX]$_2$O (X=F,I) and
(CH$_3$)$_3$SiOSi(CH$_3$)$_2$Cl (ν_a(SiO)=1100 to 1065 cm^{-1},ν_s(SiO)=
approximately 550 cm^{-1}) (590).

The structures of methylsiloxanes have been represented by combinations of the units 1.47 to 1.50

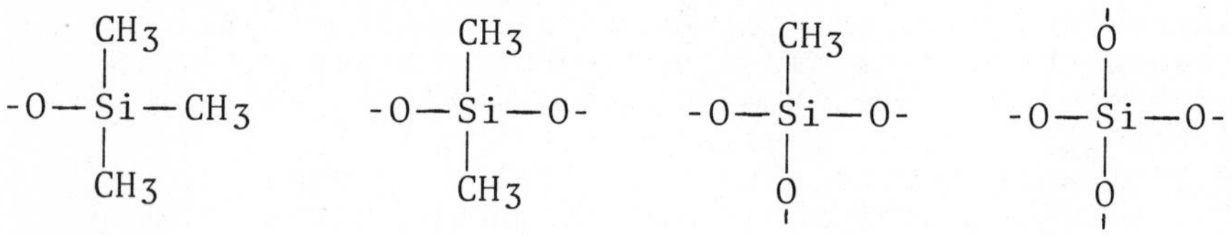

CH$_3$	CH$_3$	CH$_3$	O
|	|	|	|
-O—Si—CH$_3$	-O—Si—O-	-O—Si—O-	-O—Si—O-
|	|	|	|
CH$_3$	CH$_3$	O	O
M, M^H	D, D^H	T, T^H	Q
(1.47)	(1.48)	(1.49)	(1.50)

with the symbols M^H, D^H, or T^H representing units in
which one of the methyl groups of M, D, or T, respectively, has been replaced by a hydrogen atom. The
ν_a(SiO) modes of cyclic (593-596) and linear (593,597)
methylsiloxanes give rise to very intense infrared
bands that become broader as the number of silicon
atoms increases. This is illustrated in Figure 1.26
(593). The ν_a(SiO) modes give rise to relatively weak
intensity Raman bands (594-596). Additional vibrational data have been reported for cyclic alkylsiloxanes
(597-600). In Figure 1.27, the ν(SiO) mode frequencies
are compared for various methylsiloxanes (592). Several compounds have been prepared in which the (CH$_3$)$_3$SiO
unit is bonded through the oxygen atom to another metal
atom. The ν(SiO) (or ν_a(SiOM)) mode for these compounds
has been assigned from 1065 to 900 cm^{-1} (150,601-610).
The ν(MO) (or ν_s(SiOM)) mode is found over a broader
frequency range and has been assigned for derivatives
of mercury (approximately 490 cm^{-1}) (602), thallium
(410 to 380 cm^{-1}) (602), germanium (approximately 685
cm^{-1}) (606), lead (500 to 425 cm^{-1}) (150,603), bismuth
(approximately 575 cm^{-1}) (603), gold (478 cm^{-1}) (605),
and rhenium (475 cm^{-1}) (607).

The vibrational data reported for alkylgermanox-
anes and alkylstannoxanes are more limited than those
for the corresponding silicon derivatives. It is difficult to assign the ν(GeO) modes for the alkylgerman-
oxanes since they are expected to be in the same region
as the ρ_r(CH$_3$) modes (611). For [(CH$_3$)$_3$Ge]$_2$O, it was
originally proposed that a band at 870 cm^{-1} was most
probably associated mainly with the ν_a(GeO) mode (611).
In more recent and complete studies of (R$_3$Ge)$_2$O, the

ν_a(GeO) and ν_s(GeO) modes have been assigned at 794 and 467 cm^{-1}, respectively, for R=CH_3 (612), 853 and 440 cm^{-1}, respectively, for R=C_2H_5 (613), and 841 and 400 cm^{-1}, respectively, for R=*n*-C_4H_9 (612). Although infrared bands from 870 to 730 cm^{-1} have been attributed to both the ρ_r(CH_3) and ν(GeO) modes of $[(CH_3)_2GeO]_n$, it has been suggested that the bands at 848, 860, and 868 cm^{-1} are predominantly ν(GeO) in character for n=3, 4, and higher polymer, respectively (611). The infrared spectra have been reported and assigned for $[(CH_3)_3GeO\text{-}MR_3]_2$ (M=Ga,R=CH_3,C_6H_5;M=In,R=CH_3) (329).

Bands due to the ν(SnO) modes of alkylstannoxanes are expected at the lower end of the ρ_r(CH_3) mode frequency range. Although originally assigned from approximately 650 to 575 cm^{-1} (611), the ν_a(SnO) modes have more recently been assigned at higher frequencies for $[(CH_3)_3Sn]_2O$ (737 cm^{-1}) (574,614), $[(C_2H_5)_3Sn]_2O$ (740 cm^{-1}) (574), $[(n\text{-}C_3H_7)_3Sn]_2O$ (760 cm^{-1}) (574), and $[(n\text{-}C_4H_9)_3Sn]_2O$ (770 cm^{-1}) (574). In addition, the ν_s(SnO) mode has been assigned at 415 cm^{-1} for $[(CH_3)_3Sn]_2O$ (614). Bis(trialkyltin)oxides react with water at from 0 to 5°C to give the corresponding trialkyltin hydroxides (574). The infrared spectrum of $(CH_3)_3SnOH$, illustrated in Figure 1.28, shows this compound to have different structures in the solid state and solution (CCl_4) (573). The frequency shifts of the ν(OH) mode and, to a greater degree, the ν(SnO) mode, are most noteworthy. Molecular weight determinations in carbon tetrachloride, chloroform, and benzene solutions and infrared data have led to the conclusion that solid $(CH_3)_3SnOH$ has a linear, polymeric structure (1.51) while in solution it has a dimeric structure

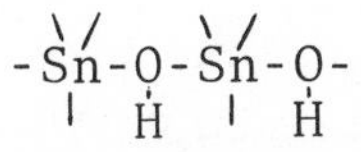

(1.51)

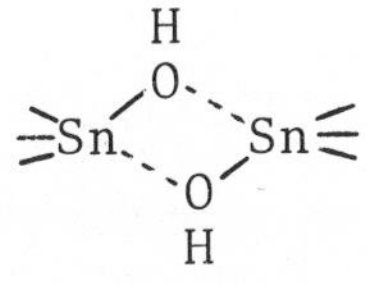

(1.52)

(1.52) with bridging through the oxygen atom of the OH ligands. An X-ray study of $(CH_3)_3SnOH$ has since confirmed the presence of the polymeric solid-state structure; the nearly planar C_3Sn units are inclined at approximately 15° to the plane normal to the Sn-O axis and almost equadistant from the OH ligands (615). Although $(CH_3)_3SnOH$ is stable at room temperature, the

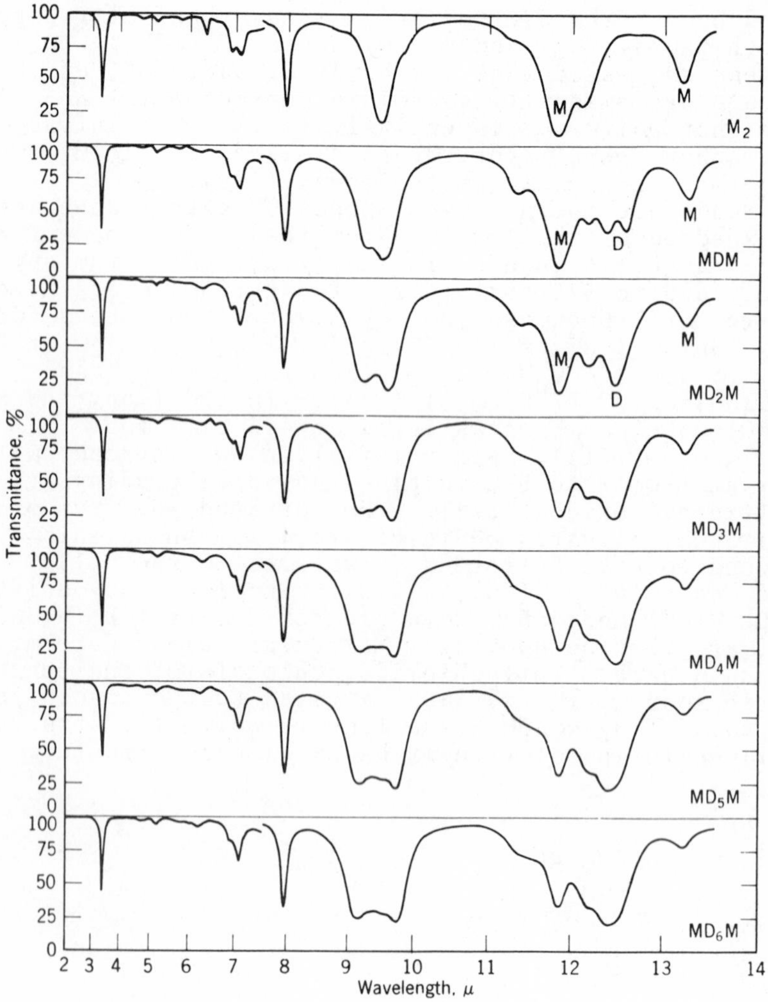
M
M
M2
M
D
M
MDM
M
M
D
MD2M
MD3M
MD4M
MD5M
MD6M
Transmittance, %
100
75
50
25
0
Wavelength, μ
2 3 4 5 6 7 8 9 10 11 12 13 14

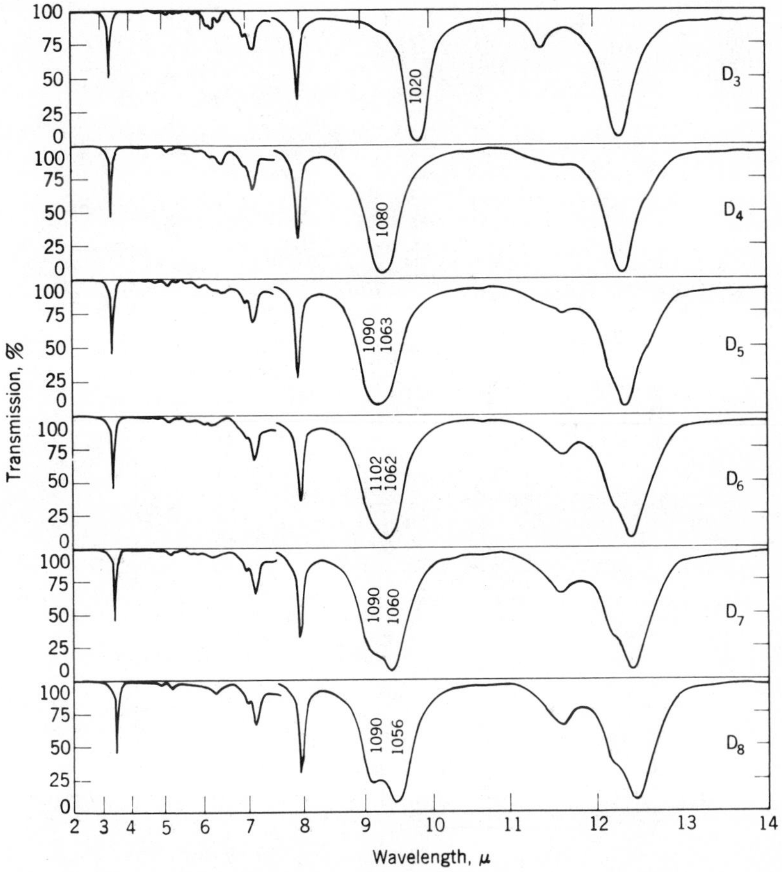

Figure 1.26. Infrared spectra of linear polymethylsiloxanes (first seven spectra) and cyclic dimethylsiloxanes (last six spectra) (593).

corresponding ethyl, *n*-propyl, and *n*-butyl derivatives are in equilibrium with the corresponding bis(trialkyltin) oxides and water,

$$2R_3SnOH \rightleftharpoons R_3SnOSnR_3 + H_2O$$

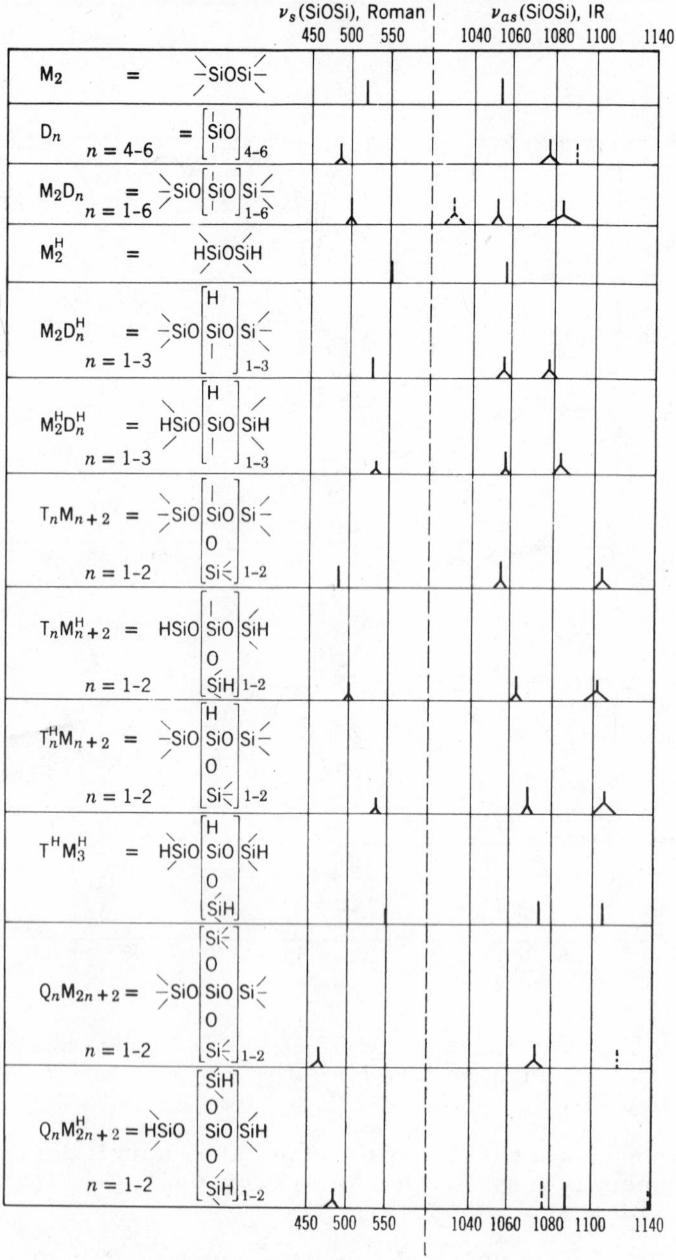

Figure 1.27. Frequencies and intensity of ν(SiOSi) bands for various polysiloxanes (⋮ denotes the band partly observed) (592)

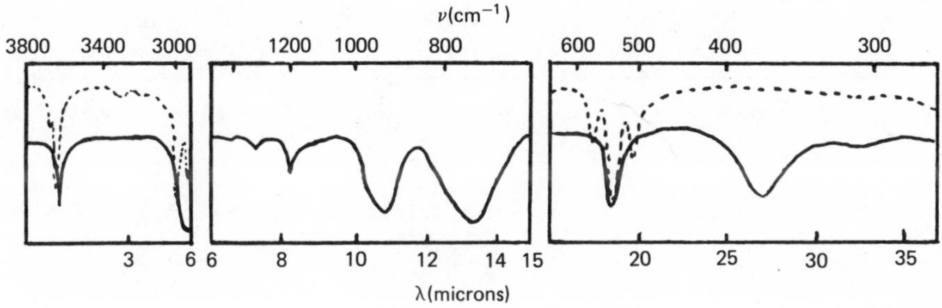

Figure 1.28. Infrared spectra of $(CH_3)_3SnOH$ as Nujol or hexachlorobutadiene mulls (——) and CCl_4 solution (---) (573).

In an earlier study (391), a band attributed to the ν_a(SnO) mode of $[(C_2H_5)_3Sn]_2O$ (770 cm^{-1}) was reported to be present with diminished intensity in the infrared spectrum of $(C_2H_5)_3SnOH$. Although these data were originally interpreted as showing that the hydroxide was the hydrated oxide (391), they have now been attributed to the equilibrium mentioned above between the hydroxide and the oxide (574). In support of this conclusion is the fact that the 770 cm^{-1} band almost disappears in the infrared spectrum of a sample prepared at low temperature by grinding the oxide with an excess of water in an inert atmosphere (574). Structures involving both bridging hydroxide and nitrate ligands have been proposed for $R_3Sn(OH)NO_3$ ($R=CH_3,C_2H_5$) (566). Although bis(trialkyltin) oxides are monomeric, polymeric structures have been proposed for mono- and dialkyltin oxides (616,617). The ν_a(SnO) mode of these polymeric compounds has been assigned from 576 to 548 cm^{-1} in the infrared spectra. The higher frequency of this mode in bis(trialkyltin) oxides (770 to 735 cm^{-1}) as compared to the value in the polymeric mono- and dialkyltin oxides has been attributed to the presence of $p\pi \rightarrow d\pi$ overlap from the oxygen atom lone pairs of electrons to the tin atoms in bis(trialkyltin) oxides (616,617) and/or to a hybridization difference between the tin atoms of the bis(trialkyltin) oxides and the mono- and dialkyltin oxides (617).

The metal-oxygen bond order in R_3MO (M=P,As,Sb) is greater than one because of the possibility of back

TABLE 1.41
Calculated Frequency Ranges (cm^{-1}) and Observed Frequencies (cm^{-1}) for the ν(MO) Modes of $(Alkyl)_3MO$ Derivatives (M=P,As,Sb)

	R_3PO[a,b]		R_3AsO[a,c]		R_3SbO[d]	
	ν(P=O)	ν(P-O)	ν(As=O)	ν(As-O)	ν(Sb=O)	ν(Sb-O)
R	1175	840	965	695	803	550
CH_3	1148		868		645	
C_2H_5	1166		885		678	
n-C_3H_7	1172		888		650	
n-C_4H_9	1169		892		650	
n-C_5H_{11}					650	

[a]Frequency range calculated using bond length from Ref. 619.
[b]Data for $(CH_3)_3PO$ from Refs. 18,442,619, and 620; those for $(C_2H_5)_3PO$, $(n\text{-}C_3H_7)_3PO$, and $(n\text{-}C_4H_9)_3PO$ from Refs. 621,622, and 623, respectively.
[c]Data for $(CH_3)_3AsO$ from Refs. 619 and 624; data for the other R_3AsO compounds from Ref. 625.
[d]Data for $(CH_3)_3SbO$ from Ref. 626; data for the other R_3SbO compounds from Ref. 627.

donation of the oxygen *p*π electrons into the empty *d*π orbitals of phosphorus, arsenic, or antimony. Using Gordy's rule (618), which relates the force constant of a diatomic system to its bond order, bond length, and the electronegativity of the two constituent atoms, a force constant can be obtained for an isolated metal-oxygen bond, assuming that the bond order is either one or two. These data can then be substituted into the harmonic oscillator equation to determine the approximate frequency range expected for the ν(MO) modes. Frequency ranges calculated for the ν(MO) modes together with assignments of these modes in several alkyl derivatives are compared in Table 1.41. Normal coordinate analyses have been reported for $(CH_3)_3MO$ (M=P (619),As (619,624),Sb (626)). Using the K(MO) force constants from these calculations and Gordy's rule, the PO, AsO, and SbO bond orders were calculated to be 1.96, 1.60, and 1.20, respectively. The methyl torsional modes have been assigned for $(CH_3)_3PO$ and $(CD_3)_3PO$ (18). In Figure 1.29, the infrared and Raman spectra are illustrated for $(CH_3)_3PO$ and $(CH_3)_3AsO$ (619). In addition to the ν(SbO) mode assignments listed in Table 1.41 for $(alkyl)_3SbO$

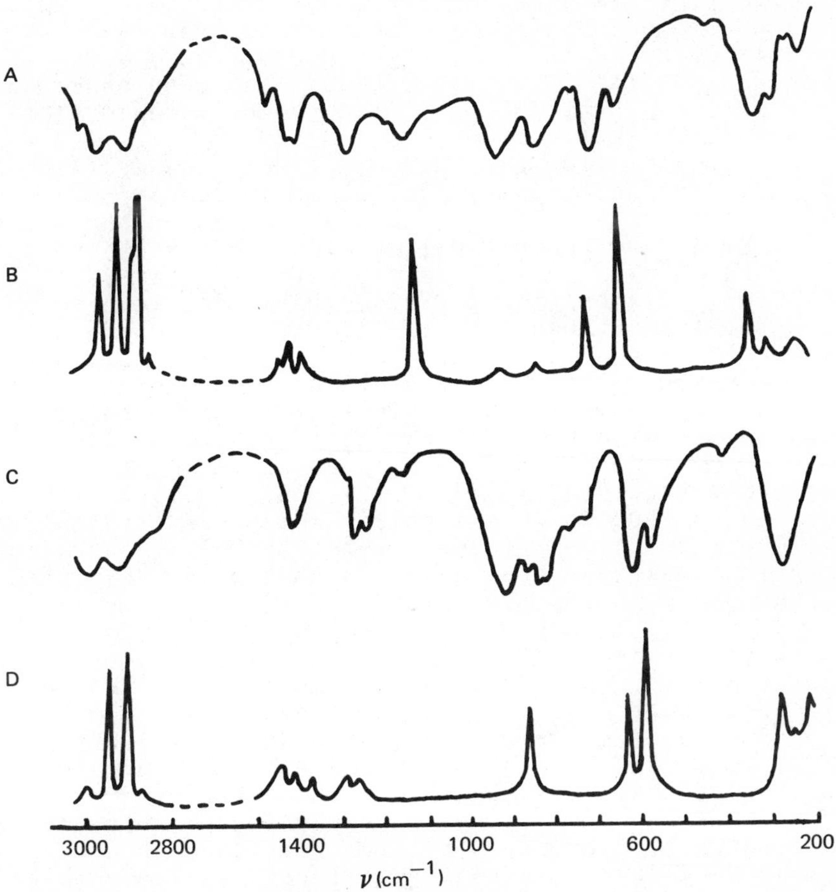

Figure 1.29. The (A) infrared and (B) Raman spectra of $(CH_3)_3PO$ and the (C) infrared and (D) Raman spectra of $(CH_3)_3AsO$ (619).

derivatives, additional bands have appeared in the infrared spectra of CCl_4 solutions of $(C_2H_5)_3SbO$ (478 cm^{-1}) and the other $(alkyl)_3SbO$ (470 cm^{-1}) derivatives (other than $(CH_3)_3SbO$), and have been attributed to ν(SbO) modes arising from molecular association of the $(alkyl)_3SbO$ molecules (627). The fact that $(C_2H_5)_3SbO$ is monomeric in $CHCl_3$ is attributed to an association of the type $(C_2H_5)_3SbO\text{-}HCCl_3$, which cannot take place in CCl_4 (627). The ν(MO) mode frequencies of $(alkyl)_3MO$

(M=P,As,Sb) have been correlated with an expression involving the total mass of the molecule and the masses and electronegativities of the M and oxygen atoms (628). The ν_a(SbO) mode of the Sb-O-Sb skeleton for $(R_3SbX)_2O$ ($R=CH_3$, X=Cl, Br, NO_3; $R=C_2H_5$, n-C_3H_7, n-C_4H_9, X=Cl) and $[(CH_3)_3Sb]_2OX$ ($X=SO_4, SeO_4, CrO_4, C_2O_4$) has been assigned to an infrared band in the 790 to 775 cm^{-1} region for the methyl derivatives (64,118,430,629) and the 792 to 734 cm^{-1} region for the other alkyl derivatives (118). Although originally formulated as an ionic compound (349), more recently it has been suggested on the basis of infrared data that $[(CH_3)_3Sb(ClO_4)]_2O$ has a covalent structure with a five-coordinate antimony atom (630).

Complete vibrational studies have been reported for several methyl and ethyl derivatives of phosphoric acid (H_3PO_4), arsenic acid (H_3AsO_4), and the corresponding anions. Selected vibrational data for these compounds are compared in Table 1.42. Three characteristic infrared bands that arise from the OH ligand under conditions of strong hydrogen bonding have been observed in the 2800 to 2500 cm^{-1}, 2375 to 1900 cm^{-1}, and 1750 to 1600 cm^{-1} regions for compounds with the M(=O)OH (M=P,As) unit (638). For $CH_3P(OH)O_2^-$, only two of these bands have been observed as shoulders on the solvent bands (631). Two similar bands have been observed in the infrared spectrum of $CH_3P(=O)(OH)_2$, with the third observed as a shoulder on the high-frequency side of a water band at 1640 cm^{-1} (631). One interpretation of these data is that the bands arise from strong Fermi resonance between the ν(OH) mode and twice the δ(MOH) modes (639). The ν(P=O) mode in the infrared spectrum of $CH_3P(=O)(OH)_2$ shifts to higher frequencies and becomes sharper as the concentration of this compound in water decreases (631). These changes have been attributed to hydrogen bonding effects. While in high concentrations, the most probable interaction is between the $CH_3P(=O)(OH)_2$ molecules, in dilute solutions, the most probable interaction is between $CH_3P(=O)(OH)_2$ and solvent water molecules. Normal coordinate calculations for $CH_3P(=O)(OH)_2$, $CH_3P(OH)O_2^-$, and $CH_3PO_3^{2-}$ show the extent of coupling between the internal modes of each of these compounds (640). The coupling is most severe for the ν(PC), ν(P-OH), and ν(PO) modes from 1060 to 750 cm^{-1}. Normal coordinate calculations have also been reported for several of the arsenic acid derivatives included in Table 1.42 (575,636). It has been suggested (631) that $Na_2(alkylPO_3)$ and $Na[(alkyl)_2PO_2]$ derivatives can be distinguished from each other by the fact that

TABLE 1.42
Selected Vibrational Assignments (cm^{-1}) for Methyl and Ethyl Derivatives of Phosphoric and Arsenic Acid, and the Corresponding Anions of These Derivatives

Compound	ν(OH)	δ(MOH)	ν(M=O)	ν_a(MO)	ν_s(MO)	ν(M-OH)	ν(MC)	Deforms.	Refs.
$CH_3P(=O)(OH)_2$	2900 2100	1250	1140			1005 950	758	492 443 330	631
$CH_3P(OH)O_2^-$	2700 2200	1250		1150	1060	925	763	507 462 320	631
$CH_3PO_3^{2-}$				1050	972		750	522 498 336	631,632
$CD_3PO_3^{2-}$				1060	969		720	510 495 320	632
$(CH_3)_2PO_2^-$				1128	1040		738 700	315 275	633
$CH_3As(=O)(OH)_2$	2845 2370	1220	913			767	636	364 265 230	575,634
$CH_3As(OH)O_2^-$	2585	1220		844	844	725	630	369 337 266	575,634
$CH_3AsO_3^{2-}$				820	820		621	368 265	575,635
$(CH_3)_2As(=O)OH$	2850 2320	1315	869			747	654 611	365 301 262 210	575,576 636
$(CH_3)_2AsO_2^-$				830	830		636 605	352 267 214	575,637
$C_2H_5As(=O)(OH)_2$	2800 2315		912			775	606	396 367 331 269	634
$C_2H_5AsO_3^{2-}$				861	834		597	420 337 284 235	635
$(C_2H_5)_2As(=O)OH$		1309	854			738	641 590	402 352 296 243 235	576
$(C_2H_5)_2AsO_2^-$				830	830		607 571	381 323 299 271	637

the ν_a(PO) and ν_s(PO) modes of the former compounds are observed at approximately 1100 and 1000 cm^{-1}, respectively, while the corresponding modes of the latter compounds are observed at approximately 1150 and 1050 cm^{-1}, respectively. Vibrational data and assignments have been reported for $[(CH_3)_2MO_2M'(CH_3)_2]_2$ (M'=Al,Ga, In;M=P,As) (641) and $(R_3SnO_2PR'_2)_n$ (R=CH_3,*n*-C_4H_9;R'= CH_3,C_6H_{13}) (642) in which the $(alkyl)_2MO_2^-$ (M=P,As) groups bridge the metal atoms (1.53). Infrared data

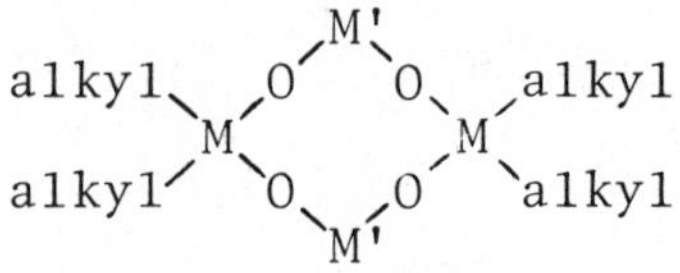

(1.53)

indicate that a four-coordinate, tetrahedral skeletal structure is found for $[(CH_3)_3AsOH]X$ (X=NO_3,ClO_4,HSO_4) while a five-coordinate, trigonal bipyramidal skeletal structure is found for $(CH_3)_3As(OH)Cl$; in the same study, however, the structures of $(CH_3)_3As(OH)Br$ and $(CH_3)_3As(OD)Br$ could not be determined (421). Infrared and Raman data show that while the reaction product of HCl and $(CH_3)_2As(=O)OH$ is $(CH_3)_2As(=O)OH\cdot HCl$, the reaction of $(C_2H_5)_2As(=O)OH$ with HCl produces $[(C_2H_5)_2As(OH)_2]Cl$ (576). The infrared mull spectrum of $(CH_3)_4SbOH$ indicates a trigonal bipyramidal skeletal structure and an axial OH ligand (431). The ν_a(SbC) mode has been assigned (566 cm^{-1}) for $(CH_3)_3Sb(OH)_2$ (430). Additional vibrational data have been reported for alkyl phosphonic acid halides (405,409,411,438-442).

The ν(SeO) mode of the dialkylselenoxides R_2SeO (R=CH_3,C_2H_5,*n*-C_3H_7,*n*-C_4H_9) has been assigned to a very strong intensity infrared band at 820 cm^{-1} for the methyl derivative and approximately 805 cm^{-1} for the higher alkyl derivatives (643,644). The ν_a(SeO) and ν_s(SeO) modes of the dialkylselenones R_2SeO_2 (R=CH_3, C_2H_5,*n*-C_3H_7,*n*-C_4H_9) have been assigned from approximately 918 to 914 cm^{-1} and approximately 889 to 872 cm^{-1}, respectively (643). The solid-state and aqueous solution Raman spectra and KBr pellet infrared spectra have been reported for the sodium and potassium salts of $RSeO_2^-$ (645) and also for RSe(=O)OH (R=CH_3,C_2H_5) (646). More than one rotational isomer was found for the ethyl derivatives in solution. The Raman spectra of $[RSe(OH)_2]Cl$ (R=CH_3,C_2H_5) in aqueous solutions and the solid-state as well as the infrared spectra of these compounds as Nujol mulls or suspended in KCl pellets have been reported and assigned (577). Two rotational isomers were found for the ethyl derivative in solution. Strong hydrogen bonding leading to associated species was found for RSe(=O)OH (646) and $[RSe(OH)_2]Cl$ ($R=CH_3$, C_2H_5) (577) in both the solid state and solution.

An X-ray study has shown that solid $(CH_3)_2Au(OH)_2$ is a tetramer with two oxygen atoms and two carbon atoms arranged in a square-planar manner about each gold atom to give a Au_4O_4 ring (578). Vibrational data indicate that the solid-state structure is retained when this compound is dissolved in organic solvents (578). Vibrational data reported for tetrameric $(CH_3)_3PtOH$ include Raman polarization measurements of a single crystal that have provided an unambiguous determination of the symmetry species of the observed bands (462).

c. Peroxides

The $\nu(OO)$ mode for the peroxide derivatives $[(CH_3)_3Si]_2OO$ and $[(C_2H_5)_3Si]_2OO$ has been assigned at 945 and 918 cm^{-1}, respectively (647). The same study included assignments of the $\nu_a(SiO)$ and $\nu_s(SiO)$ modes of $[(CH_3)_3Si]_2OO$ (1259 and 768 cm^{-1}, respectively) and $[(C_2H_5)_3Si]_2OO$ (1240 and 762 cm^{-1}, respectively). For the derivatives $R_3SiOOM(C_6H_5)_3$ ($M=Si, R=CH_3, C_2H_5, n\text{-}C_3H_7, n\text{-}C_4H_9; M=Ge, R=CH_3, C_2H_5, n\text{-}C_4H_9$) a characteristic band from 935 to 890 cm^{-1} has been attributed to the $\nu(OO)$ mode, while bands in the 797 to 770 cm^{-1} and 638 to 635 cm^{-1} regions have been assigned to $\nu(SiO)$ and $\nu(GeO)$ modes, respectively (648). The $\nu_s(SnO)$ mode has been reported (550 cm^{-1}) for $[(C_2H_5)_3Sn]_2OO$ (649).

5. Mercaptides, Sulfides, and Disulfides

Vibrational data reported for alkyl-metal mercaptides, $R_mM(SH)_n$, are very limited. The $\nu(SH)$ and $\nu(MS)$ modes have been assigned at 2590 and 582 cm^{-1}, respectively, for $(CH_3)_2BSH$ (650) and at 2584 and 454 cm^{-1}, respectively, for $(CH_3)_3SiSH$ (349). On dissolving $(CH_3)_3SiSH$ in dimethylacetamide or dimethylformamide, the $\nu(SH)$ mode frequency decreases by 60 and 46 cm^{-1}, respectively, because of hydrogen bonding between the $(CH_3)_3SiSH$ and solvent molecules (651).

A planar Hg_3S skeleton is indicated by the vibrational spectrum of $(CH_3Hg)_3S^+$ with the $\nu_a(HgS)$ mode assigned at 332 cm^{-1} and the $\nu_s(HgS)$ mode assigned at

TABLE 1.43
The ν(MS) Mode Assignments (cm^{-1}) for Several $(Alkyl)_3MS$ Derivatives (M=P,As,Sb)

R	ν(PS)[a]	ν(AsS)[b]	ν(SbS)[c]
CH_3	567	473	431
C_2H_5	552	476	439
n-C_3H_7	596 583	487 485	439
n-C_4H_9	596	487	440
n-C_5H_{11}	599 588	484	

[a]Data for $(CH_3)_3PS$ from Refs. 18,442, 657 and 658; data for the remaining $(alkyl)_3PS$ derivatives from Ref. 658.
[b]Data for $(alkyl)_3AsS$ derivatives from Ref. 151.
[c]Data for $(alkyl)_3SbS$ derivatives from Refs. 627 and/or 654.

281 cm^{-1} (583). Both the ν_a(MS) and ν_s(MS) modes of the MS_2 skeleton have been assigned for $[(CH_3)_2B]_2S$ (594 and 545 cm^{-1}, respectively) (650), $[(CH_3)_3Si]_2S$ (488 and 437 cm^{-1}, respectively) (90,652,653), $(CH_3)_3SiSGeH_3$ (461 and 401 cm^{-1}, respectively) (653), and $[(CH_3)_3Sn]_2S$ (366 and 320 cm^{-1}, respectively) (614,654,655). The infrared and Raman spectra of $(R_2SnS)_3$ (R=CH_3,C_2H_5,*n*-C_4H_9) are consistent with a cyclic Sn_3S_3 skeleton of C_{3v} or C_s but not D_{3h} symmetry (655,656).

As was true for the corresponding oxides, the metal-sulfur bond in $(alkyl)_3MS$ (M=P,As,Sb) has double-bond character because of $p\pi \rightarrow d\pi$ overlap from filled sulfur *p*-orbitals to empty metal *d*-orbitals. The ν(MS) mode assignments given for several of these alkyl derivatives are summarized in Table 1.43. The methyl torsional mode has been assigned for $(CH_3)_3PS$ and $(CD_3)_3PS$ (18). Vibrational data have also been presented for various alkyl thiophosphonic halides (405,408,409,438-442). A single crystal X-ray study of $[(C_2H_5)_2P(=S)]_2$ shows the presence of a *trans* structure in which there is a P-P bond and equivalent phosphorus atoms with two ethyl groups and a sulfur atom bonded to each (659). The solid-state as well as the liquid-state and CS_2 and cyclohexane solution infrared and Raman spectra of $[(C_2H_5)_2P(=S)]_2$ have been interpreted in terms of this structure (660). A similar structure has been proposed for $[(CH_3)_2P(=S)]_2$ (661,662). The solid-state infrared spectrum and the solution infrared and nmr spectra of $(CH_3)_4As_2S_2$ show a structure different from that of the corresponding phosphorus derivative (663). An equilibrium (1.54 and 1.55) has been proposed in solution

$$\underset{\underset{\displaystyle S}{\|}}{(CH_3)_2As}SAs(CH_3)_2 \rightleftharpoons (CH_3)_2AsSSAs(CH_3)_2$$

(1.54) (1.55)

that shifts to the right as the solution temperature is raised. This is noted in Figure 1.30, in which the solution infrared spectrum is illustrated at three different temperatures (663). With increasing temperature, the intensity of the ν(As=S) band (488 cm^{-1}) decreases while the intensity of the ν(AsS) bands (399 and 365 cm^{-1}) increase. In the solid state, it has been concluded that only species 1.54 is present. Infrared and Raman data and assignments have been reported for $Na[(CH_3)_2AsS_2]\cdot 2H_2O$ and the complexes $[(CH_3)_2AsS_2M(CO)_m]_n$ (M=Mn,Re;m=4,n=1;m=3,n=polymer) (664).

Some derivatives of the trithiodiborolane system (1.56) have been prepared and characterized. For $X=CH_3$,

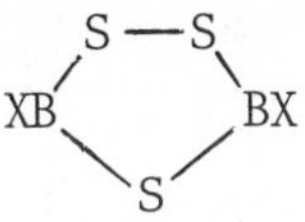

(1.56)

Cl, Br, and I, the ν(SS) mode has been assigned at 479, 450, 436, and 427 cm^{-1}, respectively (665).

6. Alkoxides, Organoperoxides, and Ethers

The most characteristic feature of the vibrational spectra of alkyl-metal alkoxides is the relatively strong infrared intensity of the ν(CO) mode (approximately 1200 to 950 cm^{-1}). The ν(MO) mode is more difficult to assign because of possible overlap with the ν(MC) mode.

Solid CH_3ZnOCH_3 is tetrameric with the zinc atoms occupying the corners of a regular tetrahedron and the oxygen atoms occupying the corners of an interpenetrating but smaller tetrahedron (666). Vibrational data have been reported for methyl- and ethylzinc methoxide as well as the corresponding cadmium derivatives (667).

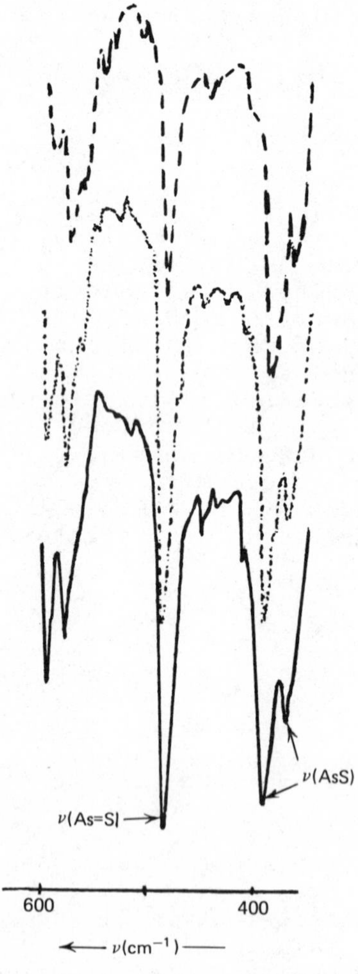

Figure 1.30. Temperature dependence of the infrared spectrum of $(CH_3)_4As_2S_2$ in CCl_4 and 1,1,2,2-tetrachloroethane at 30°C (——), 60°C (···), and 75°C (---) (663).

The trimeric, oxygen-bridged structure proposed for R_2AlOCH_3 ($R=CH_3, C_2H_5$) and the dimeric structure proposed for $(C_2H_5)_2AlOC_2H_5$ (584,668) are consistent with vibrational assignments for these and the corresponding compounds with deuterated alkoxide ligands

(669-671). A trimeric structure with a puckered M_3O_3 ring has been proposed on the basis of vibrational data reported and assigned for R_2MOR' (M=Ga,In;R and R'=CH_3, CD_3) (667). Vibrational data have also been reported for solid $(CH_3)_2TlOR$ (R=CH_3,CD_3) (667). A single crystal X-ray study of $(CH_3)_2TlOCH_3$ has shown it to have a tetrameric structure similar to that described above for CH_3ZnOCH_3 (672).

The vibrational spectra of alkyl-silicon (588,673-676) and -germanium (612,613,677) alkoxides are consistent with the presence of terminal alkoxide groups. Although a great quantity of vibrational data have appeared for alkyl-tin alkoxides (678-687), it has been noted (688) that conclusive structural data are still lacking for these derivatives. A polymeric structure has been found in a recent X-ray crystallographic study of $(CH_3)_3SnOCH_3$ with planar C_3Sn units linked by methoxy groups to form infinite zigzag -O-Sn-O-Sn- chains (689). Bands at approximately 1060 and 1035 cm^{-1} have been assigned to the ν_a(CO) and ν_s(CO) modes, respectively, for several alkyl-tin alkoxides (679,681). For $(n\text{-}C_4H_9)_2Sn(OCH_3)_2$, the intensity of these bands is concentration dependent; the band at the lower frequency disappears at a concentration of 0.01M (684). These data suggest that the higher frequency band at approximately 1060 cm^{-1} is due to the presence of terminal methoxy ligands and that the lower frequency band at approximately 1035 cm^{-1} is due to bridging methoxy ligands. It has therefore been concluded on the basis of these infrared data, as well as molecular weight measurements, that $(n\text{-}C_4H_9)_2Sn(OR)_2$ (R=CH_3,$n\text{-}C_3H_7$,$n\text{-}C_4H_9$) are dimeric in the pure liquid state (684). This conclusion is supported by the absence of the lower-frequency band in the infrared spectrum of $(n\text{-}C_4H_9)_2Sn[O(t\text{-}C_4H_9)]_2$, for which a monomeric structure has been proposed (684). For $[(\text{acetylacetone})(X)(Y)SnOCH_3]_2$ (X=CH_3,C_2H_5,$n\text{-}C_4H_9$,Cl,Br,I;Y=Cl,Br,I), which contain a Sn_2O_2 ring formed by bridging methoxy ligands, plots have been made of the ν(CO) and ν(SnO) mode frequencies against the $\Sigma\sigma^*$ for the X and Y substituents (690). These data are summarized in Figure 1.31 and show that the X and Y substituents play a predominant role in the variation of the ν(CO) and ν(SnO) mode frequencies. Alkyl-lead alkoxides show a tendency to form polymeric, oxygen-bridged structures (150,691,692). The strength of association for the series $(CH_3)_2Pb(OR)_2$ (R=CH_3,C_2H_5, $i\text{-}C_3H_7$,$t\text{-}C_4H_9$) seem to decrease with increasing size of the R group since the melting points decrease, while the

$[(CH_3O)(C_5H_7O_2)SnYX]_2$		H_3C-O (cm^{-1})	$Sn\overset{O}{\underset{O}{\diagdown\!\diagup}}Sn$ (cm^{-1})	$\Sigma\sigma^*$
Y	X			
Cl	Cl	976	531	2.1
Br	Br	972	523	2.0
I	I	997	506	1.7
CH_3	Cl	1013	490	1.05
CH_3	Br	1013	490	1.0
C_2H_5	Br	1022	484	0.9
$n\text{-}C_4H_9$	Cl	1018	481	0.87

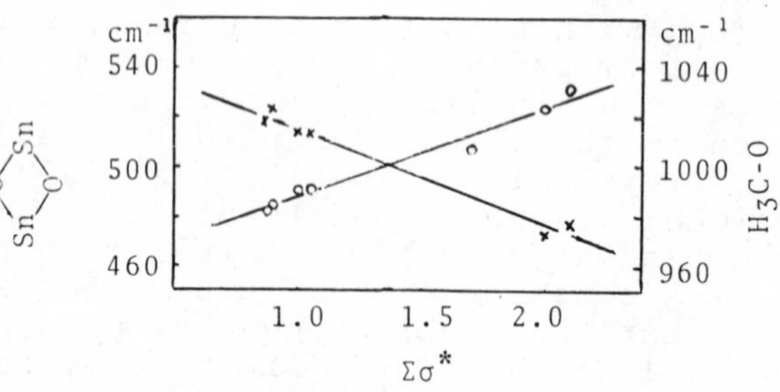

Figure 1.31. Relationships between the sum of Hammett's constant σ* and H₃C-O stretching frequencies (x), and Sn-O ring vibration frequencies (o) (690).

solubility in organic solvents, volatility, and $\nu(CO)$ mode frequencies all increase on going from the methyl to the *tert*-butyl derivative (692). The nmr data for $(CH_3)_2Pb(OCH_3)_2$ in methanol suggest octahedral coordination about the lead atom with two sites occupied by solvent molecules; molecular weight and nmr data for $(CH_3)_2Pb[O(t\text{-}C_4H_9)]_2$ in benzene suggest a monomeric structure with a four-coordinate lead atom. Vibrational data have been reported and assigned for alkylsiloxacycloalkanes (676) and alkylgermoxacycloalkanes (677) (1.57) as well as for 1-oxa-2-sila- (693,694), 1-oxa-2-

$$(alkyl)_2M\overset{O}{\underset{O}{\diagdown\!\diagup}}(CH_2)_n \quad M=Si,Ge$$

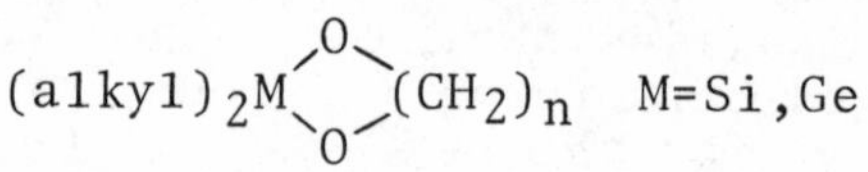

(1.57)

germa- (613,694), and 1-oxa-2-stannacyclopentanes (693, 694) (1.58).

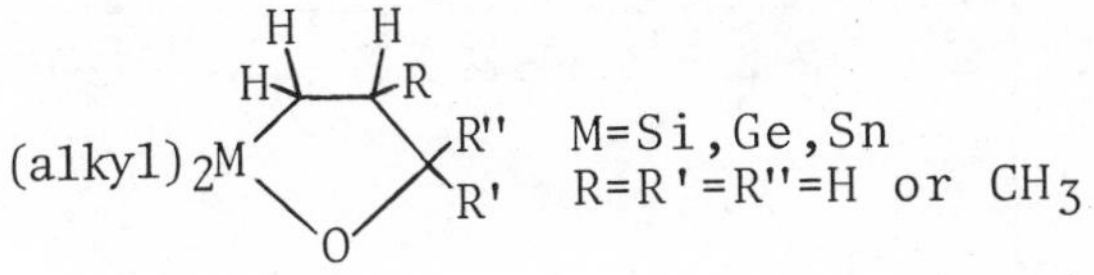

(1.58)

The complex infrared spectrum of $(CH_3)_4POCH_3$ rules out the presence of the ionic structure $[(CH_3)_4P]^+OCH_3^-$ (695). Monomeric, covalent structures are consistent with the infrared and Raman spectra of $(CH_3)_nAs(=O)$-$(OCH_3)_{3-n}$ (n=0 to 2) (696). Covalent structures with trigonal bipyramidal skeletons have been found for $(CH_3)_4SbOR$ ($R=CH_3,C_2H_5$,*i*-C_3H_7,*t*-C_4H_9) (431). On the basis of the solid-state Raman spectrum and the nmr spectrum in benzene, a dimeric structure with both terminal and bridging methoxy ligands (1.59) has been pro-

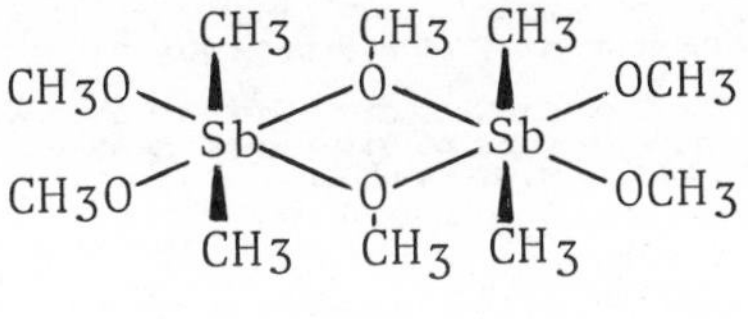

(1.59)

posed for $(CH_3)_2Sb(OCH_3)_3$ in the solid state and solution (697). Selected vibrational assignments for alkylmetal alkoxides are summarized in Table 1.44.

Vibrational data for organometallic organoperoxides are limited to organosilicon derivatives. For R_nSi-$(OOR')_{4-n}$ ($n=3,R=CH_3,C_2H_5$,*n*-C_4H_9,R'=*t*-C_4H_9,$(CH_3)_2(C_6H_5)C$; $n=2,R=CH_3$,R'=*t*-C_4H_9,$(CH_3)_2(C_6H_5)C$) (698) and $(CH_3)_3Si$-$OO(CH_3)_n(C_6H_5)_{3-n}$ (n=0 to 2) (648), a weak intensity infrared band from approximately 920 to 900 cm^{-1} has been assigned to the ν(OO) mode. With increasing organoperoxide substitution on silicon, the frequency and intensity of this band have been observed to increase (698). Also, two overlaping bands in the 940 to 900 cm^{-1} region have been observed for several multiply-substituted organoperoxide silanes (698).

The infrared and Raman spectra have been assigned for liquid (*t*-$C_4H_9)_2Be\cdot OR_2$ ($R=CH_3,C_2H_5$) and compared with the spectra of the constituents (699). While

TABLE 1.44
Selected Vibrational Assignments (cm^{-1}) for Alkyl-Metal Alkoxides

Compound	ν(CO)	ν(MO)	ν(MC)	Refs.
CH_3ZnOCH_3	1059 1020	395 341		667
CH_3ZnOCD_3	1024 990	383 342		667
$C_2H_5ZnOCH_3$	1059 1020	395 341		667
$C_2H_5ZnOCD_3$	1024 990	383 324		667
CH_3CdOCH_3	1048 1005	345 287		667
CH_3CdOCD_3	1008 981	340 277		667
$(CH_3)_2AlOCH_3$	1022 1000	646 438 383	692 600 570	670,671
$(CH_3)_2AlOCD_3$	1000 982	628 432 372	687 599 566	671
$(C_2H_5)_2AlOCH_3$	1019 992	540 460 393	655 632	670
$(C_2H_5)_2AlOC_2H_5$	1054 985	640 380	652 634 481 462	584,670
$(CH_3)_2GaOCH_3$	1040 1020	558 327	547	671,672
$(CH_3)_2GaOCD_3$	1009 990	530 315	548	671,672
$(C_2H_5)_2GaOCH_3$	1040 1001	512 326	534	671
$(t\text{-}C_4H_9)_2GaOCH_3$	1090 1060	485 389 350	540 528	148
$(CH_3)_2InOCH_3$	1040 1030	470 281	490	671,672
$(CH_3)_2InOCD_3$	1010 1000	463 271	493	671,672
$(C_2H_5)_2InOCH_3$	1047 1008	455 273	468	671
$(CH_3)_2TlOCH_3$	1040	340 231		667
$(CH_3)_2TlOCD_3$	1000	224		667
$CH_3Si(OCH_3)_3$[a]	1089	748 628	790	673,675
$(CH_3)_2Si(OCH_3)_2$	1104 1092	730[b] 688[b]	b	673,675
$(CH_3)_3SiOCH_3$[a]	1090	719	687 600	588,673
$(CH_3)_2Ge(OCH_3)_2$	1049	579[b]	b	612
$(CH_3)_3GeOCH_3$	1061	567[b]	b	612
$(CH_3)_2Ge(OC_2H_5)_2$	1053			612
$(CH_3)_3GeOC_2H_5$	1101 1064	653	614 609 571	612,677
$(C_2H_5)_2Ge(OCH_3)_2$	1062 1053	609[b] 561[b]	b	613

$(C_2H_5)_3GeOCH_3$	1062	590[b] 539[b]	b	613
$(C_2H_5)_2Ge(OC_2H_5)_2$	1066 1059	654[b] 603[b] 560[b]	b	613
$(C_2H_5)_2Ge[O(n\text{-}C_8H_{17})]_2$	1069 1049	659[b] 602[b] 561[b]	b	613
$(C_2H_5)_3GeO(n\text{-}C_8H_{17})$	1072	589[b] 540[b]	b	612
$(n\text{-}C_4H_9)_2Ge(OCH_3)_2$	1049	566[b]	b	612
$(n\text{-}C_4H_9)_3GeOCH_3$	1056	577[b]	b	612
$CH_3Sn(OCH_3)_3$	1073 1022	648	535 491	681
$(CH_3)_2Sn(OCH_3)_2$	1066 1047	642 604	561 522	679,681
$(CH_3)_2Pb(OCH_3)_2$	1059 1048	550-450[b]	b	692
$(CH_3)_3PbOCH_3$	1041	496[b] 487[b]	b	173,691
$(CH_3)_2Pb[O(t\text{-}C_4H_9)]_2$	1190 1177			692
$(CH_3)_2Sb(OCH_3)_3$			570 520	697

[a]Assignments made by different groups for this compound are contradictory.
[b]The ν(MO) and ν(MC) modes are reported in the same frequency region.

complexation did not significantly affect the *tert*-butyl modes, the ν(BeC) modes were lowered by 30 to 40 cm^{-1}. Complexation also produced a 30 to 50 cm^{-1} decrease in the ν(COC) mode frequencies of the ether molecules. A band at approximately 650 cm^{-1} was assigned mainly to the ν(BeO) mode, although this band also contains contributions from the ν(BeC) and δ(COC) modes. The infrared and Raman spectra of $CH_3MgX \cdot 2O(C_2H_5)_2$ (X=Br,I) and $CD_3MgBr \cdot 2O(C_2H_5)_2$ have been investigated in the solid state at 90°K and in the liquid state at 300°K (281). The two crystalline forms identified in the study of solid $CH_3MgBr \cdot 2O(C_2H_5)_2$ at 90°K were attributed to the presence of different conformations of the coordinated ether molecules. In a complete infrared and Raman assignment of polycrystalline samples of $C_2H_5MgX \cdot 2O(C_2H_5)_2$ (X=Br,I), $C_2H_5MgBr \cdot 2O(C_2D_5)_2$, and $RMgBr \cdot 2O(C_2H_5)_2$ ($R=C_2D_5,CD_3CH_2,CH_3CD_2$) at 90°K (700), the ν(MgC), ν_a(MgO), ν_s(MgO), ν(MgBr), and ν(MgI) modes were assigned at 485, 317, 301, 248, and 224 cm^{-1}, respectively. The infrared and Raman spectra have been investigated for $C_2H_5MgCl \cdot O(C_2H_5)_2$ (282). The structure of liquid $C_2H_5MgCl \cdot O(C_2H_5)_2$ at 300°K can be described in terms of cyclic dimers with chlorine atoms bridging the magnesium atoms. The cyclic dimers also seem to be the predominant species in ether solutions. The crystalline solid at 90°K, however, appears to have a different structure with both bridging and terminal ethyl groups. The infrared spectra of $RSrI \cdot 2THF$ ($R=CH_3,C_2H_5,n\text{-}C_3H_7$), $n\text{-}C_4H_9SrI \cdot 3THF$, and $C_2H_5BaI \cdot THF$ show bands due to coordinated THF (e.g. 1033±1 cm^{-1}) and low-frequency bands (e.g. 385 to 372 cm^{-1} for the strontium compounds) which may arise from the ν(MC) or ν(MO) modes (701).

Complete infrared and Raman assignments made for $(CH_3)_3Al \cdot OR_2$ ($R=CH_3,CD_3$) (670) are consistent with the originally proposed structure of C_s symmetry (1.60)

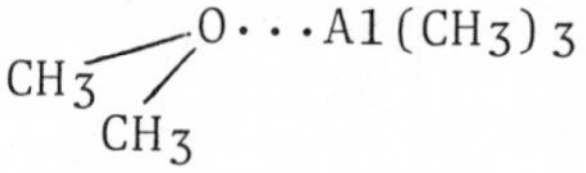

(1.60)

(702). Infrared data have also been reported for 15 complexes of trimethylaluminum, triethylaluminum, and methyl or ethyl aluminum chlorides or bromides with dimethyl ether, diethyl ether, methyl ethyl ether, or tetrahydrofuran (128,702). The infrared spectra of

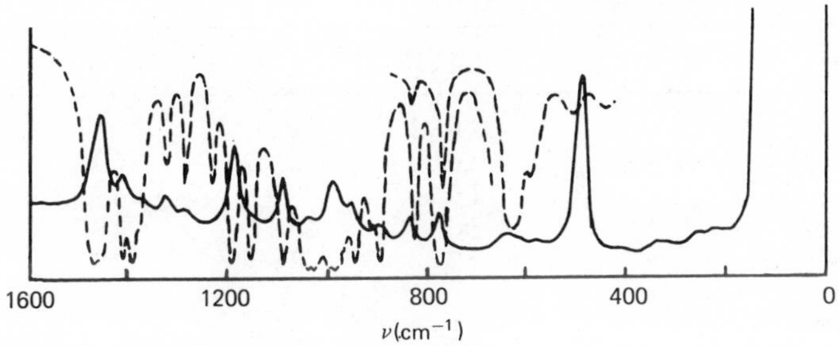

Figure 1.32. Infrared (---) and Raman (——) spectra of $(C_2H_5)_3Al \cdot O(C_2H_5)_2$ (128).

liquid films of these complexes were compared with the spectra of the uncomplexed dialkyl ethers in CCl_4 (703), and pure tetrahydrofuran in the region of the ν(COC) modes (702). There was a noticable decrease in the frequency of the ν(COC) modes of the ethers on coordination that increased as the alkyl groups on the aluminum atom were replaced by halogen atoms.

The infrared spectra at 100°K and 300°K and a single crystal X-ray study have been reported for $Li_4Cr_2(CH_3)_8 \cdot 4THF$ (704). The point symmetry is D_{4h} with a quadruple CrCr bond, stabilized by the lithium atoms assumed to exist in the $(CH_3)_8Cr_2$ unit. Each lithium atom is bonded to four methyl groups (two from each chromium atom) and an oxygen atom from a tetrahydrofuran molecule. In the infrared spectrum at 300°K, bands at 465, 422, and 421 cm^{-1} were assigned to the ν_a(CrC) mode. The infrared spectrum and a single crystal X-ray study have also been reported for the related compound $Li_3Cr(CH_3)_6 \cdot 3Dioxane$, in which both oxygen atoms of the dioxane molecules are coordinated to the lithium atoms. In addition to being coordinated to two oxygen atoms, each lithium atom is also coordinated to two methyl groups from the octahedral $(CH_3)_6Cr$ complexes (89). Table 1.45 compares the ν(COC) mode assignments made for several uncomplexed ethers and the complexes of these ethers as well as the ν(MO) mode assignments made for the alkyl-metal ether complexes. It should be noted in reference to the data in Table 1.45 that two contradictory ν(COC) mode assignments have been used for diethyl ether and the aluminum and beryllium complexes of this ether. Figure 1.32 illustrates the infrared and Raman spectra of $(C_2H_5)_3Al \cdot O(C_2H_5)_2$(128).

TABLE 1.45
The ν(COC) and ν(MO) Mode Assignments (cm^{-1}) for Alkyl-Metal Ether Complexes

Compound	ν_a(COC)	ν_s(COC)	ν(MO)	Refs.
$(CH_3)_2O$	1094	920		703
$(t\text{-}C_4H_9)_2Be \cdot O(CH_3)_2$	1060	893	648	699
$(CH_3)_3Al \cdot O(CH_3)_2$	1046	891	473	702
$(CH_3)_2AlCl \cdot O(CH_3)_2$	1031	882	489	702
$(C_2H_5)_3Al \cdot O(CH_3)_2$	1049	886	472	702
$(C_2H_5)_2O$[a]	1120	926		703
	1085	842		699
$(t\text{-}C_4H_9)_2Be \cdot O(C_2H_5)_2$	1050	795	652	699
$CH_3MgBr \cdot 2O(C_2H_5)_2$	1044	837	304 389	281
$CH_3MgI \cdot 2O(C_2H_5)_2$	1039	836	293 280	281
$C_2H_5MgBr \cdot 2O(C_2H_5)_2$	1042	838	317 301	700
$C_2H_5MgBr \cdot 2O(C_2D_5)_2$	1178	702	306 293	700
$C_2H_5MgI \cdot 2O(C_2H_5)_2$	1039	836	314 295	700
$(CH_3)_3Al \cdot O(C_2H_5)_2$	1036	898	462	702
$(CH_3)_2AlCl \cdot O(C_2H_5)_2$	1023	893	515 488	702
$CH_3AlCl_2 \cdot O(C_2H_5)_2$	1009	884	534	702
$(C_2H_5)_3Al \cdot O(C_2H_5)_2$	1035	894	457	702
$(C_2H_5)_2AlCl \cdot O(C_2H_5)_2$	1027	890	515	702
$C_2H_5AlCl_2 \cdot O(C_2H_5)_2$	1010	884	532	702
$(C_2H_5)_2AlBr \cdot O(C_2H_5)_2$	1021	889	516	702
$C_2H_5AlBr_2 \cdot O(C_2H_5)_2$	1006	881		702
$(CH_3)(C_2H_5)O$	1120	925		703
$(CH_3)_3Al \cdot O(CH_3)(C_2H_5)$	1042	833	456	702
$(C_2H_5)_3Al \cdot O(CH_3)(C_2H_5)$	1042	829	458	702
Tetrahydrofuran (THF)	1073	913		702
$Li_4Cr_2(CH_3)_8 \cdot 4THF$	1043	912	515	704
$(CH_3)_3Al \cdot THF$	1017	869	455	702
$(C_2H_5)_3Al \cdot THF$	1018	865	455	702

[a]The data for $(C_2H_5)_2O$ from Ref. 699 were used in the study of the beryllium complex while the data from Ref. 703 were used in the study of the aluminum complex.

7. Alkylsulfides and Dialkylsulfides

The ν(CS) mode of alkyl-metal alkylsulfides appears in a narrower frequency range (approximately 715 to 690 cm^{-1}) and with less intensity than the ν(CO) mode of the corresponding alkoxide derivatives. The infrared and Raman spectra for CH_3HgSCH_3 (705,706) are illustrated in Figure 1.33 (705). The facts that $(CH_3)_2Tl$-SCH_3 is dimeric in benzene (707) and that in the solid-state infrared spectrum the $\nu_a(TlC)$ (524 cm^{-1}) and $\nu_s(TlC)$ (469 cm^{-1}) modes both appear with nearly equal intensity and at lower frequencies than the corresponding modes in $(CH_3)_2TlClO_4$ suggest a nonlinear C-Tl-C skeleton and a covalent interaction between the thallium and bridging sulfur atoms in the solid state (708). Molecular weight and infrared data suggest a similar structure for $(CH_3)_2TlSC_6H_5$ (344). Selected vibrational assignments for alkyl-metal alkylsulfides are compared in Table 1.46. Additional vibrational data have been reported for several alkylsulfide derivatives of silicon (710), germanium (711), and tin (712).

The infrared and Raman spectra of $(t\text{-}C_4H_9)_2Be{\cdot}SR_2$ ($R{=}CH_3,C_2H_5$) have been compared with the spectra of the individual, uncomplexed species (713). No significant change was observed in the frequencies of the dialkylsulfides on complexation. The frequencies of the ν(BeC) modes of $(t\text{-}C_4H_9)_2Be$, however, decreased from 80 to 90 cm^{-1} on complexation. A characteristic band at 590 cm^{-1} has been assigned to the ν(BeS) mode in both complexes. On mixing $(CH_3)_2S$ with aqueous solutions of CH_3HgNO_3 or $(CH_3)_2AuNO_3$, the complexes $CH_3Hg{\cdot}S(CH_3)_2^+$ ($\nu_a(CS)$=729 cm^{-1},$\nu_s(CS)$=675 cm^{-1},ν(HgS)=302 cm^{-1}) (561) and $(CH_3)_2Au[S(CH_3)_2]_2^+$ ($\nu_a(CS)$=740 cm^{-1},$\nu_s(CS)$=687 cm^{-1}, ν(AuS)=293 cm^{-1}) (460) were formed, respectively. Although the frequencies of the ν(CS) modes in $(CH_3)_3Al{\cdot}S$-$(CH_3)_2$ ($\nu_a(CS)$=738 cm^{-1},$\nu_s(CS)$=685 cm^{-1}) (714) are slightly lower than those for uncomplexed $(CH_3)_2S$ ($\nu_a(CS)$=742 cm^{-1},$\nu_s(CS)$=690 cm^{-1}) (715), the differences are not as dramatic as those found for the ν(CO) modes in the analogous ether complexes. No assignment was given for the ν(AlS) mode of $(CH_3)_3Al{\cdot}S(CH_3)_2$.

8. Carboxylates, Thiocarboxylates, and Dithiocarboxylates

Four types of structures are possible for organometallic carboxylates (1.61 to 1.64). The most characteristic features of the vibrational spectra for these derivatives include the appearance of bands associated with the

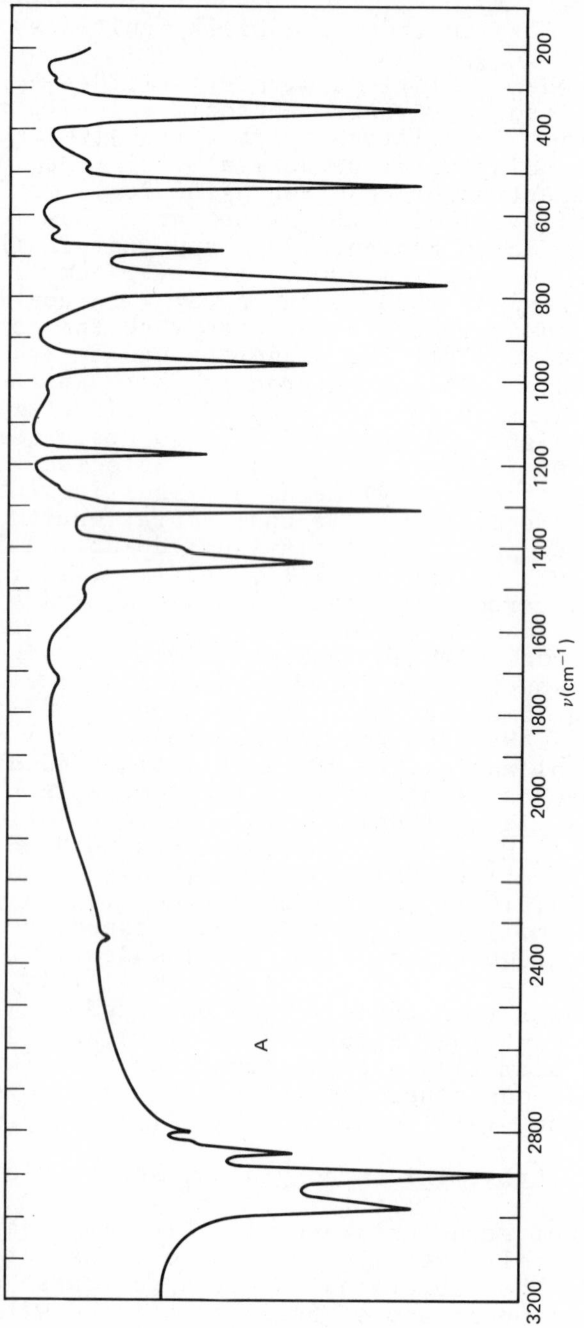
200
400
600
800
1000
1200
1400
1600
1800
2000
2400
2800
3200
ν(cm⁻¹)
A

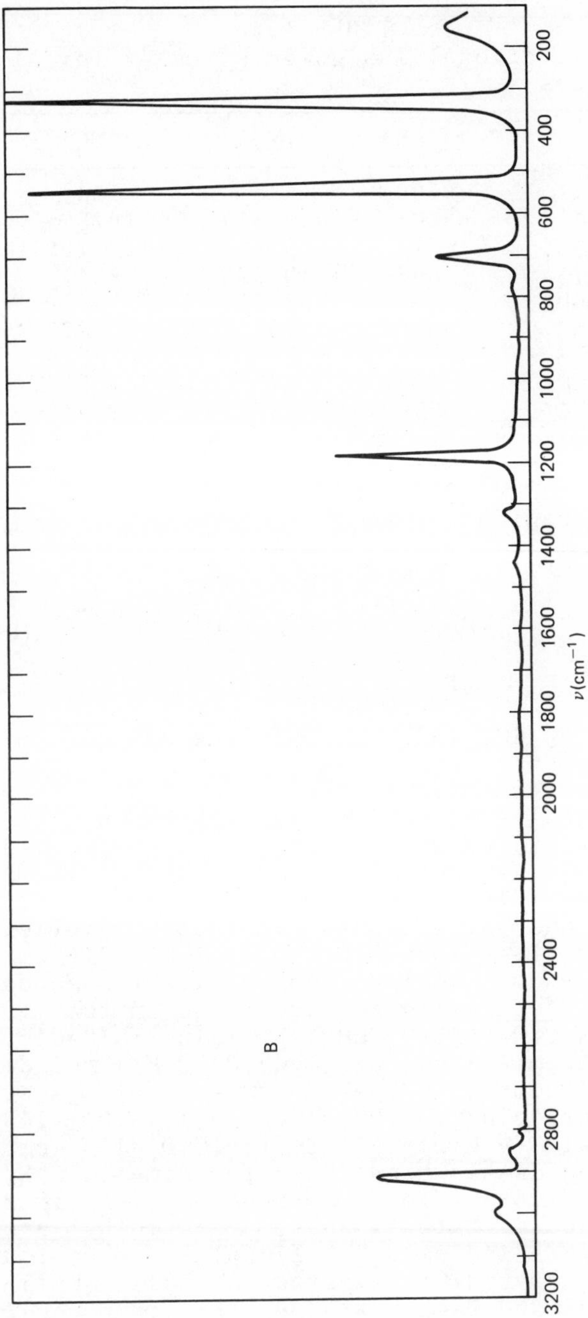

Figure 1.33. Liquid-phase (A) infrared spectrum and (B) Raman spectrum of CH_3HgSCH_3 (705).

TABLE 1.46
Selected Vibrational Assignments (cm^{-1}) for Alkyl-Metal Alkylsulfides

Compound	ν(CS)	ν(MS)	ν(MC)	Refs.
CH_3HgSCH_3	700	333	533	705,706
$CH_3B(SCH_3)_2$	708	1029 517	1073	650
$(CH_3)_2BSCH_3$	712	574	1129 1087	650
$CH_3Si(SCH_3)_3$	692	540 522	724	652
$(CH_3)_2Si(SCH_3)_2$	697	512 431	777 670	652
$(CH_3)_3SiSCH_3$	690[a]	461	690[a] 629	652
$(CH_3)_3GeSCH_3$	700	385	602 564	709
$(CH_3)_2Sn(SCH_3)_2$	694	347	541 518	655
$(CH_3)_3SnSCH_3$	699	339	531 512	655,709
$(CH_3)_2AsSCH_3$		382	578	663

[a]The ν(CS) and ν_a(SiC) modes are expected in the same frequency region.

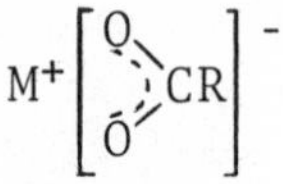
(1.61)

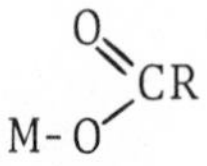
(1.62)

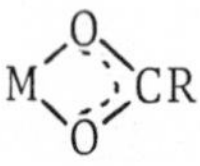
(1.63)

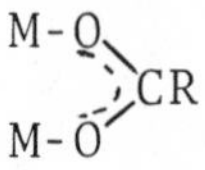
(1.64)

ν_a(CO) and ν_s(CO) modes and, in the case of covalent derivatives, the ν(MO) mode.

Cyclic, trimeric structures with $Al_3O_6C_3$ rings are found for $(CH_3)_2AlO_2CCH_3$ and $CH_3AlBrO_2CCH_3$ (716), while dimeric structures with $M_2O_4C_2$ rings are formed for $(CH_3)_2GaO_2CR$ (R=H,CH_3,CD_3) (716,717), $(C_2H_5)_2MO_2CC_2H_5$ (M=Al,Ga,In) (716,718), and $C_2H_5AlClO_2CC_2H_5$ (716) with the ν_a(CO) and ν_s(CO) modes for these trimeric and dimeric derivatives found in the 1585 to 1515 cm^{-1} and 1493 to 1450 cm^{-1} regions, respectively. Monodentate, ester like acetate ligands are found in the monomeric derivatives $(CH_3)_2Al(O_2CCH_3)$·pyridine (ν(C=O)=1672 cm^{-1}, ν(C-O)=1315 cm^{-1}) and $(C_2H_5)_2Ga(O_2CC_2H_5)$·pyridine (ν(C=O)=1635 cm^{-1},ν(C-O)=1268 cm^{-1} (716). A structure with both bridging (ν_a(CO)=1565 cm^{-1},ν_s(CO)=1482 cm^{-1}) and terminal (ν(C=O)=1656 cm^{-1},ν(C-O)=1266 cm^{-1}) propionate ligands has been found for dimeric $C_2H_5Ga(O_2CC_2H_5)_2$ (1.65) (716). A linear, polymeric structure with chelated acetate ligands, *trans* methyl groups and six-coordinate indium atoms (1.66) has been found (719) in a single crystal X-ray study of solid $(CH_3)_2InO_2CCH_3$. The infrared and Raman data for this compound (ν_a(CO)=

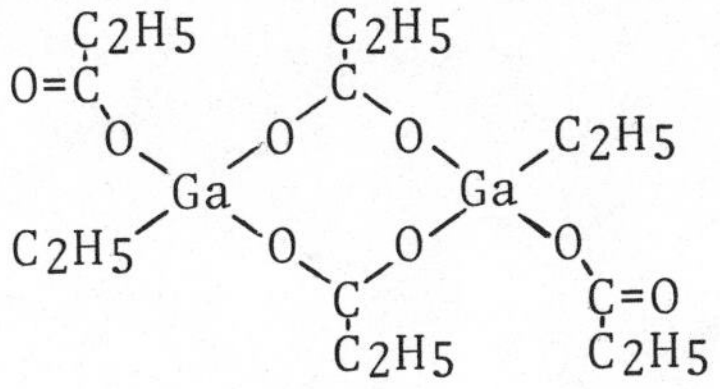

(1.65)

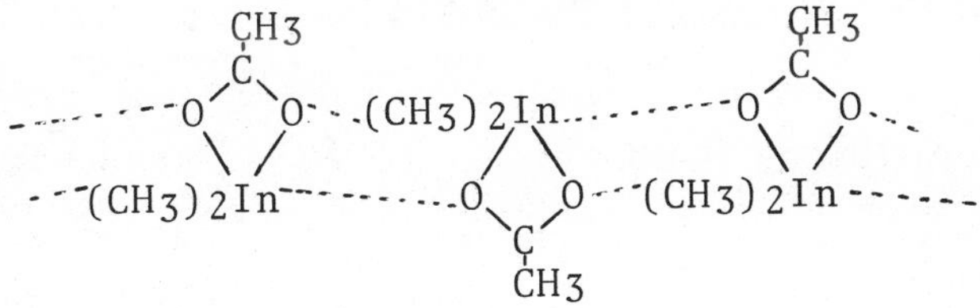

(1.66)

1535 cm^{-1}, $\nu_s(CO)$=1455 cm^{-1}, $\nu(InO)$=325 cm^{-1}) (720,721) are consistent with this structure. A similar polymeric structure has been found in single crystal X-ray and vibrational studies of solid $(C_2H_5)_2InO_2CCH_3$ ($\nu_a(CO)$= 1520 cm^{-1}, $\nu_s(CO)$=1465 cm^{-1}) (721,722). It was originally proposed that the position of the $\nu_a(CO)$ (1541 cm^{-1}) and $\nu_s(CO)$ (1418 cm^{-1}) modes in the solid-state infrared spectrum of $(CH_3)_2TlO_2CCH_3$ was consistent with the presence of bridging acetate ligands to give either a dimeric structure with a cyclic $Tl_2O_4C_2$ skeleton or a linear polymeric structure (46). A recent X-ray crystallographic study (723), however, has shown the solid-state structure to be the same as that found for the corresponding indium compound (1.66) with chelating acetate ligands. Low-frequency bands at 166 and 105 cm^{-1} have been assigned to the $\nu(TlO)$ modes in $(CH_3)_2Tl$-O_2CCH_3 (724). A dimeric structure with bridging *iso*-butyrate ligands is supported by the infrared spectrum of a dilute $CHCl_3$ solution of $(C_2H_5)_2TlO_2C(i\text{-}C_3H_7)$ (725). The $\nu_a(CO)$ and $\nu_s(CO)$ modes of this compound were assigned at 1542 and 1405 cm^{-1}, respectively, in the solid state (725). In the infrared spectrum of $CH_3Tl(O_2CCH_3)_2$, not only have bands expected for bridging acetate ligands been observed (1530 and 1428 cm^{-1}) but additional bands have also been observed at 1613 and 1375 cm^{-1} (724). These last two bands are in the position expected for a monodentate, ester like acetate ligand. Therefore, a polymeric solid-state structure with both bridging and monodentate acetate ligands (1.67) has

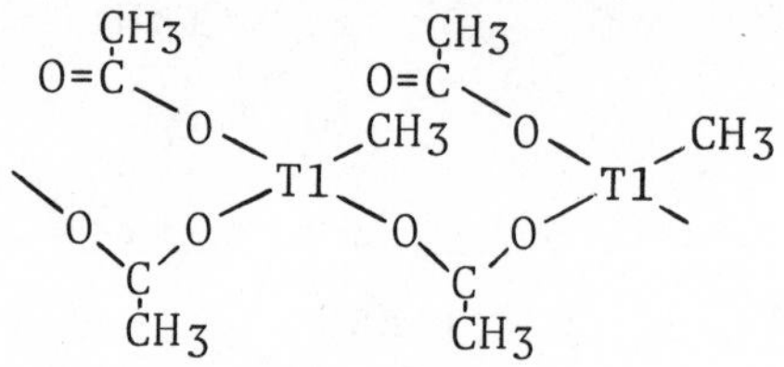

(1.67)

been proposed with a band at 287 cm^{-1} assigned to the ν(TlO) mode of the monodentate acetate ligand and bands at 165 and 114 cm^{-1} assigned to the ν(TlO) modes of the bridging acetate ligands (724). Two sets of bands have also been observed in the solution infrared spectra of $RTl[O_2C(i\text{-}C_3H_7)]_2$ ($R=CH_3, C_2H_5$) (725). The former pair (approximately 1525 and 1405 cm^{-1}) has been attributed to the presence of a bridging *iso*-butyrate ligand. The separation of the latter two bands at approximately 1580 and 1400 cm^{-1}, while not large enough for a monodentate *iso*-butyrate ligand, is consistent with the presence of a chelating *iso*-butyrate ligand. Therefore, a dimeric structure with both chelating and bridging *iso*-butyrate ligands (1.68) has been proposed for the

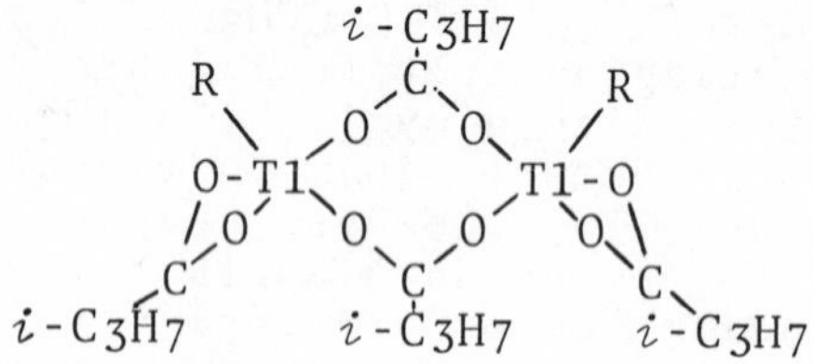

(1.68)

solution structure of these complexes (725). In the solid state, however, other structures cannot be ruled out, especially since the solution band at approximately 1580 cm^{-1} shifts to approximately 1610 cm^{-1} in the solid state (725). The infrared and Raman spectra have been discussed for dimethyl- and diethylaluminum, gallium and indium derivatives of squaric acid, $H_2C_4O_4$ (726).

Structures with monodentate, ester like acetate ligands have been proposed for $(CH_3)_nSi(O_2CCH_3)_{4-n}$ (ν(C=O)=1765, 1748, 1732, and 1725 cm^{-1}, and ν(C-O)= 1200, 1220, 1235, and 1267 cm^{-1} for n=0, 1, 2, and 3, respectively) (727) and $(CH_3)_3GeO_2CCH_3$ ($\nu(C{=}O)=1715$ cm^{-1}, $\nu(C{-}O)=1285$ cm^{-1}) (728,729). This structural conclusion was based on the difference between the frequencies for the $\nu(CO)$ modes of these silicon and germanium compounds and the frequencies of the corresponding modes for solid sodium acetate ($\nu_a(CO)=1578$ cm^{-1}, $\nu_s(CO)=1414$ cm^{-1}) (730). The infrared spectra (3500 to 1100 cm^{-1}) for $(CH_3)_nSi(O_2CCH_3)_{4-n}$ (n=0 to 3) are illustrated in Figure 1.34 (727).

The similarity between the carboxylate $\nu(CO)$ mode frequencies in the infrared spectra of several solid, ionic, sodium carboxylates to those found in the corresponding solid $(CH_3)_3Sn$ (727), $(CH_3)_2SnCl$ (727), and $(CH_3)_3Pb$ (731) carboxylates originally led to the conclusion that these tin and lead derivatives had ionic solid-state structures. This conclusion, however, was questioned in the case of $(CH_3)_nSn(O_2CCH_3)_{4-n}$ (n=2,3) since the $\nu(CO)$ mode frequencies for a bridging or chelating acetate ligand would also be very similar to those observed for an ionic acetate group (732). This point has also been made in another infrared study of several trialkyltin and trimethyllead acetate derivatives (729). It was concluded in this study that solid trialkyltin and -lead carboxylates are coordination polymers with five-coordinate metal atoms in which the coordination sphere of the planar C_3M units is completed by bonding to an oxygen atom from two different bridging acetate ligands. On dissolving in nonpolar solvents, a monomeric compound results with the frequencies of the $\nu(CO)$ modes similar to those found in organic esters. More recent data have been reported not only for trialkyltin acetates but also for trialkyltin chloroacetates (733-740). As a general rule, these compounds form linear coordination polymers in the solid state with five-coordinate tin atoms bridged by the acetate oxygen atoms. An exception is a soluble form of $(CH_3)_3$-SnO_2CCH_3, for which a cyclic polymeric structure has been proposed in the solid state rather than the linear polymeric structure proposed for the insoluble form of this compound (734). In the liquid state or solution, trialkyltin acetates are, in general, monomeric with four-coordinate tin atoms and ester like acetate ligands. The soluble form of $(CH_3)_3SnO_2CCH_3$ shows a tendency to associate in $CHCl_3$ solution (734), while the chloroacetates $(CH_3)_3SnO_2CH_nCl_{3-n}$ (n=0 to 2), although monomeric in $CHCl_3$, also show a slight degree of association in CCl_4 (735). The presence of two $\nu(C{=}O)$ bands for $R_3SnO_2CCH_nCl_{3-n}$ (R=CH_3,*n*-C_4H_9;n=1,2) is due to the exis-

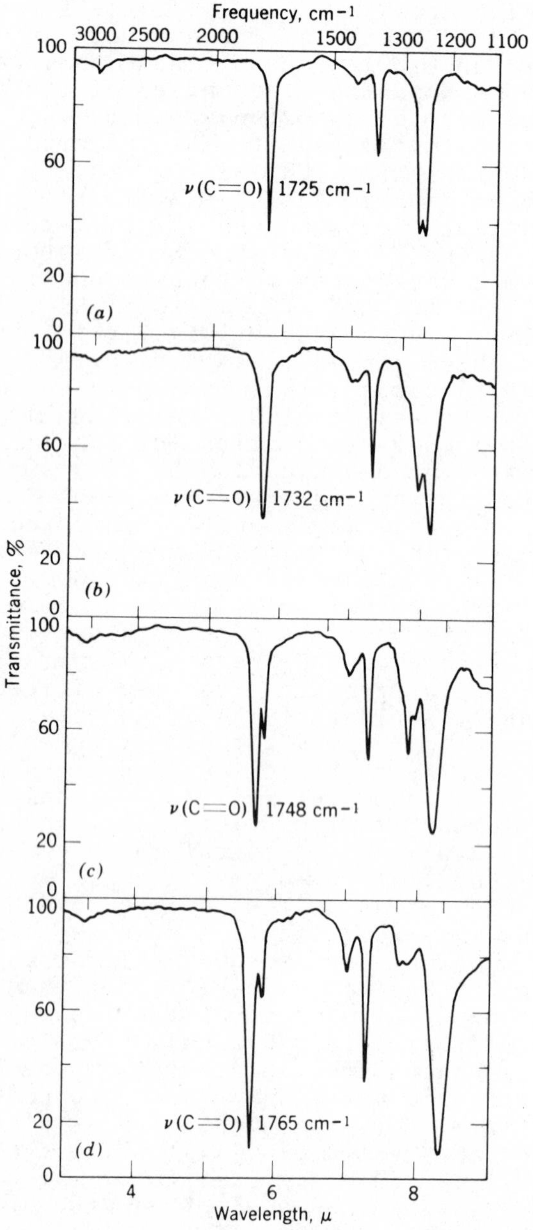

Figure 1.34. Infrared spectra of methylacetoxysilanes (727).

tence of two conformational isomers (735,739). The dialkylchlorotin carboxylates $(CH_3)_2Sn(Cl)O_2CR$ ($R=CH_3, CF_3, C_2H_5, n\text{-}C_3H_7, CF_2Cl, CH_2Cl, CHCl_2, CH_2Br, CH_2I$) (737) and $R_2Sn(Cl)O_2CCH_3$ ($R=C_2H_5, n\text{-}C_3H_7, n\text{-}C_4H_9$) (741) also have a polymeric solid-state structure with bridging carboxylate ligands. Molecular weight measurements and the fact that the $\nu_a(CO)$ mode frequency is the same in the solid-state and solution spectra indicate that the fluorinated carboxylate compounds retain a polymeric character in solution (737). Although the nonfluoronated carboxylate derivatives are apparently monomeric in solution, and the $\nu_a(CO)$ modes appear at higher frequencies in solution than in the solid state, the frequency shifts are smaller than expected for ester like solution structures with four-coordinate tin atoms. Therefore, while the $\nu_a(CO)$ mode of $(CH_3)_2SnO_2CCH_3$, which has an ester like structure with a four-coordinate tin atom in solution, appears at 1658 cm^{-1} ($CHCl_3$ solution) and 1570 cm^{-1} (solid state) (735,736), the same mode for $(CH_3)_2Sn(Cl)O_2CCH_3$ has been assigned at 1598 cm^{-1} ($CHCl_3$ solution) and 1550 cm^{-1} (solid state) (737). On the basis of these vibrational data as well as nmr data, it has been concluded that in solution the nonfluorinated carboxylate derivatives have a monomeric, chelated structure (1.69).

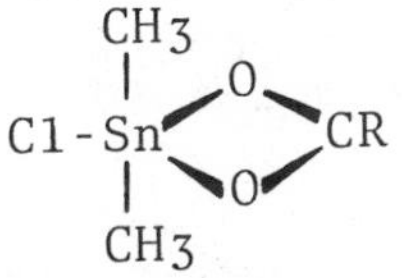

(1.69)

The infrared spectra of $R_2Sn(O_2CR)_2$ ($R=CH_3, C_2H_5, n\text{-}C_3H_7, n\text{-}C_4H_9$) as crystalline films or neat liquids indicate the presence of bridging carboxylate ligands and six-coordinate tin atoms (742,743). Molecular weight data show these compounds to be monomeric in benzene while the appearance of both the $\nu_a(SnC)$ and $\nu_s(SnC)$ modes in the infrared spectra of the methyl and ethyl derivatives indicate that the C-Sn-C skeletons may not be linear. An unsymmetrically chelated structure (1.70) has therefore been proposed for the dialkyltin acetates in solution (742,743). The fact that the infrared $\nu(CO)$ mode frequencies decrease on going from the solid state to solution indicates that $(CH_3)_2Sn(Cl)$-$(O_2CR)_2$ ($n=0, R=CH_2Cl, C_6H_5; n=1, R=CH_2Cl, CH_3$) have struc-

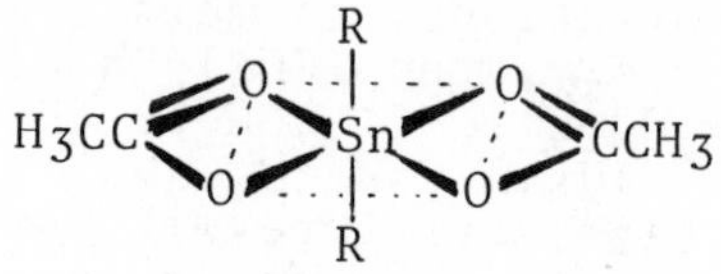

(1.70)

tures with bridging carboxylate ligands in the solid state and chelating carboxylate ligands in solution (744). Infrared data are also available for several *n*-butyltin tricarboxylates; the triacetate and tripropionate derivatives are monomeric in camphor solutions (745).

As was true for the corresponding acetate, infrared data indicate that the solid insoluble form of $(CH_3)_3SnO_2CH$ has a linear, polymeric structure with five-coordinate tin atoms and bridging formate ligands while the soluble form of solid $(CH_3)_2SnO_2CH$ most likely has a cyclic, polymeric structure (734). In $CHCl_3$ solution, the soluble form of $(CH_3)_3SnO_2CH$ also exists in a cyclic, associated form in equilibrium with the monomer (734). Linear polymers have also been proposed for solid R_3SnO_2CH ($R=C_2H_5,n\text{-}C_3H_7$) (746). In organic solvents, infrared bands associated with terminal formate ligands appear. A low linear polymer with terminal, ester like formate ligands was therefore proposed for the solution structures of R_3SnO_2CH ($R=C_2H_5,n\text{-}C_3H_7$) (746).

Infrared bands from 306 to 283 cm^{-1} have been assigned to the ν(SnO) mode for Nujol mull, liquid film, or cyclohexane solution samples of R_3SnO_2CH ($R=CH_3,C_2H_5,n\text{-}C_3H_7$) (746). The ν_a(SnO) and ν_s(SnO) modes have also been assigned from 355 to 258 cm^{-1} and 252 to 208 cm^{-1}, respectively, for R_3SnO_2CR ($R=H,CH_3,CH_2Cl,C_2H_5$) in the solid state (733). The low frequencies of the ν(SnO) modes in these trialkyltin carboxylates relative to those found in the corresponding trialkyltin alkoxides suggest that the tin-carboxylate bonds are relatively weak.

Both monodentate and chelating carboxylate ligands are found in methyl-antimony carboxylates. The vibrational spectra of the melt and CCl_4 solutions of $(CH_3)_4SbO_2CCH_3$ suggest the presence of a monodentate acetate ligand ($\nu(C{=}O)=1614\ cm^{-1}$, $\nu(C\text{-}O)=1371\ cm^{-1}$), five-coordinate antimony atom and trigonal bipyramidal skeletal structure (747). For the same compound in the solid state, the acetate ligand becomes chelating ($\nu_a(CO)=1590\ cm^{-1}$, $\nu_s(CO)=1405\ cm^{-1}$) to produce a six-coordinate antimony atom and octahedral skeletal structure (747). Similar results have been reported for $(CH_3)_4Sb$ formate, trifluoroacetate, propionate, pivalate, and benzoate (747). The vibrational spectra of $(CH_3)_3Sb(O_2CR)_2$ (R=H,CH_3,C_2H_5,*n*-C_3H_7,C_6H_5) indicate the presence of a trigonal bipyramidal skeleton with axial, monodentate carboxylate ligands (64). From the infrared spectrum of solid $(CH_3)_3Sb(O_2CCH_3)_2$, the ν(C=O) modes were assigned at 1637 and 1600 cm^{-1} and the ν(C-O) modes were assigned at 1287 and 1274 cm^{-1} (64). The ν(SbO) modes for $(CH_3)_3Sb(O_2CCH_3)_2$ both appear at approximately 279 cm^{-1}, as confirmed by the shift of this band to 268 cm^{-1} in $(CH_3)_3Sb(O_2CCD_3)_2$ (277). Molecular weight and conductance measurements as well as solid-state and solution ($CHCl_3$ and CCl_4) infrared data have been reported for several five-coordinate trimethylantimony derivatives of fluoro-, chloro-, bromo-, and cyanoacetic acids (748). A linear relationship has been found between the ν(CO) mode frequencies of these derivatives and the pK and Taft σ* constants for the parent acids. Based on infrared and Raman data, an octahedral skeletal structure with two monodentate and one chelating acetate ligand (1.71) has been proposed

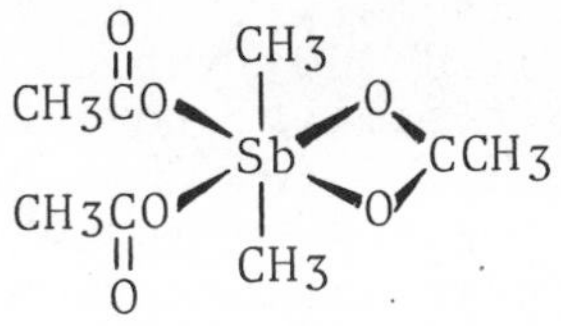

(1.71)

for $(CH_3)_2Sb(O_2CCH_3)_3$ (697). Through a comparison with solid-state infrared data previously given for $(CH_3)_nSb$-$(O_2CCH_3)_{5-n}$ (n=3,4), infrared bands at 1640 cm^{-1} (Nujol mull) or 1660 cm^{-1} (benzene solution) for $(CH_3)_2Sb$-$(O_2CCH_3)_3$ have been assigned to the ν(C=O) modes of the monodentate acetate ligands, while bands at 1568 cm^{-1} (Nujol mull) or 1567 cm^{-1} (benzene solution) have been assigned to the ν_a(CO) mode of the chelating acetate ligand. Also, the ν_a(SbO) and ν_s(SbO) modes were assigned at 296 and 261 cm^{-1}, respectively, for the chelating acetate ligand (697).

The infrared spectrum of $(CH_3)_3Ta(O_2CCH_3)_2$ in the 1560 to 1400 cm^{-1} region is consistent with the presence of bidentate acetate ligands, while bands at 530 and 450 cm^{-1} have been assigned to ν(TaC) and ν(TaO) modes, respectively (749). Solution infrared data are consistent with a dimeric structure and bridging squarate ligands for $(CH_3)_3TaC_4O_4$ (749).

Dialkylaluminum thiocarboxylates are dimeric with the vibrational spectra indicating the presence of puckered $Al_2O_2S_2C_2$ rings, while the gallium and indium derivatives are monomeric with chelating thiocarboxylate ligands forming MOSC rings (750). The derivatives $R_2AlOSCCH_3$ ($R=CH_3,C_2H_5$) are dimeric with the ν(CO) and ν(CS) modes of the bridging thioacetate ligands observed in the 1476 to 1460 cm^{-1} and 748 to 736 cm^{-1} regions, respectively. The same modes appear in the 1495 to 1480 cm^{-1} and 712 to 696 cm^{-1} ranges, respectively, for the chelating thioacetate ligands in $R_2GaOSCCH_3$ ($R=CH_3$, C_2H_5) and $(C_2H_5)_2InOSCCH_3$. Pyridine (Py) cleaves the Ga-O bond of $(CH_3)_2GaOSCCH_3$ to form $(CH_3)_2GaOSCCH_3 \cdot Py$ in which the monodentate thioacetate ligand is sulfur-bonded, resulting in a shift of the ν(C=O) and ν(C-S) modes to 1652 and 642 cm^{-1}, respectively (750). Although the ν(MO) and ν(MS) modes have been assigned for these dialkylgroup IIIb thiocarboxylates, they are mixed with other modes (750). In Figure 1.35, the infrared and Raman spectra are illustrated for $(C_2H_5)_2In$-$OSCCH_3$ (750).

In $(CH_3)_3Sb(OSCR)_2$ ($R=CH_3,C_6H_5$), coordination is through the sulfur atoms with the frequencies of the ν(C=O) modes for solid $(CH_3)_3Sb(OSCCH_3)_2$ (1639 and 1634 cm^{-1}) (751,752) very similar to those of the corresponding modes in $(CH_3)_3Sb(O_2CCH_3)_2$. The ν(SbS) modes for $(CH_3)_3Sb(OSCCH_3)_2$ are assigned at a high frequency (380 cm^{-1}) relative to the assignments of the corresponding ν(SbO) mode (279 cm^{-1}) for $(CH_3)_3Sb(O_2CCH_3)_2$. The ν(SbS) mode has also been assigned for $(CH_3)_3Sb(X)$-$OSCCH_3$ (approximately 380 cm^{-1}) and $(CH_3)_3Sb(X)OSCC_6H_5$ (approximately 360 cm^{-1}) (X=Cl,Br) (752).

The infrared and Raman spectra for the dithiocarboxylate derivatives $(CH_3)_3MS_2CR$ ($M=Si,Ge,Sn,R=CH_3;M=Pb$, $R=C_2H_5$) are consistent with a structure in which only one of the sulfur atoms is coordinated to the metal atom (1.72), as in organic dithioacetates (753).

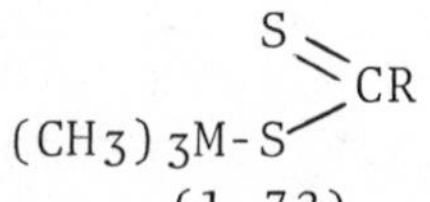

(1.72)

9. β-Diketonate, Ketonate, Acylate, and Related Ligands

Several bonding modes have been found for the β-diketonate ligand in metal complexes (1.73 to 1.76). The oxygen-bonded enol structure (1.73) is that most commonly found.

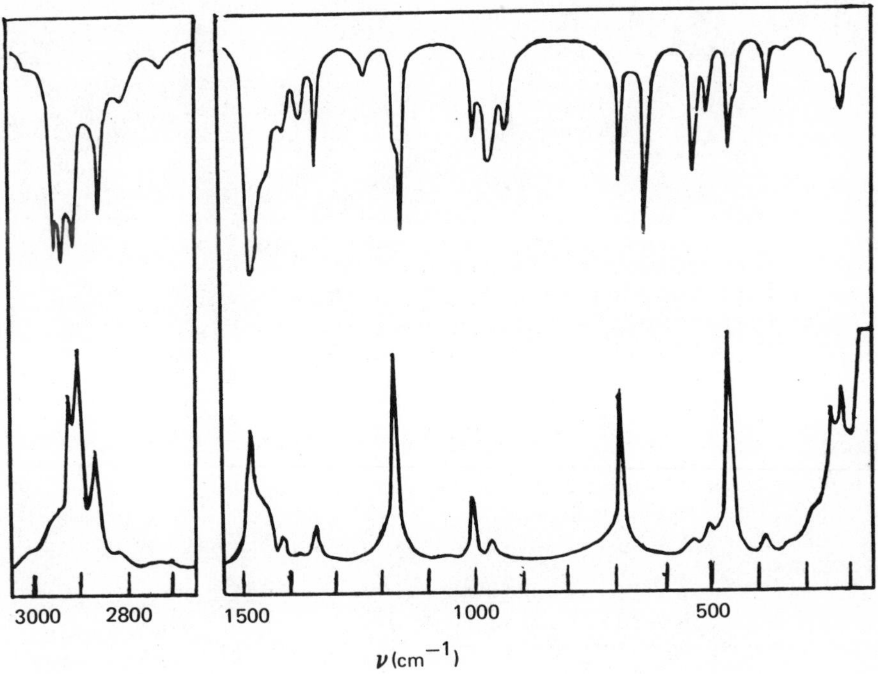

Figure 1.35. Infrared (top) and Raman (bottom) spectra of $(C_2H_5)_2InOSCCH_3$ (750). These spectra were incorrectly attributed to $(CH_3)_2GaOSCCH_3$ in Ref. 750 (308).

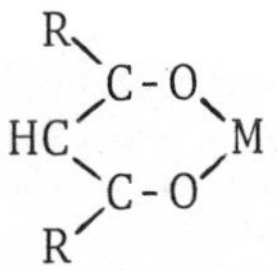

(1.73)

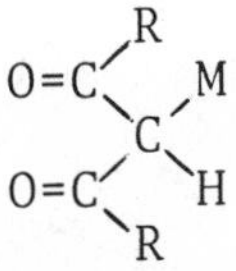

(1.74)

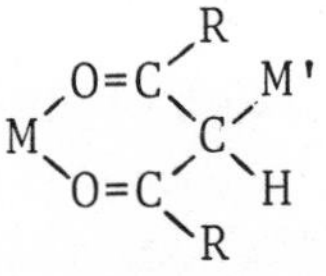

(1.75)

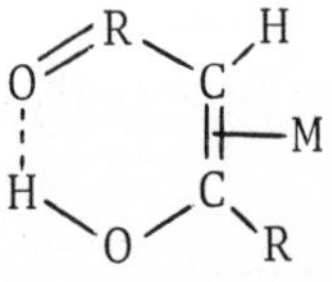

(1.76)

Compounds with the carbon-bonded keto structure (1.74) have been discussed in a recent review article (754). Several compounds have been characterized in which the diketonate ligand is carbon-bonded to one metal atom and oxygen-bonded to another (1.75) (755). The bonding scheme shown in 1.76 has been proposed for an acetyl-acetonate (Acac) ligand in one complex on the basis of a complete infrared study (756).

The vibrational spectra of liquid $(CH_3)_2GaAcac$ (562), illustrated in Figure 1.36, are typical of those observed for compounds with chelating acetylacetonate ligands. For inorganic complexes with a chelating acetylacetonate ligand, there has been disagreement as to the assignment of the vibrational spectra, especially with respect to the $\nu(CO)$ and $\nu(CC)$ modes associated with the cyclic MO_2C_3 skeleton. Although the two very strong intensity infrared bands in the 1600 to 1550 cm^{-1} and 1535 to 1500 cm^{-1} regions were originally assigned to the $\nu_a(CC)$ and $\nu_s(CO)$ modes, respectively, more recent studies involving ^{13}C and ^{18}O substitution (757, 758), normal coordinate analyses (759), and Raman polarization measurements (562,760) have led to a reversal of these assignments. The $\nu_s(CC)$ mode has been assigned to a medium intensity infrared band and a very strong intensity Raman band from 1280 to 1240 cm^{-1}, while the $\nu_a(CO)$ mode has been assigned from 1380 to 1350 cm^{-1}. The assignments of these modes are usually complicated by the presence of methyl deformation modes that appear in the same frequency region. In Table 1.47, assignments are summarized for the $\nu(CO)$, $\nu(CC)$, and $\nu(MO)$ modes of alkyl-metal derivatives with chelating acetylacetonate ligands.

Although vibrational data for both $(CH_3)_2GaAcac$ (562) and $(CH_3)_2InAcac$ (50) are consistent with the presence of a covalent structure, infrared, Raman, and pmr data for $(CH_3)_2TlAcac$ have been interpreted in terms of a dissociated or weakly chelated solution structure and an ionic or weakly coordinated solid-state structure with a linear C-Tl-C skeleton (761). A single crystal X-ray study of $(CH_3)_2TlAcac$ (723) has shown that the acetylacetonate ligand is chelating in the solid state and that the $(CH_3)_2TlAcac$ molecules are held together by further Tl-O bonds to form infinite linear polymers analogous to the structure of $(CH_3)_2InO_2CCH_3$ (1.66). Furthermore, the C-Tl-C angle is 170°.

Some controversy has occurred over the structure of $(CH_3)_2Sn(Acac)_2$. A single crystal X-ray study has shown that the atoms about the tin atom form a nearly perfect octahedron with the methyl groups *trans* to each other (766). On the basis of the large dipole moment in benzene and cyclohexane solutions, however, a structure in which the methyl groups are *cis* to each other was proposed (767). In a Raman study of $(CH_3)_2Sn(Acac)_2$ as benzene solutions, nonoriented crystalline samples, and oriented single crystals, it has been concluded that this compound has the *trans* structure in both the solid

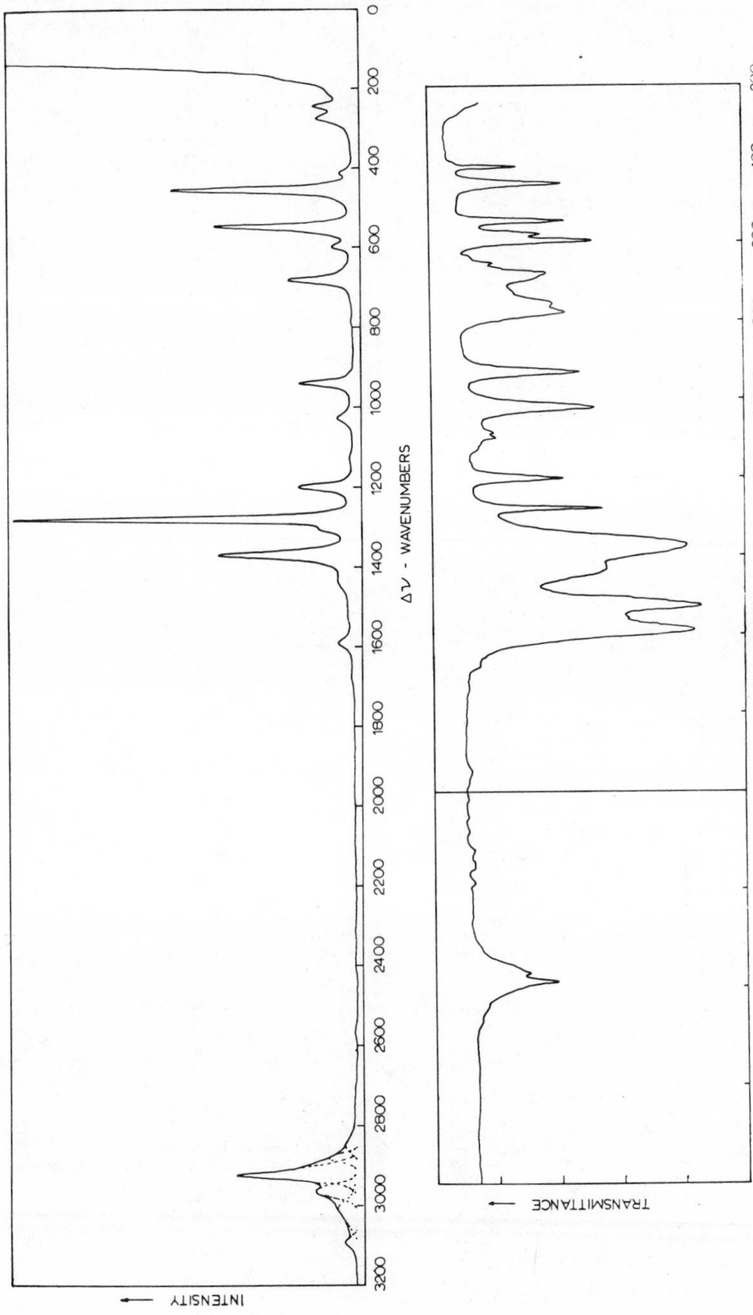

Figure 1.36. Raman spectrum (top) and infrared spectrum (bottom) of $(CH_3)_2$*GaAcac (562).*

TABLE 1.47
Characteristic Frequencies (cm^{-1}) for Alkyl-Metal Acetylacetonates with Chelating Acetylacetonate Ligands

Compound[a]	$\nu_s(CO)$	$\nu_a(CO)$	$\nu_a(CC)$	$\nu_s(CC)$	$\nu(MO)$	Refs.
$(CH_3)_2GaAcac$	1589			1280	453	562
$(CH_3)_2InAcac$	1600			1248	413	50
$(CH_3)_2TlAcac$	1566				400	761
$CH_3SnCl(Acac)_2$	1577 1558	1360	1529	1277	445 436	762
$CH_3SnBr(Acac)_2$	1575 1558	1359	1528	1283	445 436	762
$CH_3SnI(Acac)_2$	1570 1562	1361	1531	1277	440 432	762
$(CH_3)_2Sn(Acac)_2$	1566 1559	1370	1515	1259	406	762,763
$C_2H_5SnBr(Acac)_2$	1583 1560	1357	1531	1277	440 433	762
$(C_2H_5)_2Sn(Acac)_2$	1572	1381	1511	1255	404	762
$(CH_3)_2Pb(Acac)_2$	1575	1370	1509	1241	391	762
$(CH_3)_2SbCl_2Acac$	1563		1527		435 423	764,765
$(CH_3)_4SbAcac$	1587		1511 1503		387	764,765
$(C_2H_5)_2SbCl_2Acac$	1565		1527			764
$(C_2H_5)_4SbAcac$	1585		1506 1499			764
$(CH_3)_2AuAcac$[b]	1590		1520	1264	444	565
$(CH_3)_2Ta(Acac)_2$	1555	1422	1525	1280	440	749

[a]Acac, acetylacetonate anion.
[b]The original assignments hasve been changed to conform to more recent data for the $\nu(CO)$ and $\nu(CC)$ modes.

state and solution (763). To explain the dipole moment data, it was suggested that the SnO_2 plane may not be coplanar with the remainder of the acetylacetonate ring in solution. A dipole moment might then be expected even though the methyl groups are in the *trans* positions (763). Octahedral skeletons with the R and X ligands *trans* to one another have been proposed for $RSnX(Acac)_2$ ($R=CH_3,X=CH_3,Cl,Br,I;R=C_2H_5,X=C_2H_5,Br$) and $(CH_3)_2Pb$-$(Acac)_2$ (762).

Octahedral skeletons have been found for R_nSb-$Cl_{4-n}Acac$ ($R=CH_3,C_2H_5;n=1$ to 4) (764,765) with the proposal that the R groups preferentially occupy the same plane as the acetylacetonate oxygen atoms and the chloride ligands preferentially occupy the axial positions. Such a structure has been confirmed in a single crystal X-ray study of CH_3SbCl_3Acac (768).

The following are among the vibrational spectral changes observed in going from compounds with an oxygen-bonded, enol acetylacetonate group to those with a carbon-bonded, keto acetylacetonate group (769,770): (1) the $\nu(CH)$ mode frequency arising from the hydrogen atom on the γ-carbon atom of the acetylacetonate ligand decreases by approximately 100 to 150 cm^{-1}, (2) the frequencies of the $\nu_s(CO)$ and $\nu_a(CO)$ modes are shifted upward, appearing at 1700 to 1650 cm^{-1} and 1650 to 1610 cm^{-1}, respectively, and also appear with very strong infrared intensity. Table 1.48 includes selected vibrational assignments for compounds with only carbon-bonded or both carbon- and oxygen-bonded acetylacetonate ligands, while in Figure 1.37, the infrared spectra are illustrated for $Na[Pt(Acac)_2Cl_2]$ and its deuterated analogs, all of which contain two carbon-bonded acetylacetonate ligands (770).

Infrared spectra have been reported and assigned for $M[Pt(Acac)_2Cl]_n$ ($M=(VO)^{II}$,Co(II),Ni(II),Cu(II), Zn(II),Pd(II),n=2;M=Fe(III),n=3) in which one of the acetylacetonate ligands is oxygen-bonded to the platinum(II) atom and the other is carbon-bonded to the platinum(II) atom and oxygen-bonded to the M atom (1.75) (755). Stable metal isotopes were used in this study to assign the $\nu(MO)$ modes. Among the conclusions reached in this study were the following: (1) the $\nu(CO)$ modes are between the frequencies found for complexes in which the acetylacetonate ligand is only oxygen-bonded or carbon-bonded, (2) the $\nu(CC)$ mode frequencies are higher than those found in complexes with a carbon-bonded acetylacetonate ligand, and (3) the $\nu(MO)$ mode frequencies (360 to 230 cm^{-1}) are lower than those found in

TABLE 1.48
Selected Infrared Data (cm^{-1}) for the Characteristic Modes of Compounds with Carbon-Bonded, or Carbon-Bonded and Chelated Acetylacetonate Ligands

	Modes Characteristic of C-Bonded Ligands			Modes Characteristic of Chelated Ligands			
Compound[a]	ν_s(CO)	ν_a(CO)	ν(MC)	ν_s(CO)	ν_a(CC)	ν(MO)	Refs.
$(HgCl)_2Acac$	1690		566				771
$Na[Pt(Acac)_2Cl_2]$	1652	1626	567				770
$K[Pt(Acac)_2Cl]$	1699	1657	572	1570	1523	638 457	755,772
$K[Pt(Acac)_3]$	1681	1652	565 542	1568	1519	677 453	772
$Pd(Acac)_2X$[b]	1665	1632	540	1572	1523		773
$K_2[Ir(Acac)_7Cl]\cdot H_2O$	1678	1634	c	1560	1518	c	754

[a] Acac, acetylacetonate anion.
[b] X = $(C_6H_5)_3P$, pyridine, and $HN(C_2H_5)_2$.
[c] Although bands were observed from 600 to 400 cm^{-1}, no assignments were made for the ν(MC) or ν(MO) modes.

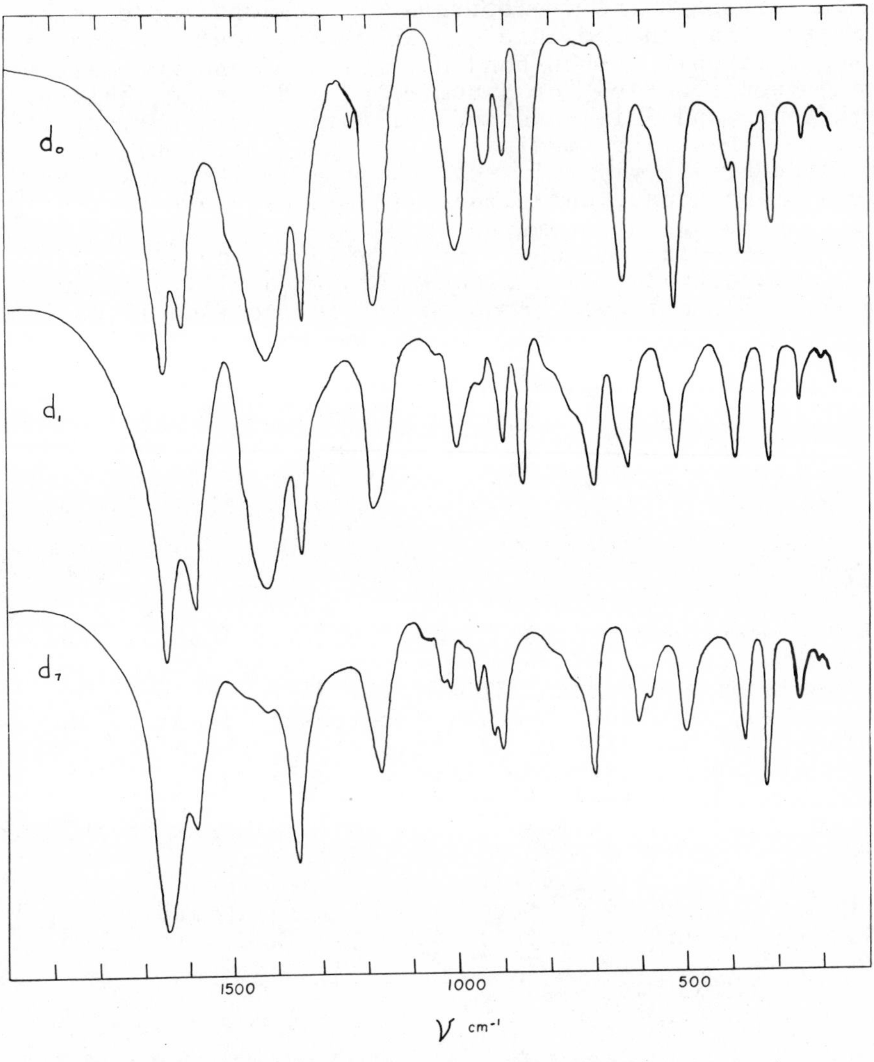

Figure 1.37. Infrared spectra of sodium dichloro(γ-acetylacetonato)platinum(II) dihydrate and its deuterated analogs (770).

analogous complexes in which the acetylacetonate ligand is only oxygen-bonded (490 to 240 cm^{-1}). These conclusions suggest that the C-O bonds retain their keto character although the π-electrons are somewhat more delocalized than in complexes in which the acetylacetonate ligand is only carbon-bonded. Infrared and nmr data have been reported for a number of β-diketone, thio-β-diketone, and β-iminoketone derivatives of $(CH_3)_3Pt(IV)$ (774). These data indicate that in $[(CH_3)_3PtL]_2$ the ligands bridge via the γ-carbon atom for L=β-diketone or β-iminoketone, while for L=thio-β-diketone, bridging is via the sulfur atoms.

A structure in which the platinum atom is bonded to the acetylacetonate ligand through a platinum-olefin bond (1.77) has been proposed for (HAcac)(Acac)PtCl (756).

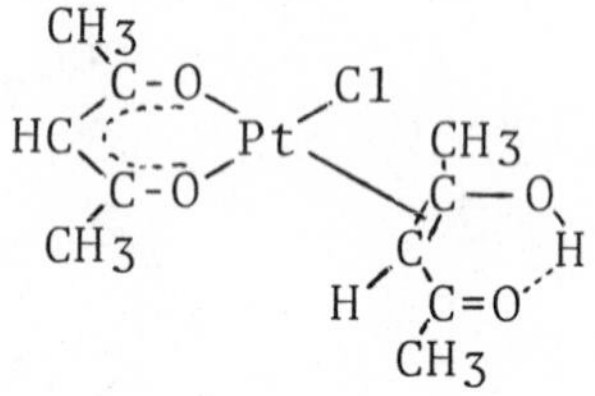

$$(1.77)$$

This conclusion was based on the observation that the infrared spectrum of this compound can be interpreted in terms of a superposition of the infrared spectrum of $AcacPtCl_2$, in which the acetylacetonate ligand is chelating, on that of the enol form of acetylacetone (756). The following bands were reported to be characteristic of the olefinic type of acetylacetone coordination: ν(C=O), 1627 cm^{-1}; ν(C=C), 1550 cm^{-1}; O-H···O stretching, 2905 cm^{-1}, and O-H···O bending, 1450 cm^{-1}.

Alkyl-metal derivatives of various ketones have been characterized in terms of the keto structure (1.78), the enol structure (1.79), or an equilibrium

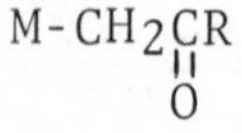

$$M-CH_2\underset{\overset{\|}{O}}{C}R \qquad\qquad M-O\underset{\overset{\|}{CH_2}}{C}R$$

$$(1.78) \qquad\qquad (1.79)$$

involving both the keto and enol forms. The keto form of trialkylmetal ketonate derivatives of silicon, germanium, and tin has been characterized by an infrared band from approximately 1700 to 1670 cm^{-1}, which is due

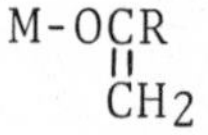

to the ν(C=O) mode; the enol form of the corresponding derivatives has been characterized by an infrared band from approximately 1655 to 1620 cm^{-1}, which is due to the ν(C=C) mode (775-778).

Infrared data have been listed for the acylate complexes CH$_3$C(=O)SiR$_2$R' (R=R'=CH$_3$;R=n-C$_4$H$_9$,R'=H,F;ν(C=O)= 1650 to 1645 cm^{-1}) (779). The complexes CH$_3$(=O)Ni-[P(CH$_3$)$_3$]$_2$X (X=Cl,Br,I) (1.80) have been prepared and

$$
\begin{array}{c}
(CH_3)_3P \diagdown \quad \diagup X \\
Ni \\
CH_3C \diagup \quad \diagdown P(CH_3)_3 \\
\diagdown O
\end{array}
$$

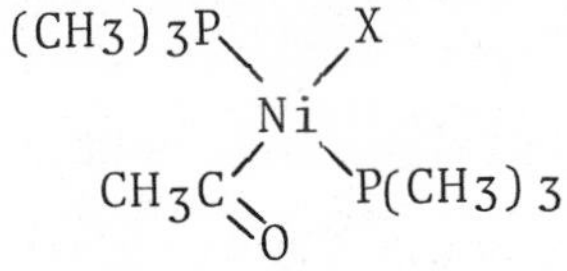

(1.80)

characterized; the ν(C=O) mode was assigned in the 1650 to 1635 cm^{-1} range while the ν(NiC) mode was assigned at 538, 537, and 530 cm^{-1} for X=Cl, Br, and I, respectively (780).

10. Nitrogen Bases and Miscellaneous Lewis-Base Adducts and Chelates

a. *Ammonia Complexes, Amines, Alkyl Amines, and Aryl Amines*

Vibrational data have been reported for ammonia complexes that were isolated as stable solids (563,781-783) or that formed on dissolving alkyl-metal salts in liquid ammonia (784). Ammonia, like water, is a useful solvent for Raman spectroscopy because of the simplicity of its spectrum and the low intensity of its Raman spectrum (784). The δ_a(NH$_3$) and δ_s(NH$_3$) modes of ammonia give weak intensity Raman bands at 1640 and 1066 cm^{-1}, respectively. Three bands appear in the ν(NH$_3$) region; the ν_a(NH$_3$) mode has been assigned at 3380 cm^{-1}, while Fermi resonance between the first overtone of the 1640 cm^{-1} band and the ν_s(NH$_3$) mode gives two intense Raman bands at 3300 and 3212 cm^{-1}. The ν(MC) modes are 12 to 19 cm^{-1} lower in alkyl-metal ammonia complexes than in the corresponding alkyl-metal aquo complexes, indicating that the metal-nitrogen bond is slightly stronger than the metal-oxygen bond (784). Vibrational assignments characteristic of alkyl-metal ammonia complexes are summarized in Table 1.49. The infrared and Raman spectra of [(CH$_3$)$_3$Pt(NH$_3$)$_3$]Cl are illustrated in Figure 1.38 (783).

TABLE 1.49
Vibrational Assignments (cm^{-1}) for Alkyl Organometallic Ammonia Complexes

Compound	Phase	$\nu(NH_3)$ or $\nu(ND_3)$	$\delta(NH_3)$ or $\delta(ND_3)$	$\rho(NH_3)$ or $\rho(ND_3)$	$\nu(MN)$	$\nu(MC)$	Refs.
$CH_3HgNH_3^+$	Solid-state IR of ClO_4^- salt.	3342 3295	1620 1292	730	585[a]	571	781
	Raman of CH_3HgI in liquid NH_3.				458[a]	547	784
$CH_3HgND_3^+$	Solid-state IR of ClO_4^- salt.	2495 2380	980		505	560	781
$(CH_3)_3SnNH_3^+$	Solid-state IR of Br^- salt.	3220 3150 3090	1600	625	503	541	782
$(CH_3)_3Sn(NH_3)_2^+$	Solid-state IR of Br^- salt.	3270 3150 3090	1600	625		544	782
$(CH_3)_2Au(NH_3)_2^+$	Raman of $[(CH_3)_2AuI]_2$ in liquid NH_3.				459	572	784
$(CH_3)_3Pt(NH_3)_3^+$	Raman of Cl^- salt in H_2O.	3368[b] 3270 3194 3166[b]	1636 1300	765[b] 691[b]	410[b] 390 377[b]	584	563,783
	Raman of $[(CH_3)_3PtI]_4$ in liquid NH_3.				403	582	784
$(CH_3)_3Pt(ND_3)_3^+$	Solid-state IR of Cl^- salt.	2506 2454 2430 2307			364	585	563,783

[a]The $\nu(HgN)$ mode assignments in Refs. 781 and 784 differ. Infrared bands reported at 468 and 450 cm^{-1} in Ref. 781 were attributed to the ClO_4^- ion.
[b]Bands observed in the infrared mull spectrum.

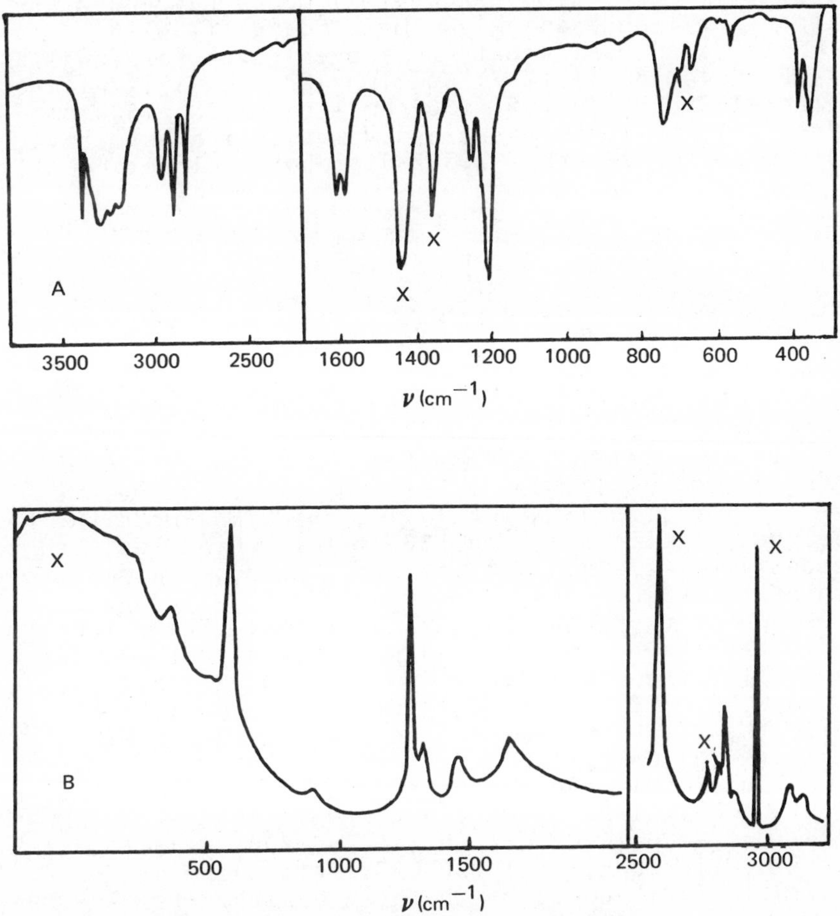

Figure 1.38. (A) Infrared spectrum of $[(CH_3)_3Pt(NH_3)_3]Cl$ *as a Nujol mull and (B) Raman spectrum of* $[(CH_3)_3Pt(NH_3)_3]Cl$ *in water. The x represents bands due to Nujol in the infrared spectrum and bands due to Hg in the Raman spectrum (783).*

The infrared and Raman spectra of 1:1 complexes between $(t\text{-}C_4H_9)_2Be$ and R_3N or R_3P ($R=CH_3, C_2H_5$) have been reported and assigned with the ν(NC) mode frequencies observed to shift to lower frequencies on complexation and the ν(PC) mode frequencies observed to shift to higher frequencies on complexation (785).

Infrared spectra have been assigned for $[(CH_3Hg)_nNH_{4-n}]X$ (n=1 to 4, X=ClO_4, F) and the related N-deuterated complexes with the aid of normal coordinate analyses (781). The $\nu_a(MN)$ and $\nu_s(MN)$ modes have been assigned for $[(CH_3)_3Si)_2N]_2M$ (M=Zn,Cd,Hg) (786).

Infrared assignments have been given for $(CH_3)_2BNH_2$ ($\nu_a(BN)$=1447 cm^{-1}) (787) and $(CH_3)_nB[N(CH_3)_2]_{3-n}$ (n=1, $\nu_a(BN)$=1521 cm^{-1}, $\nu_s(BN)$=1364 cm^{-1}; n=2, $\nu(BN)$=1525 cm^{-1}) (787-789), while infrared and Raman assignments have been made for 2-methyl-1,3,2-diazaboracyclohexane (1.81)

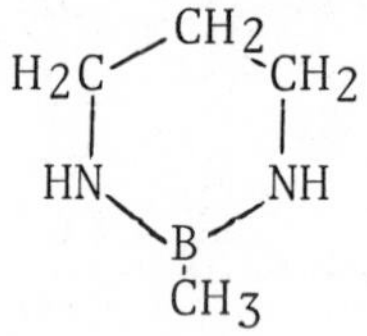

(1.81)

(789). The $\nu(AlN)$ mode has been assigned for $(CH_3)_2Al-N(CH_3)_2$ (509 cm^{-1}), which has a monomeric structure (790). For $[(CH_3)_2MN{=}P(CH_3)_3]_2$, the $\nu(MN)$ mode has been assigned at 655, 583, and 536 cm^{-1} for M=Al, Ga, and In, respectively (791). The infrared, Raman, and pmr spectra of trimeric $(c\text{-}C_3H_5)_2M(NC_2H_4)$ (M=Al,Ga) suggest the presence of nonplanar M_3N_3 rings (165). The $\nu(InC)$ modes have been assigned and possible structures discussed for the ethylenediamine (en) complexes $(CH_3)_2In(en)O_2CCH_3$ (792) and $[(CH_3)_2In(en)]X$ (X=Cl,I) (321), while the $\nu(TlC)$ modes have been assigned for $[(CH_3)_2Tl-NR_2]_2$ (R=CH_3,C_2H_5) (793).

The $\nu(SiN)$ mode has been assigned for $(C_2H_5)_3SiNH_2$ (359), $(CH_3)_3SiNHR$ (R=CH_3,C_2H_5,C_6H_5) (794), $(CH_3)_nSi(NR_2)_{4-n}$ (R=CH_3,C_2H_5, n=1 to 3) (795,796), and $[(CH_3)_3Si]_2NX$ (X=H,Cl,Br,CH_3,BBr_3) (90,797,798); there are some contradictions, however, in these assignments (795,796). Although the $\nu_a(SiN)$ mode for 13 *para*-substituted N,N-bis(trimethylsilyl)anilines, $[(CH_3)_3Si]_2N(p\text{-}XC_6H_4)$, has been assigned in a relatively narrow frequency range (975 to 968 cm^{-1}), the frequency range for the $\nu_s(SiN)$ mode is fairly broad (603 to 417 cm^{-1}); the influence of the mass of X on this range has been discussed (799). Vibrational coupling made it impossible to assign the $\nu(SiN)$ mode of N-trimethylsilyl pyrrole (800). Assignments have been reported for $[(CH_3)_3Si]_2NN[Si(CH_3)_3]_2$ ($\nu(SiN)$=925, 579, 411 cm^{-1}; $\nu(NN)$=1039 cm^{-1}) (799) and the $\nu_a(SiNM)$ and $\nu_s(SiNM)$ modes of $(CH_3)_3Si[N(C_6H_5)M-$

$(CH_3)_3]$ (M=Si,Ge,Sn) (786). The infrared and Raman spectra of $[(CH_3)_3SiN]_3$ are consistent with a planar Si_3N_3 skeleton of D_{3h} symmetry (596). The characteristic vibrations have been assigned from the infrared and Raman spectra of $R_3GeNR'_2$ (R=CH_3,C_2H_5;R'=CH_3,n-C_4H_9; ν(GeN)=580 to 550 cm^{-1}), $(R_3Ge)_2NR'$ (R=CH_3,C_2H_5;R'=H, C_6H_5;ν_a(GeN)=800 to 790 cm^{-1},ν_s(GeN)=600 to 570 cm^{-1}) and two triethylgermylpyrazoles (801). Infrared data have also been reported for $(CH_3)_2Ge[N(C_2H_5)_2]_2$ and $(CH_3)_2Ge(CH_3NCH_2CH_2NCH_3)$ (796). The infrared and Raman spectra have been reported and assigned for $[(n\text{-}C_4H_9)Sn\text{-}]_2NC_2H_5$, $(n\text{-}C_4H_9)_3SnNRR'$ (R=R'=CH_3,C_2H_5,n-C_3H_7,n-C_4H_9, C_6H_5;R=C_2H_5,R'=C_6H_5) and $(n\text{-}C_4H_9)_3SnNR_2$ (R=CH_3,C_2H_5, n-C_4H_9) (802), while the ν_a(SnN) and ν_s(SnN) modes have been assigned for $(R_3Sn)_2NH$ (R=n-C_3H_7 (803),i-C_3H_7 (804), n-C_4H_9 (804), neophyl (804), neopentyl (804), C_6H_{11} (805,806)). Some contradictions have appeared in the ν(SnN) mode assignments made in other studies (796,807, 808).

The ν_a(SbN) mode for $(R_3SbCl)_2NH$ has been assigned to a strong intensity infrared band at 741, 758, 752, and 770 cm^{-1} for R=CH_3, C_2H_5, n-C_3H_7, and n-C_4H_9, respectively (118).

Vibrational data, including assignments for the ν(MC) modes, have been reported for $RTi(NR'_2)_3$ (R=CH_3, C_2H_5;R'=CH_3,CD_3,C_2H_5,n-C_3H_7,i-C_3H_7,n-C_4H_9,t-C_4H_9,C_6H_5, $C{\equiv}CC_6H_5$) (809) and $(CH_3)_3Ta[N(CH_3)_2]_2$ (749).

b. *Pyridine, 2,2'-Bipyridine, and 1,10-Phenanthroline Complexes*

The limited data for alkyl-metal complexes of pyridine, 2,2'-dipyridine, and 1,10-phenanthroline deal mainly with the ν(MC) mode assignments. In several, although by no means all instances, coordination by these ligands results in a decrease in the ν(MC) mode frequencies relative to their values in the uncomplexed alkyl-metal derivatives (111,314,321,388,467,708,810-813). Data illustrating this point are presented in Table 1.50. In some methyl and ethyl derivatives, the $\rho_r(CH_3)$ or $\rho_r(CH_2)$ mode frequencies, respectively, have also been observed to decrease on coordination with nitrogen bases (111,314,321).

Coordination of pyridine to heavy metals results in the replacement of the two Raman active A_1 ring modes at 991 and 1031 cm^{-1} with an intense A_1 Raman band at approximately 1019 cm^{-1} and a weaker intensity band at approximately 1050 cm^{-1}; this contrasts with the infrared spectra of these complexes in which the only

TABLE 1.50
Comparison of the ν(MC) Mode Assignments (cm^{-1}) for Complexes with and without Pyridine, 2,2'-Bipyridine, or 1,10-Phenanthroline

Compound[a]	ν(MC)	Refs.
$(C_2H_5)_2Zn$	561 484	111
$(C_2H_5)_2ZnPy$	501 450	111
$[(CH_3)_2InCl]_2$	563 492	314
$(CH_3)_2In(Cl)Bipyr$	529 481	321
$(CH_3)_2In(Cl)Phen$	522 483	321
$[(CH_3)_2InI]_2$	548 480	314
$(CH_3)_2In(I)Py$	522 482	314
$(CH_3)_2In(I)Bipyr$	539 480	321
$(C_2H_5)_3SnCl$	518 488	812
$(C_2H_5)_3Sn(Cl)Py$	521 481	812
$(C_2H_5)_2SnCl_2$	530 495	812
$(C_2H_5)_2SnCl_2(Py)_2$	530 480	812
$(C_2H_5)_2SnCl_2(Bipyr)$	529 481	812
$(C_2H\ )_2SnCl_2(Phen)_2$	525 470	812
$C_2H_5SnCl_3$	515	812
$C_2H_5SnCl_3(Py)_2$	497	812
$C_2H_5SnCl_3(Bipyr)_2$	504	812
$C_2H_5SnCl_3(Phen)_2$	507	812
CH_3NbBr_4	500	467
CH_3NbBr_4Bipyr	470	467

[a]Py, pyridine; Bipyr, 2,2'-bipyridine; Phen, 1,10-phenanthroline.

significant differences between the spectra of free and coordinated pyridine occur below 650 cm^{-1} (784). The 1031 and 991 cm^{-1} bands have been assumed to involve C-N and C-C stretching and C-C-C, C-N-C, and N-C-C bending (784). Raman data indicate that while CH_3HgI reacts with neat pyridine (Py) to give solutions of the neutral iodo complex, pyridine solutions of CH_3HgClO_4 produce CH_3HgPy^+ (ν(HgN)=204 cm^{-1}) (784).

The ν(GaN) mode of $(CH_3)_2Ga(OSCCH_3)Py$ has been assigned at 496 cm^{-1} (750). Although $(CH_3)_2InX$ (X=Cl,I) are dimeric in benzene, molecular weight and vibrational data indicate that reaction of the chloride with pyridine (50,314) or 1,10-phenanthroline (Phen) (321) and the chloride and iodide with 2,2'-bipyridine (Bipyr) (314) results in cleavage of the di-μ-halide bridges to produce monomeric $(CH_3)_2In(X)L$ (X=Cl,L=Py,Bipyr, Phen;X=I,L=Bipyr). Although polymeric in the solid state, $(CH_3)_2InO_2CCH_3$ reacts with pyridine in organic solvents to produce $(CH_3)_2In(O_2CCH_3)Py$ in which the indium atom is five-coordinate and with 2,2'-bipyridine and 1.10-phenanthroline to produce $[(CH_3)_2InO_2CCH_3]_2L$ (L=Bipyr,Phen); the nmr and vibrational spectra are discussed for these derivatives (792). Infrared and Raman data for the pyridine and 1,10-phenanthroline complexes of $(CH_3)_2TlClO_4$ have led to the conclusion that the C-Tl-C skeletons are slightly bent (342). This has been confirmed in a single crystal X-ray study of $(CH_3)_2Tl(Phen)ClO_4$ that also showed the presence of a weak interaction between the $(CH_3)_2Tl^+$ and ClO_4^- ions.

The metal-carbon and metal-halide stretching modes have been assigned in an infrared study of several 2,2'-bipyridine and 1,10-phenanthroline complexes of alkyl-tin halides (388). It has been suggested in this study that the ν(SnC) modes appear below 200 cm^{-1} and are mixed, in the case of the bromides and iodides, with the ν(Sn-halide) modes. Mossbauer and far-infrared data indicate the presence of a *trans* $(CH_3)_2Sn$ configuration in the 1:1 adducts formed between ten substituted 1,10-phenanthrolines and two substituted 2,2'-bipyridines with $(CH_3)_2SnCl_2$ (815). It was also concluded that no unequivocal ν(SnN) mode assignments could be made without the aid of further data which can be supplied by metal and/or nitrogen isotopic substitution (815).

Solid-state Raman and infrared data have been reported for the square planar derivatives *cis*-$(CH_3)_2Au$-(X)Py (X=Cl,SCN) (810). Dissolving dimeric $(CH_3)_2AuX$ in neat pyridine produced Raman and nmr spectra characteristic of $(CH_3)_2Au(X)Py$ (X=Cl,I,SCN) with the ν(AuN) mode for all three compounds assigned at approximately 195 cm^{-1} (784). Raman and nmr data indicate that on dissolving tetrameric $(CH_3)_3PtI$ in neat pyridine, the complex $(CH_3)_3Pt(I)(Py)_2$ (ν(PtN)= 210 cm^{-1}) is formed (794).

c. Monodentate Oxygen-Donor Lewis Bases

Vibrational data have been discussed for dimethylsulfoxide (DMSO) adducts of $(CH_3)_3Al$, $(C_2H_5)_3Al$, and $(C_2H_5)_2AlCl$ (816). Both vibrational and nmr data indicate that $(CH_3)_2In(O_2CCH_3)DMSO$ ($\nu(SO)=1020\ cm^{-1}$) has a monomeric structure with *trans* methyl groups (1.82) in

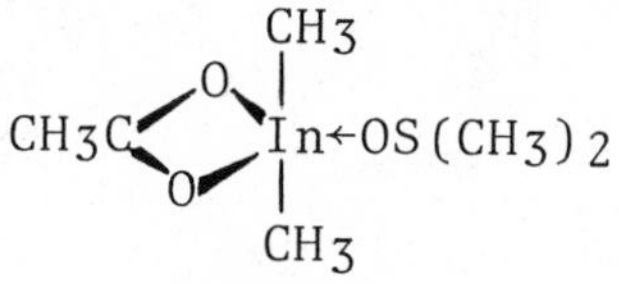

(1.82)

both the solid state and DMSO solution (792).

Infrared and nmr data have been reported for the DMSO, pyridine N-oxide (PyO), triphenylphosphine oxide (Ph_3PO), and triphenylarsine oxide (Ph_3AsO) adducts $(CH_3)_nMX_{4-n}L_m$ (n=3,X=Br,M=Sn,m=1,L=PyO,Ph_3PO,Ph_3AsO, DMSO;n=3,X=Cl,M=Pb,m=1,L=PyO,Ph_3PO,Ph_3AsO;n=2,X=Cl,M=Sn, m=2,L=PyO,Ph_3PO,Ph_3AsO,DMSO) (817). The presence of the $\nu_a(SnC)$ but not the $\nu_s(SnC)$ mode in the infrared spectra led to the conclusion that the dimethyl derivatives have a *trans* methyl, octahedral skeletal structure while the trimethyl derivatives have a trigonal bipyramidal skeletal structure. For $(CH_3)_2SnCl_2L_2$ (L=PyO, Ph_3PO,Ph_3AsO,DMSO), it was also noted that $\Delta\nu(SO) > \Delta\nu(PO) > \Delta\nu(NO) > \Delta\nu(AsO)$ with values of +109, +39, +38, and 0 cm^{-1}, respectively, where $\Delta\nu(MO) = \nu(MO)$ for the free ligand minus $\nu(MO)$ for the complexed ligand (817). Using infrared data, structures involving either *trans* methyls and *trans* chlorines (384) or *trans* methyls and *cis* chlorines (818) have been proposed for $(CH_3)_2Sn$-$Cl_2(DMSO)_2$. A single crystal X-ray study has shown that in the solid state the molecule has the latter structure with *trans* methyls and *cis* chlorines (819). Raman spectra have been reported for single crystals of $(CH_3)_2Sn$-$Cl_2(DMSO)_2$, unoriented samples, and solutions in DMSO (820). The similarity in the $\nu(SnC)$ region of the spectra for the single crystals and DMSO solutions indicate similar bonding in both phases. This study also included values of the components of the derived polarizability tensor for symmetric Sn-C stretching. An infrared study has been reported of the $\nu(NO)$ mode frequencies of pyridine N-oxide derivatives in which R_3M (R=

CH_3,M=Si,Ge,Sn;R=C_2H_5,M=Si) are not bonded to the oxygen atom of the pyridine N-oxide ligand but to the 4-position of the pyridine N-oxide ring (821). There is an increase in the ν(NO) mode frequency for the organometallic derivatives when compared to that of the unsubstituted pyridine N-oxide or the 4-substituted *tert*-butyl pyridine N-oxide ligand due to the electron acceptor ability of the organometallic derivatives, which increases the amount of ring π-back bonding.

Complexes containing the $[(CH_3)_3SbL]_2O^{2+}$ (L=PyO, Ph_3PO,PhAsO,DMSO,DPSO,DMA) cations have been isolated and characterized using infrared and conductance data (630). Infrared data have also been reported for neutral complexes of diorganoantimony trichlorides with PyO, Ph_3P, and DMSO (427).

Infrared data have been listed and some assignments made for the complexes formed between $CH_3W(=O)Cl_3$ and Ph_3P, Ph_3As, and DMSO (468).

d. Amino Acids and Other Biologically Important Ligands

The structures have been discussed for 16 trimethyltin and tricyclohexyltin α- and β-amino acids and dipeptides, and tri(*n*-butyl)tin glycinate using vibrational, mass spectral, and Mossbauer data (822).

Detailed discussions have appeared on changes in the Raman (particularly Raman difference) (823-825) and, in part, infrared spectra (825) of several nucleosides and nucleotides on binding with CH_3Hg^+. Structures were proposed for these complexes on the basis of the observed spectral changes. Raman difference spectroscopy makes it possible to detect very small frequency shifts and/or intensity changes in broad bands. The difference spectrum at a pH of 7 for 50 mM uridine + 50 mM CH_3Hg^+ versus 50 mM uridine is illustrated in Figure 1.39 (823). Vibrations arising from the CH_3Hg^+ unit such as the ν(HgC) and $\delta_s(CH_3)$ modes at 560 and 1207 cm^{-1}, respectively, appear as positive features in the difference spectrum. New bands from the mercury-nucleoside complex and characteristic of neither the CH_3Hg^+ nor nucleoside (e.g., at 1291, 1044, and 605 cm^{-1}) also appear as positive features. Small shifts in the uridine bands give rise to derivative features such as that centered at 786 cm^{-1}.

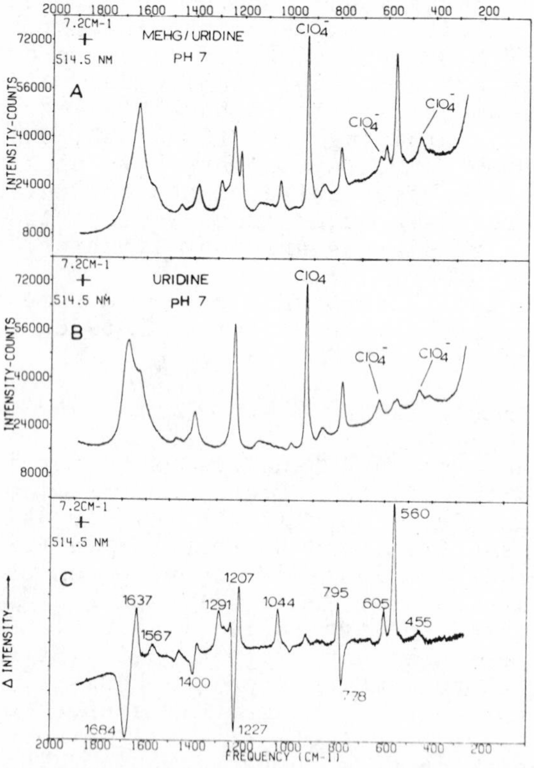

Figure 1.39. Raman difference spectrum: (A) 50 mM uridine + 50 mM CH_3HgClO_4; (B) 50 mM uridine (both solutions 0.1 M in $NaClO_4$); (C) difference spectrum (A-B). Solution pH 7. Scan conditions: 0.25 A intervals, 10 sec counting time (823).

e. Miscellaneous Chelating Ligands

Limited infrared data have been used to help in characterizing the structure of the oxinate (Ox) derivative $(C_2H_5)_2InOx$ for which a dimeric structure with five-coordinate indium atom has been proposed as well as for $(CH_3)_2InDtc$ and $RIn(Dtc)_2$ (Dtc=$SSCN(CH_3)_2$,R=CH_3, C_2H_5) that contain four- and five coordinate indium atoms, respectively (826).

Both Mossbauer and infrared data have been used to characterize the structures of dimethyltin and dimethyllead phthalate (827). The structures of the 1:1 adducts formed between N,N'-ethylenebis(salicylideneiminate)-Ni(II) (NiSalen) and $(CH_3)_nSnCl_{4-n}$ (n=2,3) have been studied using solid state infrared and other spectral data (828). A structure with *trans* methyl and *cis* chloride groups was proposed for $(CH_3)_2SnCl_2$(NiSalen) (1.83). Vibrational data, including ν(SnS) mode as-

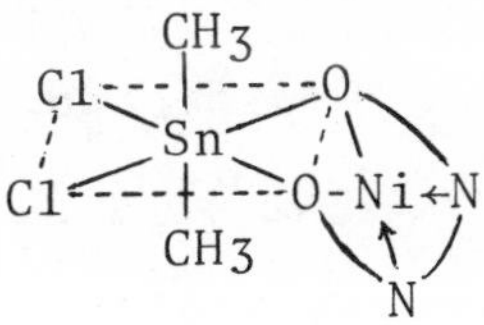

(1.83)

signments (353 to 291 cm^{-1}), have been reported (829) for the dicyanoethylene-1,2-dithiolate (Mnt) complexes $(CH_3)_2$SnMnt, $[(C_6H_5)_4As](CH_3SnMntX)$ (X=Cl,Br,I), and $[(C_2H_5)_4N]_2[(CH_3)_2Sn(Mnt)_2]$. Infrared data have been used to help determine the configuration about the tin atom for several alkyl-tin oxinates (830). The structures of $(CH_3)_2SnL$ (L^{2-}=tridentate ONO and SNO ligands) have been investigated by vibrational, pmr, and electronic spectroscopy (831). Several alkyl-tin complexes with multidentate, anionic Schiff base ligands have been investigated in the solid state using ^{119}Sn Mossbauer and infrared spectroscopy (832).

Infrared, uv, and nmr data show the oxinate complexes $R_nSbCl_{4-n}Ox$ (R=CH_3,C_2H_5,*n*-C_3H_7;n=1,2,4) to have six-coordinate antimony atoms and chelating oxinate ligands in benzene solution (833). The $RSbCl_3Ox$ and R_2SbCl_2Ox complexes also have six-coordinate skeletal structures in neat ethanol or $CHCl_3$ solutions, although there is partial or complete rupture of the Sb-N bond in polar solvents. The $R_3Sb(Cl)Ox$ complexes apparently have five-coordinate skeletal structures (833). The presence of three infrared bands in the ν(SbC) mode region of $(CH_3)_3Sb$(*o*-phenylenedioxide) indicates that the C_3Sb skeleton is not planar (834). Some features of the infrared spectra have been discussed for $[(CH_3)_3PtSalnr]_2$ and $(CH_3)_3PtSalnrX$ (Salnr=N-substituted salicylaldiminate anion;X=3,5-lutidine,$(C_6H_5)_3P$) (835).

Both the infrared and Raman spectra have been assigned for the 1,2-bis(diphenylphosphino)ethane complexes

$[(CH_3)_2Au[P(C_6H_5)_2CH_2]_2]X$ ($X=Cl, (CH_3)_2AuCl_2$) (810), while infrared data have been reported for $(CH_3)_2In(O_2CCH_3)[P(C_6H_5)_2CH_2]_2$ (792), $CH_3NbX_4[P(C_6H_5)_2CH_2]_2$ ($X=Cl, Br$) (467), $CH_3WCl_5[P(C_6H_5)_2CH_2]_2$ (468), and $CH_3W(=O)Cl_3[P(C_6H_5)_2CH_2]_2$ (468).

11. Oxo- and Halo-Acid Salts

The organometallic derivatives of oxo- and halo-acid anions are frequently ionic salts. This section treats those derivatives in which a covalent interaction is found between the metal atom and the acido anion.

a. *Nitrates and Related Ligands with Nitrogen-Oxygen Bonds*

Covalent nitrates are unidentate (1.84), bidentate chelating (1.85), or bidentate bridging (1.86). Vibra-

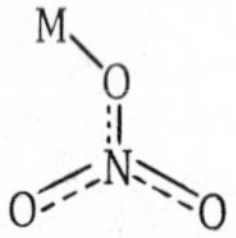

(1.84)

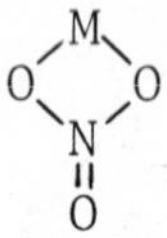

(1.85)

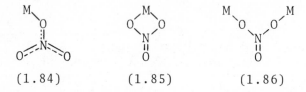

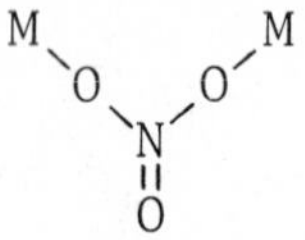

(1.86)

tional assignments for covalent nitrate derivatives are summarized in Table 1.51.

The Raman spectrum shows CH_3HgNO_3 to be undissociated in benzene (561), while similar data in water show the equilibrium

$$CH_3HgNO_3 + H_2O \rightleftharpoons CH_3HgH_2O^+ + NO_3^-$$

to be present (561,836).

The infrared and Raman spectra of the colorless liquid $(CH_3)_3GeNO_3$ show the presence of a unidentate nitrate group and a four-coordinate germanium atom (837). The structures of the methyl-tin nitrates $(CH_3)_nSn(NO_3)_{4-n}$ ($n=1$ to 3) are rather complex. Using infrared data, it was originally proposed that the tin atom of $(CH_3)_3SnNO_3$ is four-coordinate and that the nitrate group is unidentate (840). In a later study by the same authors (782), however, it was proposed that the nitrate group bridges the tin atoms to give a polymeric structure with five-coordinate tin atoms and planar C_3Sn units. A polymeric structure has also been proposed for $(CH_3)_3SnNO_3$ on the basis of several vibrational studies (566,813,838) and confirmed in a single

TABLE 1.51
Selected Vibrational Assignments (cm^{-1}) for Methyl Organometallic Nitrates

Compound[a]	Type of Nitrate Bonding	$\nu(NO_2)$ Asym.	$\nu(NO_2)$ Sym.	$\nu(NO)$[b]	$\rho_t(NO_3)$	$\delta_s(NO_2)$	$\nu(MO)$	Refs.
$RHgNO_3$	Monodentate	1502	1285	1000	820	750	292	561,836
R_3GeNO_3	Monodentate	1585	1248	940	778	698	520	837
R_3SnNO_3Py	Monodentate	1465	1290	1011	821	718	227	813
$R_3SnNO_3(Bipyr)_{0.5}$	Monodentate	1470	1285	1015	815	735	230	813
R_3SnNO_3	Bidentate bridging	1282 1270	1032	1484	809	728	220	813
$R_2Sn(NO_3)_2(Py)_2$	Monodentate	1480	1285	1000	818	731	280 245	813
$R_2Sn(NO_3)_2Bipyr$	Monodentate	1470	1292 1280	1010	812	735	234	813
$R_2Sn(NO_3)_2$	Bidentate chelating	1285 1255	1010 998	1555 1536	815 813	710 700	283	813
$RSn(NO_3)_3(Py)_2$	Monodentate	1510	1275	970	795	732 608	302	813
$RSn(NO_3)_3Bipyr$	Monodentate	1540 1512	1305 1280	991 965	795	735	295	813
$RSn(NO_3)_3$	Bidentate chelating	1240	974	1605	775	685	304 226	838
$R_3Sb(NO_3)_2$	Monodentate	1530	1290 1275 1245	965	795	728 708	275	64,277, 839

[a] $R = CH_3$.
[b] The NO bond has partial double-bond character for the monodentate nitrate group and full double-bond character for the bidentate nitrate group.

crystal X-ray study (841) that also showed a slight distortion of the C_3Sn skeleton from planarity and non-equivalent Sn-O bond lengths. The infrared spectrum of $(CH_3)_3SnNO_3 \cdot H_2O$ has been interpreted as consistent with a nonplanar C_3Sn skeleton (566,782). Using infrared data, a structure with a tetrahedral skeleton and two monodentate nitrate groups was proposed for $(CH_3)_2Sn(NO_3)_2$ (842). Others have interpreted the vibrational data as consistent with the presence of bidentate bridging (782,813) or chelating (813) nitrate groups. An X-ray study has now shown that the solid-state structure contains asymmetrically chelating nitrate groups (843). Polymeric structures with strongly coordinating hydroxyl and nitrate groups have been proposed for $R_2Sn(OH)NO_3$ ($R=CH_3,C_2H_5$) (566). The $\nu(NO_3)$ and $\nu(SnC)$ mode frequencies of $R_2Sn(OH)NO_3$ and $(R_2SnNO_3)_2O$ ($R=n\text{-}C_3H_7,n\text{-}C_4H_9$) are very similar; for the *n*-butyl derivatives in $CHCl_3$ the equilibrium

$$(n\text{-}C_4H_9)_2Sn(OH)NO_3 \rightleftharpoons [(n\text{-}C_4H_9)_2SnNO_3]_2O + H_2O$$

has been proposed (844). A structure with a seven-coordinate tin atom and three bidentate, chelating nitrate groups has been proposed for $CH_3Sn(NO_3)_3$ on the basis of vibrational data (838). Infrared data have been tabulated for $R_nPb(NO_3)_{4-n}$ (n=2,3) and $[R_2Pb(RNO)_2](NO_3)_2$ ($R=CH_3,C_2H_5$) (846). It has been noted in this study that the infrared data for $(C_2H_5)_nPb(NO_3)_{4-n}$ (n=2,3) do not agree with infrared data previously reported (847) for these derivatives. Infrared and Raman data and assignments for solid $R_3M[ON(O)CR'R'']$ ($M=Sn,Pb$; $R=CH_3,C_2H_5,n\text{-}C_4H_9$; $R'=R''=H,CH_3$) confirm a local symmetry of D_{3h} for the R_3Sn groups and a symmetry of C_{2v} for the $-ON(O)CR'R''$ ligands (848).

Although an ionic structure was originally proposed for $(CH_3)_3Sb(NO_3)_2$ (430), subsequent studies have shown the presence of covalent nitrate groups and a trigonal bipyramidal skeleton (64,277,839). While a covalent structure has also been found for $[(CH_3)_3SbNO_3]_2O$, an ionic structure has been found for $[(CH_3)_4Sb]^+NO_3^-$ (64).

The Raman spectrum of an aqueous solution of $(CH_3)_3PtNO_3$ shows it to be extensively dissociated (564).

b. Sulfates and Related Ligands with Sulfur-Oxygen Bonds

The Raman spectrum of an aqueous $(CH_3Hg)_2SO_4$ solution up to 0.8M in concentration is consistent with the equilibrium

$$CH_3HgSO_4^- + H_2O \rightleftharpoons CH_3HgH_2O^+ + SO_4^{2-}$$

with no detectable amount of $(CH_3Hg)_2SO_4$ present (849). The $\nu(SO_4)$ modes for the coordinated sulfate group were assigned at 1160, 1040, and 958 cm^{-1} (a fourth mode expected at approximately 1100 cm^{-1} was obscured by other bands) while the $\nu(HgO)$ mode arising from the coordinated sulfate group was assigned at 273 cm^{-1}.

Although $(R_2M)_2SO_4$ ($R=CH_3,C_2H_5$;M=Ga,In,Tl) dissociate in water to give R_2M^+ and sulfate ions, covalent compounds are indicated by the vibrational spectra of these compounds in ths solid state (850).

The infrared spectrum of solid $[(CH_3)_3Sn]_2SO_4$ can be interpreted in terms of either an ionic structure with planar $(CH_3)_3Sn^+$ cations or a polymeric structure in which every oxygen atom on each sulfate group is coordinated to a tin atom (1.87), thus approximately pre-

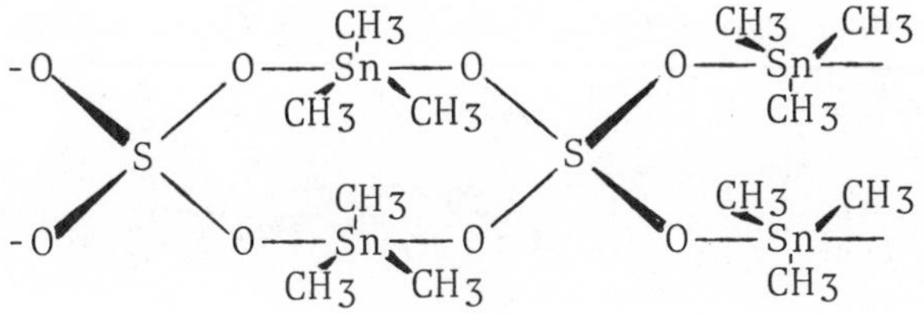

(1.87)

serving the T_d symmetry of the sulfate group (383). The fact that only the $\nu_a(SO_4)$ (1090 cm^{-1}) and $\delta_a(SO_4)$ (630 cm^{-1}) modes are infrared active makes it necessary to preserve the T_d symmetry of this group, while the infrared activity of the $\nu_a(SnC)$ (522 cm^{-1}) mode and not the $\nu_s(SnC)$ mode (520 cm^{-1}) indicates that the C_3Sn skeleton is planar. In Figure 1.40 the infrared spectra (1300 to 700 cm^{-1}) are illustrated for $[(CH_3)_3Sn]_2SO_4$ and a methanol adduct of this compound (383). Although an ionic structure has also been proposed for $[(C_2H_5)_3Sn]_2SO_4$ on the basis of its infrared spectrum (391), it is also possible to interprete these data in terms of a polymeric structure. In more recent reports of the infrared and Raman spectra of $(R_3Sn)_2SO_4$ ($R=CH_3$, C_2H_5, *n*-C_3H_7, *i*-C_3H_7), the polymeric structure has been favored (851,852). Based on Mossbauer and supportive infrared and physicochemical data, a monomeric structure was proposed for $[(n\text{-}C_4H_9)_3Sn]_2SO_4$ in which a chelating sulfate group is bonded to two nonplanar $(n\text{-}C_4H_9)_3Sn$ units (1.88) (853). This interpretation

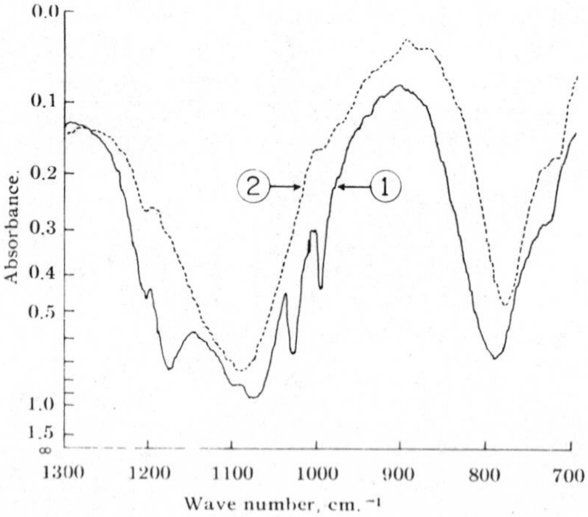

Figure 1.40. *The infrared spectra of (1) bis(trimethyltin) sulfate methanol adduct and (2) bis(trimethyltin) sulfate in Nujol mulls (383).*

$$(n\text{-}C_4H_9)_3Sn\underset{O}{\overset{O}{<}}S\underset{O}{\overset{O}{>}}Sn(n\text{-}C_4H_9)_3$$

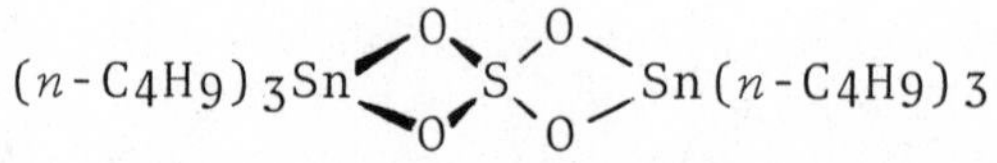

(1.88)

of these data has been challenged in favor of the polymeric structure proposed for other bis(trialkyltin) sulfates (854). In a more recent study, mass spectral, in addition to more detailed Mossbauer and low-frequency infrared data, have been used by the original authors to support their originally proposed monomeric structure (855). The infrared spectrum of $(CH_3)_2SnSO_4$ is consistent with either an ionic structure or a polymeric structure with six-coordinate tin atoms (1.89) (384), as

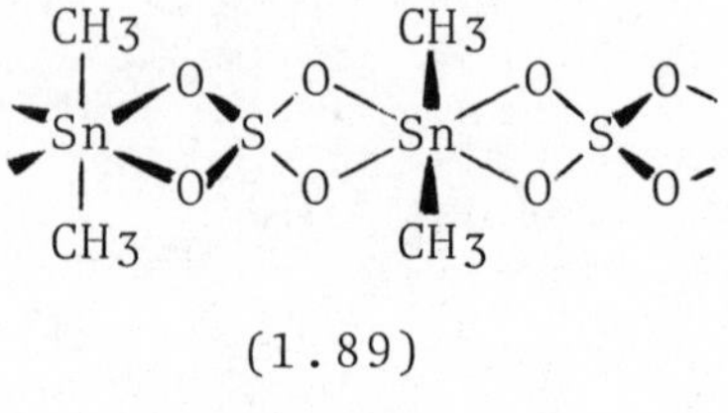

(1.89)

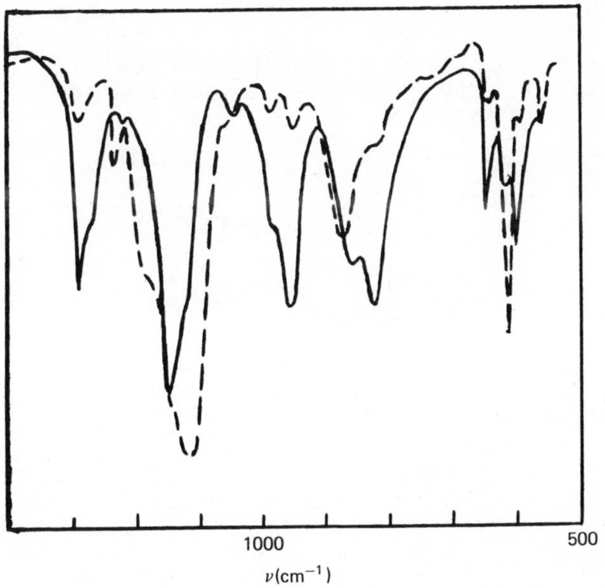

Figure 1.41. The initial infrared spectrum of 0.25% $(CH_3)_3SbSO_4$ in a KBr pellet (——) and the infrared spectrum of a similar sample after 45-minute exposure to air at 40% humidity (---) (430).

are similar data for R_2SnSO_4 ($R=C_2H_5$, *n*-C_3H_7, *i*-C_3H_7, *n*-C_4H_9) (852). The infrared spectra of $[(CH_3)_3Pb]_2SO_4$ and $(CH_3)_2PbSO_4$, however, have been interpreted in terms of an ionic structure (856).

The infrared spectrum (1400 to 500 cm^{-1}) of $(CH_3)_3SbSO_4$ is illustrated in Figure 1.41 along with that of a sample of this compound that had absorbed water from the air (430). The infrared spectrum of $(CH_3)_3SbSO_4$ is more complex than that of $[(CH_3)_3Sn]_2SO_4$ (Figure 1.40). A polymeric structure in which the sulfate groups bridge the antimony atoms has been used to explain the infrared spectra of $(CH_3)_3SbSO_4$ (430,839) and $(C_2H_5)_3SbSO_4$ (430). The $\nu_a(SO_2)$ and $\nu_s(SO_2)$ modes of the oxygen atoms not involved in bonding in $(CH_3)_3Sb$ SO_4 have been assigned at 1285 and 1145 cm^{-1}, respectively, while the corresponding modes from the two oxygen atoms of each sulfate that bridge the antimony atoms have been assigned at 950 and 825 cm^{-1}, respectively (839).

The Raman spectrum of an aqueous solution of $[(CH_3)_3Pt]_2SO_4$ indicates complete dissociation with no evidence for covalently bonded sulfate ligands (564).

The infrared spectra have been reported for the sulfite complexes R_2SnSO_3 ($R=CH_3$ (857), C_2H_5 (857), n-C_3H_7, n-C_4H_9) (852) and R_2PbSO_3 ($R=CH_3, C_2H_5, n$-C_4H_9) (858). The sulfite ligand in the tin derivatives acts simultaneously as a chelating and bridging ligand (1.90).

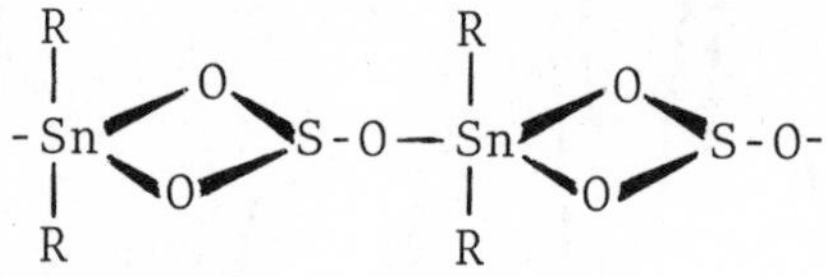

(1.90)

The $\nu(SO_3)$ modes have been assigned to infrared bands from 985 to 825 cm^{-1} for the tin derivatives and 1195 to 1185 cm^{-1} for the lead derivatives. The infrared spectra have also been discussed for $[(CH_3)_3Sn]_2SO_3 \cdot H_2O$ and $[(C_2H_5)_3Sn]_2SO_3$ (857).

Sulfur-bridged structures (1.91) have been proposed

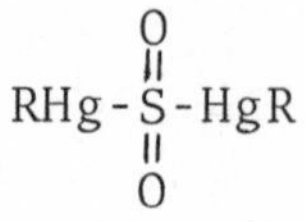

(1.91)

for the sulfonyl derivatives $(RHg)_2SO_2$ with the $\nu_a(S=O)$ and $\nu_s(S=O)$ modes assigned at 1245 and 1085 cm^{-1}, respectively, for $R=CH_3$ (859) and 1251 and 1094 cm^{-1}, respectively, for $R=C_2H_5$ (860). The polymeric sulfonyl derivative $[(CH_3)_3SnSO_2]_n$, which has been described in terms of either one of two structures (1.92 and 1.93),

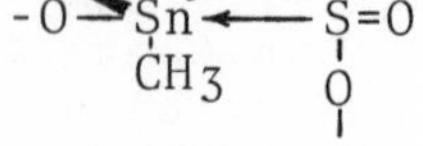

(1.92)

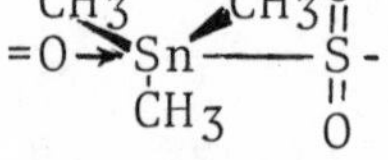

(1.93)

shows a strong intensity infrared band that has been assigned to the $\nu(SO)$ mode of the coordinated S-O bond (990 cm^{-1}) and a band in the region expected for a terminal $\nu(S=O)$ mode (1100 cm^{-1}) (861). Sulfonyl derivatives have been formed through cleavage of the metal-metal bonds in $(CH_3)_3SnMn(CO)_5$ (861) and $(CH_3)_3MFe(CO)_2$-C_2H_5 (M=Ge,Sn) (862) by sulfur dioxide. Although attempts were made to interprete the infrared spectra of the resulting products in terms of possible structures, it was noted that final structural elucidation must await X-ray crystallographic studies.

A Raman study has been reported of $CH_3HgSO_3CH_3$ in both the solid state and in aqueous solutions of different concentrations (849). A monomeric solid-state structure has been proposed in which the methylsulfonate ligand is bonded through one oxygen atom to the mercury atom; the ν(HgO) mode was assigned at 234 cm^{-1}. Although a 1M aqueous solution of $CH_3HgSO_3CH_3$ is completely dissociated, at a concentration of 4.7M, the equilibrium

$$CH_3HgSO_3CH_3 + H_2O \rightleftharpoons CH_3HgH_2O^+ + CH_3SO_3^-$$

was found. Spectroscopic data, including in many cases detailed infrared and Raman data and assignments, have been used to characterize the structures of $R_2MSO_3CH_3$ ($R=CH_3,C_2H_5$;M=Al,Ga,In,Tl) (717,863,864). The aluminum (ν(AlO)=440 cm^{-1}) and gallium (ν(GaO)=405 cm^{-1}) derivatives are dimeric or trimeric in benzene with the metal atoms bridged by two oxygen atoms of bidentate $CH_3(O=)SO_2^-$ ligands (717,864). Polymeric solid-state structures with weakly bridging and essentially ionic, tridentate methylsulfonate ligands are found for the indium and thallium derivatives (863,864). In aqueous solution, the indium and thallium derivatives dissociate into R_2M^+ and $CH_3SO_3^-$ ions (864). A monomeric structure has been proposed for $(CH_3)_2InSO_3C_6H_5$ vapor (865) and solid $(CH_3)_2TlSO_3C_2H_5$ (866). Mossbauer, infrared, and partial Raman data have been discussed for $(CH_3)_3SnSO_3X$ ($X=CH_3,CF_3$,F (867)) (868) and $(CH_3)_mSnCl_n(SO_3F)_p$ (m=2, n=0,p=2;m=2,n=p=1;m=n=1,p=2;m=1,n=2,p=1) (867). Since the anion spectra are more complex than expected for an ionic group with C_{3v} symmetry, it has been concluded that these groups are either monodentate or bidentate (867,868). A bidentate group has been found in a single crystal X-ray study of $(CH_3)_2Sn(SO_3F)_2$ (869). It has been concluded that the other methyl-tin alkylsulfonate and chlorosulfonate derivatives also have polymeric structures with bridging, bidentate sulfonate ligands (867,868). Spectral evidence supports ionic structures for the alkylsulfonates $(CH_3)_3PbSO_3CH_3$ and $(CH_3)_2Pb$-$(SO_3CH_3)_2$ (856).

Covalent, monomeric structures with Hg-S bonds have been found for the alkylsulfinate derivatives CH_3Hg-SO_2CH_3 (859) and $C_2H_5HgSO_2C_2H_5$ (860). In various hydrocarbon solvents, however, the spectroscopic evidence indicates that $C_2H_5HgSO_2C_2H_5$ adopts a chelated, oxygen-bonded structure (1.94) although the presence of the

$$C_2H_5Hg \overset{\displaystyle O}{\underset{\displaystyle O}{<>}} SC_2H_5$$

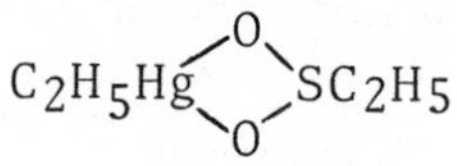

(1.94)

rapid equilibrium

$$C_2H_5Hg\text{-}O\underset{\displaystyle O}{\overset{|}{S}}C_2H_5 \;\rightleftharpoons\; C_2H_5Hg\text{-}O\overset{\displaystyle O}{\overset{\|}{S}}C_2H_5$$

could not be ruled out (860). Dimeric structures with cyclic $M_2O_4S_2$ skeletons are present in $(CH_3)_2MSO_2CH_3$ (M=Al,Ga,In,Tl) (865,866,870,871), $(CH_3)_2MSO_2C_6H_5$ (M=Al, Ga) (717), $(C_2H_5)_nCl_{2-n}GaSO_2C_6H_5$ (M=Al,Ga) (717), $(C_2H_5)_nCl_{2-n}GaSO_2C_2H_5$ (n=1,2) (872), and $(CH_3)_2InSO_2C_2H_5$ (865). Characteristic of these compounds are the $\nu_a(SO_2)$, $\nu_s(SO_2)$, and $\nu(CS)$ modes at approximately 1050 to 990 cm^{-1}, 1000 to 940 cm^{-1}, and 715 to 660 cm^{-1}, respectively. The $\nu(MO)$ modes have been assigned for the dimeric compounds $(CH_3)_2AlSO_2CH_3$ (561 to 515 cm^{-1}) (870), $(CH_3)_2GaSO_2CH_3$ (455 to 385 cm^{-1}) (871), $(CH_3)_2In$-SO_2CH_3 (435 to 375 cm^{-1}) (871), $(CH_3)_2TlSO_2CH_3$ (400 cm^{-1}) (866), and $(C_2H_5)_nCl_{2-n}GaSO_2C_2H_5$ (n=1,2) (500 to 450 cm^{-1}) (872). In Figure 1.42, the infrared and Raman spectra are illustrated for $(C_2H_5)_2GaSO_2C_2H_5$ (872).

Ester like structures with five-coordinate central metal atoms and trigonal bipyramidal skeletons have been proposed for $(CH_3)_3MO_2SCH_3$ (M=Si,Ge) on the basis of nmr, mass spectral, infrared, and Raman data (873). Similar data have been used to characterize the structures of $(CH_3)_2Ge(SO_2CH_3)_2$ (874) and $CH_3Ge(SO_2CH_3)_3$ (875). Infrared and Raman data for R_3SnSO_2R (R=CH_3,C_2H_5,n-C_3H_7, n-C_4H_9) are consistent with polymeric structures and five-coordinate tin atoms (1.95) (851 852,876). Polymeric structures with six-coordinate tin or lead atoms (1.96) have been proposed for $R_2Sn(SO_2R)_2$ (R=CH_3,C_2H_5, n-C_3H_7,i-C_3H_7,n-C_4H_9) (852,877) and $R_2Pb(SO_2R)_2$ (R=CH_3, C_2H_5,n-C_4H_9) (859). The $\nu_a(SO_2)$ modes for these polymeric species give rise to either one or two very strong intensity infrared bands from 1000 to 935 cm^{-1}.

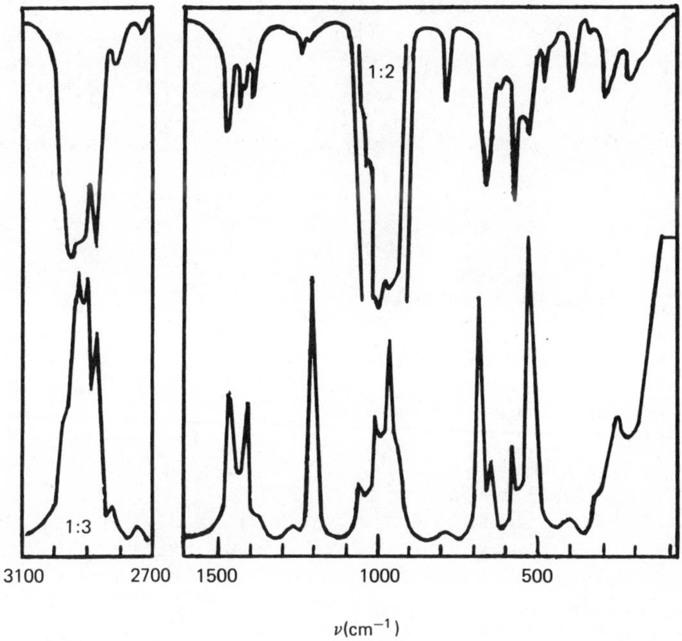

Figure 1.42. Infrared (top) and Raman (bottom) spectra of $[(C_2H_5)_2Ga(O_2SC_2H_5)]_2$ *(872).*

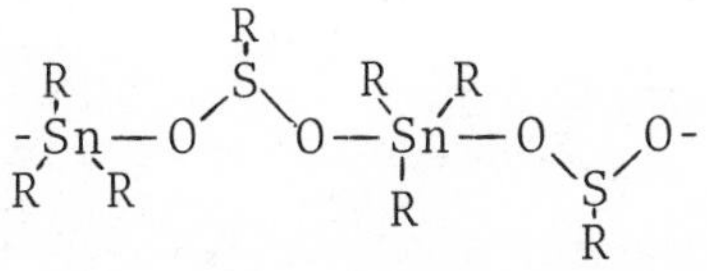

(1.95)

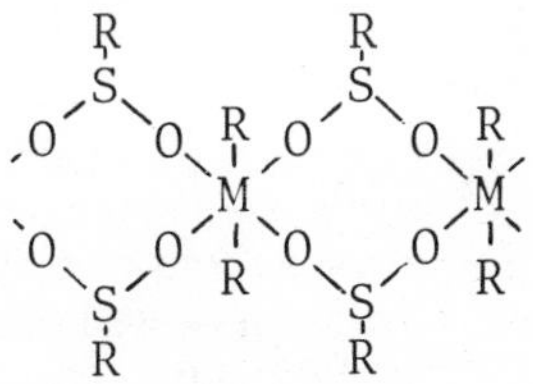

(1.96)

c. Selenates and Related Ligands with Selenium-Oxygen Bonds

The infrared spectra of the selenate compound $[(CH_3)_3Si]_2SeO_4$ are consistent with a monomeric, covalent structure (878). The selenite compound $[(CH_3)_3Si]_2SeO_3$ as a similar structure (878). Spectral data indicate that both $(CH_3)_3SbSeO_4$ and $[((CH_3)_3Sb)_2O]SeO_4$ are polymeric with five-coordinate antimony atoms (629).

d. Carbonates

The infrared spectrum previously attributed to $(CH_3)_3SnCO_3$ (384) has now been shown to be that of $[(CH_3)_2Sn]_2O(CO_3)$ (526). The infrared spectrum of solid $[(CH_3)_2Sn]_2O(CO_3)$ along with Raman data below 900 cm^{-1} are consistent with a polymeric structure containing bridging carbonate ligands and Sn_2O_2 rings (526). The new interpretation of the data previously attributed to $(CH_3)_2SnCO_3$ raises questions concerning the infrared data reported (391) for a compound formulated as $(C_2H_5)_2SnCO_3$. Infrared bands at 450, 375, and 250 cm^{-1} have been assigned to ν(SbO) modes in $(CH_3)_3SbCO_3$ (839).

e. Halo-Acid and Miscellaneous Oxo-Acid Salts

The infrared spectra of $(CH_3)_3SnX$ ($X{=}BF_4,AsF_6,SbF_6$) have been interpreted in terms of planar C_3Sn units bridged by the ligand fluorine atoms (879). Attempts to prepare $(CH_3)_2SnX_2$ ($X{=}BF_4,PF_6,AsF_6$) gave mixtures containing large amounts of $(CH_3)_2SnF_2$, while this was the exclusive product when $X{=}SiF_6$ (384).

Vibrational data have been reported for $(CH_3)_2M$-O_2PH_2 (M=Al,Ga,In) (880). The vibrational spectra indicate that the gallium atoms in the oxo-halo derivatives $[(CH_3)_2GaO_2PCl_2]_2$ (881) and $[(C_2H_5)_2GaO_2PX_2]_2$ (X=F,Cl) (882) are bridged by the oxygen rather than halogen atoms to form cyclic $Ga_2P_2O_4$ skeletons.

The infrared and ^{119}Sn Mossbauer spectra suggest that $R_2Sn(O_2PH_2)$ ($R{=}CH_3,C_2H_5,n\text{-}C_4H_9$) have six-coordinate, octahedral tin atoms with *trans* alkyl groups and bidentate, oxygen-bridged hypophosphite ligands; that R_2Sn-PO_3X ($R{=}CH_3,C_2H_5,n\text{-}C_4H_9$,X=H;$R{=}CH_3$,X=OH,F) have polymeric structures with nonlinear R_2Sn units; and that $[(CH_3)_2Sn$-$]_3(PO_4)_2$ has a five-coordinate tin atom with at least one three-coordinate oxygen atom per formula unit (883).

A structure with essentially ionic perchlorate groups is indicated by infrared data reported for $(CH_3)_2TlClO_4$ (342). The presence of at least four $\nu(ClO_4)$ bands in the infrared spectrum of $(CH_3)_3SnClO_4$ (1200,1112,998, and 908 cm^{-1}) indicates that the symmetry of the perchlorate group is C_{2v} and can be explained by a bridged structure (1.97) (884). The complex $(CH_3)_3Ta(ClO_4)_2$ is a nonconductor and a dimer in acetonitrile, and exhibits strong intensity infrared bands (1100,1015 and 910 cm^{-1}) characteristic of a bidentate, bridging perchlorate ligand (749).

$$\begin{array}{c} \underset{|}{CH_3} \ \underset{|}{CH_3} \qquad\qquad \underset{|}{CH_3} \\ -Sn-O \diagdown \quad \diagup O-Sn- \\ CH_3 \quad \diagup Cl \diagdown \quad CH_3 \ \ CH_3 \\ O \diagup \quad \diagdown O \end{array}$$

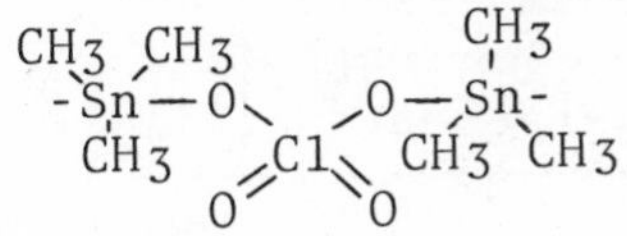

(1.97)

A polymeric structure with five-coordinate antimony atoms is consistent with the infrared spectrum of
[((CH$_3$)$_3$Sb)$_2$O]CrO$_4$ (629). The ν(SnC) modes have been
assigned for (CH$_3$)$_2$SnMoO$_4$ (827).

12. Metal-Metal Bonds and Complexes

Although the ν(MM') modes of heteronuclear skeletons and
some skeletal modes in metal cluster compounds are infrared active, the infrared bands arising from these
modes are weaker and less easily identified than those
in the corresponding Raman spectra. Raman spectroscopy
has therefore proved more useful than infrared spectroscopy in the study of the modes arising from metal-metal
interactions. This is especially true for homonuclear
compounds with dimetallic or cluster skeletons in which
the symmetric ν(MM) modes are infrared inactive. The
Raman active modes arising from metal-metal interactions
give rise to relatively intense signals due to the large
change in the polarizability during these vibrations.

a. Homonuclear

The ν(MM) mode assignments for several (alkyl)$_n$M$_2$
derivatives are summarized in Table 1.52. The extent of
skeletal mode coupling for these derivatives decreases
as the mass of the metal atom increases. This is illustrated in Figure 1.43 for the Raman active skeletal
modes of (CH$_3$)$_6$M$_2$ (M=Si,Ge,Sn,Pb) (886).
The vibrational spectra of (CH$_3$)$_4$P$_2$ (888) and
(CH$_3$)$_4$As$_2$ (889) show both to exist exclusively as the
trans conformer in the solid state and as a mixture of
both the *trans* and *gauche* conformers in the liquid
state. These studies have also found that an increase
in the M-M bond length favors a greater percentage of
the *trans* form. Therefore, while (CH$_3$)$_4$P$_2$ was found to
be a 60-40% mixture of the *gauche* and *trans* isomers, respectively, a 40-60% mixture of the *gauche* and *trans*
isomers, respectively, was found for (CH$_3$)$_4$As$_2$. From
the Raman spectrum of (CH$_3$)$_4$P$_2$ vapor, it has also been
concluded that both the *gauche* and *trans* isomers are

TABLE 1.52
The ν(MM) Mode Assignments (cm^{-1}) of $(Alkyl)_nM_2$ Complexes

Compound	ν(MM) Mode	Refs.
$(CH_3)_6Si_2$	404	885,886
$(CH_3)_6Ge_2$	273	885,886
$(CH_3)_6Sn_2$	190	885,886
$(CH_3)_6Pb_2$	116	886,887
$(CH_3)_4P_2$	455[a] 429[b]	888
$(CH_3)_4As_2$	271[a] 254[b]	889
$(CH_3)_2Se_2$	286	890,891
$(CD_3)_2Se_2$	275	890
$(CH_3)_2Te_2$	188	892
$(C_2H_5)_4P_2$	424	893
$(n\text{-}C_4H_9)_4P_2$	419	893
$(n\text{-}C_4H_9)_2Se_2$	293	894

[a]Due to the *trans* conformer.
[b]Due to the *gauche* conformer.

are present and that the *gauche* form is predominant with a concentration of approximately 60% (895). A structure with puckered P_4 rings has been proposed for $(RP)_4$ (R= C_2H_5,n-C_3H_7,i-C_3H_7,i-C_4H_9) with the ν_a(PP) and ν_s(PP) modes assigned from 488 to 467 cm^{-1} and 409 to 391 cm^{-1}, respectively (896). The ν(AsAs) modes of $C_2H_3As_3$ (1.98)

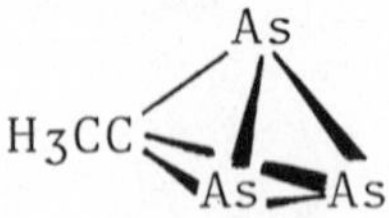

(1.98)

have been assigned to Raman bands at 300 (A_1) cm^{-1} and at 252 and 246 (E) cm^{-1} while the ν(AsC) modes have been assigned to infrared bands at 546 and 522 cm^{-1} (A_1 + E) (897). A normal coordinate analysis indicates that the Te-Te bond of $(CH_3)_2Te_2$ is weaker than the Se-Se bond of $(CH_3)_2Se_2$ (892).

In several compounds the presence of metal-metal interactions is not as obvious as in the compounds discussed above since the metal atoms are bridged by other ligands. If metal-metal interactions are present, it is likely that they will give rise to relatively strong intensity, low-frequency Raman bands. Therefore, Raman intensities are important in assigning the vibrational data of polymetallic compounds with bridging ligands.

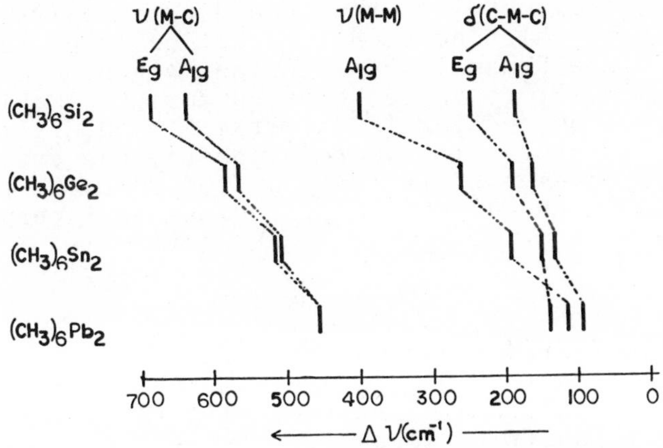

Figure 1.43. Raman active skeletal modes for $(CH_3)_6M_2$ (M=Si,Ge, Sn,Pb).

Absolute Raman intensities have been measured for the totally symmetric modes of $(t\text{-}C_4H_9Li)_4$ (154). The eigenvectors calculated for $(t\text{-}C_4H_9Li)_4$ were used to transform the Raman intensities into bond polarizability derivatives that were used to estimate bond orders. This calculation suggested that the extent of the Li-Li bonding is less than 5% of the total bonding-electron density of the Li_4C_4 cage.

A polymeric structure has been found for solid $(CH_3)_2Be$ (74). A strong intensity Raman band at 455 cm^{-1} has been assigned to the ν(BeBe) mode of this compound.(80). For CH_3BeBH_4, which has a dimeric, methyl-bridged structure, the ν(BeBe) mode has been assigned to a Raman band at 357 cm^{-1} (80). The higher frequency of the ν(BeBe) mode in polymeric $(CH_3)_3Be$ than in dimeric CH_3BeBH_4 has been interpreted as implying that electron density has been shifted somewhat from the bridge bonds to the middle of the Be-Be region in $(CH_3)_2Be$ (80). The ν(BeB) mode of dimeric CH_3BeBH_4 has been assigned at 740 cm^{-1} (80).

The possibility of an Al-Al interaction has been discussed for the methyl-bridged dimer of $(CH_3)_3Al$ (82, 83,898,899). A K(AlAl) force constant of 1.059 mdynes/A was used in a normal coordinate analysis of $(CH_3)_6Al_2$ that included all 26 atoms; a Raman band at 312 cm^{-1} was assigned as primarily due to the ν(AlAl) mode (84). The need of the K(AlAl) force constant, however, does not necessarily reflect the presence of such an interaction since similar metal-metal interaction force constants have been included in normal coordinate analyses for compounds in which no such interaction is probable (900).

Absolute Raman intensities and eigenvector calculations similar to those reported for $(t\text{-}C_4H_9Li)_4$ have also been reported for $[(CH_3)_3PtX]_4$ (X=OH,Cl,I) (462). Weak Pt-Pt interactions were proposed, the extent of which decreases as X is changed from OH to Cl to I. This is also the direction of increasing internuclear distance.

b. Heteronuclear

The infrared and Raman spectra of the 1:1 complexes formed between $(t\text{-}C_4H_9)_2Be$ and R_3M ($R=CH_3,C_2H_5$;M=N,P) have been reported and assigned with the ν(NC) mode observed to shift to lower frequencies on complexation and the ν(PC) mode observed to shift to higher frequencies on complexation (785).

Low-frequency infrared and Raman assignments have been reported for $(CH_3Hg)_3As$ (ν_a(HgAs)=240 cm^{-1}, ν_s(HgAs)=210 cm^{-1}) and $[(CH_3Hg)_4As]X$ ($X=NO_3,BF_4,PF_6$) (ν_a(HgAs)=243 cm^{-1},ν_s(HgAs)=109 cm^{-1}) (901).

Low-frequency vibrational assignments for $[(CH_3)_3\text{-}Si]_nM$ derivatives are summarized in Table 1.53. Vibrational assignments have also been reported for $(CH_3)_2H\text{-}PBH_2$ (907) and several members of the series $[(alkyl)_3\text{-}M]_nM'X_m$ (M=Si,Ge,Sn,P,As;M'=main group element;X=H,D, halide,alkyl group;n=1 to 3;m=0 to 4) (507,614,908-921). The position of the ν(MM') mode for $(alkyl)_3MM'$ (M=P,As, Sb;M'=Se,Te) has been interpreted in terms of multiple bonding between the group Vb element and the selenium or tellurium atom (627,628,922,923). The infrared and Raman spectra of $(R_2SnSe)_3$ show the Sn_3Se_3 ring to have a symmetry of C_s or C_{3v} but not D_{3h} (656).

Trialkyl derivatives of the group IVb and Vb elements have been used frequently as ligands in transition-metal complexes. The ν(MM') mode assignments for several compounds with group IVb element-transition-metal bonds are summarized in Table 1.54. Successive replacement of methyl groups with chlorine atoms in $(CH_3)_3SnMn(CO)_5$ and $(CH_3)_3SnMo(CO)_3C_5H_5$ has been observed, in general, to slightly increase the ν(SnM) mode frequency (931). This has been explained in terms of a competition between the increasing mass of the $(CH_3)_nCl_{3-n}Sn$ unit and an increasing tin-transition

TABLE 1.53
Skeletal Mode Assignments (cm^{-1}) for $[(CH_3)_3Si]_nM$ Compounds

Compound	$\nu_a(SiM)$	$\nu_s(SiM)$	$\delta(Si_nM)$	$\delta_a(SiC_3)$	$\delta_s(SiC_3)$	$\rho(SiC_3)$	Refs.
$[(CH_3)_3Si]_2Se$	369vs,IR	363vs,R	92vw,R	232m,R	172vs,R	160s,R	902
$[(CH_3)_3Si]_2Te$	323vs,IR	330vs,R	80vw,R	227s,IR	167s,R	155,R	902
$[(CH_3)_3Si]_3In$	311m,R	307s,R	111m,R	224w,R	165sh,R	157,R	903
$[(CH_3)_3Si]_3Tl$	286sh,R	296,R	112m,R	222w,R	161sh,R	152vs,R	904
$[(CH_3)_3Si]_3P$	461m,R	380vs,R	102s,R	225m,IR 232s,R	265m,IR 276w,IR	177vs,R	905
$[(CH_3)_3Si]_3As$	356m,R	341vs,R	89w,R	210s,R 248m,IR	210s,R 249m,IR	155sh,R 175sh,R	905
$[(CH_3)_3Si]_3Sb$	319s,IR	318vs,R	71s,R	229s,IR	193s,IR 187vs,R	145sh,R 162vs,R	905
$[(CH_3)_3Si]_4Si$	457m,IR	328s,R	101w,IR	223s,IR	222m,R 262s,IR	172vs,R	906
$[(CH_3)_3Si]_4Ge$	360vs,IR	319vs,R	89vs,IR	240m,IR	202w,R 213vs,IR	170vs,R	906
$[(CH_3)_3Si]_4Sn$	328vs,IR	311vs,R	73w,IR	230w,IR	180sh,R 196vs,IR	158vs,R	906

TABLE 1.54
Group IVb-Transition-Metal Stretching Mode Frequencies (cm^{-1}) for Transition-Metal Complexes with Trialkyl-Group IVb Element Ligands

Compound	ν(MM') Mode	Frequency	Refs.
$(CH_3)_3SiMn(CO)_5$	Si-Mn	297 (R)	924
$(CH_3)_3SiCo(CO)_4$	Si-Co	295 (R)	925,926
$(CH_3)_3GeCr(CO)_3C_5H_5$	Ge-Cr	119 (IR)	927
$(CH_3)_3GeMo(CO)_3C_5H_5$	Ge-Mo	184 (R)	927
$(CH_3)_3GeW(CO)_3C_5H_5$	Ge-W	177 (R)	927
$(CH_3)_3GeMn(CO)_5$	Ge-Mn	191 (R)	924,928
$(CH_3)_3GeRe(CO)_5$	Ge-Re	169 (R)	924,928
$(CH_3)_3GeCo(CO)_4$	Ge-Co	192 (R)	929
$(C_2H_5)_3GeCo(CO)_4$	Ge-Co	188 (R)	929
$(CH_3)_3SnCr(CO)_3C_5H_5$	Sn-Cr	183 (IR)	927
$(CH_3)_3SnMo(CO)_3C_5H_5$	Sn-Mo	172 (IR)	930,931
$(CH_3)_3SnW(CO)_3C_5H_5$	Sn-W	168 (IR)	931
$(CH_3)_3SnMn(CO)_5$	Sn-Mn	179 (IR,R)	924,928, 930,931
$(CH_3)_3SnRe(CO)_5$	Sn-Re	147 (R)	928
$(CH_3)_3SnFe(CO)_2C_5H_5$	Sn-Fe	185 (IR)	930,931
$(CH_3)_3SnCo(CO)_4$	Sn-Co	176 (IR,R)	925,929, 931
$(C_2H_5)_3PbMn(CO)_5$	Pb-Mn	167 (IR,R)	932

metal interaction in proceeding from the fully methylated to the fully chlorinated compounds. The conclusions reached in this study support those reached in earlier studies using infrared data in the ν(CO) mode region (933-936). The infrared spectra of $(CH_3)_nSnCl_{3-n}Mo(CO)_3C_5H_5$ (n=1 to 3) are illustrated in Figure 1.44 (931). The ν(SnCo) modes have been assigned at approximately 170 cm^{-1} in the infrared spectra of $RSn[Co(CO)_4]_3$ ($R=CH_3,C_2H_5,n\text{-}C_4H_9$) (937).

Infrared and Raman data including boron-group Vb element stretching mode assignments have been reported for the compounds formed between $(CH_3)_3M$ (M=P,As,Sb) and the Lewis acids BF_3, BCl_3, BBr_3, B_2H_6, and $(CH_3)_3B$ (938). The ν(MM') mode assignments have been reported for several compounds with group Vb-transition-metal bonds (467,939-946). Among these has been the use of stable nickel isotopes to assign the ν(NiP) modes in $Ni[P(C_2H_5)_3]_2X_2$ (X=Cl,Br) (942,943). The ν(PtTe) modes have been assigned from 197 to 177 cm^{-1} for $(R_2Te)_2Pt_2Cl_4$ ($R=C_2H_5,n\text{-}C_3H_7$) (947).

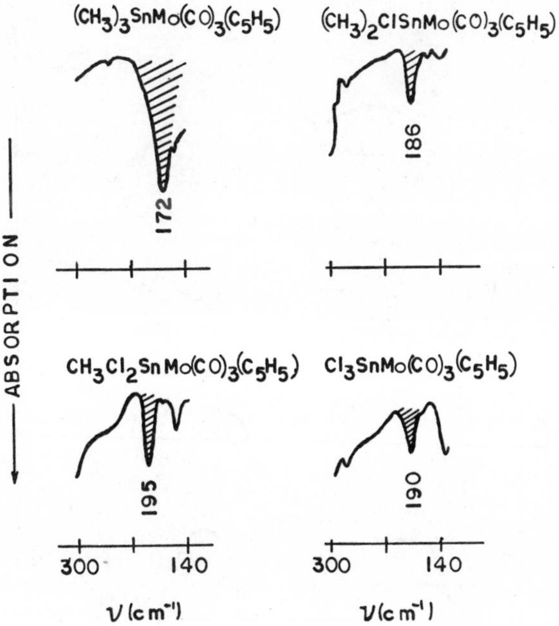

Figure 1.44. The infrared spectra (140 to 300 cm^{-1}) of $(CH_3)_{3-n}Cl_nSnMo(CO)_3C_5H_5$ (n=0 to 3). The shaded bands indicate the ν(SnMo) mode (931).

F. HALOGENATED ALKYL COMPOUNDS

1. Main Group Elements

a. Perhaloalkyl Complexes

The reaction of alkali metal atoms with CCl_4 at high dilution in argon deposited at 15°K produced not only trichloromethyl radicals and dichlorocarbene but also two carbenoids characterized as CCl_3M and a dichlorocarbene-metal halide complex (1.99) (948). These

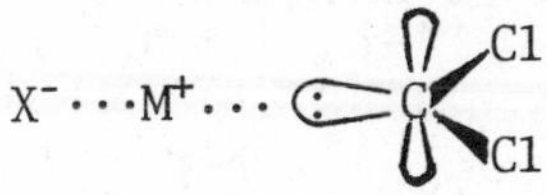

(1.99)

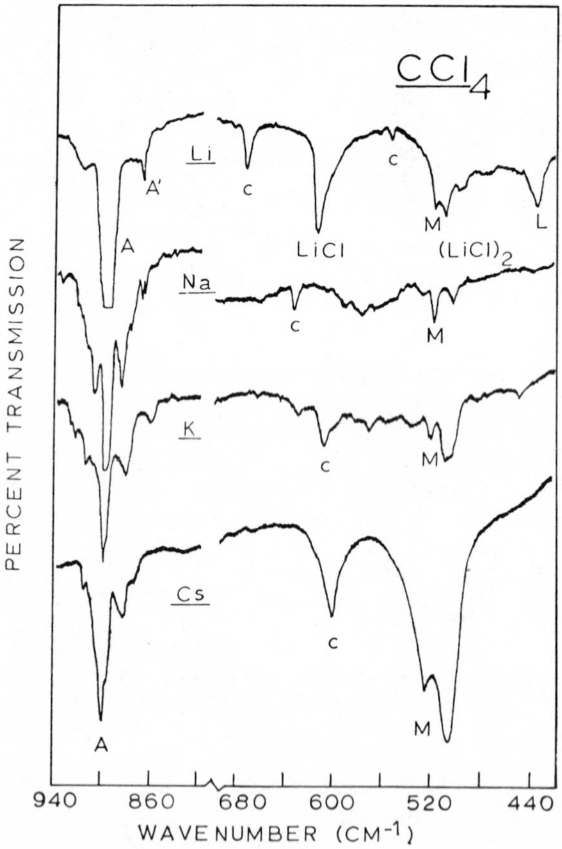

Figure 1.45. Infrared spectra in the 940 to 860 cm^{-1} and 680 to 440 cm^{-1} regions following codeposition of ^6Li, Na, K, and Cs atoms at 14°K with CCl_4 at high dilution in argon (948).

carbenoids have been studied using infrared spectra of the matrix species. In Figure 1.45, the infrared spectra (940 to 860 cm^{-1} and 680 to 400 cm^{-1}) are illustrated for the products obtained by cocondensing ^6Li, Na, K, or Cs atoms at 14°K with CCl_4 at high dilutions in argon (948). The *A* band (898 cm^{-1}) has been assigned to the ν_3 mode of CCl_3 (949) while the weaker intensity *A'* band (869 cm^{-1}) is the naturally abundant ^{13}C counterpart of the 898 cm^{-1} band. An intense doublet at

745.7 and 743.7 cm^{-1}, and a weaker doublet at 719.5 and 716.7 cm^{-1}, which have been omitted from Figure 1.45, have been assigned to the ν_3 and ν_1 modes, respectively, of CCl_2 (949-951). The *L* and *M* bands were attributed to $LiCCl_3$ (and analogously for the other alkali metal atoms) and were favored at high lithium atom concentrations (952). The *C* band that was previously assigned to the ν_1 mode of CCl_3 (949) has been reassigned as arising from the carbenoid-alkali metal chloride complex (948). The weaker intensity *C* band in the ^6Li spectrum may be the ^6Li-Cl mode in the complex carbene species (948). Lithium atom matrix reactions with CBr_4 produce sharp bands at 641 and 596 cm^{-1} that have been assigned to the ν_3 and ν_1 modes, respectively, of CBr_2 (953) while a band at 582 cm^{-1}, previously attributed to the CBr_3 species (953), has been reassigned as the bromine counterpart of the 674 cm^{-1} *C* band. The $\nu(CBr_3)$ modes have been assigned to infrared bands at 640 and 622 cm^{-1} for $(CBr_3)_2Hg$ (954).

Table 1.55 summarizes vibrational assignments reported for $(CF_3)_nM$ derivatives. There is a noticable frequency decrease of the $(CF_3)_3M$ modes as M is changed from P to As to Sb (958). Normal coordinate analyses for $(CF_3)_3M$ (M=P,As,Sb) indicate that the K(MC) force constants are slightly smaller than those of the corresponding methyl derivatives (958). These calculations also show several of the modes to be strongly coupled, although the extent of coupling decreases as the mass of M increases. In Figure 1.46, the gas-phase infrared and liquid-phase Raman spectra are illustrated for $(CF_3)_3As$ (958). Infrared data have also been reported for $(C_2F_5)_2Te$ (959).

b. Perhaloalkyl Complexes with Other Functional Groups

The $\nu(CF_3)$ modes have been assigned for $(CH_3)_3MCF_3$ (M=Ge,1194 and 1098 cm^{-1} (960);M=Sn,1158 and 1071 cm^{-1} (961)). Similar bands have been observed for $(CH_3)_3Pb-C_2F_5$ at 1198, 1085, and 1070 cm^{-1} (962). Vibrational data and some assignments have also been reported for $C_6H_5HgCX_2Y$ (X=Y=Cl=Br;X=Cl,Y=Br;X=Br,Y=Cl) (963), R_3Sn-CX_3 ($R=CH_3,C_2H_5,n-C_4H_9$;X=Cl,Br,I) (964-966), $(CH_3)_2PCF_3$ (967), and CH_3TeR ($R=CF_3,C_2F_5$) (959).

The $\nu(MH)$ and $\nu(MD)$ mode assignments summarized in Table 1.56 for perfluoromethyl hydrides and deuterides appear at higher frequencies, while the $\nu(MC)$ mode assignments, also included in Table 1.56, appear at lower frequencies than those found for the corresponding

TABLE 1.55
Vibrational Assignments (cm^{-1}) for Perhalomethyl Derivatives of the Main Group Elements

Mode	Compound						
	$(CF_3)_2Hg$[a]	$(CF_3)_4Ge$[b]	$(CF_3)_4Sn$[c]	$(CF_3)_3P$[d]	$(CF_3)_3As$[d]	$(CF_3)_3Sb$[d]	$(CF_3)_2Te$[e]
$\nu_a(CF_3)$	1160 1143	1255 1170	1238 1150	1235 1189	1219 1189	1194 1148	1178 1143
$\nu_s(CF_3)$	1072 1049	1102		1158 1129	1152 1114	1129 1089	1077
$\delta_d(CF_3)$	524 515	525		573 559	557 537	526	
$\delta_s(CF_3)$	716 713	735	744	747	737	725	740
$\rho_r(CF_3)$	264 216			270 250	249 232	222 201	
$\nu_a(MC)$	274			470	337	269	
$\nu_s(MC)$	226			450	349	286	
$\delta(MC_3)$				169 109	144 90	130 72	

[a]References 955 and 956.
[b]Reference 956.
[c]References 956 and 957.
[d]Reference 958.
[e]References 956 and 959.

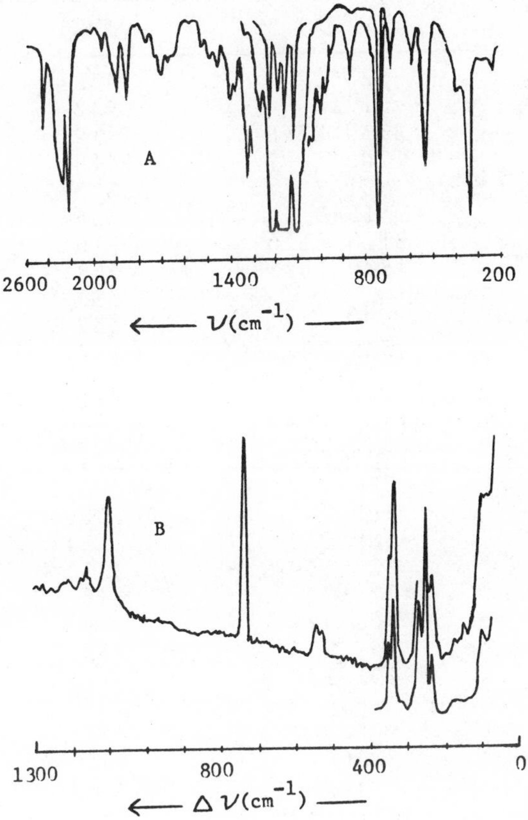

Figure 1.46. The (A) vapor-phase infrared spectrum and (B) liquid-phase Raman spectrum of $(CF_3)_3As$ (958).

methyl hydrides. Vibrational assignments for $(CF_3)_2MH$ (968), CF_3MH_3 (971) (M=P,As), and their deuterated analogs have been made on the basis of normal coordinate analyses. The ν(PH) mode of $(CF_3)_2PH$ is split in the solid-state infrared spectrum (2372 and 2342 cm^{-1}). This has been attributed to the presence of more than one molecule per unit cell (968).

A normal coordinate analysis has been included in a vibrational study of the potassium salt of $CF_3BF_3^-$ (974). The vibrational spectra of $(CF_3)_2GeX_2$ in the gas and liquid phases (975) and of CF_3GeX_3 in the liquid

TABLE 1.56
Selected Assignments (cm^{-1}) for $(CF_3)_mMX_n$ (X=H,D,Halide)

Compound	ν(MX)	ν(MC)	Refs.
$(CF_3)_2PH$	2360	439	968-970
$(CF_3)_2PD$	1712	442	968
CF_3PH_2	2339 2348	419	971,972
CF_3PD_2	1705 1697	410	971
$(CF_3)_2AsH$	2143	355 333	968
$(CF_3)_2AsD$	1549	336	968
CF_3AsH_2	2123 2116	310	971
CF_3AsD_2	1529 1523	308	971
CF_3SeH		334	973
CF_3SeD		332	973
CF_3HgCl	339 327	252	294
CF_3HgBr	233	270	294
CF_3HgI		255	294
$CF_3{}^{11}BF_3^-$	1060 732		974
$CF_3{}^{10}BF_3^-$	1100 732	1124	974
$(CF_3)_2GeF_2$[a]	755 742	362 312	975
$(CF_3)_2GeCl_2$	465 441	342 290	975
$(CF_3)_2GeBr_2$	359 232	348 343	975
$(CF_3)_2GeI_2$	325 190	325	975
CF_3GeF_3	775 754.5	340	976
CF_3GeCl_3	463 424	300	976
CF_3GeBr_3	349 234	333.5	976
CF_3GeI_3	291 173.7	310	976
$(CF_3)_2PF$	850	455	970
$(CF_3)_2PCl$	533	444	970
$(CF_3)_2PBr$	466	441	970
$(CF_3)_2PI$	419	444 441	970
CF_3PF_2	861 850	460	972
CF_3PCl_2	524 515	424	406,972
CF_3PBr_2	434 421	385	972
CF_3PI_2	389 371	411	972
$(CF_3)_2AsF$	693	347 325	977
$(CF_3)_2AsCl$	428	348 337	977
$(CF_3)_2AsBr$	325	333 312	977
$(CF_3)_2AsI$	216	332 311	977
$CF_3P(H)Br$	2318[b] 429[c]	425	978
$CF_3P(D)Br$	1689[b] 425[c]	423	978
$CF_3P(H)I$	2327[b] 383[c]	415	978
$CF_3P(D)I$	1694[b] 381[c]	412	978

[a] Data for the vapor and liquid phases in which $(CF_3)_2GeF_2$ has a monomeric structure.
[b] ν(PH) or ν(PD) mode.
[c] ν(PBr) or ν(PI) mode.

TABLE 1.57
Vibrational Assignments (cm^{-1}) for RPF_4 ($R=CH_3,CF_3,CCl_3$)

Mode	Symmetry	CH_3PF_4[a]	CF_3PF_4[b]	CCl_3PF_4[c]
$\nu(CX_3)$	B_1	2963	1162	773
	B_2	2949	1231	757
	A_1	2932	1183	713
$\delta_a(CX_3)$	B_1	1434	510	328
	B_2	1434	518	368
$\delta_s(CX_3)$	A_1	1328	754	409
$\rho_r(CX_3)$	B_1	985	229	267
	B_2	985	304	302
$\nu(PC)$	A_1	596	423	569
$\nu(PF)$	B_1	1009	986	985
	B_2	843	892	807
	A_1	932	909	918
	A_1	725	674	665

[a]Reference 424.
[b]Reference 425.
[c]Reference 982.

phase (976) (X=F,Cl,Br,I) are consistent with monomeric structures. The solid-state Raman spectrum of $(CF_3)_2Ge$-F_2, however, suggests an associated structure with Ge-F-Ge bridges (975). The vibrational spectra of $(CF_3)_2MX$ (M=P (970),As (977)) and CF_3PX_2 (X=F,Cl,Br,I) (406,972) are consistent with monomeric structures. Using infrared and Raman data, structures with a trigonal bipyramidal skeleton and axial CF_3 group(s) have been proposed for CF_3PCl_4 (976) and $(CF_3)_2PCl_3$ (980). Although microwave data for CF_3PF_4 have been reported to be consistent with a trigonal bipyramidal skeleton and an axial CF_3 group (981), gas-phase infrared and liquid-phase Raman data have been reported to be consistent with an equatorial CF_3 group (425). The infrared and Raman spectra of CCl_3PF_4 show the CCl_3 group to be in an equatorial position (982). In Table 1.57, the vibrational assignments are compared for RPF_4 ($R=CH_3,CF_3$, CCl_3) while ν(M-halide) and ν(MC) mode assignments for other perfluoromethyl-metal halides are included in Table 1.56. Vibrational data have also been reported for CCl_3HgX (X=Cl,ν(HgCl)=335 cm^{-1};X=Br,ν(HgBr)=238 cm^{-1}) (294), CBr_3HgBr (ν(CBr)=621 cm^{-1}) (954), $RSiF_3$ ($R=CF_3,C_2F_5$) (983), $RSiF_2I$ ($R=CF_3$ (984),C_2F_5 (983), $K_2[CF_3GeF_5]$ (985), and CBr_3GeBr_3 (986).

The infrared spectra of $(CF_3)_2PXH$ and $(CF_3)_2PXD$ (X=O,S) have been recorded in all three phases (987). The O-H(D) and S-H(D) stretching and bending modes together with the first overtone of the stretching modes consistently showed two bands in each region. This has been attributed to the presence of rotational isomers thought to arise from the electronic effect of intermolecular hydrogen-bonding and lone-pair repulsions.

The vapor- and solid-phase infrared spectra and liquid- and solid-phase Raman spectra of $[(CF_3)_2P]_2X$ (X=O,S,Se) have been assigned assuming nonlinear P-X-P skeletons (988). For X=O and S, the ν_a(PX) mode has been assigned at 923 and 509 cm^{-1}, respectively, and the ν_s(PX) mode at 716 and 520 cm^{-1}, respectively, while the ν(PSe) modes for $[(CF_3)_2P]_2Se$ were reported to appear in the ν(PC) mode region (456 to 422 cm^{-1}).

In one vibrational study of $(CF_3)_4P_2$ (969), it has been concluded that although only the *trans* conformer is present in the solid state, the vapor and liquid states consist of mixtures of both the *trans* and *gauche* conformers with the *trans* isomer predominating. In a second study of $(CF_3)_4P_2$ (989), however, the spectra have been interpreted as indicating that only the *trans* conformer is present in the vapor and liquid phases. In a vibrational analysis of $(CF_3)_4As_2$, it has been concluded that only the *trans* isomer is present in the vapor (989,990), liquid (989,990), and solid (990) phases. Normal coordinate analyses have been reported for $(CF_3)_4M_2$ (M=P,As) (989). The ν(MM) mode has been assigned at 485 and 205 cm^{-1} for the *trans* isomer of $(CF_3)_4P_2$ (969,989) and $(CF_3)_4As_2$ (989,990), respectively. The ν(PP) mode for the *gauche* isomer of $(CF_3)_4P_2$ has been assigned at 406 cm^{-1} in the first study (969) while in the second study (989) this band was assigned to a ν(PC) mode. The infrared and Raman spectra have also been reported for $(CF_3)_2PPF_2$ (ν(PP)=486 cm^{-1}) (991), while the infrared spectrum has been reported for H_3Si-$P(CF_3)_2$ (992), $(CH_3)_3MM'(CF_3)_2$ (M=Si,Ge;M'=P,As) (993), and $(CF_3)_2Te_2$ (994).

c. Partially Halogenated Alkyl Complexes

Infrared assignments and normal coordinate analyses have been reported for CH_2XSiH_3 (X=Cl,Br), CH_2ClGeH_3, and the corresponding deuterides (995). The vapor-state infrared spectra, low-temperature (-190°C), solid-state far-infrared spectra, and liquid-state Raman spectra have been reported for $RSiCl_2H$ (R=$CHCl_2$,CH_2Cl,CH_3) (996).

The assignment of the torsional modes for these compounds in the solid-state infrared spectra at 177, 172, and 175 cm^{-1}, respectively, gave corresponding barriers of 3.69, 3.00, and 2.25 kcal/mole, respectively. The $\nu(SiF)$ mode has been assigned for $(CH_2X)(CH_3)_2SiF$ (X=F, Cl,Br,I) (357). Complete infrared and Raman data and assignments, and normal coordinate analyses have been reported for $CH_nCl_{3-n}PO_3H_2$, $CH_nCl_{3-n}PO_3H^-$, and $CH_nCl_{3-n}PO_3^{2-}$ (n=1 (997,2 (998)).

Additional data have been reported for $C_6H_5HgCHXY$ (X,Y=Cl,Br,I) (295,999), $(CH_2X)_2Hg$ (X=Cl,Br,I) (1000), CHX_2HgBr (X=Cl,Br) (295), CHClBrHgBr (295), CH_2XSiH_3 (X=Cl,Br,I) (1001), and CH_2ClAsH_2 (1002)

2. Transition Elements

Vibrational studies for perhaloalkyl-transition metal derivatives are limited to those compounds with perfluoroalkyl groups. Based on the infrared band shapes of $CF_3Mn(CO)_5$, the $\nu_a(CF_3)$ and $\nu_s(CF_3)$ modes have been assigned at 1045 and 1063 cm^{-1}, respectively (1003, 1004). This order is the reverse of that found for non-transition metal derivatives included in Table 1.55 where the $\nu_a(CF_3)$ mode appears at a higher frequency than the $\nu_s(CF_3)$ mode. Support for the assignments made for $CF_3Mn(CO)_5$ has come from studies of other trifluoromethyl-transition metal derivatives (1005,1006) in which it has been observed that the lower-frequency band is often split, presumably because of a low molecular symmetry. This splitting is consistent with the assignment of the lower-frequency band to the degenerate mode.

The $\nu(CF_3)$ mode frequencies of $CF_3Mn(CO)_5$ (1003, 1004) are lower than those of CF_3I (1185 and 1076 cm^{-1}) and CF_3Br (1207 and 1087 cm^{-1}) (1007) with K(CF) force constants of 5.9 and 4.6 mdynes/A calculated for CF_3I and $CF_3Mn(CO)_5$, respectively (1004). The frequencies of the $\nu(CF_3)$ modes for $CF_3Fe(CO)_3C_5H_5$ (1068, 1042, 1015, and 985 cm^{-1}) and $CF_3Mo(CO)_3C_5H_5$ (1044, 1004, and 976 cm^{-1}) are also lower than for the trifluoroacyl derivatives $CF_3C(=O)Fe(CO)_2C_5H_5$ and $CF_3(=O)M(CO)_3C_5H_5$ (M=Mo,W) (1224±4, 1175±1, and 1126±1 cm^{-1}). These results have been interpreted as indicating that when the CF_3 group is bonded to a transition metal, electrons from the filled metal *d*-orbitals are donated through π-back bonding to antibonding σ* orbitals of the CF_3 group, resulting in a weakening of the C-F bonds and a lowering of the frequencies of the $\nu(CF_3)$ modes (1003,1004,1008, 1009).

In more recent studies, this interpretation has been questioned (1006,1010,1011). Rather, it has been stated that the value of the K(CF) force constant in $CF_3Mn(CO)_5$ can be understood in terms of σ-donation from an antibonding CF_3 orbital and energy stabilization due to charge effects on neighboring atoms with no need for Mn to CF_3 backbonding (1011).

The fact that the infrared $\nu_s(CF_3)$ mode frequency in several trifluoromethylplatinum complexes is dependent on the mass of the ligands coordinated to the platinum atom has been attributed to coupling between the ν(CF) and ν(PtC) modes (1012). The presence of coupling has also been offered as an explanation for the absence of reasonable intensity infrared and Raman bands that can be assigned to the ν(PtC) mode in these complexes although the corresponding mode of methylplatinum complexes shows moderate Raman intensity (1006). For the trifluoromethyl-platinum complexes, however, a very weak intensity band at approximately 300 cm^{-1} may have some Pt-C character (1006).

In several perfluoroalkyl carbonyl derivatives of rhenium, manganese, and iron, ν(CF), ν(CC), and $\delta(CF_3)$ modes have been assigned from 1350 to 1170, 920 to 800, and 730 to 715 cm^{-1}, respectively (1012). The CF_3CF_2 fragment of the *n*-C_3F_7 unit gave rise to five characteristic infrared bands in the ν(CF) mode region while the bands arising from the CF_2M fragment showed a much greater frequency variation. Additional vibrational data have been reported for *n*-C_3F_7 derivatives of manganese (1013) and cobalt (1014).

REFERENCES

1. S. Venkateswaren, *Indian J. Phys.*, *5*, 145 (1930).
2. A. J. Downs, in *Spectroscopic Methods in Organometallic Chemistry*, W. O. George, Ed., Butterworth, London, 1970, p. 1.
3. J. R. Allkins and P. J. Hendra, *Spectrochim. Acta*, *22*, 2075 (1966).
4. R. A. Kovar and G. L. Morgan, *Inorg. Chem.*, *8*, 1099 (1969).
5. G. E. Coates and A. J. Downs, *J. Chem. Soc.*, 3353 (1964).
6. J. R. Hall, L. A. Woodward, and E. A. V. Ebsworth, *Spectrochim. Acta*, *20*, 2049 (1964).
7. H. Siebert, *Z. Anorg. Allg. Chem.*, *273*, 161 (1953).

8. E. J. Rosenbaum and T. A. Ashford, *J. Chem. Phys.*, *7*, 554 (1939).
9. N. G. Pai, *Proc. Roy. Soc. (London)*, *A149*, 29 (1935).
10. A. M. W. Bakke, *J. Mol. Spectrosc.*, *41*, 1 (1772).
11. J. R. Durig and S. C. Brown, *J. Mol. Spectrosc.*, *45*, 338 (1973).
12. J. L. Bribes and R. Gaufrès, *J. Chim. Phys., Physio-chim. Biol.*, *67*, 1168 (1970).
13. J. Mink and L. Nemes, *J. Organometal. Chem.*, *28*, C39 (1971).
14. W. J. Lehmann, C. O. Wilson, Jr., and I. Shapiro, *J. Chem. Phys.*, *31*, 1071 (1959).
15. L. A. Woodward, J. R. Hall, R. N. Dixon, and N. Sheppard, *Spectrochim. Acta*, *15*, 249 (1959).
16. R. J. O'Brien and G. A. Ozin, *J. Chem. Soc. A*, 1136 (1971).
17. F. Oswald, *Z. Anal. Chem.*, *197*, 309 (1963).
18. H. Rojhantalab, J. W. Nibler, and C. J. Wilkins, *Spectrochim. Acta*, *32A*, 519 (1976).
19. E. J. Rosenbaum, D. J. Rubin, and C. R. Sandberg, *J. Chem. Phys.*, *8*, 366 (1940).
20. C. W. Young, J. S. Koehler, and D. S. McKinney, *J. Amer. Chem. Soc.*, *69*, 1410 (1947).
21. S. Sportouch, C. Lacoste, and R. Gaufrès, *J. Mol. Struct.*, *9*, 119 (1971).
22. E. R. Lippincott and M. C. Tobin, *J. Amer. Chem. Soc.*, *75*, 4141 (1953).
23. W. F. Edgell and C. H. Ward, *J. Amer. Chem. Soc.*, *77*, 6486 (1955).
24. G. A. Crowder, G. Gorin, F. H. Kruse, and D. W. Scott, *J. Mol. Spectrosc.*, *16*, 115 (1965).
25. H. Bürger and S. Biedermann, *Spectrochim. Acta*, *28A*, 2283 (1972).
26. L. A. Woodward, in *Raman Spectroscopy*, Vol. 2, H. A. Szymanski, Ed., Plenum, New York, 1970, p. 1.
27. G. Costa and G. De Alti, *Gazz. Chim. Ital.*, *87*, 1273 (1957).
28. L. Andrews, *J. Chem. Phys.*, *47*, 4834 (1967).
29. J. R. Allkins and P. J. Hendra, *Spectrochim. Acta*, *23A*, 1671 (1967).
30. W. J. Lehmann, C. O. Wilson, Jr., and I. Shapiro, *J. Chem. Phys.*, *28*, 777 (1958).
31. G. Bouquet and M. Bigorgne, *Spectrochim. Acta*, *23A*, 1231 (1967).
32. P. J. D. Park and P. J. Hendra, *Spectrochim. Acta*, *24A*, 2081 (1968).
33. S. J. W. Price and J. P. Richard, *Can. J. Chem.*, *50*, 966 (1972).

34. H. H. Eysel, H. Siebert, G. Groh, and H. J. Berthold, *Spectrochim. Acta, 26A,* 1595 (1970).
35. A. J. Downs, R. Schmutzler, and I. A. Steer, *Chem. Commun.,* 221 (1966).
36. A. J. Shortland and G. Wilkinson, *J. Chem. Soc., Dalton Trans.,* 872 (1973).
37. R. K. Sheline, *J. Chem. Phys., 18,* 602 (1950).
38. E. Weiss and E. A. C. Lucken, *J. Organometal. Chem., 2,* 197 (1964).
39. P. G. Perkins and D. H. Hall, *J. Chem. Soc. A,* 1207 (1966).
40. G. Herzberg, *Infrared and Raman Spectra of Polyatomic Molecules,* Van Nostrand, New York, 1945, p. 200.
41. J. Mink and B. Gellai, *J. Organometal. Chem., 66,* 1 (1974).
42. J. R. Durig, C. M. Player, Jr., J. Bragin, and Y. S. Li, *J. Chem. Phys., 55,* 2895 (1971).
43. A. L. Smith, *J. Chem. Phys., 21,* 1997 (1953).
44. J. W. Nibler and T. H. Cook, *J. Chem. Phys., 58,* 1596 (1973).
45. P. L. Goggin and L. A. Woodward, *Trans. Faraday Soc., 56,* 1591 (1960).
46. G. B. Deacon and J. H. S. Green, *Spectrochim. Acta, 24A,* 885 (1968).
47. V. B. Ramos and R. S. Tobias, *Inorg. Chem., 11,* 2451 (1972).
48. M. M. McGrady and R. S. Tobias, *Inorg. Chem., 3,* 1157 (1964).
49. M. G. Miles, J. H. Patterson, C. W. Hobbs, M. J. Hopper, J. Overend, and R. S. Tobias, *Inorg. Chem., 7,* 1721 (1968).
50. C. W. Hobbs and R. S. Tobias, *Inorg. Chem., 9,* 1998 (1970).
51. M. J. S. Gynane and I. J. Worrall, *Inorg. Nucl. Chem. Lett., 8,* 547 (1972).
52. C. E. Freidline and R. S. Tobias, *Inorg. Chem., 5,* 354 (1966).
53. H. Kriegsmann and S. Pischtschan, *Z. Anorg. Allg. Chem., 308,* 212 (1961).
54. H. Kriegsmann, H. Hoffmann, and S. Pischtschan, *Z. Anorg. Allg. Chem., 315,* 283 (1962).
55. P. Goggin, D. Phil. Thesis, Oxford, 1960, p. 156, cited by R. S. Tobias, *J. Organometal. Chem., 1,* 93 (1966).
56. A. J. Downs and I. A. Steer, *J. Organometal. Chem., 8,* P21 (1967).
57. H. Hope, *Acta Crystallogr., 20,* 610 (1966).

58. K. J. Wynne and J. W. George, *J. Amer. Chem. Soc.*, *91*, 1649 (1969).
59. M. T. Chen and J. W. George, *J. Amer. Chem. Soc.*, *90*, 4580 (1968).
60. J. A. Creighton, G. B. Deacon, and J. H. S. Green, *Aust. J. Chem.*, *20*, 583 (1967).
61. G. B. Deacon and R. A. Jones, *Aust. J. Chem.*, *15*, 554 (1962).
62. R. Baumgärtner, W. Sawodny, and J. Goubeau, *Z. Anorg. Allg. Chem.*, *333*, 171 (1964).
63. M. Arshad, A. Beg, and M. S. Siddiqui, *Can. J. Chem.*, *43*, 608 (1965).
64. M. Shindo and R. Okawara, *J. Organometal. Chem.*, *5*, 537, (1966).
65. G. W. Rice and R. S. Tobias, *Inorg. Chem.*, *15*, 489 (1976).
66. G. W. Rice and R. S. Tobias, *Inorg. Chem.*, *14*, 2402 (1975).
67. K. Hoffmann and E. Weiss, *J. Organometal. Chem.*, *37*, 1 (1972).
68. G. W. Rice and R. S. Tobias, *Chem. Commun.*, 994 (1975).
69. D. N. Shigorin, T. V. Talalaeva, K. A. Kocheskov, and A. N. Rodinov, *Izv. Akad. Nauk SSSR, Ser. Fiz.*, *22*, 1110 (1959).
70. D. N. Shigorin, *Spectrochim. Acta*, *14*, 198 (1959).
71. T. L. Brown and M. T. Rogers, *J. Amer. Chem. Soc.*, *79*, 1859 (1957).
72. R. West and W. Glaze, *J. Amer. Chem. Soc.*, *83*, 3580 (1961).
73. J. Goubeau and K. Walter, *Z. Anorg. Allg. Chem.*, *322*, 58, (1963).
74. A. I. Snow and R. E. Rundle, *Acta Crystallogr.*, *4*, 348 (1951).
75. E. Weiss, *J. Organometal. Chem.*, *2*, 314 (1964).
76. G. E. Coates, M. L. H. Green, and K. Wade, *Organometallic Compounds*, Vol. 1, 3rd ed., Methuen, London, 1967, p. 105.
77. P. Krohmer and J. Goubeau, *Z. Anorg. Allg. Chem.*, *369*, 238 (1969).
78. T. Holm, *Acta Chem. Scand.*, *19*, 1819 (1965).
79. G. E. Parris and E. C. Ashby, *J. Organometal. Chem.*, *72*, 1 (1974).
80. L. J. Allamandola and J. W. Nibler, *J. Amer. Chem. Soc.*, *98*, 2096 (1976).
81. E. C. Ashby and H. S. Prasad, *Inorg. Chem.*, *14*, 2869 (1975).
82. P. H. Lewis and R. E. Rundle, *J. Chem. Phys.*, *21*, 986 (1953).

83. T. Ogawa, K. Hirota, and T. Miyazawa, *Bull. Chem. Soc. Japan*, *38*, 1105 (1965).
84. T. Ogawa, *Spectrochim. Acta*, *24A*, 15 (1968).
85. A. P. Gray, *Can. J. Chem.*, *41*, 1511 (1963).
86. K. Mach, *J. Organometal. Chem.*, *2*, 410 (1964).
87. J. Yamamoto and C. A. Wilke, *Inorg. Chem.*, *10*, 1129 (1971).
88. E. Weiss and R. Wolfrum, *Chem. Ber.*, *101*, 35 (1968).
89. J. Krausse and G. Marx, *J. Organometal. Chem.*, *65*, 215 (1974).
90. H. Kriegsmann, *Z. Electrochem.*, *61*, 1088 (1957).
91. D. C. McKean, G. Davidson, and L. A. Woodward, *Spectrochim. Acta*, *26A*, 1815 (1970).
92. F. Glockling, S. R. Stobart, and J. J. Sweeney, *J. Chem. Soc., Dalton Trans.*, 2029 (1973).
93. W. Mowat, A. Shortland, G. Yagupsky, N. J. Hill, M. Yagupsky, and G. Wilkinson, *J. Chem. Soc., Dalton Trans.*, 533 (1972).
94. S. Moorhouse and G. Wilkinson, *J. Chem. Soc., Dalton Trans.*, 2187 (1974).
95. J. Z. Nyathi, J. M. Ressner, and J. D. Smith, *J. Organometal. Chem.*, *70*, 35 (1974).
96. S. Numata, H. Kurosawa, and R. Okawara, *J. Organometal. Chem.*, *70*, C21 (1974).
97. M. R. Collier, M. F. Lappert, and R. Pearce, *J. Chem. Soc., Dalton Trans.*, 445 (1973).
98. K. Mertis, D. H. Williamson, and G. Wilkinson, *J. Chem. Soc., Dalton Trans.*, 607 (1975).
99. H. Schmidbaur, J. Adlkofer, and W. Buchner, *Angew. Chem., Inter. Ed. Engl.*, *12*, 415 (1973).
100. H. Schmidbaur and R. Franke, *Angew. Chem., Inter. Ed. Engl.*, *12*, 416 (1973).
101. H. Schmidbaur and R. Franke, *Inorg. Chim. Acta*, *13*, 79 (1975).
102. J. Ellermann, H. Schössner, A. Haag, and H. Schödel, *J. Organometal. Chem.*, *65*, 33 (1974).
103. Y. Matsumura and R. Okawara, *Inorg. Nucl. Chem. Lett.*, *5*, 449 (1969).
104. Y. Matsumura and R. Okawara, *Inorg. Nucl. Chem. Lett.*, *7*, 113 (1970).
105. Y. Matsumura and R. Okawara, *J. Organometal. Chem.*, *25*, 439 (1970).
106. C. H. Van Dyke, E. W. Kifer, and G. A. Gibbon, *Inorg. Chem.*, *11*, 408 (1972).
107. W. Sawodny, *Z. Anorg. Allg. Chem.*, *368*, 284 (1969).
108. T. J. Barton and C. L. McIntosh, *Chem. Commun.*, 861 (1972).

109. S. E. Rudakova and Yu. A. Pentin, *Opt. Spectrosk.*, *18*, 339 (1965); *20*, 353 (1966); *21*, 240 (1966).
110. J. H. S. Green, *Spectrochim. Acta*, *24A*, 137 (1968).
111. S. Inoue and T. Yamada, *J. Organometal. Chem.*, *25*, 1 (1970).
112. J. Mink and Yu. A. Pentin, *J. Organometal. Chem.*, *23*, 293 (1970).
113. J. L. Bribes and R. Gaufrès, *Spectrochim. Acta*, *27A*, 2133 (1971).
114. J. Mink, L. Bursics, and G. Vegh, *J. Organometal. Chem.*, *34*, C4 (1972).
115. W. J. Lehmann, C. O. Wilson, Jr., and I. Shapiro, *J. Chem. Phys.*, *28*, 781 (1958).
116. J. Chouteau, G. Davidovics, F. d'Amato, and L. Savidan, *C. R. Acad. Sci., Paris*, *260*, 2759 (1965).
117. A. E. Borisov, N. V. Novikova, N. A. Chumaevskii, and E. B. Shkirtil, *Dokl. Akad. Nauk SSSR*, *173*, 855 (1967).
118. R. L. McKenney and H. H. Sisler, *Inorg. Chem.*, *6*, 1178 (1967).
119. J. A. Jackson and R. J. Nielson, *J. Mol. Spectrosc.*, *14*, 320 (1964).
120. M. I. Batuev, A. D. Petrov, V. A. Ponomarenko, and A. D. Matveeva, *Izv. Akad. Nauk SSSR, Otd. Khim. Nauk*, 1070 (1956).
121. L. A. Leites, Yu. P. Egorov, G. Ya. Zueva, and V. A. Ponomarenko, *Izv. Akad. Nauk SSSR, Otd. Khim. Nauk*, 2132 (1961).
122. W. R. Cullen, G. B. Deacon, and J. H. S. Green, *Can. J. Chem.*, *43*, 3193 (1965).
123. P. Taimsalu and J. L. Wood, *Trans. Faraday Soc.*, *59*, 1754 (1963).
124. C. R. Dillard and J. R. Lawson, *J. Opt. Soc. Amer.*, *50*, 1271 (1960).
125. C. A. Wilkie, *J. Organometal. Chem.*, *32*, 161 (1971).
126. J. Mink and Yu. A. Pentin, *Acta Chim. Acad. Sci. Hung.*, *65*, 273 (1970).
127. Y. Takashi, *J. Organometal. Chem.*, *8*, 225 (1967).
128. E. G. Hoffmann, *Z. Electrochem.*, *64*, 616 (1960).
129. H. Dietrich, *Acta Crystallogr.*, *16*, 681 (1963).
130. T. L. Brown, R. L. Gerteis, D. A. Bafus, and J. A. Ladd, *J. Amer. Chem. Soc.*, *86*, 2135 (1964); T. L. Brown, J. A. Ladd, and C. N. Newman, *J. Organometal Chem.*, *3*, 1 (1965).
131. T. L. Brown, in *Advances in Organometallic Chemistry*, Vol. III, F. G. A. Stone and R. West, Eds., Academic, New York, 1965, p. 374.

132. G. E. Coates and P. D. Roberts, *J. Chem. Soc. A*, 2651 (1968).
133. W. Strohmeier, K. Hümpfner, K. Miltenberger, and F. Seifert, *Z. Electrochem.*, *63*, 537 (1959).
134. N. Atam, H. Müller, and K. Dehnicke, *J. Organometal. Chem.*, *37*, 15 (1972).
135. O. Yamamoto, *Bull. Chem. Soc. Japan*, *35*, 619 (1962).
136. R. Wolfrum, G. Sauermann, and E. Weiss, *J. Organometal. Chem.*, *18*, 27 (1969).
137. R. L. Gerteis, R. E. Dickerson, and T. L. Brown, *Inorg. Chem.*, *3*, 872 (1964).
138. J. K. Brown and N. Sheppard, *Trans. Faraday Soc.*, *50*, 1164 (1954).
139. K. Ulbricht and V. Chvalovsky, *J. Organometal. Chem.*, *12*, 105 (1968).
140. R. A. Cummins, *Aust. J. Chem.*, *16*, 985 (1963).
141. R. A. Cummins, *Aust. J. Chem.*, *18*, 985 (1965).
142. R. J. Cross and F. Glockling, *J. Organometal. Chem.*, *3*, 146 (1965).
143. M. G. H. Wallbridge, *Spectrochim. Acta*, *20*, 1829 (1964).
144. E. G. Hoffmann, *Bull. Chem. Soc. France*, 1467 (1963).
145. P. J. Oliver and L. G. Stevens, *J. Inorg. Nucl. Chem.*, *24*, 953 (1962).
146. J. R. Sanders and E. C. Ashby, *J. Organometal. Chem.*, *25*, 277 (1970).
147. D. Seybold and K. Dehnicke, *J. Organometal. Chem.*, *11*, 1 (1968).
148. H.-U. Schwering, E. Jungk, and J. Weidlein, *J. Organometal. Chem.*, *91*, C4 (1975).
149. E. A. Chernyshev, Dissertation, 1963, cited by Yu. P. Egorov, *Izvest. Akad. Nauk SSSR, Otd. Khim. Nauk*, 1553 (1960).
150. E. Amberger and R. Hönigschmid-Grossich, *Chem. Ber.*, *98*, 3795 (1965).
151. R. A. Zingaro, R. E. McGlothlin, and R. M. Hedges, *Trans. Faraday Soc.*, *59*, 798 (1963).
152. J. Müller and W. Holzinger, *Angew. Chem., Inter. Ed. Engl.*, *14*, 760 (1975).
153. M. Weiner, G. Vogel, and R. West, *Inorg. Chem.*, *1*, 654 (1962).
154. W. M. Scovell, B. Y. Kimura, and T. G. Spiro, *J. Coord. Chem.*, *1*, 107 (1971).
155. G. E. Coates, P. D. Roberts, and A. J. Downs, *J. Chem. Soc. A*, 1085 (1967).
156. J. Mounier, *J. Organometal. Chem.*, *38*, 7 (1972).
157. W. Kruse, *J. Organometal. Chem.*, *42*, C39 (1972).
158. P. J. Davidson, M. F. Lappert, and R. Pearce, *J. Organometal. Chem.*, *57*, 269 (1973).

159. G. M. Whitesides, E. J. Panek, and E. R. Stedronsky, *J. Amer. Chem. Soc., 94,* 232 (1972).
160. R. Zerger, W. Rhine, and G. Stucky, *J. Amer. Chem. Soc., 96,* 6048 (1974).
161. K.-H. Thiele, S. Wilcke, and M. Ehrhardt, *J. Organometal. Chem., 14,* 13 (1968).
162. J. D. Roberts and V. C. Chambers, *J. Amer. Chem. Soc., 73,* 5030 (1951).
163. A.-F. Shihada and K. Dehnicke, *J. Organometal. Chem., 24,* 45 (1970).
164. A. H. Cowley and T. A. Furtsch, *J. Amer. Chem. Soc., 91,* 39 (1969).
165. J. Müller, K. Margiolis, and K. Dehnicke, *J. Organometal. Chem., 46,* 219 (1972).
166. D. A. Sanders, P. A. Scherr, and J. P. Oliver, *Inorg. Chem., 15,* 861 (1976).
167. K. Margiolis and K. Dehnicke, *J. Organometal. Chem., 33,* 147 (1971).
168. B. Busch and K. Dehnicke, *J. Organometal. Chem., 67,* 237 (1974).
169. A. H. Cowley, J. L. Mills, T. M. Loehr, and T. V. Long, II, *J. Amer. Chem. Soc., 93,* 2150 (1971).
170. W. H. Green, A. B. Harvey, and J. A. Greenhouse, *J. Chem. Phys., 54,* 850 (1971).
171. R. W. Mitchell, L. J. Kuzma, R. J. Pirkle, and J. A. Merritt, *Spectrochim. Acta, 25A,* 819 (1969).
172. S. Chan, H. Goldwhite, H. Keyzer, and R. Tang, *Spectrochim. Acta, 26A,* 249 (1970).
173. J. Laane, *Spectrochim. Acta, 26A,* 517 (1970).
174. J. Laane and R. C. Lord, *J. Chem. Phys., 48,* 1508 (1968).
175. A. B. Harvey, J. R. Durig, and A. C. Morrissey, *J. Chem. Phys., 50,* 4949 (1969).
176. M. G. Pettit, J. S. Gibson, and D. O. Harris, *J. Chem. Phys., 53,* 3408 (1970).
177. J. Laane, *J. Chem. Phys., 50,* 776 (1969).
178. J. R. Durig and J. N. Willis, Jr., *J. Mol. Spectrosc., 32,* 320 (1969).
179. J. Laane, *J. Chem. Phys., 50,* 1946 (1969).
180. J. R. Durig and J. N. Willis, Jr., *J. Chem. Phys., 52,* 6108 (1970).
181. J. R. Durig, Y. S. Li, and L. A. Carreira, *J. Chem. Phys., 58,* 2393 (1973).
182. W. N. Greenwood and J. C. Wright, *J. Chem. Soc.,* 448 (1965).
183. G. D. Oshesky and F. F. Bentley, *J. Amer. Chem. Soc., 79,* 2057 (1957).

184. G. Fogarasi, F. Török, and V. M. Vdovin, *Acta Chim. Acad. Sci. Hung.*, *54*, 277 (1967).
185. E. C. Juenge and S. Gray, *J. Organometal. Chem.*, *10*, 465 (1967).
186. D. Vedal, O. H. Ellestad, P. Klaboe, and G. Hagen, *Spectrochim. Acta*, *31A*, 355 (1975).
187. L. Verdonck and Z. Eeckhaut, *Spectrochim. Acta*, *28A*, 433 (1972).
188. W. Brüser, K.-H. Thield, P. Zdunneck, and F. Brune, *J. Organometal. Chem.*, *32*, 335 (1971).
189. I. W. Bassi, G. Allegra, R. Scordamaglia, and G. Chioccola, *J. Amer. Chem. Soc.*, *93*, 3787 (1971).
190. G. R. Davies, J. A. J. Jarvis, and B. T. Kilbourn, *Chem. Commun.*, 1511 (1971).
191. K.-H. Thiele, A. Russek, R. Opitz, B. Mohai, and W. Brüsser, *Z. Anorg. Allg. Chem.*, *412*, 11 (1975).
192. E. Köhler, W. Brüser, and K.-H. Thiele, *J. Organometal. Chem.*, *76*, 235 (1974).
193. W. L. Hase and J. W. Simons, *J. Organometal. Chem.*, *32*, 47 (1971).
194. I. V. Shevchenko, I. F. Kovalev, V. S. Dernova, M. G. Voronkov, Yu. I. Khudobin, N. A. Andreeva, and N. P. Kharitonov, *Izv. Akad. Nauk SSSR, Ser. Khim.*, 98 (1972).
195. A. L. Smith, *Spectrochim. Acta*, *24A*, 695 (1968).
196. M. Hayashi, K. Ohno, and H. Murata, *Bull. Chem. Soc. Japan*, *45*, 298 (1972).
197. M. Horák, B. Schneider, and V. Bažant, *Chem. Listy*, *52*, 2048 (1948).
198. W. R. Cullen, G. B. Deacon, and J. H. S. Green, *Can. J. Chem.*, *44*, 717 (1966).
199. J. Goubeau and D. Langhardt, *Z. Anorg. Allg. Chem.*, *338*, 163 (1965).
200. J. H. S. Green, W. Kynaston, and G. A. Rodley, *Spectrochim. Acta*, *24A*, 853 (1968).
201. W. J. Lehmann, C. O. Wilson, Jr., and I. Shapiro, *J. Chem. Phys.*, *32*, 1088 (1960).
202. W. J. Lehmann, C. O. Wilson, Jr., and I. Shapiro, *J. Chem. Phys.*, *32*, 1786 (1960).
203. W. J. Lehmann, C. O. Wilson, Jr., and I. Shapiro, *J. Chem. Phys.*, *33*, 590 (1960).
204. W. J. Lehmann, C. O. Wilson, Jr., and I. Shapiro, *J. Chem. Phys.*, *34*, 476 (1961).
205. W. J. Lehmann, C. O. Wilson, Jr., and I. Shapiro, *J. Chem. Phys.*, *34*, 783 (1961).
206. J. H. Carpenter, W. J. Jones, R. W. Jotham, and L. H. Long, *Chem. Commun.*, 881 (1968).
207. J. H. Carpenter, W. J. Jones, R. W. Jotham, and L. H. Long, *Spectrochim. Acta*, *26A*, 1199 (1970).

208. W. J. Lehmann and I. Shapiro, *Spectrochim. Acta, 17,* 396 (1961).
209. H. C. Brown and S. K. Gupta, *J. Organometal. Chem., 32,* C1 (1971).
210. H. C. Brown and G. J. Klender, *Inorg. Chem., 1,* 204 (1962).
211. E. F. Knights and H. C. Brown, *J. Amer. Chem. Soc., 90,* 5280 (1968).
212. H. C. Brown and E. Negishi, *J. Amer. Chem. Soc., 93,* 6682 (1971).
213. H. C. Brown and E. Negishi, *J. Organometal. Chem., 26,* C67 (1971).
214. E. Negishi, P. L. Burke, and H. C. Brown, *J. Amer. Chem. Soc., 94,* 7431 (1972).
215. E. Breuer and H. C. Brown, *J. Amer. Chem. Soc., 91,* 4164 (1969).
216. E. Negishi, J.-J. Katz, and H. C. Brown, *J. Amer. Chem. Soc., 94,* 4026 (1972).
217. H. C. Brown, E. Negishi, and J.-J. Katz, *J. Amer. Chem. Soc., 97,* 2791 (1975).
218. E. G. Hoffmann and G. Schomburg, *Z. Electrochem., 61,* 1101 (1957).
219. G. Schomburg and E. G. Hoffmann, *Z. Electrochem., 61,* 1110 (1957).
220. A. Storr and V. G. Wiebe, *Can. J. Chem., 47,* 673 (1969).
221. A. B. Harvey and M. K. Wilson, *J. Chem. Phys., 44,* 3535 (1966).
222. J. A. Lannon and E. R. Nixon, *Spectrochim. Acta, 23A,* 2713 (1967).
223. D. F. Ball, P. L. Goggin, D. C. McKean, and L. A. Woodward, *Spectrochim. Acta, 16,* 1358 (1960).
224. D. F. Ball, T. Carter, D. C. McKean, and L. A. Woodward, *Spectrochim. Acta, 20,* 1721 (1964).
225. M. Randic, Ph. D. Thesis, Cambridge Univ., 1958, cited by D. F. Ball, P. L. Gogin, D. C. McKean, and L. A. Woodward, *Spectrochim. Acta, 16,* 1358 (1960).
226. I. F. Kovalev, *Opt. Spectrosk., 8,* 315 (1960).
227. H. Kimmel and C. R. Dillard, *Spectrochim. Acta, 24A,* 909 (1968).
228. H. C. Beachell and B. Katlafsky, *J. Chem. Phys., 27,* 182 (1957).
229. M. Baudler and H. Grundlach, *Naturwissenschaften, 42,* 152 (1955).
230. I. V. Shevchenko, I. F. Kovalev, M. G. Voronkov, Yu. I. Khudobin, and N. P. Kharitonov, *Izv. Akad. Nauk SSSR, Otd. Khim.,* 57 (1969).
231. I. V. Shevchenko, I. F. Kovalev, M. G. Voronkov, Yu. I. Khudobin, and N. P. Kharitonov, *Zh. Fiz. Khim., 44,* 1898 (1970).

232. V. Galasso, A. Bigotto, and G. De Alti, *Z. Phys. Chem. (Frankfurt)*, *50*, 38 (1966).
233. J. R. Durig and C. W. Hawley, *J. Phys. Chem.*, *75*, 3993 (1971).
234. D. M. Cameron, W. C. Sears, and H. H. Nielsen, *J. Chem. Phys.*, *7*, 994 (1939).
235. A. B. Harvey and M. K. Wilson, *Inorg. Nucl. Chem. Lett.*, *1*, 101 (1965).
236. A. B. Harvey and M. K. Wilson, *J. Chem. Phys.*, *45*, 678 (1966).
237. C. W. Sink and A. B. Harvey, *J. Chem. Phys.*, *57*, 4434 (1972).
238. T. Shimanouchi, I. Nakagawa, J. Hiraishi, and M Ishii, *J. Mol. Spectrosc.*, *19*, 78 (1966).
239. R. J. Nielsen and J. D. Walker, *Spectrochim. Acta*, *21*, 1163 (1965).
240. E. Mayer, *J. Mol. Struct.*, *26*, 347 (1975).
241. J. L. Duncan and I. M. Mills, *Spectrochim. Acta*, *20*, 523 (1964).
242. R. E. Wilde, *J. Mol. Spectrosc.*, *8*, 427 (1962).
243. L. P. Lindeman and M. K. Wilson, *Z. Phys. Chem. (Frankfurt)*, *9*, 29 (1956).
244. J. E. Griffiths, *J. Chem. Phys.*, *38*, 2879 (1963).
245. D. F. Van de Vondel and G. P. Van der Kelen, *Bull. Soc. Chim. Belges*, *74*, 467 (1965).
246. V. A. Ponomarenko, G. Ya. Zueva, and N. S. Andreev, *Izv. Akad. Nauk SSSR*, 1758 (1961).
247. R. Mathis, J. Satge, and F. Mathis, *Spectrochim. Acta*, *18*, 1463 (1962).
248. I. W. Levin and H. Ziffer, *J. Chem. Phys.*, *43*, 4023 (1965).
249. C. R. Dillard and L. May, *J. Mol. Spectrosc.*, *14*, 250 (1964).
250. H. Kriegsmann and K. Ulbricht, *Z. Anorg. Allg. Chem.*, *328*, 90 (1964).
251. E. Amberger, *Angew. Chem.*, *72*, 494 (1960).
252. K. M. Mackay and R. Watt, *Spectrochim. Acta*, *23A*, 2761 (1967).
253. V. A. Ponomarenko, Yu. P. Egorov, and G. Ya. Vzenkova, *Izv. Akad. Nauk SSSR, Otd. Chim.*, 54 (1958).
254. D. E. Webster, *J. Chem. Soc.*, 5132 (1960).
255. V. A. Ponomarenko, G. Ya. Vzenkova, and Yu. P. Egorov, *Dokl. Akad. Nauk SSSR*, *122*, 405 (1958).
256. M. L. Maddox, N. Flitcroft, and H. D. Kaesz, *J. Organometal. Chem.*, *4*, 50 (1965).
257. R. West and E. G. Rochow, *J. Org. Chem.*, *18*, 303 (1953).

258. M. I. Batuev, A. D. Petrov, V. A. Ponomarenko, and A. D. Matveeva, *Izv. Akad. Nauk SSSR, Otd. Khim.*, 1243 (1956).
259. W. P. Newmann and K. Kühlein, *Angew. Chem.*, *77*, 808 (1965).
260. H. Westermark, *Acta Chem. Scand.*, *9*, 947 (1955).
261. R. Mathis, M. Constant, J. Satgé, and F. Mathis, *Spectrochim. Acta*, *20*, 515 (1964).
262. A. L. Smith and N. C. Angelotti, *Spectrochim. Acta*, *15*, 412 (1959).
263. H. W. Thompson, *Spectrochim. Acta*, *16*, 238 (1968).
264. V. A. Ponomarenko and Yu. P. Egorov, *Izv. Akad. Nauk SSSR, Otd. Khim.*, 1133 (1960).
265. C. J. Attridge, *J. Organometal. Chem.*, *13*, 259 (1968).
266. A. N. Egorochkin, S. Ya. Khorshev, N. S. Ostasheva, J. Satgé, P. Riviere, J. Barrau, and M. Massol, *J. Organometal. Chem.*, *76*, 29 (1974).
267. P. E. Potter, L. Pratt, and G. Wilkinson, *J. Chem. Soc.*, 524 (1964).
268. Y. Kawasaki, K. Kawakami, and T. Tanaka, *Bull. Chem. Soc. Japan*, *38*, 1102 (1965).
269. A. K. Sawyer, J. E. Brown, and E. L. Hanson, *J. Organometal. Chem.*, *3*, 464 (1965).
270. K. Kawakami, T. Saito, and R. Okawara, *J. Organometal. Chem.*, *8*, 377 (1967).
271. A. K. Sawyer and J. E. Brown, *J. Organometal. Chem.*, *5*, 438 (1966).
272. R. Gupta and B. Majee, *J. Organometal. Chem.*, *36*, 71 (1972).
273. V. Galasso, G. De Alti, and A. Bigotto, *Z. Phys. Chem. (Frankfurt)*, *57*, 132 (1968).
274. A. E. Borisov, N. V. Novikova, N. A. Chumaevskii, and E. B. Shkirtil, *Ukr. Fiz. Zh.*, *13*, 75 (1968).
275. A. L. Smith, *Spectrochim. Acta*, *16*, 87 (1960).
276. R. A. Chittenden and L. C. Thomas, *Spectrochim. Acta*, *21*, 861 (1965).
277. R. G. Goel, E. Maslowsky, Jr., and C. V. Senoff, *Inorg. Chem.*, *10*, 2572 (1971).
278. L. Y. Tan and G. C. Pimentel, *J. Chem. Phys.*, *48*, 5202 (1968).
279. R. M. Salinger and H. S. Mosher, *J. Amer. Chem. Soc.*, *86*, 1782 (1964).
280. E. C. Ashby and S. Yu, *J. Organometal. Chem.*, *29*, 339 (1971).
281. J. Kress and A. Novak, *J. Organometal. Chem.*, *99*, 199 (1975).
282. J. Kress and A. Novak, *J. Organometal. Chem.*, *99*, 23 (1975).

283. P. T. Moseley and H. M. M. Shearer, *J. Chem. Soc., Dalton Trans.*, 64 (1973).
284. D. F. Evans and I. Wharf, *J. Chem. Soc. A*, 783 (1968).
285. K. Cavanagh and D. F. Evans, *J. Chem. Soc. A*, 2890 (1969).
286. P. L. Goggin and L. A. Woodward, *Trans. Faraday Soc.*, *62*, 1432 (1966).
287. A. Büchler, W. Klemperer, and A. G. Emslie, *J. Chem. Phys.*, *36*, 2499 (1962).
288. W. Klemperer, *J. Chem. Phys.*, *25*, 1066 (1956).
289. D. Breitinger, A. Zober, and M. Newbauer, *J. Organometal. Chem.*, *30*, C49 (1971).
290. W. Klemperer, *Z. Electrochem. Soc.*, *110*, 1023 (1963).
291. W. Klemperer and L. Lindeman, *J. Chem. Phys.*, *25*, 397 (1956).
292. Z. Meić and M. Randić, *Chem. Commun.*, 1608 (1968).
293. Z. Meić and M. Randić, *J. Chem. Soc. Faraday Trans. II*, 444 (1972).
294. G. E. Coates and D. Ridley, *J. Chem. Soc.*, 166 (1964).
295. J. H. S. Green, *Spectrochim. Acta*, *24A*, 863 (1968).
296. Z. Meić and M. Randić, *Trans. Faraday Soc.*, *64*, 1438 (1968).
297. Z. Meić, *J. Mol. Struct.*, *23*, 131 (1974).
298. Z. Meić and M. Randić, *J. Mol. Spectrosc.*, *39*, 39 (1971).
299. P. R. Reed, Jr. and R. W. Lovejoy, *Spectrochim. Acta*, *26A*, 1087 (1970).
300 W. J. Jones and N. Sheppard, *Proc. Roy. Soc., Ser. A*, *304*, 135 (1968).
301. H. J. Becher, *Z. Anorg. Allg. Chem.*, *291*, 151 (1957).
302. H. J. Becher, *Z. Anorg. Allg. Chem.*, *271*, 243 (1953).
303. W. Schabacher and J. Goubeau, *Z. Anorg. Allg. Chem.*, *294*, 183 (1958).
304. W. Haubold and J. Weidlein, *Z. Anorg. Allg. Chem.*, *420*, 251 (1976).
305. F. Joy, M. F. Lappert, and B. Prokai, *J. Organometal. Chem.*, *5*, 506 (1966).
306. J. Weidlein and V. Krieg, *J. Organometal. Chem.*, *11*, 9 (1968).
307. A. W. Laubengayer and G. F. Lengnick, *Inorg. Chem.*, *5*, 503 (1966).
308. J. Weidlein, *personal communication*.
309. G. Gundersen, T. Haugen, and A. Haaland, *J. Organometal. Chem.*, *54*, 77 (1973).
310. V. Krieg and J. Weidlein, *J. Organometal. Chem.*, *21*, 281 (1970).
311. H. Schmidbaur, H.-F. Klein, and K. Eiglmeier, *Angew. Chem.*, *79*, 821 (1967).

312. H. Schmidbaur, J. Weidlein, H. F. Klein, and K. Eiglmeier, *Chem. Ber., 101,* 2268 (1968).
313. H. Schmidbaur and H.-F. Klein, *Chem. Ber., 101,* 2278 (1968).
314. H. C. Clark and A. L. Pickard, *J. Organometal. Chem., 8,* 427 (1967).
315. T. Maeda, H. Tada, K. Yasuda, and R. Okawara, *J. Organometal. Chem., 27,* 13 (1971).
316. T. Ehemann and K. Dehnicke, *J. Organometal. Chem., 71,* 191 (1974).
317. R. P. Bell and H. C. Longuet-Huggins, *Proc. Roy. Soc., Ser. A, 183,* 357 (1945).
318. H. D. Hausen, K. Mertz, E. Veigel, and J. Weidlein, *Z. Anorg. Allg. Chem., 410,* 156 (1974).
319. J. S. Poland and D. G. Tuck, *J. Organometal. Chem., 42,* 315 (1972).
320. M. J. S. Gynane, L. G. Waterworth, and I. J. Worrall, *J. Organometal. Chem., 43,* 257 (1972).
321. H. C. Clark and A. L. Pickard, *J. Organometal. Chem., 13,* 61 (1968).
322. W. Klemperer, *J. Chem. Phys., 24,* 353 (1956).
323. H. Gerding and E. Smit, *Z. Phys. Chem., B50,* 171 (1941).
324. J. Weidlein, *J. Organometal. Chem., 17,* 213 (1969).
325. T. Onishi and T. Shimanouchi, *Spectrochim. Acta, 20,* 325 (1964).
326. M. P. Groenewege, *Z. Phys. Chem. (Frankfurt), 18,* 147 (1958).
327. I. R. Beattie, T. Gilson, and G. A. Ozin, *J. Chem. Soc. A,* 813 (1968).
328. I. R. Beattie and J. R. Horder, *J. Chem. Soc. A,* 2655 (1969).
329. I. R. Beattie, T. Gilson, and P. Cocking, *J. Chem. Soc. A,* 702 (1967).
330. E. Kinsella, J. Chadwick, and J. Coward, *J. Chem. Soc. A,* 969 (1968).
331. B. Armer and H. Schmidbaur, *Chem. Ber., 100,* 1521 (1967).
332. W. Lind and I. J. Worrall, *J. Organometal. Chem., 36,* 35 (1972).
333. M. Wilkinson and I. J. Worrall, *J. Organometal. Chem., 93,* 39 (1975).
334. N. N. Greenwood, D. J. Prince, and B. P. Straughan, *J. Chem. Soc. A,* 1694 (1968).
335. W. Lind and I. J. Worrall, *J. Organometal. Chem., 40,* 35 (1972).
336. M. J. S. Gynane and I. J. Worrall, *J. Organometal. Chem., 40,* C59 (1972).

337. R. A. Kovar, H. Derr, D. Brandau, and J. O. Callaway, *Inorg. Chem.*, *14*, 2809 (1975).
338. K. Kawai, I. Kanesaka, and F. Ichimura, *Spectrochim. Acta*, *26A*, 593 (1970).
339. I. Kanesaka, M. Shindo, and K. Kawai, *Spectrochim. Acta*, *26A*, 2345 (1970).
340. M. Boleslawski, S. Pasynkiewicz, and M. Harasimowicz, *J. Organometal. Chem.*, *78*, 61 (1974).
341. I. L. Wilson and K. Dehnicke, *J. Organometal. Chem.*, *67*, 229 (1974).
342. I. R. Beattie and P. A. Cocking, *J. Chem. Soc.*, 3860 (1965).
343. M. Tanaka, H. Kurosawa, and R. Okawara, *Inorg. Nucl. Chem. Lett.*, *3*, 565 (1967).
344. H. Kurosawa, K. Yasuda, and R. Okawara, *Bull. Chem. Soc. Japan*, *40*, 861 (1967).
345. H. M. Powell and D. M. Crowfoot, *Z. Kristallogr.*, *87*, 370 (1934).
346. W. Beck and E. Schuierer, *J. Organometal. Chem.*, *3*, 55 (1965).
347. F. Klanberg and E. L. Muetterties, *Inorg. Chem.*, *7*, 155 (1968).
348. K. Licht, C. Peuker, and C. Dathe, *Z. Anorg. Allg. Chem.*, *380*, 293 (1971).
349. H. Kriegsmann, *Z. Anorg. Allg. Chem.*, *294*, 113 (1958).
350. H. Bürger, *Spectrochim. Acta*, *24A*, 2015 (1968).
351. H. Kriegsmann, *Z. Electrochem.*, *62*, 1033 (1958).
352. R. L. Collins and J. R. Nielsen, *J. Chem. Phys.*, *23*, 351 (1955).
353. J. Goubeau, H. Siebert, and M. Winterwerb, *Z. Anorg. Allg. Chem.*, *259*, 240 (1949).
354. T. Shimanouchi, I. Tsuchiya, and Y. Mikawa, *J. Chem. Phys.*, *18*, 1306 (1950).
355. J. Goubeau and H. Sommer, *Z. Anorg. Allg. Chem.*, *289*, 1 (1957).
356. H. Murata and S. Hayashi, *J. Chem. Phys.*, *19*, 1217 (1951).
357. K. Licht, P. Koehler, and H. Kriegsmann, *Z. Anorg. Allg. Chem.*, *415*, 31 (1975).
358. T. Tanaka and S. Murakami, *Bull. Chem. Soc. Japan*, *38*, 1465 (1965).
359. Yu. I. Ruskin and M. G. Vorinkov, *Collect. Czech. Chem. Commun.*, *24*, 3816 (1959).
360. H. Murata, *J. Chem. Soc. Japan, Pure Chem. Sec.*, *73*, 465 (1952).
361. S. E. Rudakova and Yu. A. Pentin, *Opt. Spektrosk.*, *18*, 592 (1965).
362. C. Peuker and K. Licht, *Z. Phys. Chem. (Leipzig)*, *248*, 103 (1971).

363. J. R. Durig, K. K. Lau, J. B. Turner, and J. Bragin, *J. Mol. Spectrosc.*, *31*, 419 (1969).
364. D. F. Van de Vondel and G. P. Van der Kelen, *Bull. Soc. Chim. Belges*, *74*, 453 (1965).
365. J. E. Griffiths, *Spectrochim. Acta*, *20*, 1335 (1964).
366. J. R. Aronson and J. R. Durig, *Spectrochim. Acta*, *20*, 219 (1964).
367. D. F. Van de Vondel, G. P. Van der Kelen, and G. Van Hooydonk, *J. Organometal. Chem.*, *23*, 431 (1970).
368. J. R. Durig, C. F. Jumper, and J. N. Willis, Jr., *J. Mol. Spectrosc.*, *37*, 260 (1971).
369. J. W. Anderson, G. K. Barker, A. J. F. Clark, J. E. Drake, and R. T. Hemmings, *Spectrochim. Acta*, *30A*, 1081 (1974).
370. J. W. Anderson, G. K. Barker, J. E. Drake, and R. T. Hemmings, *Can. J. Chem.*, *49*, 2931 (1971).
371. K. Licht and P. Koehler, *Z. Anorg. Allg. Chem.*, *383*, 174 (1971).
372. V. F. Mironov and A. L. Kravchenko, *Izv. Akad. Nauk SSSR, Ser. Khim.*, 1026 (1965).
373. J. R. Durig, P. J. Cooper, and Y. S. Li, *J. Mol. Spectrosc.*, *57*, 169 (1975).
374. C. Peuker, K. Licht, and H. Kriegsmann, *Z. Phys. Chem. (Leipzig)*, *244*, 61 (1971).
375. K. M. Mackay and R. Watt, *J. Organometal. Chem.*, *6*, 336 (1966).
376. J. R. Durig and K. L. Hellams, *Appl. Spectrosc.*, *22*, 153 (1968).
377. J. R. Durig and C. W. Hawley, *J. Chem. Phys.*, *58*, 237 (1973).
378. J. R. Durig, S. M. Craven, and J. Bragin, *J. Chem. Phys.*, *51*, 5663 (1969).
379. E. O. Schlemper and W. C. Hamilton, *Inorg. Chem.*, *5*, 995 (1966).
380. H. C. Clark, R. J. O'Brien, and J. Trotter, *J. Chem. Soc.*, 2332 (1964).
381. L. E. Levchuk, J. R. Sams, and F. Aubke, *Inorg. Chem.*, *11*, 43 (1972).
382. C. W. Hobbs and R. S. Tobias, *Inorg. Chem.*, *9*, 1037 (1970).
383. H. C. Clark and R. G. Goel, *Inorg. Chem.*, *4*, 1428 (1965).
384. H. C. Clark and R. G. Goel, *J. Organometal. Chem.*, *7*, 263 (1967).
385. M. Goldstein and W. D. Unsworth, *J. Chem. Soc. A*, 2121 (1971).
386. K. Licht, H. Geissler, P. Koehler, and K. Moueller, *Z. Chem.*, *11*, 272 (1971).

387. K. Licht, H. Geissler, P. Koehler, K. Hottmann, H. Schnorr, and H. Kriegsmann, *Z. Anorg. Allg. Chem.*, *385*, 271 (1971).
388. R. J. H. Clark, A. G. Davies, and R. J. Puddephatt, *J. Chem. Soc. A*, 1828 (1968).
389. H. Kriegsmann, C. Peufer, R. Heess, and H. Geissler, *Z. Naturforsch. A*, *24*, 778 (1969).
390. H. Geissler and H. Kriegsmann, *J. Organometal. Chem.*, *11*, 85 (1968).
391. D. H. Lohmann, *J. Organometal. Chem.*, *4*, 382 (1965).
392. L. Verdonck and G. P. Van der Kelen, *J. Organometal. Chem.*, *40*, 135 (1972).
393. H. N. Farrer, M. M. McGrady, and R. S. Tobias, *J. Amer. Chem. Soc.*, *87*, 5019 (1965).
394. M. K. Dass, J. Buckle, and P. G. Harrison, *Inorg. Chim. Acta*, *6*, 17 (1972).
395. I. R. Beattie, F. C. Stokes, and L. E. Alexander, *J. Chem. Soc., Dalton Trans.*, 465 (1973).
396. C. J. Wilkins and H. M. Haendler, *J. Chem. Soc.*, 3174 (1965).
397. J. P. Clark and C. J. Wilkins, *J. Chem. Soc. A*, 871 (1966).
398. R. C. Poller, *The Chemistry of Organotin Compounds*, Logos Press, London, 1970, p. 227.
399. T. Tanaka, *Organometal. Chem. Rev. A*, *5*, 1 (1970).
400. R. J. H. Clark, A. G. Davies, and R. J. Puddephatt, *J. Amer. Chem. Soc.*, *90*, 6923 (1968).
401. M. Mammi, V. Busetti, and A. Del Pra, *Inorg. Chim. Acta*, *1*, 419 (1967).
402. J. R. Durig and J. E. Saunders, *J. Mol. Struct.*, *27*, 403 (1975).
403. F. Steel, K. Rudolph, and R. Budenz, *Z. Anorg. Allg. Chem.*, *341*, 196 (1965).
404. H. W. Schiller and R. W. Rudolph, *Inorg. Chem.*, *11*, 187 (1972).
405. J. R. Durig, F. Block, and I. W. Levin, *Spectrochim. Acta*, *21*, 1105 (1965).
406. J. E. Griffiths, *Spectrochim. Acta*, *21*, 1135 (1965).
407. L. Maier, *Helv. Chim. Acta*, *46*, 2026 (1963).
408. C. Christol and H. Christol, *J. Chim. Phys.*, *62*, 246 (1965).
409. R. R. Holmes and M. Fild, *Spectrochim. Acta*, *27A*, 1525 (1971).
410. R. R. Holmes, G. Ting-Kuo Fey, and R. H. Larkin, *Spectrochim. Acta*, *29A*, 665 (1973).
411. R. R. Holmes and M. Fild, *Spectrochim. Acta*, *27A*, 1537 (1971).
412. G. P. Van der Kelen and M. A. Herman, *Bull. Soc. Chim. Belges*, *65*, 350 (1956).

413. D. M. Revitt and D. B. Sowerby, *Spectrochim. Acta, 26A,* 1581 (1970).
414. J. R. Durig, C. F. Jumper, and J. N. Willis, Jr., *Appl. Spectrosc., 25,* 218 (1971).
415. H. F. Shurvell, M. R. Gold, and A. R. Norris, *Can. J. Chem., 50,* 2691 (1972).
416. E. G. Claeys and G. P. Van der Kelen, *Spectrochim. Acta, 22,* 2103 (1966).
417. I. R. Beattie, K. Livingston, and T. Gilson, *J. Chem. Soc. A,* 1 (1968).
418. H. Schmidbaur, K. H. Mitschke, and J. Weidlein, *Angew. Chem., Inter. Ed. Engl., 11,* 144 (1972).
419. J. Goubeau and R. Baumgärtner, *Z. Electrochem., 64,* 598 (1960).
420. C. Woods and G. G. Long, *J. Mol. Spectrosc., 40,* 435 (1971).
421. M. H. O'Brien, G. O. Doak, and G. G. Long, *Inorg. Chim. Acta, 1,* 34 (1967).
422. A. Schmidt, *Chem. Ber., 102,* 380 (1969).
423. A. Schmidt, *Chem. Ber., 103,* 3928 (1970).
424. A. J. Downs and R. Schmutzler, *Spectrochim. Acta, 21,* 1927 (1965).
425. J. E. Griffiths, *J. Chem. Phys., 49,* 1307 (1968).
426. A. J. Downs and R. Schmutzler, *Spectrochim. Acta, 23A,* 681 (1967).
427. N. Nishii, Y. Matsumura, and R. Okawara, *J. Organometal. Chem., 30,* 59 (1971).
428. B. A. Nevett and A. Perry, *J. Organometal. Chem., 71,* 399 (1974).
429. C. Woods and G. G. Long, *J. Mol. Spectrosc., 38,* 387 (1971).
430. G. G. Long, G. O. Doak, and L. D. Freedman, *J. Amer. Chem. Soc., 86,* 209 (1964).
431. H. Schmidbaur, J. Weidlein, and K.-H. Mitschke, *Chem. Ber., 102,* 4136 (1969).
432. H. G. Nadler and K. Dehnicke, *J. Organometal. Chem., 90,* 291 (1975).
433. L. Verdonck and G. P. Van der Kelen, *Spectrochim. Acta, 29A,* 1675 (1973).
434. L. Verdonck and G. P. Van der Kelen, *Spectrochim. Acta, 31A,* 1707 (1975).
435. R. R. Holmes and M. Fild, *Inorg. Chem., 10,* 1109 (1971).
436. R. R. Holmes, G. Ting-Kuo Fey, and R. H. Larkin, *Inorg. Chem., 12,* 2225 (1973).
437. K. Dehnicke and H.-G. Nadler, *Z. Anorg. Allg. Chem., 418,* 229 (1975).
438. J. R. Durig and J. M. Casper, *J. Chem. Phys., 75,* 1956 (1971).

439. R. A. Nyquist, *Appl. Spectrosc.*, *22*, 452 (1968).
440. J. R. Durig, D. W. Wertz, B. R. Mitchell, F. Block, and J. M. Greene, *J. Phys. Chem.*, *71*, 3815 (1967).
441. J. R. Durig, B. R. Mitchell, J. S. DiYorio, and F Block, *J. Phys. Chem.*, *70*, 3190 (1966).
442. D. Köttgen, H. Stoll, R. Pantzer, A. Lentz, and J. Goubeau, *Z. Anorg. Allg. Chem.*, *389*, 269 (1972).
443. G. D. Christofferson, R. A. Sparks, and J. D. McCullough, *Acta Crystallogr.*, *11*, 782 (1958).
444. R. H. Vernon, *J. Chem. Soc.*, *117*, 86 (1920).
445. H. D. K. Drew, *J. Chem. Soc.*, 560 (1929).
446. F. Einstein, J. Trotter, and C. Williston, *J. Chem. Soc. A*, 2018 (1967).
447. L. Y. Y. Chan and F. W. B. Einstein, *J. Chem. Soc., Dalton Trans.*, 316 (1972).
448. G. C. Hayward and P. J. Hendra, *J. Chem. Soc. A*, 1760 (1969).
449. K. V. Smith and J. S. Thayer, *Inorg. Chem.*, *13*, 3021 (1974).
450. K. J. Wynne and P. S. Pearson, *Inorg. Chem.*, *10*, 1871 (1971).
451. K. J. Wynne and J. W. George, *J. Amer. Chem. Soc.*, *87*, 4750 (1965).
452. K. J. Wynne and P. S. Pearson, *Inorg. Chem.*, *10*, 2735 (1971).
453. R. H. Larkin, H. D. Stidham, and K. J. Wynne, *Spectrochim. Acta*, *27A*, 2261 (1971).
454. A. P. Gray, A. B. Callear, and F. H. C. Edgecombe, *Can. J. Chem.*, *41*, 1502 (1963).
455. H. M. van Looy, L. A. M. Rodriquez, and J. A. Gabant, *J. Polym. Sci., Part A*, *4*, 1927 (1966).
456. J. F. Hanlan and J. D. McCowan, *Can. J. Chem.*, *50*, 747 (1972).
457. R. J. H. Clark and A. J. McAlees, *J. Chem. Soc. A*, 2026 (1970).
458. R. J. H. Clark and A. J. McAlees, *Inorg. Chem.*, *11*, 342 (1972).
459. W. M. Scovell, G. C. Stocco, and R. S. Tobias, *Inorg. Chem.*, *9*, 2682 (1970).
460. W. M. Scovell and R. S. Tobias, *Inorg. Chem.*, *9*, 945 (1970).
461. D. E. Clegg and J. R. Hall, *J. Organometal. Chem.*, *22*, 491 (1970).
462. P. A. Bulliner, V. A. Maroni, and T. G. Spiro, *Inorg. Chem.*, *9*, 1887 (1970).
463., R. E. Rundle and J. H. Sturdivant, *J. Amer. Chem. Soc.*, *69*, 1561 (1947).

464. G. Donnay, L. B. Coleman, N. G. Krieghoff, and D. O. Cowan, *Acta Crystallogr.*, *B24*, 157 (1968).
465. J. R. Hall and G. A. Swile, *J. Organometal. Chem.*, *56*, 419 (1973).
466. S. E. Binns, R. H. Cragg, R. D. Gillard, B. T. Heaton, and M. F. Pilbrow, *J. Chem. Soc. A*, 1227 (1969).
467. C. Santini-Scampucci and J. G. Riess, *J. Chem. Soc. Dalton Trans.*, 2436 (1973).
468. C. Santini-Scampucci and J. G. Reiss, *J. Chem. Soc. Dalton Trans.*, 195 (1976).
469. J. Müller and K. Dehnicke, *J. Organometal. Chem.*, *10*, P1 (1967).
470. K. Dehnicke and D. Seybold, *J. Organometal. Chem.*, *11*, 227 (1968).
471. A. F. Shihada and K. Dehnicke, *J. Organometal. Chem.*, *26*, 157 (1972).
472. J. Müller and K. Dehnicke, *J. Organometal. Chem.*, *12*, 37 (1968).
473. J. Müller and K. Dehnicke, *Z. Anorg. Allg. Chem.*, *348*, 261 (1966).
474. K. Dehnicke, J. Strähle, D. Seybold, and J. Müller, *J. Organometal. Chem.*, *6*, 298 (1966).
475. J. Müller and K. Dehnicke, *J. Organometal. Chem.*, *7*, P1 (1967).
476. V. Krieg and J. Weidlein, *Z. Anorg. Allg. Chem.*, *368*, 44 (1969).
477. P. Gray and T. C. Waddington, *Trans. Faraday Soc.*, *53*, 901 (1957).
478. H. A. Papazian, *J. Chem. Phys.*, *34*, 1614 (1961).
479. K. Dehnicke and I. L. Wilson, *J. Chem. Soc., Dalton Trans.*, 1428 (1973).
480. F. Weller, I. L. Wilson, and K. Dehnicke, *J. Organometal. Chem.*, *30*, C1 (1971).
481. F. Weller and K. Dehnicke, *J. Organometal. Chem.*, *35*, 237 (1972).
482. K. Dehnicke and N. Röder, *J. Organometal. Chem.*, *86*, 335 (1975).
483. H. Bürger, *Monatch. Chem.*, *96*, 1710 (1965).
484. J. S. Thayer and D. P. Strommen, *J. Organometal. Chem.*, *5*, 383 (1966).
485. J. S. Thayer, *Organometal. Chem. Rev.*, *1*, 157 (1966).
486. E. G. Rochow, D. Seyferth, and A. C. Smith, Jr., *J. Amer. Chem. Soc.*, *75*, 3099 (1953).
487. J. S. Thayer and R. West, *Inorg. Chem.*, *3*, 889 (1964).
488. G. J. M. Van der Kerk, J. G. A. Luijten, and M. J. Janssen, *Chimia*, *16*, 10 (1962).

489. R. Barbieri, N. Bertazzi, C. Tomarchio, and R. H. Herber, *J. Organometal. Chem.*, *84*, 39 (1975).
490. H. Schmidbaur, K.-H. Mitschke, J. Weidlein, and St. Cradock, *Z. Anorg. Allg. Chem.*, *386*, 139 (1971).
491. W. Beck, W. P. Fehlhammer, P. Pollmann, and R. S. Tobias, *Inorg. Chim. Acta*, *2*, 467 (1968).
492. K. H. von Dahlen and J. Lorberth, *J. Organometal. Chem.*, *65*, 267 (1974).
493. M. Atam and U. Müller, *J. Organometal. Chem.*, *71*, 435 (1974).
494. E. Ettenhuber and K. Rühlmann, *Chem. Ber.*, *101*, 743 (1968).
495. I. Ruidisch and M. Schmidt, *J. Organometal. Chem.*, *1*, 493 (1964).
496. A. Schmidt, *Chem. Ber.*, *103*, 3923 (1970).
497. D. M. Revitt and D. B. Sowerby, *Inorg. Nucl. Chem. Lett.*, *5*, 459 (1969).
498. D. M. Revitt and D. B. Sowerby, *J. Chem. Soc., Dalton Trans.*, 847 (1972).
499. A. Schmidt, *Chem. Ber.*, *101*, 3976 (1968).
500. R. G. Goel and D. R. Ridley, *Inorg. Nucl. Chem. Lett.*, *7*, 21 (1971).
501. R. G. Goel and D. R. Ridley, *Inorg. Chem.*, *13*, 1252 (1974).
502. H. Leimeister and K. Dehnicke, *J. Organometal. Chem.*, *31*, C3 (1971).
503. J. Goubeau and H. Gräbner, *Chem. Ber.*, *93*, 1379 (1960).
504. J. Goubeau, E. Heubach, D. Paulin, and I. Widmaier, *Z. Anorg. Allg. Chem.*, *300*, 194 (1959).
505. F. Stocco, G. C. Stocco, W. M. Scovell, and R. S. Tobias, *Inorg. Chem.*, *10*, 2639 (1971).
506. J. S. Thayer and R. West, *Adv. Organometal. Chem.*, *5*, 169 (1967).
507. R. P. Hirschmann, R. N. Kniseley, and V. A. Fassel, *Spectrochim. Acta*, *21*, 2125 (1965).
508. D. Martin, *Angew. Chem., Inter. Ed. Engl.*, *3*, 311 (1964).
509. J. E. Förster, M. Vargas and H. Müller, *J. Organometal. Chem.*, *59*, 97 (1973).
510. T. Wizemann, H. Müller, D. Seybold, and K. Dehnicke, *J. Organometal. Chem.*, *20*, 211 (1969).
511. R. P. J. Cooney and J. R. Hall, *Aust. J. Chem.*, *22*, 2117 (1969).
512. J. Relf, R. P. Cooney, and H. F. Henneike, *J. Organometal. Chem.*, *39*, 75 (1972).
513. F. Weller and K. Dehnicke, *J. Organometal. Chem.*, *36*, 23 (1972).
514. K. Dehnicke, *Angew. Chem.*, *79*, 942 (1967).

515. N. Bertazzi, G. C. Stocco, L. Pellerito, and A. Silvestri, *J. Organometal. Chem.*, *81*, 27 (1974).
516. J. Goubeau and J. Reyhing, *Z. Anorg. Allg. Chem.*, *294*, 96 (1958).
517. D. B. Sowerby, *J. Inorg. Nucl. Chem.*, *22*, 205 (1961).
518. M. Wada and R. Okawara, *J. Organometal. Chem.*, *8*, 261 (1967).
519. M. A. Mullins and C. Curran, *Inorg. Chem.*, *7*, 2584 (1968).
520. N. Bertazzi, G. Alonzo, A. Silvestri, and G. Consiglio, *J. Organometal. Chem.*, *37*, 281 (1972).
521. J. M. Homan, J. M. Kawamoto, and G. L. Morgan, *Inorg. Chem.*, *9*, 2533 (1970).
522. G. C. Stocco and R. S. Tobias, *J. Coord. Chem.*, *1*, 133 (1971).
523. R. A. Forder and G. M. Sheldrick, *J. Organometal. Chem.*, *21*, 115 (1970).
524. R. A. Forder and G. M. Sheldrick, *J. Organometal. Chem.*, *22*, 611 (1970).
525. Y. M. Chow, *Inorg. Chem.*, *10*, 673 (1971).
526. R. G. Goel, H. S. Prasad, G. M. Bancroft, and T. K. Sham, *Can. J. Chem.*, *54*, 711 (1976).
527. L. H. Jones, *J. Chem. Phys.*, *25*, 1069 (1956).
528. E. E. Aynsley, N. N. Greenwood, G. Hunter, and M. J. Sprague, *J. Chem. Soc. A*, 1344 (1966).
529. H. W. Morgan, *J. Inorg. Nucl. Chem.*, *16*, 367 (1961).
530. H. Bürger and U. Goetze, *J. Organometal. Chem.*, *10*, 380 (1967).
531. J. S. Thayer, *J. Organometal. Chem.*, *9*, P30 (1967).
532. J. S. Thayer, *Inorg. Chem.*, *7*, 2599 (1968).
533. E. E. Aynsley, N. N. Greenwood, and M. J. Sprague, *J. Chem. Soc.*, 2395 (1965).
534. W. Beck and E. Schuierer, *Chem. Ber.*, *97*, 3517 (1964).
535. J. J. McBride, Jr. and H. C. Beachell, *J. Amer. Chem. Soc.*, *74*, 5247 (1952).
536. J. Müller, F. Schmock, A. Klopsch, and K. Dehnicke, *Chem. Ber.*, *108*, 664 (1975).
537. J. R. Hall and J. C. Mills, *J. Organometal. Chem.*, *6*, 445 (1966).
538. J. C. Mills, H. S. Preston, and C. H. L. Kennard, *J. Organometal. Chem.*, *14*, 33 (1968).
539. W. Morell and D. Breitinger, *J. Organometal. Chem.*, *71*, C43 (1974).
540. T. Ehemann and K. Dehnicke, *J. Organometal. Chem.*, *64*, C33 (1974).
541. K. Tanaka, H. Kurosawa, and R. Okawara, *J. Organometal. Chem.*, *30*, 1 (1971).
542. H. Bürger and G. Schirawski, *Spectrochim. Acta*, *27A*, 159 (1971).

543. M. R. Booth and S. G. Frankiss, *Spectrochim. Acta*, *26A*, 859 (1970).
544. T. A. Bither, W. H. Knoth, R. V. Lindsey, Jr., and W. H. Sharkey, *J. Amer. Chem. Soc.*, *80*, 4151 (1968).
545. J. A. Secker and J. S. Thayer, *Inorg. Chem.*, *14*, 573 (1975).
546. M. R. Booth and S. G. Frankis, in *Spectroscopic Methods in Organometallic Chemistry*, W. O. George, Ed., Butterworth, London, 1970, p. 200.
547. J. R. Durig, P. J. Cooper, and Y. S. Li, *Inorg. Chem.*, *14*, 2845 (1975).
548. E. O. Schlemper and D. Britton, *Inorg. Chem.*, *5*, 511 (1966).
549. J. R. Durig, Y. S. Li, and J. B. Turner, *Inorg. Chem.*, *13*, 1495 (1974).
550. D. Seyferth and N. Kahlen, *J. Org. Chem.*, *25*, 809 (1960).
551. J. Lorberth, *Chem. Ber.*, *98*, 1201 (1965).
552. E. O. Schlemper and D. Britton, *Inorg. Chem.*, *5*, 507 (1966).
553. Y. M. Chow and D. Britton, *Acta Crystallogr.*, *B27*, 856 (1971).
554. H. G. M. Edwards, J. S. Ingman, and D. A. Long, *Spectrochim. Acta*, *32A*, 739 (1976).
555. A. Burawoy, C. S. Gibson, and S. Holt, *J. Chem. Soc.*, 1024 (1935).
556. R. F. Phillips and H. M. Powell, *Proc. Roy. Soc. London, Ser. A*, *173*, 147 (1939).
557. J. R. Durig, A. W. Cox, Jr., and Y. S. Li, *Inorg. Chem.*, *13*, 2302 (1974).
558. W. J. Franklin, R. L. Werner, and R. A. Ashby, *Spectrochim. Acta*, *30A*, 387 (1974).
559. R. S. Tobias, *Organometal. Chem. Rev.*, *1*, 93 (1966).
560. M. Wada and R. Okawara, *J. Organometal. Chem.*, *4*, 487 (1965).
561. P. L. Goggin and L. A. Woodward, *Trans. Faraday Soc.*, *58*, 1495 (1962).
562. R. S. Tobias, M. J. Sprague, and G. E. Glass, *Inorg. Chem.*, *7*, 1714 (1968).
563. D. E. Clegg and R. J. Hall, *J. Organometal. Chem.*, *17*, 175 (1969).
564. D. E. Clegg and R. J. Hall, *Spectrochim. Acta*, *21*, 357 (1965).
565. M. G. Miles, G. E. Glass, and R. S. Tobias, *J. Amer. Chem. Soc.*, *88*, 5738 (1966).
566. K. Yasuda and R. Okawara, *J. Organometal. Chem.*, *3*, 76 (1965).

567. J. E. de Moor, G. P. Van der Kelen, and Z. Eeckhaut, *J. Organometal. Chem.*, *9*, 31 (1967).
568. K. Licht and H. Kriegsmann, *Z. Anorg. Allg. Chem.*, *323*, 190 (1963).
569. J. Rouviere, V. Tabacik, and G. Fleury, *Spectrochim. Acta*, *29A*, 229 (1973).
570. S. W. Kantor, *J. Amer. Chem. Soc.*, *75*, 2712 (1953).
571. M. Kakudo, P. N. Kasai, and T. Watase, *J. Chem. Phys.*, *21*, 1894 (1953).
572. R. S. Tobias and S. Hutcheson, *J. Organometal. Chem.*, *6*, 535 (1966).
573. R. Okawara and K. Yasuda, *J. Organometal. Chem.*, *1*, 356 (1964).
574. J. M. Brown, A. C. Chapman, R. Harper, D. J. Mowthorpe, A. G. Davies, and P. J. Smith, *J. Chem. Soc.*, *Dalton Trans.*, 338 (1972).
575. H.-V. Gründler, H.-D. Schumann, and E. Steger, *J. Mol. Struct.*, *21*, 149 (1974).
576. A. Simon and H.-D. Schumann, *Z. Anorg. Allg. Chem.*, *400*, 294 (1973).
577. R. Paetzold, H.-D. Schumann, and A. Simon, *Z. Anorg. Allg. Chem.*, *305*, 98 (1960).
578. G. E. Glass, J. H. Konnert, M. G. Miles, D. Britton, and R. S. Tobias, *J. Amer. Chem. Soc.*, *90*, 1131 (1968).
579. P. A. Bulliner and T. Spiro, *Inorg. Chem.*, *8*, 1023 (1969).
580. S. Detoni and D. Hadži, *Spectrochim. Acta*, *20*, 949 (1964).
581. L. C. Thomas and R. A. Chittenden, *Spectrochim. Acta*, *20*, 489 (1964).
582. D. Gredniḉ and F. Zado, *J. Chem. Soc.*, 521 (1962).
583. J. H. R. Clarke and L. A. Woodward, *Spectrochim. Acta*, *23A*, 2077 (1967).
584. A. Storr, K. Jones, and A. W. Laubengayer, *J. Amer. Chem. Soc.*, *90*, 3173 (1968).
585. N. Ueyama, T. Araki, and H. Tani, *Inorg. Chem.*, *12*, 2218 (1973).
586. T. Aoyagi, T. Araki, N. Oguni, M. Mikumo, and H. Tani, *Inorg. Chem.*, *12*, 2702 (1973).
587. G. S. Smith and J. L. Hoard, *J. Amer. Chem. Soc.*, *81*, 3907 (1959).
588. A. Marchand, J. Valade, M. T. Forel, M. L. Josien, and R. Calas, *J. Chim. Phys.*, *59*, 1142 (1962).
589. G. Engelhardt and H. Kriegsmann, *Z. Anorg. Allg. Chem.*, *330*, 155 (1964).
590. G. Engelhardt and H. Kriegsmann, *Z. Anorg. Allg. Chem.*, *328*, 194 (1964).

591. H. Kriegsmann, *Z. Electrochem.*, *65*, 336 (1961).
592. H. Kriegsmann, *Z. Electrochem.*, *65*, 342 (1961).
593. N. Wright and M. J. Hunter, *J. Amer. Chem. Soc.*, *69*, 803 (1947).
594. T. Alvik and J. Dale, *Acta Chem. Scand.*, *25*, 2142 (1971).
595. G. Fogarasi, H. Hacker, V. Hoffmann, and S. Dobos, *Spectrochim. Acta*, *30A*,
596. D. M. Adams and W. S. Fernando, *J. Chem. Soc.*, *Dalton Trans.*, 410 (1973).
597. R. E. Richards and H. W. Thompson, *J. Chem. Soc.*, 124 (1949).
598. C. W. Young, P. C. Servais, C. C. Currie, and M. J. Hunter, *J. Amer. Chem. Soc.*, *70*, 3758 (1948).
599. I. F. Kovalev, L. Ozolins, and M. G. Voronkov, *Dokl. Akad. Nauk SSSR*, *181*, 577 (1968).
600. H. Kriegsmann, *Z. Anorg. Allg. Chem.*, *298*, 232 (1959).
601. H. Schmidbaur, J. A. Perez-Garcia, and H. S. Arnold, *Z. Anorg. Allg. Chem.*, *328*, 105 (1964).
602. H. Schmidbaur, M. Bergfeld, and F. Schindler, *Z. Anorg. Allg. Chem.*, *363*, 73 (1968).
603. H. Schmidbaur and M. Bergfeld, *Z. Anorg. Allg. Chem.*, *363*, 84 (1968).
604. H. Schmidbaur and H. Hussek, *J. Organometal. Chem.*, *1*, 257 (1964).
605. H. Schmidbaur and M. Bergfeld, *Inorg. Chem.*, *5*, 2069 (1966).
606. H. Schmidbaur and H. Hussek, *J. Organometal. Chem.*, *1*, 235 (1964).
607. D. H. Boal and G. A. Ozin, *Can. J. Chem.*, *48*, 3026 (1970).
608. H. Schmidbaur and H. Hussek, *J. Organometal. Chem.*, *1*, 244 (1964).
609. H. Schmidbaur, *Chem. Ber.*, *97*, 842 (1964).
610. H. Schmidbaur, *J. Organometal. Chem.*, *1*, 28 (1963).
611. M. P. Brown, R. Okawara, and E. G. Rochow, *Spectrochim. Acta*, *16*, 595 (1960).
612. A. Marchand, M.-T. Forel, M. Lebedeff, and J. Valade, *J. Organometal. Chem.*, *26*, 69 (1971).
613. A. Marchand, M.-H. Soulard, M. Massol, J. Barrau, and J. Satgé, *J. Organometal. Chem.*, *63*, 175 (1973).
614. H. Kriegsmann, H. Hoffmann, and H. Geissler, *Z. Anorg. Allg. Chem.*, *341*, 24 (1965).
615. N. Kasai, K. Yasuda, and R. Okawara, *J. Organometal. Chem.*, *3*, 172 (1965).
616. R. C. Poller, *J. Inorg. Nucl. Chem.*, *24*, 593 (1962).
617. R. A. Cummins, *Aust. J. Chem.*, *18*, 98 (1965).
618. W. Gordy, *J. Chem. Phys.*, *14*, 305 (1946).

619. F. Choplin and G. Kaufmann, *Spectrochim. Acta, 26A*, 2113 (1970).
620. J. H. S. Green and H. A. Lauwers, *Bull. Soc. Chim. Belges, 79*, 571 (1970).
621. F. N. Hooge and P. J. Christen, *Rec. Trav. Chim. Pays-Bas, 77*, 911 (1958).
622. L. W. Daasch and D. C. Smith, *Anal. Chem., 23*, 853 (1951).
623. R. A. Zingaro and R. M. Hedges, *J. Phys. Chem., 65*, 1132 (1961).
624. F. Watari, *Spectrochim. Acta, 31A*, 1143 (1975).
625. A. Merijanian and R. A. Zingaro, *Inorg. Chem., 5*, 187 (1966).
626. W. Morris, R. A. Zingaro, and J. Loone, *J. Organometal. Chem., 91*, 295 (1975).
627. G. N. Chremos and R. A. Zingaro, *J. Organometal. Chem., 22*, 637 (1970).
628. G. N. Chremos and R. A. Zingaro, *J. Organometal. Chem., 22*, 647 (1970).
629. R. G. Goel, P. N. Joshi, D. R. Ridley, and R. E. Beaumont, *Can. J. Chem., 47*, 1423 (1969).
630. R. G. Goel and H. S. Prasad, *Inorg. Chem., 11*, 2141 (1972).
631. B. J. Van der Veken and M. A. Herman, *J. Mol. Struct., 15*, 225 (1973).
632. R. A. Nyquist, *J. Mol. Struct., 2*, 111 (1968).
633. R. A. Nyquist, *J. Mol. Struct., 2*, 123 (1968).
634. A. Simon and H.-D. Schumann, *Z. Anorg. Allg. Chem., 398*, 145 (1973).
635. A. Simon and H.-D. Schumann, *Z. Anorg. Allg. Chem., 393*, 23 (1972).
636. F. K. Vansant, B. J. Van der Veken, and M. A. Herman, *Spectrochim. Acta, 30A*, 69 (1974).
637. A. Simon and H.-D. Schumann, *Z. Anorg. Allg. Chem., 399*, 97 (1973).
638. J. T. Braunholtz, G. E. Hall, F. G. Mann, and N. Sheppard, *J. Chem. Soc.*, 868 (1959).
639. D. Hadži, *Pure Appl. Chem., 11*, 435 (1965).
640. J. Van der Veken and M. A. Herman, *J. Mol. Struct., 15*, 237 (1973).
641. H. Olapinski, B. Schaible, and J. Weidlein, *J. Organometal. Chem., 33*, 107 (1972).
642. R. E. Ridenour and E. E. Flagg, *J. Organometal. Chem., 16*, 393 (1969).
643. R. Paetzold and G. Bochmann, *Spectrochim. Acta, 26A*, 391 (1970).
644. R. Paetzold, U. Lindner, G. Bochmann, and P. Reich, *Z. Anorg. Allg. Chem., 352*, 295 (1967).

645. R. Paetzold, H.-D. Schumann, and A. Simon, *Z. Anorg. Allg. Chem., 305,* 78 (1960).
646. R. Paetzold, H.-D. Schumann, and A. Simon, *Z. Anorg. Allg. Chem., 305,* 88 (1960).
647. A. Simon and H. Arnold, *J. Prakt. Chem., 8,* 17 (1959).
648. V. A. Yablokov, S. Ya. Khorshev, A. P. Tarabarina, and A. N. Sunin, *Zh. Obshch. Khim., 43,* 607 (1973).
649. N. N. Vyshinskii, Yu. A. Aleksandrov, and N. K. Rudnevskii, *Izv. Akad. Nauk SSSR, Ser. Fiz., 26,* 1285 (1962)
650. H. Vahrenkamp, *J. Organometal. Chem., 28,* 181 (1971).
651. R. M. Salinger and R. West, *J. Organometal. Chem., 11,* 631 (1968).
652. K. A. Hootan and A. L. Allred, *Inorg. Chem., 4,* 671 (1965).
653. M. A. Finch and C. H. Van Dyke, *Inorg. Chem., 14,* 136 (1975).
654. Y. Matsumura and R. Okawara, *J. Organometal. Chem., 11,* 299 (1968).
655. P. G. Harrison and S. R. Stobart, *J. Organometal. Chem., 47,* 89 (1973).
656. H. Kriegamann, H. Hoffmann, and H. Geissler, *Z. Anorg. Allg. Chem., 359,* 58 (1968).
657. J. Goubeau and D. Köttgen, *Z. Anorg. Allg. Chem., 360,* 182 (1968).
658. R. A. Zingaro and R. E. McGlothlin, *J. Chem. Eng. Data, 8,* 226 (1963).
659. S. N. Dutta and M. M. Woolfson, *Acta Crystallogr., 14,* 178 (1961).
660. P. J. D. Park, G. Chambers, E. Wyn-Jones, and P. J. Hendra, *J. Chem. Soc. A,* 646 (1967).
661. A. H. Cowley and H. Steinfink, *Inorg. Chem., 4,* 1827 (1965).
662. A. H. Cowley and W. D. White, *Spectrochim. Acta, 22,* 1431 (1966).
663. R. A. Zingaro, K. J. Irgolic, D. H. O'Brien, and L. J. Edmonson, Jr., *J. Amer. Chem. Soc., 93,* 5677 (1971).
664. E. Lindner and H.-M. Ebinger, *J. Organometal. Chem., 66,* 103 (1974).
665. M. Schmidt and W. Siebert, *Chem. Ber., 102,* 2752 (1969).
666. H. M. M. Shearer and C. B. Spencer, *Chem. Commun.,* 194 (1966).
667. G. Mann, H. Olapinski, R. Ott, and J. Weidlein, *Z. Anorg. Allg. Chem., 410,* 195 (1974).
668. E. G. Hoffmann, *Ann. Chem., 629,* 104 (1960).
669. R. Tarao, *Bull. Chem. Soc. Japan, 39,* 725 (1966).
670. R. Tarao, *Bull. Chem. Soc. Japan, 39,* 2126 (1966).

671. G. Mann, A. Haaland, and J. Weidlein, *Z. Anorg. Allg. Chem.*, *398*, 231 (1973).
672. L. F. Dahl, G. L. Davis, D. L. Wampler, and R. West, *J. Inorg. Nucl. Chem.*, *24*, 357 (1962).
673. R. Forneris and E. Funck, *Z. Electrochem.*, *62*, 1130 (1958).
674. M. Hayashi, *J. Chem. Soc. Japan, Pure Chem. Sec.*, *79*, 436 (1958).
675. T. Tanaka, *Bull. Chem. Soc. Japan*, *33*, 446 (1960).
676. A. Marchand and J. Valade, *J. Organometal. Chem.*, *12*, 305 (1968).
677. M. Lebedeff, A. Marchand, and J. Valade, *C. R. Acad. Sci., Paris, Ser. C*, *267*, 813 (1968).
678. N. N. Vyshinskii and K. K. Rudnevskii, *Opt. Spectrosk.*, *10*, 297 (1961).
679. F. F. Butcher, W. Gerrard, E. F. Mooney, R. G. Rees, and H. A. Willis, *Spectrochim. Acta*, *20*, 51 (1964).
680. A. Marchand, J. Mendelsohn, and J. Valade, *C. R. Acad. Sci., Paris*, *259*, 1737 (1964).
681. J. Lorberth and M.-R. Kula, *Chem. Ber.*, *97*, 3444 (1964).
682. R. A. Cummins and J. V. Evans, *Spectrochim. Acta*, *21*, 1016 (1965).
683. J. Mendelsohn, A. Marchand, and J. Valade, *J. Organometal. Chem.*, *6*, 25 (1966).
684. J. Mendelsohn, J.-C. Pommier, and J. Valade, *C. R. Acad. Sci., Paris, Ser. C*, *263*, 921 (1966).
685. J. C. Maire and R. Ouaki, *Helv. Chem. Acta*, *51*, 1150 (1968).
686. J.-C. Pommier and J. Valade, *J. Organometal. Chem.*, *12*, 433 (1968).
687. A. C. Chapman, A. G. Davies, P. G. Harrison, and W. McFarlane, *J. Chem. Soc. C*, 821 (1970).
688. A. J. Bloodworth and A. G. Davies, in *Organotin Compounds*, Vol. 1, A. K. Sawyer, Ed., Marcel Dekker, Ney York, 1971, p. 153.
689. A. M. Domingos and G. M. Sheldrick, *Acta Crystallogr.*, *B30*, 519 (1974).
690. Y. Kawasaki, T. Tanaka, and R. Okawara, *J. Organometal. Chem.*, *6*, 95 (1966).
691. A. G. Davies and R. J. Puddephatt, *J. Chem. Soc. C*, 2663 (1967).
692. R. J. Puddephatt and G. H. Thistlethwaite, *J. Chem. Soc., Dalton Trans.*, 570 (1972).
693. A. Marchand, P. Gerval, M. Massol, and J. Barrau, *J. Organometal. Chem.*, *74*, 209 (1974).

694. A. Marchand, P. Gerval, M. Massol, and J. Barrau, *J. Organometal. Chem.*, *74*, 227 (1974).
695. H. Schmidbaur and H. Stuhler, *Angew. Chem., Inter. Ed. Engl.*, *11*, 145 (1972).
696. F. K. Vansant and B. J. Van der Veken, *J. Mol. Struct.*, *22*, 273 (1974).
697. H. A. Meinema and J. G. Noltes, *J. Organometal. Chem.*, *36*, 313 (1972).
698. Y. L. Fan and R. G. Shaw, *J. Org. Chem.*, *38*, 2410 (1973).
699. J. Mounier, *J. Organometal. Chem.*, *56*, 67 (1973).
700. J. Kress and A. Novak, *J. Organometal. Chem.*, *86*, 281 (1975).
701. B. G. Gowenlock, W. E. Lindsell and B. Singh, *J. Organometal. Chem.*, *101*, C37 (1975).
702. S. Takeda and R. Tarao, *Bull. Chem. Soc. Japan*, *38*, 1567 (1965).
703. Y. Mashiko, *J. Chem. Soc. Japan, Pure Chem. Sec.*, *79*, 470 (1958).
704. J. Krausse, G. Marx, and G. Schödl, *J. Organometal. Chem.*, *21*, 159 (1970).
705. R. A. Nyquist and J. R. Mann, *Spectrochim. Acta*, *28A*, 511 (1972).
706. N. Iwasaki, J. Tomooka, and K. Toyoda, *Bull. Chem. Soc. Japan*, *47*, 1323 (1974).
707. G. E. Coates and R. A. Whitcombe, *J. Chem. Soc.*, 3351 (1956).
708. G. D. Shier and R. S. Drago, *J. Organometal. Chem.*, *5*, 330 (1966).
709. D. F. Van de Vondel, E. V. Van den Berghe, and G. P. Van der Kelen, *J. Organometal. Chem.*, *23*, 105 (1970).
710. E. W. Abel, *J. Chem. Soc.*, 4406 (1960).
711. R. C. Mehrotra, V. D. Gupta, and D. Sukhari, *J. Organometal. Chem.*, *9*, 263 (1967).
712. E. W. Abel and D. B. Brady, *J. Chem. Soc.*, 1192 (1965).
713. J. Mounier, *J. Organometal. Chem.*, *56*, 79 (1973).
714. R. Tarao, *Bull. Chem. Soc. Japan*, *39*, 2132 (1966).
715. R. Fonteyne, *J. Chem. Phys.*, *8*, 60 (1940).
716. J. Weidlein, *Z. Anorg. Allg. Chem.*, *378*, 245 (1970).
717. G. E. Coates and R. N. Mukherjee, *J. Chem. Soc.*, 1295 (1964).
718. J. Weidlein, *J. Organometal. Chem.*, *19*, 253 (1969).
719. F. W. B. Einstein, M. M. Gilbert, and D. G. Tuck, *J. Chem. Soc., Dalton Trans.*, 248 (1973).
720. J. J. Habeeb and D. G. Tuck, *J. Chem. Soc., Dalton Trans.*, 243 (1973).
721. H. D. Hausen and H. U. Schwering, *Z. Anorg. Allg. Chem.*, *398*, 119 (1973).

722. H. D. Hausen, *J. Organometal. Chem.*, *39*, C37 (1972).
723. Y. M. Chow and D. Britton, *Acta Cryst.*, *B31*, 1929 (1975).
724. H. Kurosawa and R. Okawara, *J. Organometal. Chem.*, *19*, 253 (1969).
725. H. Kurosawa and R. Okawara, *J. Organometal. Chem.*, *10*, 211 (1967).
726. H.-U. Schwering, H. Olapinski, E. Jungk, and J. Weidlein, *J. Organometal. Chem.*, *76*, 315 (1974).
727. R. Okawara, D. E. Webster, and E. G. Rochow, *J. Amer. Chem. Soc.*, *82*, 3287 (1960).
728. T. N. Srivastava and M. Onyszchuk, *Can. J. Chem.*, *41*, 1244 (1963).
729. M. J. Janssen, J. G. A. Luijten, and G. J. M. Van der Kerk, *Rec. Trav. Chim. Pays-Bas*, *82*, 90 (1963).
730. K. Ito and H. J. Bernstein, *Can. J. Chem.*, *34*, 170 (1956).
731. R. Okawara and H. Sato, *J. Inorg. Nucl. Chem.*, *16*, 204 (1961).
732. I. R. Beattie and T. Gilson, *J. Chem. Soc.*, 2585 (1961).
733. R. E. Hester, *J. Organometal. Chem.*, *23*, 123 (1970).
734. P. B. Simons and W. A. G. Graham, *J. Organometal. Chem.*, *8*, 479 (1967).
735. P. B. Simons and W. A. G. Graham, *J. Organometal. Chem.*, *10*, 457 (1967).
736. C. Poder and J. R. Sams, *J. Organometal. Chem.*, *19*, 67 (1969).
737. C. S. C. Wang and J. M. Shreeve, *J. Organometal. Chem.*, *38*, 287 (1972).
738. R. A. Cummins and P. Dunn, *Aust. J. Chem.*, *17*, 185 (1964).
739. R. A. Cummins, *Aust. J. Chem.*, *17*, 594 (1964).
740. M. Vilarem and J. C. Maire, *C. R. Acad. Sci., Paris, Ser. C*, *262*, 480 (1966).
741. M. Wada, M. Shindo, and R. Okawara, *J. Organometal. Chem.*, *1*, 95 (1963).
742. Y. Maeda, C. R. Dillard, and R. Okawara, *Inorg. Nucl. Chem. Lett.*, *2*, 197 (1966).
743. Y. Maeda and R. Okawara, *J. Organometal. Chem.*, *10*, 247 (1967).
744. A. D. Cohen and C. R. Dillard, *J. Organometal. Chem.*, *25*, 421 (1970).
745. H. H. Anderson, *Inorg. Chem.*, *3*, 912 (1964).
746. R. Okawara and M. Ohara, *J. Organometal. Chem.*, *1*, 360 (1964).
747. H. Schmidbaur, K.-H. Mitschke, and J. Weidlein, *Z. Anorg. Allg. Chem.*, *386*, 147 (1971).

748. R. G. Goel and D. R. Ridley, *J. Organometal. Chem., 38,* 83 (1972).
749. C. Santini-Scampucci and G. Wilkinson, *J. Chem. Soc., Dalton Trans.,* 807 (1976).
750. J. Weidlein, *J. Organometal. Chem., 32,* 181 (1971).
751. Y. Matsumura, M. Shindo, and R. Okawara, *Inorg. Nucl. Chem. Lett., 3,* 219 (1967).
752. J. Otera and R. Okawara, *J. Organometal. Chem., 17,* 353 (1969).
753. S. Kato, A. Hori, H. Shiotani, M. Mizuta, N. Hayashi, and T. Takakuwa, *J. Organometal. Chem., 82,* 223 (1974).
754. D. Gibson, *Coord. Chem. Rev., 4,* 225 (1969).
755. Y. Nakamura and K. Nakamoto, *Inorg. Chem., 14,* 63 (1975).
756. G. T. Behnke and K. Nakamoto, *Inorg. Chem., 7,* 2030 (1968).
757. S. Pinchas, B. L. Silver, and I. Laulicht, *J. Chem. Phys., 46,* 1506 (1967).
758. H. Junge and H. Musso, *Spectrochim. Acta, 24A,* 1219 (1968).
759. G. T. Behnke and K. Nakamoto, *Inorg. Chem., 6,* 433 (1967).
760. E. E. Ernstbrunner, *J. Chem. Soc. A,* 1558 (1970).
761. H. Kurosawa, K. Yasuda, and R. Okawara, *Inorg. Nucl. Chem. Lett., 1,* 131 (1965).
762. Y. Kawasaki, T. Tanaka, and R. Okawara, *Spectrochim. Acta, 22,* 1571 (1966).
763. V. B. Ramos and R. S. Tobias, *Spectrochim. Acta, 29A,* 953 (1973).
764. H. A. Meinema and J. G. Noltes, *J. Organometal. Chem., 16,* 257 (1969).
765. H. A. Meinema, A. Mackor, and J. G. Noltes, *J. Organometal. Chem., 37,* 285 (1972).
766. G. A. Miller and E. O. Schlemper, *Inorg. Chem., 12,* 677 (1973).
767. C. Z. Moore and W. H. Nelson, *Inorg. Chem., 8,* 138 (1969).
768. N. Kanehisa, Y. Kai, and N. Kasai, *Inorg. Nucl. Chem. Lett., 8,* 375 (1972).
769. J. Lewis, R. F. Long, and C. Oldham, *J. Chem. Soc.,* 6740 (1965).
770. G. T. Behnke and K. Nakamoto, *Inorg. Chem., 6,* 440 (1967).
771. F. Bonati and G. Minghetti, *Angew. Chem., Inter. Ed. Engl., 7,* 629 (1968).
772. G. T. Behnke and K. Nakamoto, *Inorg. Chem., 7,* 330 (1968).

773. S. Baba, T. Ogura, and S. Kawaguchi, *Inorg. Nucl. Chem. Lett.*, *7*, 1195 (1971).
774. J. R. Hall and G. A. Swile, *J. Organometal. Chem.*, *47*, 195 (1973).
775. I. F. Lutsenko, Yu. I. Baukov, O. V. Dudukina, and E. N. Kramarova, *J. Organometal. Chem.*, *11*, 35 (1968).
776. I. F. Lutsenko, Yu. I. Baukov, I. Yu. Belavin, and A. N. Tvorogov, *J. Organometal. Chem.*, *14*, 229 (1968).
777. I. F. Lutsenko, Yu. I. Baukov, and I. Yu. Belavin, *J. Organometal. Chem.*, *24*, 359 (1970).
778. C. R. Krüger and E. G. Rochow, *J. Organometal. Chem.*, *1*, 476 (1964).
779. E. M. Dexheimer and L. Spialter, *J. Organometal. Chem.*, *107*, 229 (1976).
780. H. F. Klein, *Angew. Chem.*, *Inter. Ed. Engl.*, *12*, 5 (1973).
781. N. Q. Dao and D. Breitinger, *Spectrochim. Acta*, *27A*, 905 (1971).
782. H. C. Clark, R. J. O'Brien, and A. L. Pickard, *J. Organometal. Chem.*, *4*, 43 (1965).
783. D. E. Clegg and J. R. Hall, *Spectrochim. Acta*, *23A*, 263 (1967).
784. H. Hagnauer, G. C. Stocco, and R. S. Tobias, *J. Organometal. Chem.*, *46*, 179 (1972).
785. J. Mounier, B. Mula, and A. Potier, *J. Organometal. Chem.*, *105*, 289 (1976).
786. H. Bürger, W. Sawodny, and U. Wannagat, *J. Organometal. Chem.*, *3*, 113 (1965).
787. H. J. Becher, *Spectrochim. Acta*, *19*, 575 (1963).
788. J. W. Dawson, P. Kritz, and K. Niedenzu, *J. Organometal. Chem.*, *5*, 13 (1966).
789. J. W. Dawson, P. Fritz, and K. Niedenzu, *J. Organometal. Chem.*, *5*, 211 (1966).
790. R. E. Hall and E. P. Schram, *Inorg. Chem.*, *8*, 270 (1969).
791. W. Wolfsberger and H. Schmidbaur, *J. Organometal. Chem.*, *17*, 41 (1969).
792. J. J. Habeeb and D. G. Tuck, *Can. J. Chem.*, *52*, 3950 (1974).
793. B. Walther and K. Thiede, *J. Organometal. Chem.*, *32*, C7 (1971).
794. H. Bürger and U. Goetze, *Monat. Chem.*, *99*, 155 (1968).
795. H. Bürger and W. Sawodny, *Spectrochim. Acta*, *23A*, 2827 (1967).
796. C. H. Yoder and J. J. Zuckerman, *Inorg. Chem.*, *5*, 2055 (1966).

797. R. E. Bailey and R. West, *J. Organometal. Chem., 4,* 430 (1965).
798. K. Seppelt and H. H. Eysel, *Z. Anorg. Allg. Chem., 384,* 147 (1971).
799. K. Witke, P. Reich, and H. Kriegsmann, *Z. Anorg. Allg. Chem., 380,* 164 (1971).
800. H. Bürger and K. Burczyk, *Z. Anorg. Allg. Chem., 381,* 176 (1971).
801. A. Marchand, M. Riviere-Baudet, J. Satge and M.-H. Soulard, *J. Organometal. Chem., 107,* 33 (1976).
802. A. Marchand, C. Lemerle, M.-T. Forel, and M.-H. Soulard, *J. Organometal. Chem., 42,* 353 (1972).
803. R. E. Highsmith and H. H. Sisler, *Inorg. Chem., 8,* 996 (1969).
804. H.-J. Götze, *J. Organometal. Chem., 47,* C25 (1973).
805. O. Schmitz-DuMont, H.-J. Götze, and H. Götze, *Z. Anorg. Allg. Chem., 366,* 180 (1969).
806. O. Schmitz-DuMont and H.-J. Götze, *Z. Anorg. Allg. Chem., 371,* 38 (1969).
807. R. E. Hester and K. Jones, *Chem. Commun.,* 317 (1966).
808. H. Schumann and S. Ronecker, *J. Organometal. Chem., 23,* 451 (1970).
809. H. Bürger and H. J. Neese, *J. Organometal. Chem., 20,* 129 (1969).
810. G. C. Stocco and R. S. Tobias, *J. Amer. Chem. Soc., 93,* 5057 (1971).
811. G. D. Shier and R. S. Drago, *J. Organometal. Chem., 6,* 359 (1966).
812. E. V. Van den Berghe, L. Verdonck, and G. P. Van der Kelen, *J. Organometal. Chem., 16,* 497 (1969).
813. D. Potts, H. D. Sharma, A. J. Carty, and A. Walker, *Inorg. Chem., 13,* 1205 (1974).
814. T. L. Blundell and H. M. Powell, *Chem. Commun.,* 54 (1967).
815. W. D. Honnick, M. C. Hughes, C. D. Schaffer, Jr., and J. J. Zuckerman, *Inorg. Chem., 15,* 1391 (1976).
816. J. Meunier and M.-T. Forel, *Spectrochim. Acta, 29A,* 487 (1973).
817. V. G. K. Das and W. Kitching, *J. Organometal. Chem., 13,* 523 (1968).
818. T. Tanaka, *Inorg. Chim. Acta, 1,* 217 (1967).
819. N. W. Isaacs and C. H. L. Kennard, *J. Chem. Soc. A,* 1257 (1970).
820. V. B. Ramos and R. S. Tobias, *Spectrochim. Acta, 30A,* 181 (1974).
821. M. A. Weiner, *J. Organometal. Chem., 23,* C20 (1970).
822. Y. K. Ho and J. J. Zuckerman, *Inorg. Chem., 12,* 1552 (1973).

823. S. Mansy, T. E. Wood, J. C. Sprowles, and R. S. Tobias, *J. Amer. Chem. Soc., 96,* 1762 (1974).
824. S, Mansy and R. S. Tobias, *J. Amer. Chem. Soc., 96,* 6874 (1974).
825. S. Mansy and R. S. Tobias, *Inorg. Chem., 14,* 287 (1975).
826. T. Maeda and R. Okawara, *J. Organometal. Chem., 39,* 87 (1972).
827. N. W. G. Debye, D. E. Fenton, and J. J. Zuckerman, *J. Inorg. Nucl. Chem., 34,* 352 (1972).
828. L. Pellerito, R. Cefalu, A. Gianguzza, and R. Barbieri, *J. Organometal. Chem., 70,* 303 (1974).
829. E. S. Bretschneider and C. W. Allen, *Inorg. Chem., 12,* 623 (1973).
830. K. Kawakami and R. Okawara, *J. Organometal. Chem., 6,* 249 (1966).
831. R. Cefalù, R. Bosco, F. Bonati, F. Maggio, and R. Barbieri, *Z. Anorg. Allg. Chem., 376,* 180 (1970).
832. J. N. R. Ruddick and J. R. Sams, *J. Organometal. Chem., 60,* 233 (1973).
833. H. A. Meinema, E. Rivarola, and J. G. Noltes, *J. Organometal. Chem., 17,* 71 (1969).
834. M. Shindo and R. Okawara, *Inorg. Nucl. Chem. Lett., 5,* 77 (1969).
835. J. R. Hall and G. A. Swile, *Aust. J. Chem., 28,* 1507 (1975).
836. J. H. R. Clarke and L. A. Woodward, *Trans. Faraday Soc., 62,* 3022 (1966).
837. D. Potts and A. Walker, *Can. J. Chem., 49,* 2171 (1971).
838. J. R. Ferraro, D. Potts, and A. Walker, *Can. J. Chem., 48,* 711 (1970).
839. H. C. Clark and R. G. Goel, *Inorg. Chem., 5,* 998 (1966).
840. H. C. Clark and R. J. O'Brien, *Inorg. Chem., 2,* 740 (1963).
841. P. Au, M.Sc. Thesis, University of Western Ontario, London, Ontario, Canada, cited by D. Potts, H. D. Sharma, A. J. Carty, and A. Walker, *Inorg. Chem., 13,* 1205 (1974).
842. C. C. Addison, W. B. Simpson, and A. Walker, *J. Chem. Soc.,* 2360 (1964).
843. J. Hilton, E. K. Nunn, and S. C. Wallwork, *J. Chem. Soc., Dalton Trans.,* 173 (1973).
844. K. Yasuda, Y. Matsumoto, and R. Okawara, *J. Organometal. Chem., 6,* 528 (1966).
845. G. S. Brownlee, S. C. Nyburg, and T. J. Szymański, *Chem. Commun.,* 1073 (1971).
846. K. C. Williams and D. W. Imhoff, *J. Organometal. Chem., 42,* 107 (1972).

847. D. Potts and A. Walker, *Can. J. Chem.*, *47*, 1621 (1969).
848. J. Lorberth, J. Pebler, and G. Lange, *J. Organometal. Chem.*, *54*, 177 (1972).
849. J. H. R. Clarke and L. A. Woodward, *Trans. Faraday Soc.*, *64*, 1041 (1968).
850. H. Olapinski and J. Weidlein, *J. Organometal. Chem.*, *54*, 87 (1973).
851. E. Lindner, U. Kunze, G. Ritter, and A. Haag, *J. Organometal. Chem.*, *24*, 119 (1970).
852. U. Kunze, E. Lindner, and J. Koola, *J. Organometal. Chem.*, *38*, 51 (1972).
853. C. H. Stapfer, K. L. Leung, and R. H. Herber, *Inorg. Chem.*, *9*, 970 (1970).
854. R. E. B. Garrod, R. H. Platt, and J. R. Sams, *Inorg. Chem.*, *10*, 424 (1971).
855. R. H. Herber and C. H. Stapfer, *Inorg. Nucl. Chem. Lett.*, *7*, 617 (1971).
856. U. Stahlberg, R. Gelius, and R. Müller, *Z. Anorg. Allg. Chem.*, *355*, 230 (1967).
857. U. Kunze and H. P. Völker, *Chem. Ber.*, *107*, 3818 (1974).
858. F. Huber and F. J. Padberg, *Z. Anorg. Allg. Chem.*, *351*, 1 (1967).
859. N. A. D. Carey and H. C. Clark, *Can. J. Chem.*, *46*, 649 (1968).
860. P. J. Pollick, J. P. Bibler, and A. Wojcicki, *J. Organometal. Chem.*, *16*, 201 (1969).
861. N. A. D. Carey and H. C. Clark, *Can. J. Chem.*, *46*, 643 (1968).
862. R. E. J. Bichler and H. C. Clark, *J. Organometal. Chem.*, *23*, 427 (1970).
863. H. Olapinski and J. Weidlein, *J. Organometal. Chem.*, *35*, C53 (1972).
864. H. Olapinski, J. Weidlein, and H.-D. Hausen, *J. Organometal. Chem.*, *64*, 193 (1974).
865. A. T. T. Hsieh, *J. Organometal. Chem.*, *27*, 293 (1971).
866. A. G. Lee, *J. Chem. Soc. A*, 467 (1970).
867. P. A. Yeats, J. R. Sams, and F. Aubke, *Inorg. Chem.*, *11*, 2634 (1972).
868. P. A. Yeats, J. R. Sams, and F. Aubke, *Inorg. Chem.*, *10*, 1877 (1971).
869. F. H. Allen, J. A. Lerbscher, and J. Trotter, *J. Chem. Soc. A*, 2507 (1971).
870. V. Gaiser, J. Weidlein, and E. Lindner, *J. Organometal. Chem.*, *56*, C1 (1973).
871. J. Weidlein, *J. Organometal. Chem.*, *24*, 63 (1970).
872. J. Weidlein, *Z. Anorg. Allg. Chem.*, *366*, 22 (1969).
873. E. Lindner and K. Schardt, *J. Organometal. Chem.*, *81*, 145 (1974).

874. E. Lindner and K. Schardt, *J. Organometal. Chem.*, *82*, 73 (1974).
875. E. Lindner and K. Schardt, *J. Organometal. Chem.*, *82*, 81 (1974).
876. G. Vitzthum, U. Kunze, and E. Lindner, *J. Organometal. Chem.*, *21*, P38 (1970).
877. E. Lindner, U. Kunze, G. Vitzthum, G. Ritter, and A. Haag, *J. Organometal. Chem.*, *24*, 131 (1970).
878. R. Paetzold and D. Knaust, *Z. Anorg. Allg. Chem.*, *368*, 196 (1969).
879. H. C. Clark and R. J. O'Brien, *Inorg. Chem.*, *2*, 1020 (1963).
880. B. Schaible, K. Roessel, J. Weidlein, and H. D. Hausen, *Z. Anorg. Allg. Chem.*, *409*, 176 (1974).
881. B. Schaible and J. Weidlein, *J. Organometal. Chem.*, *35*, C7 (1972).
882. J. Weidlein and B. Schaible, *Z. Anorg. Allg. Chem.*, *386*, 176 (1971).
883. T. Chivers, J. H. G. Van Roode, J. N. R. Ruddick, and J. R. Sams, *Can. J. Chem.*, *51*, 3702 (1973).
884. B. J. Hathaway and A. E. Underhill, *J. Chem. Soc.*, 3091 (1961).
885. M. P. Brown, E. Cartmell, and G. W. A. Fowles, *J. Chem. Soc.*, 506 (1960).
886. B. Fontal and T. G. Spiro, *Inorg. Chem.*, *10*, 9 (1971).
887. R. J. H. Clark, A. G. Davies, R. J. Puddephatt, and W. McFarlane, *J. Amer. Chem. Soc.*, *91*, 1334 (1969).
888. J. R. Durig and J. S. DiYorio, *Inorg. Chem.*, *8*, 2796 (1969).
889. J. R. Durig and J. M. Casper, *J. Chem. Phys.*, *55*, 198 (1971).
890. W. H. Green and A. B. Harvey, *J. Chem. Phys.*, *49*, 3586 (1968).
891. S. G. Frankiss, *J. Mol. Spectrosc.*, *3*, 89 (1969).
892. C. W. Sink and A. B. Harvey, *J. Mol. Struct.*, *4*, 203 (1969).
893. E. Steger and K. Stopperka, *Chem. Ber.*, *94*, 3029 (1961).
894. K. G. Allum, J. A. Creighton, J. H. S. Green, G. J. Minkoff, and L. J. S. Prince, *Spectrochim. Acta*, *24A*, 927 (1968).
895. J. R. Durig and R. W. MacNamee, *J. Mol. Struct.*, *17*, 426 (1973).
896. R. L. Amster, N. B. Colthup, and W. A. Henderson, *Spectrochim. Acta*, *19*, 1841 (1963).
897. J. Ellermann and H. Schössner, *Angew. Chem., Inter. Ed. Engl.*, *13*, 601 (1974).
898. R. G. Vranka and E. L. Amma, *J. Amer. Chem. Soc.*, *89*, 3121 (1967).

899. K. A. Levison and P. G. Perkins, *Discuss. Faraday Soc.*, *47*, 183 (1969).
900. F. J. Farrell, V. A. Maroni, and T. G. Spiro, *Inorg. Chem.*, *8*, 2638 (1969).
901. D. Breitinger and G. P. Arnold, *Inorg. Nucl. Chem. Lett.*, *10*, 517 (1974).
902. H. Bürger, U. Goetze, and W. Sawodny, *Spectrochim. Acta*, *24A*, 2003 (1968).
903. H. Bürger and U. Goetze, *Angew. Chem., Inter. Ed. Engl.*, *8*, 140 (1969).
904. A. G. Lee, *Spectrochim. Acta*, *25A*, 1841 (1969).
905. H. Bürger, U Goetze, and W. Sawodny, *Spectrochim. Acta*, *26A*, 671 (1970).
906. H. Bürger, U. Goetze, and W. Sawodny, *Spectrochim. Acta*, *26A*, 685 (1970).
907. J. R. Durig, B. A. Hudgens, Y. S. Li, and J. D. Odom, *J. Chem. Phys.*, *61*, 4890 (1974).
908. R. D. George, K. M. Mackay, and S. R. Stobart, *J. Chem. Soc. A*, 3250 (1970).
909. J. W. Anderson and J. E. Drake, *J. Chem. Soc., Dalton Trans.*, 951 (1972).
910. P. L. Goggin, R. J. Goodfellow, S. R. Haddock, and J. G. Eary, *J. Chem. Soc., Dalton Trans.*, 647 (1972).
911. H. Bürger, W. Kilian, and K. Burczyk, *J. Organometal. Chem.*, *21*, 291 (1970).
912. H. Bürger and W. Kilian, *J. Organometal. Chem.*, *18*, 299 (1969).
913. H. Bürger and W. Kilian, *J. Organometal. Chem.*, *26*, 47 (1971).
914. G. Engelhardt, P. Reich, and H. Schumann, *Z. Naturforsch.*, *22b*, 352 (1967).
915. I. R. Beattie and G. A. Ozin, *J. Chem. Soc. A*, 370 (1970).
916. G. A. Ozin, *J. Chem. Soc. A*, 1307 (1970).
917. A. Balls, N. N. Greenwood, and B. P. Straughan, *J. Chem. Soc. A*, 753 (1968).
918. R. D. George, K. M. Mackay, and S. R. Stobart, *J. Chem. Soc. A*, 3250 (1970).
919. J. R. Durig, B. A. Hudgens, and J. D. Odom, *Inorg. Chem.*, *13*, 2306 (1974).
920. H. Bürger, J. Cichon, R. Demuth, J. Grobe, and F. Höfler, *Spectrochim. Acta*, *30A*, 1977 (1974).
921. H. Schumann and U. Arbenz, *J. Organometal. Chem.*, *22*, 411 (1970).
922. J. R. Durig, J. S. DiYorio, and D. W. Wertz, *J. Mol. Spectrosc.*, *28*, 444 (1968).
923. J. Quinchon, M. Le Sech, and E. G. Trochimowski, *Bull. Soc. Chim. Fr.*, 735 (1961).

924. R. A. Burnham and S. R. Stobart, *J. Chem. Soc., Dalton Trans.*, 1269 (1973).
925. O. Kahn and M. Bigorgne, *C. R. Acad. Sci., Paris, Ser. C*, *266*, 792 (1968).
926. J. R. Durig, S. J. Meischen, S. J. Hannum, R. R. Hitch, S. K. Gondal, and C. T. Sears, *Appl. Spectrosc.*, *25*, 182 (1971).
927. D. J. Cardin, S. A. Keppie, and M. F. Lappert, *Inorg. Nucl. Chem. Lett.*, *4*, 365 (1968).
928. A. Terzis, T. C. Strekas and T. G. Spiro, *Inorg. Chem.*, *13*, 1346 (1974).
929. G. F. Bradley and S. R. Stobart, *J. Chem. Soc., Dalton Trans.*, 264 (1974).
930. N. A. D. Carey and H. C. Clark, *Chem. Commun.*, 292 (1967).
931. N. A. D. Carey and H. C. Clark, *Inorg. Chem.*, *7*, 94 (1968).
932. M. J. Ware and A. G. Cram, private commun., cited by T. G. Spiro, *Prog. Inorg. Chem.*, *11*, 1 (1970).
933. H. R. H. Patil and W. A. G. Graham, *Inorg. Chem.*, *5*, 1401 (1966).
934. P. J. Patmore and W. A. G. Graham, *Inorg. Chem.*, *5*, 1405 (1966).
935. P. J. Patmore and W. A. G. Graham, *Inorg. Chem.*, *5*, 1586 (1966).
936. W. Jetz, P. B. Simons, J. A. J. Thompson, and W. A. G. Graham, *Inorg. Chem.*, *5*, 2217 (1966).
937. L. F. Wuyts and G. P. Van der Kelen, *Spectrochim. Acta*, *32A*, 689 (1976).
938. D. C. Mente and J. L. Mills, *Inorg. Chem.*, *14*, 1862 (1975).
939. D. A. Duddell, P. L. Goggin, R. J. Goodfellow, M. G. Norton, and J. G. Smith, *J. Chem. Soc. A*, 545 (1970).
940. P. J. D. Park and P. J. Hendra, *Spectrochim. Acta*, *25A*, 227 (1969).
941. H. Schmidbaur and A. Shiotani, *Chem. Ber.*, *104*, 2821 (1971).
942. K. Nakamoto, K. Shobatake, and B. Hutchinson, *Chem. Commun.*, 1451 (1969).
943. K. Shobatake and K. Nakamoto, *J. Amer. Chem. Soc.*, *92*, 3332 (1970).
944. P. L. Goggin and J. R. Knight, *J. Chem. Soc., Dalton Trans.*, 1489 (1973).
945. C. F. Shaw and R. S. Tobias, *Inorg. Chem.*, *12*, 965 (1973).
946. H. K. Klein and H. H. Karsch, *Inorg. Chem.*, *14*, 473 (1975).

947. D. M. Adams and P. J. Chandler, *J. Chem. Soc. A*, 588 (1969).
948. D. A. Hatzenbühler, L. Andrews, and F. A. Carey, *J. Amer. Chem. Soc.*, *97*, 187 (1975).
949. L. Andrews, *J. Chem. Phys.*, *48*, 972 (1968).
950. L. Andrews, *J. Chem. Phys.*, *48*, 979 (1968).
951. L. Andrews, *Tetrahedron Lett.*, 1423 (1968).
952. L. Andrews and T. G. Carver, *J. Phys. Chem.*, *72*, 1743 (1968).
953. L. Andrews and T. G. Carver, *J. Chem. Phys.*, *48*, 896 (1968).
954. R. Robson and I. E. Dickson, *J. Organometal. Chem.*, *15*, 7 (1968).
955. A. J. Downs, *J. Chem. Soc.*, 5273 (1963).
956. R. J. Lagow, L. L. Gerchman, R. A. Jacobs, and J. A. Morrison, *J. Amer. Chem. Soc.*, *97*, 518 (1975).
957. R. A. Jacob and R. L. Lagow, *Chem. Commun.*, 104 (1973).
958. H. Bürger, J. Chichon, J. Grobe, and F. Höfler, *Spectrochim. Acta*, *28A*, 1275 (1972).
959. M. L. Denniston and D. R. Martin, *J. Inorg. Nucl. Chem.*, *37*, 1871 (1975).
960. N. Hota and C. J. Willis, cited by H. C. Clark and J. H. Tsai, *J. Organometal. Chem.*, *7*, 515 (1967).
961. H. C. Clark and C. J. Willis, *J. Amer. Chem. Soc.*, *82*, 1888 (1960).
962. H. D. Kaesz, J. R. Phillips, and F. G. A. Stone, *J. Amer. Chem. Soc.*, *82*, 6228 (1960).
963. D. Seyferth and J. M. Burlitch, *J. Organometal. Chem.*, *4*, 127 (1965).
964. A. G. Davies and T. N. Mitchell, *J. Chem. Soc. C*, 1896 (1969).
965. T. Chivers and B. David, *J. Organometal. Chem.*, *13*, 177 (1968).
966. D. Seyferth and F. M. Armbrecht, Jr., *J. Organometal. Chem.*, *16*, 249 (1969).
967. M. A. A. Beg and H. C. Clark, *Can. J. Chem.*, *40*, 393 (1962).
968. H. Bürger, J. Chichon, J. Grobe, and R. Demuth, *Spectrochim. Acta*, *29A*, 47 (1973).
969. J. D. Witt, J. W. Thompson, and J. R. Durig, *Inorg. Chem.*, *12*, 811 (1973).
970. R. C. Dobbie and B. P. Straughan, *J. Chem. Soc., Dalton Trans.*, 2754 (1973).
971. H. Bürger, J. Cichon, R. Demuth, and J. Grobe, *Spectrochim. Acta*, *29A*, 943 (1973).
972. J. D. Brown, R. C. Dobbie, and B. P. Straughan, *J. Chem. Soc., Dalton Trans.*, 1691 (1973).

973. C. F. Marsden and G. M. Sheldrick, cited by H. Burger, J. Cichon, J. Grobe, and R. Demuth, *Spectrochim. Acta, 29A,* 943 (1973).
974. J. F. Jackovitz, C. E. Falletta, and J. C. Carter, *Appl. Spectrosc., 27,* 209 (1973).
975. H. Bürger and R. Eujen, *Spectrochim. Acta, 31A,* 1655 (1975).
976. H. Bürger and R. Eujen, *Spectrochim. Acta, 31A,* 1645 (1975).
977. R. Demuth, *Z. Anorg. Allg. Chem., 418,* 149 (1975).
978. R. C. Dobbie, P. D. Gosling, and B. P. Straughan, *J. Chem. Soc., Dalton Trans.,* 2368 (1975).
979. J. E. Griffiths, *J. Chem. Phys., 41,* 3510 (1964).
980. J. E. Griffiths and A. L. Beach, *J. Chem. Phys., 44,* 2686 (1966).
981. E. A. Cohen and C. D. Cornwell, *Inorg. Chem., 7,* 398 (1968).
982. R. R. Holmes and M. Fild, *J. Chem. Phys., 53,* 4161 (1970).
983. K. G. Sharp and T. D. Coyle, *Inorg. Chem., 11,* 1259 (1972).
984. J. L. Margrave, K. G. Sharp, and P. W. Wilson, *J. Inorg. Nucl. Chem., 32,* 1817 (1970).
985. H. C. Clark and C. J. Willis, *J. Amer. Chem. Soc., 84,* 898 (1962).
986. M. D. Curtis and P. Wolber, *Inorg. Chem., 11,* 431 (1972).
987. R. C. Dobbie and B. P. Straughan, *Spectrochim. Acta, 27A,* 255 (1971).
988. R. C. Dobbie, M. J. Hopkinson, and B. P. Straughan, *J. Mol. Struct., 23,* 141 (1974).
989. H. Bürger, J. Cichon, R. Demuth, J. Grobe, and F. Höfler, *Z. Anorg. Allg. Chem., 396,* 199 (1973).
990. J. W. Thompson, J. D. Witt, and J. R. Durig, *Inorg. Chem., 12,* 2124 (1973).
991. H. W. Schiller and R. W. Rudolph, *Inorg. Chem., 10,* 2500 (1971).
992. L. Maya and A. B. Burg, *Inorg. Chem., 14,* 698 (1975).
993. S. Ansari, J. Grobe, and P. Schmid, *J. Fluorine Chem., 2,* 281 (1973).
994. T. N. Bell, B. J. Pullman, and B. O. West, *Aust. J. Chem., 16,* 722 (1963).
995. K. Ohno and H. Murata, *Bull. Chem. Soc. Japan, 45,* 3333 (1972).
996. J. R. Durig and C. W. Hawley, *J. Chem. Phys., 59,* 1 (1973).

997. B. J. Van der Veken and M. A. Herman, *J. Mol. Struct.*, *28*, 371 (1975).
998. B. J. Van der Veken, *J. Mol. Struct.*, *25*, 75 (1975).
999. D. Seyferth and H. D. Simmons, *J. Organometal. Chem.*, *6*, 306 (1966).
1000. Y. Imai and K. Aida, *Spectrochim. Acta*, *28A*, 517 (1972).
1001. J. M. Bellama and A. G. MacDiarmid, *J. Organometal. Chem.*, *18*, 275 (1969).
1002. A. L. Reingold and J. A. Bellama, *J. Organometal. Chem.*, *102*, 437 (1975).
1003. F. A. Cotton and J. A. McCleverty, *J. Organometal. Chem.*, *4*, 490 (1965).
1004. F. A. Cotton and R. M. Wing, *J. Organometal. Chem.*, *9*, 511 (1967).
1005. M. P. Johnson, *Inorg. Chim. Acta*, *3*, 232 (1969).
1006. T. G. Appleton, M. H. Chisholm, H. C. Clark, and L. E. Manzer, *Inorg. Chem.*, *11*, 1786 (1972).
1007. P. R. McGee, F. F. Cleveland, A. G. Meister, and C. E. Decker, *J. Chem. Phys.*, *21*, 242 (1953).
1008. R. B. King and M. B. Bisnette, *J. Organometal. Chem.*, *2*, 15 (1964).
1009. H. C. Clark and J. H. Tsai, *J. Organometal. Chem.*, *7*, 515 (1967).
1010. W. A. G. Graham, *Inorg. Chem.*, *7*, 315 (1968).
1011. M. B. Hall and R. F. Fenske, *Inorg. Chem.*, *11*, 768 (1972).
1012. E. Pitcher and F. G. A. Stone, *Spectrochim. Acta*, *18*, 585 (1962).
1013. J. B. Wilford, P. M. Treichel, and F. G. A. Stone, *J. Organometal. Chem.*, *2*, 119 (1964).
1014. P. M. Treichel and G. P. Werber, *J. Organometal. Chem.*, *7*, 157 (1967).

II

Noncyclic, Unsaturated Organometallic Derivatives

A. VINYL COMPOUNDS

The infrared and Raman spectra of $(CH_2=CH)_4Si$ are illustrated in Figure 2.1 while the frequency ranges associated with the vinyl group are summarized in Table 2.1. In addition to the bands associated with the fundamental modes of the vinyl group, several combination bands are observed in the infrared spectrum from approximately 2000 to 1850 cm^{-1}.

The $\nu(CH_2)$ and $\nu(CH)$ modes are relatively intense in the infrared and Raman spectra. The $\nu(C=C)$ mode shows weak to moderate infrared intensity but very strong Raman intensity. The infrared intensity of the $\nu(C=C)$ mode of the $M-CH=CH_2$ (M=C,Si,Ge,Sn) unit has been reported to decrease as the mass of M increases (2). The Raman intensity of the $\nu(C=C)$ mode for $(CH_3)_3MCH=CH_2$ (M=C,Si,Ge,Sn) has also been reported to decrease as the mass of M increases; this has tentatively been attributed to the effect of $p\pi \rightarrow d\pi$ bonding from the filled vinyl π-orbitals to empty metal *d*-orbitals. The higher frequency of the $\nu(C=C)$ mode in free ethylene (1623 cm^{-1}) (4) than in vinyl derivatives has also been attributed in part to the presence of $p\pi \rightarrow \pi d$ bonding in the vinyl derivatives (5,6). Therefore, while in the di(tetrahydrofuran) and mono(1,2-dimethoxyethane) adducts of $CH_2=CHTiCl_3$ the $\nu(C=C)$ mode has been assigned at 1590 cm^{-1} (7), in $(\eta^5-C_5H_5)_2Ti(CH=CH_2)Cl$, where $p\pi \rightarrow d\pi$ bonding is unlikely since the empty titanium *d*-orbitals are engaged in bonding with the $\eta^5-C_5H_5$ rings, the $\nu(C=C)$ mode has been assigned at 1635 cm^{-1} (8).

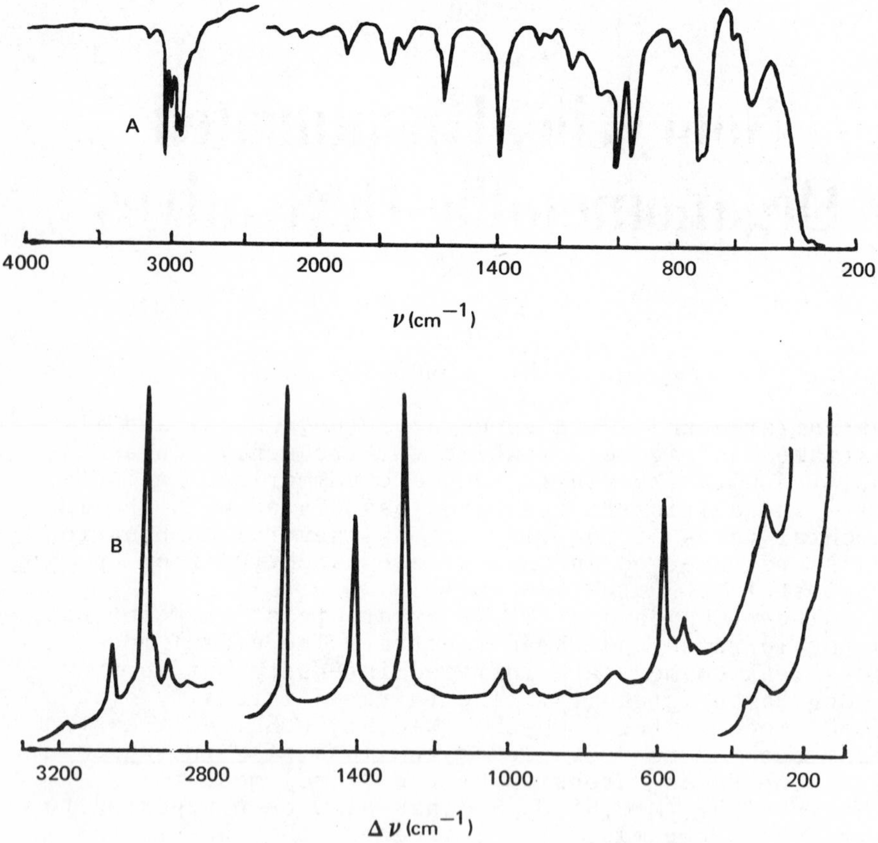

Figure 2.1. Liquid-phase (A) infrared and (B) Raman spectra of $(CH_2{=}CH)_4Si$ *(1).*

Although the $\delta_s(CH_2)$ and $\delta(CH)$ modes have consistently been assigned to from moderate to strong intensity infrared and Raman bands, there is less general agreement on the assignment of the other C-H deformation modes, namely, $\rho_t(CH_2)$, $\rho_w(CH_2)$, $\rho_r(CH_2)$, and $\pi(CH)$. In Table 2.2, the vinyl mode assignments are compared for the $(CH_2{=}CH)_nM$ derivatives of the main group elements while the skeletal mode assignments for several of these derivatives, as well as for $(CH_2{=}CH)_mMX_n$ (X=H, halogen), are summarized in Table 2.3.

TABLE 2.1
Frequency Ranges (cm^{-1}) for the Normal Modes of the Vinyl Group

Mode	Frequency Range
$\nu_a(CH_2)$	3100-2900
$\nu_s(CH_2)$	
$\nu(CH)$	
$\nu(C{=}C)$	1630-1565
$\delta(CH_2)$	1425-1385
$\delta(CH)$	1280-1235
$\rho_t(CH_2)$	1025-835
$\rho_w(CH_2)$	
$\rho_r(CH_2)$	
$\pi(CH)$	730-450

Two complete infrared and Raman studies have been reported for $(CH_2{=}CH)_3B$. In the first, which included data for the liquid and vapor phases, it was concluded that there is strong vibrational evidence for conjugation between the empty $p\pi$ orbital on the sp^2 hybridized boron and the π-orbitals of the vinyl groups (11). In the second, which included data for all three phases, no discussion was given of the possibility of π-bonding between the boron atom and the vinyl groups (12). It was concluded in this second study, however, that the vapor and liquid phases consist of two conformers: a planar form of C_{3v} symmetry and a nonplanar form in which the vinyl groups are twisted out of the plane. In the solid state, only the planar form was found. The position of the $\nu(C{=}C)$ mode (1629 to 1910 cm^{-1}) in the infrared spectra of $(CH_2{=}CH)_{3-n}BX_n$ (n=1 to 3,X=F,Cl) (24) and $(CH_2{=}CR)BX_2$ ($R{=}H,OC_4H_9;X{=}OH,CH_3$) (25) has been presented as evidence for the presence of a π-interaction between the boron atom and the vinyl group(s). In more recent studies of $CH_2{=}CHBF_2$ that included infrared, Raman, microwave, and dipole moment data, it has been concluded that there is no evidence for or against the presence of π-overlap in the B-C bonds (16,17). The infrared spectrum of $(CH_2{=}CH)_3Ga$ in CCl_4 has been reported (27). This compound is a dimer in solution and may exist as a dimer or higher polymer in the pure liquid or solid states.

There is the possibility of a π-interaction between empty *d*-orbitals on the group IVb metals and the π-orbitals of the vinyl group. A vibrational study of

TABLE 2.2
Vinyl Mode Assignments (cm^{-1}) for $(CH_2{=}CH)_nM$ Derivatives of the Main Group Elements

	Compound[a]									
Mode	R_2Zn[b]	R_2Hg[c]	R_3B[d]	R_4Si		R_4Ge[e]	R_4Sn		R_4Pb[e]	R_3Pb
$\nu_a(CH_2)$	3000	3051	3080	3055[e]	2981[f]	3052	3041[e]	3025[g]	3038	3095
$\nu_s(CH_2)$	2900	2944	2967	2952	3062	2947	2936	2910	2931	2970
$\nu(CH)$	2930	2985	2990	2973	3014	2981	2977	2955	2975	3005
$\nu(C{=}C)$	1565	1603	1604	1592	1591	1595	1583	1588	1580	1595
$\delta_s(CH_2)$	1390	1402	1423	1406	1404	1399	1395	1397	1389	1392
$\delta(CH)$	1258	1263	1301	1273	1269	1265	1257	1248	1227	1262
$\rho_t(CH_2)$	990	1019	1017	1013	1008	1008	999	1000	989	982
$\rho_w(CH_2)$	952	937	970	957	958	949	950	948	942	948
$\rho_r(CH_2)$		1019	1088	1013	1062	1008	999		989	918
$\pi(CH)$		495 456	748	524		514	513		453	

[a] $R = CH_2{=}CH$.
[b] Reference 9.
[c] Reference 10.
[d] References 11 and 12.
[e] Reference 13.
[f] Reference 1.
[g] References 9 and 14.

TABLE 2.3
Skeletal Mode Assignments (cm^{-1}) for Vinyl Derivatives of the Main Group Elements

Compound	ν(MC)	ν(MX)	δ(MCC)	$\delta_a(MX_3)$	$\delta_s(MX_3)$	$\rho_r(MX_3)$	Refs.
$(CH_2{=}CH)_2Hg$	541		291				10
	513						
$CH_2{=}CHHgCl$	541	343	295				15
$CH_2{=}CHHgBr$	538	228	292				15
$CH_2{=}CHHgI$	527	188	287				15
$(CH_2{=}CH)_3B$	1186[a]		327				11,12
	1156[b]		317				
	651[c]						
$CH_2{=}CHBF_2$	765	1375	221				16,17
		1318					
$(CH_2{=}CH)_4Si$	725		522[d]				1,13
	578		490[d]				
	541		382[e]				
			336[e]				
			305[e]				
$CH_2{=}CHSiH_3$	714	2161	282	944	920	623	18
		2150		920		431	
$CH_2{=}CHSiD_3$	640	1575	262	681	724	509	18
		1553		672		374	
		1545					
$CH_2{=}CHSiF_3$	687	970	287	360	433	217	19
		890				170	
$(CH_2{=}CH)_3SiF$		870					20
$(CH_2{=}CH)_4Ge$	600		355				14
	561		322				

TABLE 2.3 (continued)

Compound	$\nu(MC)$	$\nu(MX)$	$\delta(MCC)$	$\delta_a(MX_3)$	$\delta_s(MX_3)$	$\rho_r(MX_3)$	Refs.
$CH_2=CHGeH_3$	639	2090	267	887	834	543	21
				876		430	
$CH_2=CHGeD_3$	639	1497		637	601	454	21
				629		328	
$CH_2=CHGeCl_3$	621	428	340	173	173	155	22
		404					
$CH_2=CHGeBr_3$	613	276	308	122	122	91	22
		256					
$(CH_2=CH)_4Sn$	531		309				13,14
	490		304				
$(CH_2=CH)_3SnCl$	544	351					14
	525						
	480						
$(CH_2=CH)_2SnCl_2$	557	366					14
	528						
	490						
$(CH_2=CH)_4Pb$	495		280				13
	479		252				
$CH_2=CHPF_2$	742	829					23

[a]Due to the $\nu_a(^{10}BC)$ mode.
[b]Due to the $\nu_a(^{11}BC)$ mode.
[c]Due to the $\nu_s(BC)$ mode.
[d]Out-of-plane mode.
[e]In-plane mode.

$(CH_2{=}CH)_4Si$ included only weak evidence for such an interaction (1). Similarly, no convincing evidence could be found for such an interaction in a vibrational study of $(CH_2{=}CH)_6M_2$ (M=Si,Ge,Sn) (28). Infrared assignments have been made for $CH_2{=}CHSi(CH_3)_nCl_{3-n}$ (n=0 to 3) (29), $CH_2{=}CHSi(CH_3)_n(OC_2H_5)_{3-n}$ (n=0 to 2), and $(CH_2{=}CH)_{3-n}Sn(CH_3)_nX$ (n=0,1,3;X=I,SCN,O_2CCH_3) (30), while the infrared spectra have been illustrated for $CH_2{=}CHM(C_6H_5)_3$ (M=Si,Ge,Sn,Pb) and $(CH_2{=}CH)_2M(C_6H_5)_2$ (M=Si,Ge,Sn) (31). The infrared spectrum of $(CH_2{=}CH)_3SnO_2CH$ in CCl_4 is consistent with an equilibrium between the monomeric and dimeric species (32). The solid-state infrared spectrum of this compound suggests the presence of nonplanar C_3Sn groups (32). In the infrared spectra of $(CH_2{=}CH)_3SnO_2CR$, the positions of the ν(CO) modes for R=CH_3, C_2H_5, and CH_2Cl are similar to those of the corresponding carboxylate ions and therefore indicate a polymeric solid-state structure in which the $(CH_2{=}CH)_3Sn$ units are associated through symmetrically bridged carboxylate groups (33,34). The larger separation of the ν(CO) modes for R=$CHCl_2$ and CF_3 implies the presence of monomeric solid-state structures with ester like carboxylate ligands (33,34). These conclusions are supported by the $CHCl_3$ solution data where there is a marked increase in the separation of the ν(CO) mode frequencies on dissolution for R=CH_3 and CH_2Cl but very little change for R=$CHCl_2$ and CF_3 (34). Molecular weight data have shown that for R=CF_3 and CH_2Cl, monomeric structures are present in $CHCl_3$ although a trimeric structure is found for R=CH_3 (33). The complexity of the infrared spectra in the ν(SnC) mode region for these carboxylates suggests that the three vinyl groups are not all in the same equatorial plane (33,34). The ν(C=C) mode for $(R_2C_2SnX_2)_2$ (R=O_2CCH_3;X=Cl,Br,I) (2.1) has been assigned from

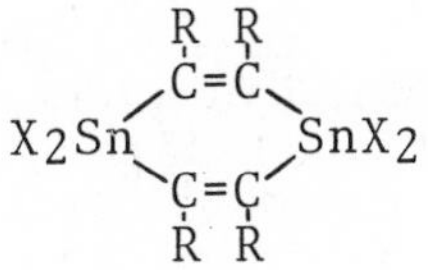

(2.1)

1550 to 1533 cm^{-1} in the infrared spectra and at approximately 1570 cm^{-1} in the Raman spectra. The coordination number of the tin atoms is six because of intermolecular coordination via the carbonyl groups; an infinite coordination polymer results. While the *trans* isomers of $(CH_3)_3M[CH=C(CH_3)H]$ $(M=Si,Ge,Sn)$ show characteristic infrared bands from 1620 to 1605 cm^{-1} and 987 to 980 cm^{-1}, the *cis* isomers have a characteristic infrared band at 1610 to 1605 cm^{-1} but no bands in the 1050 to 925 cm^{-1} region (36). The *trans* isomer of $(C_6H_5)(CH_3)_2SiCH=CHSn(C_6H_5)_3$ exhibits a characteristic infrared band at 1012 cm^{-1} that has been assigned to the olefinic $\pi(CH)$ mode (37).

The infrared and Raman assignments for $K_6[Co_2-(CN)_{10}C_2H_2]\cdot 4H_2O$ and its totally deuterated analog show both compounds to have a *trans* structure (2.2) with the

$$\underset{(NC)_5Co}{\overset{H}{\diagdown}}C=C\underset{H}{\overset{Co(CN)_5}{\diagup}}$$

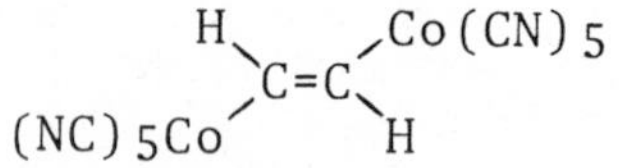

(2.2)

$\nu(C=C)$ modes assigned at 1518 and 1506 cm^{-1}, respectively (38). Partial infrared data have been reported for vinyl derivatives of titanium (8), iron (39), rhenium (39), and platinum (40).

Vibrational data for halovinyl derivatives include some tentative infrared assignments for perfluoro-2-propenyl mercury compounds ($\nu(C=C)=1710$ cm^{-1}) (41) and $(CF_2=CF)_nBX_{3-n}$ $(n=1,X=F,Cl;n=2,X=F;n=3)$ ($\nu(C=C)=1725$ to 1677 cm^{-1}) (42,43). For the *cis* isomers of $(ClCH=CH)_2Hg$, $ClCH=CHHgX$ $(X=Cl,Br)$, $(ClCH=CH)_2TlCl$, and $(ClCH=CH)_nSnCl_{4-n}$ $(n=2,3)$, the $\delta(CH)$ and $\pi(CH)$ modes have been assigned in the 1275 to 1260 cm^{-1} and 920 to 915 cm^{-1} ranges, respectively, while for the *trans* isomers these modes have been assigned in the 1157 to 1140 cm^{-1} and 950 to 935 cm^{-1} ranges, respectively. For $ClCH=CHSiCl_3$ and $CCl_2=CClSi(CH_3)_3$, the $\nu(C=C)$ mode has been assigned at 1560 cm^{-1} (45) and 1545 cm^{-1} (46), respectively. Infrared data and some assignments have also been reported for several chlorinated vinyl complexes of platinum (40) as well as for $[(CF_3)_3Sn)(CF_3)C=C(CF_3)][Fe(CO)_2(\eta^5-C_5H_5)]$ (47).

Additional vibrational data are discussed at the end of Sec. B of this chapter for vinyl derivatives in which the $C=C$ bond is π-bonded to another metal atom to give compounds in which the vinyl group bridges two metal atoms by means of both a σ- and π-bond.

B. π-BONDED ETHYLENE AND RELATED MONOALKENE COMPOUNDS

Ethylene and similar olefins commonly form π-bonds with transition metals. One of the most studied of these derivatives has been Zeise's salt, $K[(C_2H_4)PtCl_3]\cdot H_2O$.

The first infrared studies of Zeise's salt (cf. Figure 2.2 (48)) and other olefin complexes focused on the assignment of the olefin ν(C=C) mode, which is expected at a lower frequency in the complex than in the free olefin. The change in the ν(C=C) mode frequency on complexation should give a measure of the strength of the metal-olefin bond. Although most workers, on the basis of infrared data alone, have consistently assigned the ν(C=C) mode of Zeise's salt and other platinum-ethylene complexes to a weak band at 1530 to 1500 cm^{-1} (48-54) as compared to a value of 1623 cm^{-1} for this mode in free ethylene (4), one group preferred to assign this band to the $\delta_s(CH_2)$ mode (55). They considered that the platinum atom of Zeise's salt formed a three-membered ring on coordination with ethylene resulting in partial or complete loss of the double-bond character of ethylene. Since the $\delta_s(CH_2)$ mode of ethylene oxide had been assigned at 1498 cm^{-1}, it seemed reasonable to make a similar assignment for Zeise's salt. This conclusion was originally questioned since both K[(*cis*-CH_3CH=$CHCH_3$)$PtCl_3$] (56) and other substituted olefins (57) that do not contain a CH_2 group exhibit an infrared band at approximately 1505 cm^{-1}. Also, a band at 1504 cm^{-1} for K[(CH_2=$CHCH_3$)$PtCl_3$]·H_2O showed a small isotopic shift on deuteration, although a large shift is expected if this band arose from a $\delta_s(CH_2)$ mode.

The matter therefore seemed settled until the Raman spectrum of Zeise's salt was recorded (cf. Figure 2.3 (59)). Since a Raman band at 1240 cm^{-1} (which is absent in the infrared spectrum) is much stronger than the 1515 cm^{-1} Raman band, and because the ν(C=C) mode is expected to show strong Raman intensity, the 1240 cm^{-1} band was assigned to the ν(C=C) mode with the 1515 cm^{-1} band reassigned to the $\delta_s(CH_2)$ mode (59). Based on these assignments and the apparently large change in the ν(C=C) mode frequency on coordination of the ethylene, it might at first appear that the platinum-ethylene bond is much stronger than had been thought previously. On comparing these Raman data for Zeise's salt with Raman data for the deuterated analog of Zeise's salt (59), other platinum-olefin complexes (60), [C_2H_4Ag]BF_4 (60), and $C_2H_4Fe(CO)_4$ (61) this is seen to be an oversimplification.

In any analysis of the data for metal-ethylene complexes, account must be taken of the effect of coupling between the ν(C=C) and $\delta_s(CH_2)$ modes, which is an important factor in both free and complexed ethylene (48, 60). Since the ν(C=C) mode is not pure, the frequency of the band that is predominantly ν(C=C) in character is

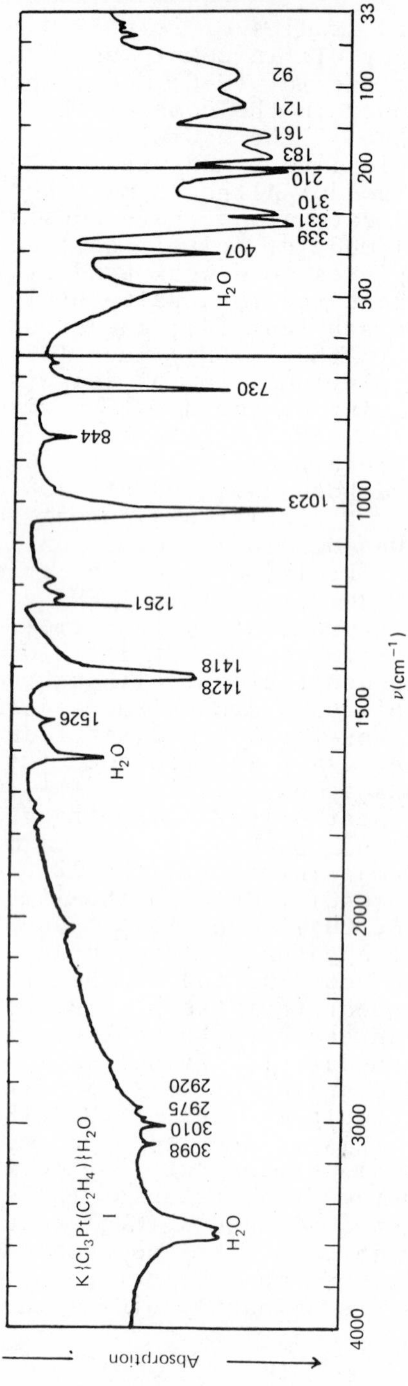

Figure 2.2. Infrared spectrum of Zeise's salt from 4000 to 33 cm^{-1} *(48).*

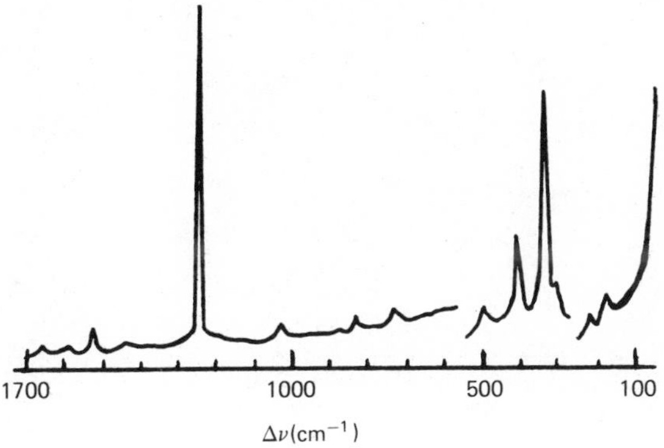

Figure 2.3. Raman spectrum (1700 to 100 cm^{-1}) of an aqueous solution of K[$C_2H_4PtCl_3$]·H_2O at 48°C (59).

influenced not only by the strength of the metal-ethylene bond but also by the extent to which the ν(C=C) and $\delta_s(CH_2)$ modes are coupled.

In the following discussion, reference will be made to the ν(C=C) and $\delta_s(CH_2)$ mode assignments summarized in Figure 2.4 for $CH_2=CH_2$, $CD_2=CD_2$, and various metal-ethylene complexes. The decrease in the ν(C=C) mode frequency on coordination would bring the ν(C=C) and $\delta_s(CH_2)$ modes closer in frequency and increase the extent to which they couple. The coupling would then produce an increase in the frequency of the higher-frequency band and a decrease in the frequency of the lower-frequency band. A relatively strong metal-ethylene interaction could force the ν(C=C) mode to a position where there is a cross over in the nature of the bands assigned as predominantly ν(C=C) and $\delta_s(CH_2)$. This is apparently what has happened to K[$C_2H_4PtCl_3$] and $C_2H_4Fe(CO)_4$ for which bands at approximately 1500 cm^{-1} now become predominantly $\delta_s(CH_2)$ in character while bands at approximately 1200 cm^{-1} have been assigned to the ν(C=C) mode. The silver-olefin bond is considered weaker than the platinum-olefin bond (60) and the shift in the ν(C=C) mode of ethylene on coordination in [C_2H_4Ag]BF_4 is not large enough to produce the cross over; the ν(C=C) mode is therefore assigned at 1579 cm^{-1}. Figure 2.4 also shows that the frequency separation of the two modes is greater in $CD_2=CD_2$ than in $CH_2=CH_2$;

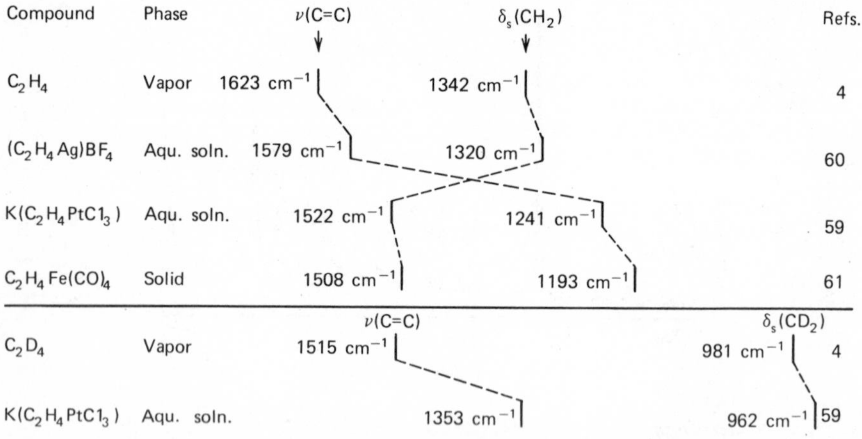

Compound	Phase	ν(C=C)	δ_s(CH$_2$)	Refs.
C$_2$H$_4$	Vapor	1623 cm^{-1}	1342 cm^{-1}	4
(C$_2$H$_4$Ag)BF$_4$	Aqu. soln.	1579 cm^{-1}	1320 cm^{-1}	60
K(C$_2$H$_4$PtCl$_3$)	Aqu. soln.	1522 cm^{-1}	1241 cm^{-1}	59
C$_2$H$_4$Fe(CO)$_4$	Solid	1508 cm^{-1}	1193 cm^{-1}	61

		ν(C=C)	δ_s(CD$_2$)	
C$_2$D$_4$	Vapor	1515 cm^{-1}	981 cm^{-1}	4
K(C$_2$H$_4$PtCl$_3$)	Aqu. soln.	1353 cm^{-1}	962 cm^{-1}	59

Figure 2.4. The ν(C=C) and δ_s(CH$_2$) mode assignments for ethylene-d$_0$, ethylene-d$_4$, and metal-ethylene complexes.

therefore, coupling, though still present, is less extensive in CD$_2$=CD$_2$ and a cross over is less likely on coordination. Therefore, a better measure of the lowering of the ν(C=C) mode frequency of ethylene on coordination in Zeise's salt is obtained by comparing frequency values for the deuterated complexes in which the ν(C=C) mode frequency decreases by approximately 10.5% on complex formation.

Similarly, in a complete infrared and Raman study of K[*cis*-CH$_3$CH=CHCH$_3$PtCl] and K[*trans*-CH$_3$CH=CHCH$_3$PtCl$_3$], it has been concluded that Raman bands at 1242 and 1230 cm^{-1} in the former compound and 1263 cm^{-1} in the latter compound arise from the ν(C=C) mode, while the bands at 1503 and 1526 cm^{-1}, respectively, are due to C-H deformation modes (62). This contrasts with the infrared study mentioned previously (56) in which the ν(C=C) mode was assigned at approximately 1500 cm^{-1}. Since, however, there is no δ_s(CH$_2$) mode in [(CH$_3$)$_2$C=C(CH$_3$)$_2$PtCl$_2$]$_2$, an infrared band at 1500 cm^{-1} for this compound has been assigned to the ν(C=C) mode as compared to a value of 1670 cm^{-1} for this mode in the uncoordinated ligand (60). These data indicate a 10% decrease in the ν(C=C) mode frequency on coordination, which compares favorably with similar data discussed above for free CD$_2$=CD$_2$ and deuterated Zeise's salt. Therefore, it has been concluded that in totally substituted olefin complexes such as $[(CH_3)_2C{=}C(CH_3)_2PtCl_2]_2$ the 1500 cm^{-1} band is mainly $\nu(C{=}C)$ in character.

Controversy has not only arisen over the assignment of the $\nu(C{=}C)$ mode of Zeise's salt but also over the assignment of the platinum-ethylene skeletal modes. The platinum-ethylene interaction has been pictured in terms of two model extremes (2.3 and 2.4). Actually, the bonding description is somewhere between these two

(2.3) (2.4)

extremes. For both models, three platinum-ethylene skeletal modes (exclusive of the twisting mode) can be drawn (2.5 to 2.7). Although modes 2.5 and 2.7 can be

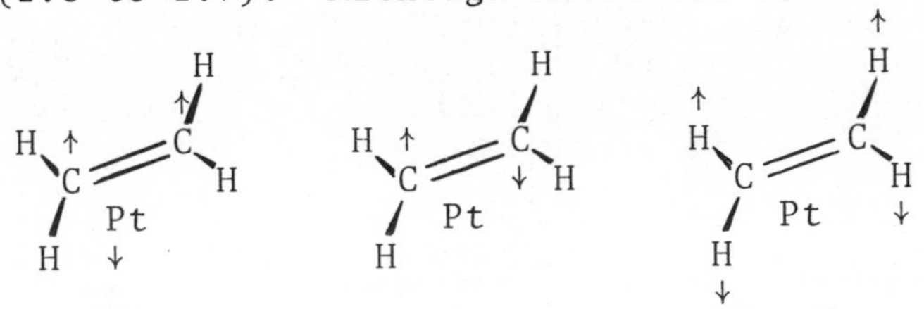

(2.5) (2.6) (2.7)

described as platinum-ethylene stretching and tilting modes, respectively, irrespective of which model is used, mode 2.6 can be described as a tilting mode if the first model is used or as an antisymmetric platinum-ethylene stretching mode if the second model is used. If the first model is closer to reality, mode 2.6 would be expected at a lower frequency than mode 2.5. If, however, the second model is more realistic, mode 2.6 might be expected to be at a frequency much closer to or even higher than that of mode 2.5. One group has adopted the first model in assigning the low-frequency data for Zeise's salt with the $\nu(Pt\text{-}C_2H_4)$ mode assigned to an infrared band at 407 cm^{-1} and the other platinum-ethylene skeletal modes assigned to bands below 210 cm^{-1} (48). An infrared band at 493 cm^{-1} was attributed to a lattice water band for Zeise's salt monohydrate (48) or, since this band appeared with diminished intensity in anhydrous

Zeise's salt, to a combination or overtone band (53). Several other groups have preferred the second model for Zeise's salt (52,59,63,64) and have assigned the two expected $\nu(Pt-C_2H_4)$ modes to bands at 407 cm^{-1} (which is polarized in the Raman spectrum and therefore the $\nu_s(Pt-C_2H_4)$ mode (59)) and 493 cm^{-1}. The 493 cm^{-1} band has been attributed to a fundamental mode by these workers since it appears in both the infrared (52,63) and Raman (59,64) spectra of the monohydrate and anhydrous forms of Zeise's salt and its bromo analog indicating that it is not due to water of crystallization. It has also been argued (64) that this cannot be an overtone or combination band since Zeise's dimer, $(C_2H_4PtCl_2)_2$, shows a band at 490 cm^{-1} (59,64) although the low-frequency spectra of Zeise's salt and Zeise's dimer are otherwise very different.

The low-frequency infrared spectra have been assigned for $C_2H_4PtCl_2L$ (L=NH_3,CH_3NH_2,$(CH_3)_2NH$,NC_5H_4R (R=H,CH_3,Cl,etc.)) (63). The $\nu_a(Pt-C_2H_4)$ and $\nu_s(Pt-C_2H_4)$ modes have been assigned at 489 and 404 cm^{-1}, respectively, for K[*cis*-CH_3CH=$CHCH_3PtCl_3$] and 493 and 386 cm^{-1}, respectively, for K[*trans*-CH_3CH=$CHCH_3PtCl_3$] (62). The low-frequency infrared and Raman spectra of $C_2H_4Fe(CO)_4$ have been assigned ($\nu(Fe-C_2H_4)$=356 cm^{-1} (A_1), Fe-C_2H_4 tilt=400 and 305 cm^{-1}) while infrared bands at 399 and 393 cm^{-1} for $[(C_2H_4)_2RhCl]_2$ and $[(C_2H_4)_2RhBr]_2$, respectively, have been assigned to the $\nu(Rh-C_2H_4)$ mode (65).

Complete assignments have been made of the infrared (52,53,59) and Raman (59) spectra of Zeise's dimer and the infrared spectra of $C_2H_4Pt(Cl)L$ (L=Acac (66), glycino (54)) and the palladium analog of Zeise's dimer (53). An infrared analysis has been made of the metal-sensitive vibrations in complexes formed between three different dipeptides and Zeise's salt (67). Solid-state and solution infrared and Raman data for MX_2L (M=Pt,Pd; X=Cl,Br,I;L=$(C_6H_5)_nP(CH_2CH_2CH{=}CH_2)_{3-n}$,n=0 to 2) indicate that only one olefin is coordinated to the M atom (67). The infrared and Raman spectra of *trans*-$C_2H_4PtCl_2$(4-R-Py) (R=NH_3,CH_3,C_2H_5,CH_2OH,H,Cl,Br,$C_2H_5CO_2$,COOH,$COCH_3$,CN) show that the R group has no influence on the $\nu(C{=}C)$ or platinum-ethylene mode frequencies but strongly influences the position of the $\nu(PtN)$ mode, which has a minimum value for R=H (69). Less complete infrared data, including the $\nu(C{=}C)$ mode assignment, have been reported for C_2H_4Ag (70), $(C_2H_4)_nCu$ (n=1 to 3) (70), $(C_2H_4)_3Co$ (70), $(C_2H_4)_3Ni$ (70,71), $(C_2H_4)_3Pd$ (72), $[C_6H_3(CH_3)_3]M(CO)_2C_2H_4$ (M=Cr,Mo) (73), η^5-$C_5H_5Mn(CO)_2C_2H_4$ (74), and $[(C_2H_4)_2RhCl]_2$ (75). Raman data have been reported for several alkene complexes of silver(I) (76). The infrared spectra of ethylene chemisorbed on silica-supported palladium and platinum catalysts show bands characteristic of a π-bonded species at approximately 3000 and 1500 cm^{-1} and a σ-bonded species MCH_2CH_2M at 2875 and 2790 cm^{-1} (77).

The ν(CO) and ν(C=C) modes have been assigned for partially and totally halogenated monoolefin complexes of $Fe(CO)_4$ (78). The infrared spectra have been analyzed for several iridium complexes of $CF_2{=}CF_2$ (79). These iridium complexes exhibited a strong intensity band at approximately 800 cm^{-1}, which correlates with a band at 778 cm^{-1} in free $CF_2{=}CF_2$, which is mainly ν(CF) in character. Also, two or three strong intensity bands were observed between 1170 and 1000 cm^{-1}. Lastly, most of these iridium complexes showed a relatively intense band between 1600 and 1340 cm^{-1}, which was assigned as mainly ν(C=C) in character; the ν(C=C) mode of free $CF_2{=}CF_2$ has been assigned at 1872 cm^{-1}.

Approximate normal coordinate analyses for the C_2H_4Pt and C_2H_4Fe fragments of Zeise's salt and $C_2H_4Fe(CO)_4$, respectively, show that Pt^{II} interacts more strongly with $CH_2{=}CH_2$ than Fe^0 (80). Additional approximate normal coordinate calculations have been reported for Zeise's salt (48,52,59), Zeise's dimer (53), and its palladium analog (53). A normal coordinate analysis for the hypothetical complex $C_2H_4TlH_2O^+$ has been reported to show in principle the possibility of kinetic coupling between internal ligand modes and the skeletal modes of an organometallic complex (81).

Compounds are known in which a monoolefin acts simultaneously in a σ- and π-bonded fashion. This appears to be true in $(CH_3)_3MCH{=}CH_2 \cdot CuCl$ (M=Si,Sn) and $R_2M(CH{=}CH_2)_2 \cdot 2CuCl$ (M=Si,Sn,R=CH_3;M=Sn,R=*n*-C_4H_9) (82). The ν(C=C) mode frequency in these vinyl derivatives (1595 to 1578 cm^{-1}) has been reported to decrease to 1508 to 1490 cm^{-1} when the copper atom coordinates to the C=C bond. It has been proposed that these complexes are probably dimerized or polymerized through chloride bridges across the copper atoms. The ν(C=C) mode of $(CH_3)_2Si(CH{=}CH_2)_2$ was lowered by approximately 100 cm^{-1} on complex formation with $Fe(CO)_3$ (2.8) (83). Similar complexes, with both σ- and π-bonded monoolefins have been reported when vinyl derivatives of iron, tungsten, or rhenium react with iron nonacarbonyl. The ν(C=C) mode of these complexes has been assigned from approximately 1430 to 1396 cm^{-1} (39). Although a structure has been considered for $I_3In[CH_2{=}CHP(C_6H_5)_2]$ in which both

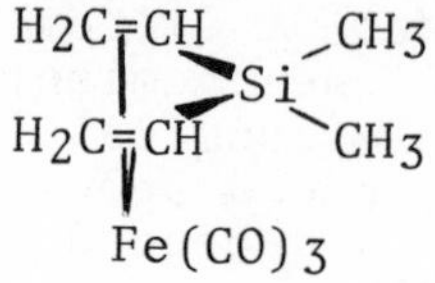

(2.8)

the phosphorus atom and olefin C=C bond are coordinated to the indium atom, it has been concluded on the basis of infrared data that this compound is polymeric rather than chelated via olefin C=C bonds (84).

C. ACETYLENIC AND OTHER ALKYNE COMPOUNDS

As was true for alkenes, alkynes can form σ- and/or π-bonds with metals. The most characteristic feature of the vibrational spectra of σ-type derivatives is the signal due to the $\nu(C{\equiv}C)$ mode. This mode gives rise to a relatively strong intensity signal in both the infrared and Raman spectra, as illustrated for $(CH_3C{\equiv}C)_3As$ in Figure 2.5 (85).

The infrared data for the $\nu(C{\equiv}C)$ mode of several group Ia alkyne derivatives are summarized in Table 2.4 (86). The facts that this mode is lower in frequency in the derivatives than in the free alkynes and that the frequency decreases from lithium to cesium are interpreted to indicate an increase in the importance of structure 2.10 in a resonance between 2.9 and 2.10.

$$R\text{-}C{\equiv}C^{-} \quad \leftrightarrow \quad R\text{-}\bar{C}{=}C$$

(2.9) (2.10)

In an infrared study of the the group IIa derivatives $(C_6H_5C{\equiv}C)_2M$ (M=Ca,Sr,Ba), the $\nu(C{\equiv}C)$ mode frequency is also lower in the complexes than in $C_6H_5C{\equiv}CH$ and decreases from the calcium (2036 cm^{-1}) to the strontium (2023 cm^{-1}) to the barium (2017 cm^{-1}) derivative, although it has been noted that coupling of the $\nu(C{\equiv}C)$ and $\nu(CH)$ modes in $C_6H_5C{\equiv}CH$ makes a comparison with the complexes difficult (87). The intensity of the $\nu(C{\equiv}C)$ mode is reduced in these derivatives relative to that found in $C_6H_5C{\equiv}CH$; the same is ture for $C_6H_5C{\equiv}CNa$. An infrared band at 324 cm^{-1} was assigned to a $\nu(CaC{\equiv})$ mode of $(C_6H_5C{\equiv}C)_2Ca$. It has been concluded that solid $(C_6H_5C{\equiv}C)_2Ca$ has a polymeric structure similar to that of $(CH_3)_2Be$ (1.4) but that it dissolves as a low polymer

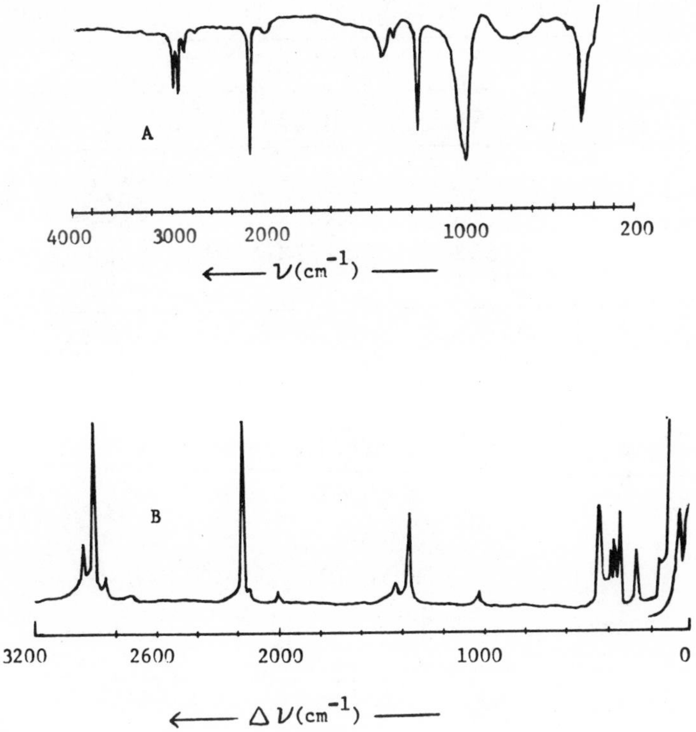

Figure 2.5. The CCl_4 solution (A) infrared spectrum and solid-state (B) Raman spectrum of $(CH_3C{\equiv}C)_3As$ (85).

which slowly depolymerizes in solution to produce mainly a dimer. The isomorphism shown by X-ray powder photographs of $(C_6H_5C{\equiv}C)_2M$ (M=Sr,Ba) with $(C_6H_5C{\equiv}C)_2Ca$ indicates that the former two compounds are also polymeric in the solid state.

The properties of the group Ib derivatives RC≡CM (R=alkyl, aryl groups;M=Cu (88),Ag (88-91)) have been interpreted as consistent with a polymeric structure (2.11). It has been concluded that the position of the ν(C≡C) mode for the copper (88,93,94) and silver (88, 90,93) compounds relative to its value in the corresponding free disubstituted alkynes (88,93) supports this proposed structure. Therefore, the ν(C≡C) mode in $CH_3C{\equiv}CCu$ (1958 cm^{-1}) and $C_6H_5C{\equiv}CCu$ (1926 cm^{-1}) is approximately 180 cm^{-1} lower than in free $CH_3C{\equiv}CH$

TABLE 2.4
The ν(C≡C) Mode Assignments (cm^{-1}) for Alkyne Complexes of the Group IA Elements

	Alkyne		
M	HC≡C-	CH_3C≡C-	C_6H_5C≡C-
H	1974	2142	2107
Li		2053	2036
Na	1867	2032	2018
K	1858	2123	2000
Rb	1851	2020	1990

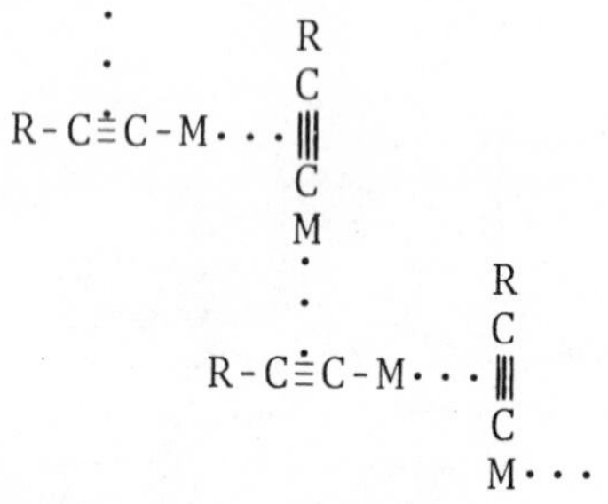

(2.11)

(2142 cm^{-1}) (94) and C_6H_5C≡CH (2107 cm^{-1}) (95) and much lower than noted for any of the previously discussed derivatives that had no metal-alkyne π interaction. Raman bands for CH_3C≡CCu and C_6H_5C≡CCu at 435 and 422 cm^{-1}, respectively, have been assigned to the ν(CuC≡) mode while Raman bands at 262 and 308 cm^{-1}, respectively, have been tentatively assigned to modes arising from the π-type copper-alkyne interaction (93). The small shift in the ν(C≡C) mode frequency of CH_3C≡CAg (2063 cm^{-1}) and C_6H_5C≡CAg (2052 cm^{-1}) relative to the corresponding positions of this mode in the free alkynes is indicative of the relative weakness of the silver-alkyne π interaction (88,93). No Raman bands that could be attributed to silver-alkyne stretching modes were observed (93). Vibrational data have also been reported for *p*-XC_6H_4C≡CCu (X=CH_3,Cl,NO_2) and *n*-$CH_3(CH_2)_nC$≡CCu (n=3,5) (93) as well as for the reaction products of alkyneCu(I) and alkyne-Ag(I) compounds with bases such as trialkyl- and triarylphosphines. A tetrameric structure, illustrated in

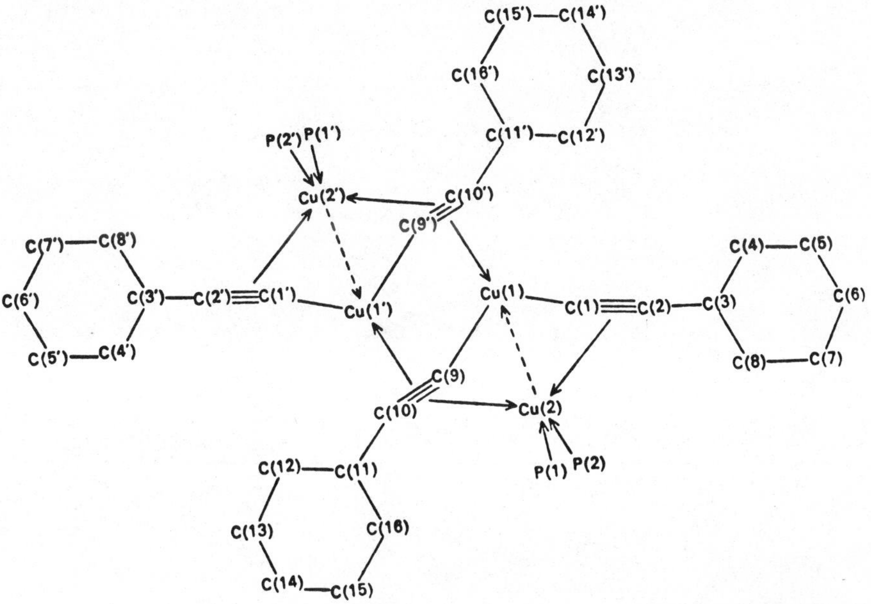

Figure 2.6. Structure found for solid $C_6H_5C{\equiv}CCuP(CH_3)_3$ (96).

Figure 2.6, has been found for solid $C_6H_5C{\equiv}CCuP(CH_3)_3$ (96). The presence of two different types of phenylacetylene groups, as shown by the single crystal X-ray study, is supported by the appearance of two ν(C≡C) bands (2045 and 2019 cm^{-1}) in the infrared spectrum (88). More than one ν(C≡C) band has also been observed in the infrared spectra of other alkyne$CuPR_3$ (R=alkyl, aryl) compounds (88). In addition, there is a noticeable increase in the ν(C≡C) mode frequency of the phosphine (88,97) and other (97) complexes over the value found in the uncomplexed RC≡CCu derivatives. The one ν(C≡C) band observed in the infrared spectrum of solid $C_6H_5C{\equiv}CAgP(CH_3)_3$ (2075 cm^{-1}) (88) is consistent with the structure found in a single crystal X-ray study of this compound (98) and illustrated in Figure 2.7. In boiling benzene, this compound becomes dimeric (89).

Molecular weight data have led to the conclusion that the group IIb compounds $(RC{\equiv}C)_2M$ (M=Zn,Cd;R=C_6H_5, C_6H_{11}) have a polymeric structure (2.12) (99). This structure is supported by the ν(C≡C) mode assignments

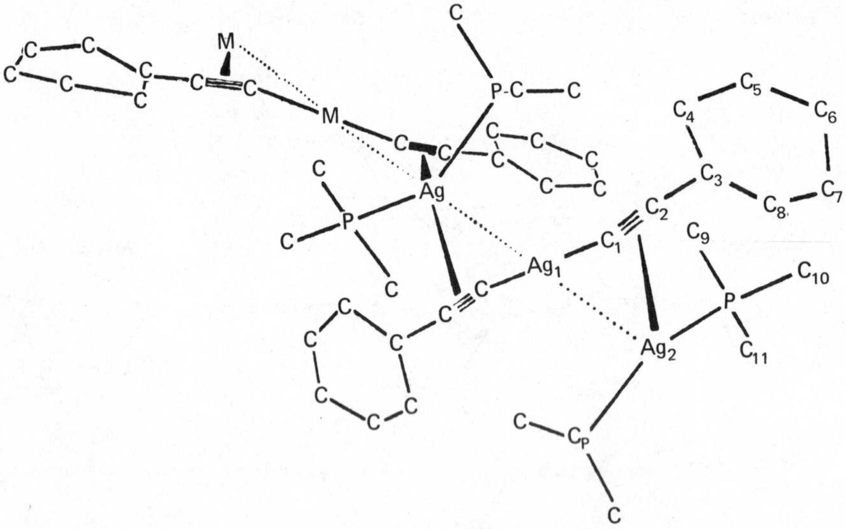

Figure 2.7. Structure found for solid $C_6H_5C{\equiv}CAgP(CH_3)_3$ (98).

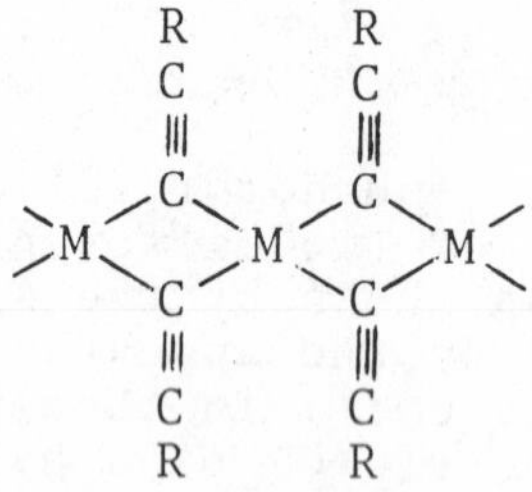

(2.12)

of 2060 and 2050 cm^{-1} for $(C_6H_5C{\equiv}C)_2Zn$ and $(C_6H_5C{\equiv}C)_2Cd$, respectively, and approximately 2070 cm^{-1} for the corresponding bis(1-octyne)zinc and -cadmium compounds (as compared to 2110 cm^{-1} for free 1-octyne), which are in the range found for the corresponding mode in $C_6H_5C{\equiv}CM(CH_3)_2$ (M=Al,Ga,In) that are dimeric and have a similar structure with bridging alkyne groups (99). The range of the $\nu(C{\equiv}C)$ mode frequencies found for $(RC{\equiv}C)_2Hg$ (R=alkyl or aryl group) (2188 to 2142 cm^{-1}) indicates that these compounds have a monomeric structure with nonbridging alkyne groups (88).

Complete infrared and Raman assignments have been reported for $HC{\equiv}CBX_2$ (X=F,Cl) (100), while complete infrared assignments have been made for $CH_3C{\equiv}CBF_2$ (101). Partial infrared data and assignments have been given for $(C_6H_5C{\equiv}C)_nB(C_6H_5)_{3-n}$ (n=1,2) (102). Although alkyne derivatives of boron have a monomeric structure, several of the alkyne derivatives of the other group IIIb metals have an associated structure. A dimeric structure (2.13) has been proposed for $C_6H_5C{\equiv}CM(CH_3)_2$

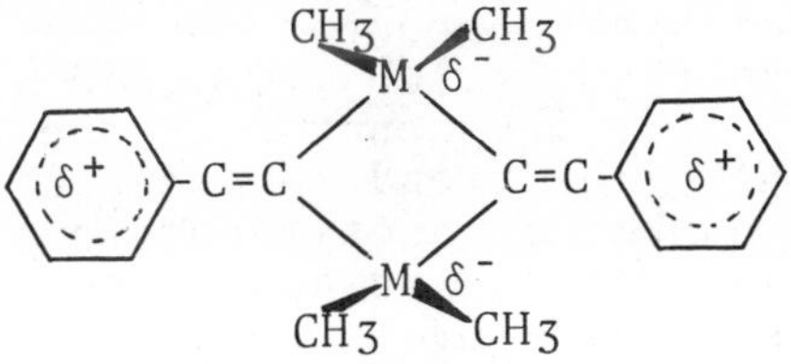

(2.13)

(M=Al,Ga,In) with the alkyne bridges becoming somewhat allenic in character (99). This structure is disrupted in donor solvents such as tetrahydrofuran (99), as indicated by the fact that in benzene the $\nu(C{\equiv}C)$ mode frequency of these derivatives has been assigned at approximately 2050 cm^{-1} while in tetrahydrofuran it increases to 2110 cm^{-1}, near the region observed for terminal phenylacetylene groups (i.e., $C_6H_5C{\equiv}CSi(CH_3)_3$, $\nu(C{\equiv}C)$= 2160 cm^{-1} (103) and $C_6H_5C{\equiv}CPb(C_6H_5)_3$, $\nu(C{\equiv}C)$=2130 cm^{-1} (104)). According to cryoscopic and conductivity measurements, $RC{\equiv}CTl(CH_3)_2$ (R=CH_3, $\nu(C{\equiv}C)$=2095 cm^{-1}; R=C_6H_5, $\nu(C{\equiv}C)$=2068 cm^{-1}) and $M[(RC{\equiv}C)_4Tl]$ (M=K, $(C_6H_5)_4P$,R=CH_3, $\nu(C{\equiv}C)$=2138 cm^{-1};M=Na,$(C_6H_5)_4P$,R=C_6H_5, $\nu(C{\equiv}C)$=approximately 2116 cm^{-1}) are monomeric (105). Monomeric structures have also been inferred from the position of the $\nu(C{\equiv}C)$ mode in $(C_6H_5C{\equiv}C)(CH_3)TlO_2CCH_3$ (2151 cm^{-1}) and $[(C_6H_5C{\equiv}C)(CH_3)Tl]B(C_6H_5)_4$ (2160 cm^{-1}) (106).

The physical properties, molecular weight data, and position of the $\nu(C{\equiv}C)$ mode (2045 cm^{-1}) for $C_6H_5C{\equiv}CSc(\eta^5\text{-}C_5H_5)_2$ suggest a bridged structure of at least a dimer similar to that found for related aluminum, gallium, and indium derivatives (107). Although infrared data have been reported for $(C_6H_5C{\equiv}C)_3Sc$ ($\nu(C{\equiv}C)$=2050 cm^{-1}), no discussion has been presented of the probable structure of this compound (108).

Vibrational assignments for σ-type alkyne derivatives of the group IVb and Vb elements are given in

Table 2.5. These data show that, in general, the ν(C≡C) mode frequency is higher in the complex than in the free alkyne (where, for example, in HC≡CH, ν(C≡C)= 1974 cm^{-1} (124) and in $CH_3C{\equiv}CH$, ν(C≡C)=2124 cm^{-1} (86)) and decreases in compounds of a given stoichiometry as the mass of the central atom increases. The ν(MC≡) mode frequencies follow the same trend. Also, the ν(C≡C) mode frequency generally decreases for a given central atom as alkyne groups are replaced with methyl groups. The same trends have been observed for several of the perfluoroalkyne compounds $(RC{\equiv}C)_{4-n}M(CH_3)_n$ (M=Si, Ge,Sn;n=1 to 4;R=CF_3,C_2F_5,$(CF_3)_2CF$), in addition to which the ν(C≡C) mode frequency increases as the size of R increases (125). The trend in the ν(C≡C) mode frequencies as the number of acetylene groups bonded to the central atom changes does not appear to support the presence of back bonding from the π system of the acetylene group to empty, central-atom *d*-orbitals (110). A normal coordinate analysis of $HC{\equiv}CMH_3$ (M=C,Si,Ge), however, indicates that there is *p*π→*d*π bonding from the α-carbon atom of the acetylene to the silicon and, to a lesser extent, the germanium atom (126). The mean amplitudes of vibration have been calculated for $HC{\equiv}CMH_3$ (M=C,Si,Ge) (127). Less complete vibrational data have been reported for $X_3SiC{\equiv}CSiX_3$ (X=H (128),Cl (129),CH_3 (112,129),OCH_3 (129);ν(C≡C)=2142 to 2108 cm^{-1}), $(CH_3)_3GeC{\equiv}CGe(CH_3)_3$ (130), $(C_6H_5)(CH_3)_2SiC{\equiv}CSn(C_6H_5)_3$ (ν(C≡C)=2085 cm^{-1}) (34), $[RC{\equiv}CSi(CH_3)_2]_2O$ (R=H,CH_3, $(CH_3)_2COH$) (131), and various other trialkylsilicon (103) and triethyl- and triphenyllead alkynes (104).

Alkyne derivatives of the transition elements are found with both purely σ-type and purely π-type structures. The vibrational spectra of these two structural extremes differ in two major respects. The first is that while the ν(C≡C) mode in σ-type derivatives appears with from strong to moderate infrared and Raman intensity, the ν(C≡C) mode is often observed only in the Raman spectrum for π-type derivatives. Second, the changes observed in the ν(C≡C) mode frequency of the alkyne on coordination are much larger for the π-type than for the σ-type structures. The ν(C≡C) mode has been assigned for the σ-type alkyne derivatives $(C_6H_5C{\equiv}C)_2M(\eta^5\text{-}C_5H_5)_2$ (M=Ti (2070 cm^{-1}) (132,133),Zr (2073 cm^{-1}) (133),Hf (2083 cm^{-1}) (133)), $[(C_6H_5)_4E]_3Na$-$[Co(C{\equiv}CH)_6]$ (1905 cm^{-1}) $[(C_6H_5)_4E]_3M[Co(C{\equiv}CC_6H_5)_6]$ (2045 cm^{-1}) (E=P,As;M=Na,K) (134), and HC≡CNi (1947 cm^{-1}) (135). For the product of the reaction of $Ni(CO)_4$ with $(C_6H_5C{\equiv}C)_2Ti(\eta^5\text{-}C_5H_5)_2$ (2.14), the ν(C≡C) mode frequency

TABLE 2.5
Vibrational Assignments (cm^{-1}) for Alkyne Derivatives of the Group IVb and Group Vb Elements

Compound	ν(CR)	ν(C≡C)	δ(C≡CR)	ν(MC≡)	δ(MC≡C)	δ(≡CMC≡)	ν(MX)[a]	Refs.
$(HC{\equiv}C)_4Si$	3298 3315	2053 2062	687 695	534 708	357 392.5	103 105.5		109,110
$(HC{\equiv}C)_4Ge$	3298 3316	2057 2062	683	507 523	346 356	91 98		109,110
$(HC{\equiv}C)_4Sn$	3311	2043	686	447 or 504[b]	290	45 or 97		109,110
$(HC{\equiv}C)_3P$	3323	2061	637 690	615 646	266 424	101 120		111
$(HC{\equiv}C)_3As$	3317	2053	651 681	517 526	231 379 399	89 100		111
$(HC{\equiv}C)_3Sb$	3312	2033	662 682	450 477	219 325 341	72 94		111
$HC{\equiv}CSi(CH_3)_3$	3287	2035	679	555			650 697	110,112
$DC{\equiv}CSi(CH_3)_3$	2590	1909	540	547			647 692	110,112
$(HC{\equiv}C)_2Si(CH_3)_2$	3284	2041	626 684	595	300 377 385 548	315	701 794	110,113,114
$(HC{\equiv}C)_3SiCH_3$	3282	2046	595 603 699	605 658	304 375 541	105 456	756	110,115

TABLE 2.5 (continued)

Compound	ν(CR)	ν(C≡C)	δ(C≡CR)	ν(MC≡)	δ(MC≡C)	δ(≡CMC≡)	ν(MX)[a]	Refs.
HC≡CGe$(CH_3)_3$	3290	2030	662	497	316		581 616	110,115
DC≡CGe$(CH_3)_3$	2569	1898		486	305		583 619	110
(HC≡C)$_2$Ge$(CH_3)_2$	3283	2041	671 686	521 538	334 359 492	387 419	598 625	110,113
(HC≡C)$_3$GeCH$_3$	3282	2046	577 687 693	522 562	280 353 500	98 429	624	110,113
HC≡CSn$(CH_3)_3$	3380	2005	662	432	258		517 538	110,115
(HC≡C)$_2$Sn$(CH_3)_2$	3276	2016	675 687	454	272 293 432	99 138 223?	532 552	
(HC≡C)$_3$SnCH$_3$	3305	2018	667		262 353 437		540	110,113
HC≡CSiH$_3$	3314	2056	684[c] 668[d]	679[c] 657[d]	220		2193	116,117
HC≡CSiD$_3$	3315	2055	683	679	206		1595	116
HC≡CGeH$_3$	3315	2060	673	530	216		2117 2120	118
HC≡CGeD$_3$	3318	2053	673	518	203		1525	118
IC≡CSi$(CH_3)_3$	731	2100	320	366	237		622 698	119
BrC≡CSi$(CH_3)_3$	783	2126	349	372	248		624 699	119
ClC≡CSi$(CH_3)_3$	906	2143	405	383	261		628 699	120

$ClC{\equiv}CGe(CH_3)_3$	860	2137	369	352	220		563 600	120
$ClC{\equiv}CSn(CH_3)_3$	830	2120	345	345			518 538	120
$ClC{\equiv}CPb(CH_3)_3$	790	2098					483	120
$(CH_3C{\equiv}C)_4Si$	1045	2190	415	384	276	63 87		85
$(CH_3C{\equiv}C)_4Ge$	1019	2192	401	371 440	259	82		85
$(CH_3C{\equiv}C)_4Pb$	992	2170	365	331	208 252	50 84		85
$(CH_3C{\equiv}C)_3P$	962 992	2193	411 530 565	641 677	283 364 436	65 77		85
$(CH_3C{\equiv}C)_3As$	1012 999	2192	374 392 446	450 615	221 266 349	71		85
$(CH_3C{\equiv}C)_3Sb$	996	2156	191 257 335	406 454	101 120 143			85
$CH_3C{\equiv}CSi(CH_3)_3$	1026	2186	455	393	268		635 695	115
$CH_3C{\equiv}CGe(CH_3)_3$	1003	2180	465	400			574 604	115
$CH_3C{\equiv}CSn(CH_3)_3$	991	2160	360	340			515 538	115
$CH_3C{\equiv}CPF_2 \cdot BH_3$		2220		645				121
$CH_3C{\equiv}CPF_2$		2193		593				121
$CH_3C{\equiv}CPF_4$		2240		672				121

TABLE 2.5 (continued)

Compound	ν(CR)	ν(C≡C)	δ(C≡CR)	ν(MC≡)	δ(MC≡C)	δ(≡CMC≡)	ν(MX)[a]	Refs.
$(CF_3C{\equiv}C)_3P$	1187 1222	2225	435 454 482	871 883	254 263 328	88 136		122
$(CF_3C{\equiv}C)_3As$	1185 1219	2218	272 383 405	851 852.5	230 237 259	85 132		122
$(CF_3C{\equiv}C)_3Sb$	1183	2201	332 359	229 240	200 204 211	75 128.5		
$(t\text{-}C_4H_9C{\equiv}C)_3P$	947	2175 2207		576 602				123
$(t\text{-}C_4H_9C{\equiv}C)_3As$	919	2155 2189		495 542				123
$(t\text{-}C_4H_9C{\equiv}C)_3Sb$	912	2149 2176		479 550				123
$(t\text{-}C_4H_9C{\equiv}C)_3P{=}O$	967	2186 2212 2233					1235	123
$(t\text{-}C_4H_9C{\equiv}C)_3P{=}S$	958	2177 2200 2218					689	123

[a]These assignments are for other than the ν(MC≡) mode.
[b]The presence of two bands has been attributed to Fermi resonance.
[c]Assignment from Ref. 116.
[d]Assignment from Ref. 117.

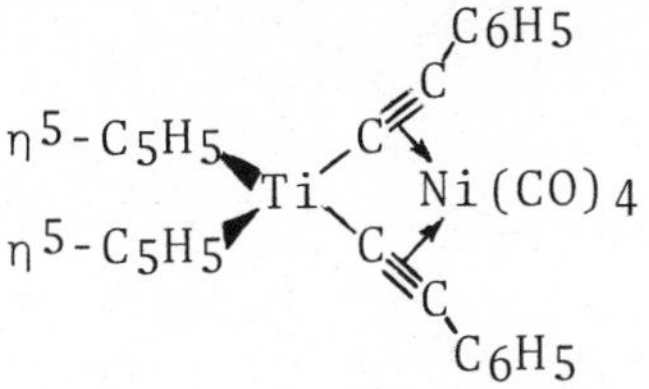

(2.14)

(1850 cm^{-1}) is approximately 220 cm^{-1} lower than in $(C_6H_5C{\equiv}C)_2Ti(\eta^5\text{-}C_5H_5)_2$ because of the π interaction of the nickel atom with the two phenylacetylene groups (136).

While bands at 2109, 2093, 2043, and 2025 cm^{-1} are in the region expected for the ν(PtH) mode and the ν(C≡C) mode of the σ-bonded, alkyne ligand in *trans*-$C_6F_5C{\equiv}CPt(H)[P(C_2H_5)_3]_2$, these bands are replaced by a new band at 1687 cm^{-1} which has been assigned to the ν(C≡C) mode of a π-bonded alkyne molecule in the related complex $C_6F_5C{\equiv}CHPt[P(C_6H_5)_3]_2$ (137). Similarly, the ν(C≡C) mode has been assigned at approximately 2100 cm^{-1} for the σ-bonded alkyne complexes $RC{\equiv}CMo(CO)_3$-$(\eta^5\text{-}C_5H_5)$ ($R{=}C_4H_9,C_6H_5$) (138) and 1613 cm^{-1} for the π-bonded acetylene derivative $HC{\equiv}CHMo(\eta^5\text{-}C_5H_5)_2$ (139).

Replacement of a CO ligand in $RC{\equiv}CFe(CO)_2(\eta^5\text{-}C_5H_5)$ ($R{=}C_4H_9,C_6H_5$) with a $(C_6H_5)_3P$ ligand results in a decrease in the ν(C≡C) mode frequency from 2125 and 2105 cm^{-1}, respectively, to 2095 and 2085 cm^{-1}, respectively, suggesting the presence of π back bonding from the iron *d*-orbitals to π* orbitals of the alkyne group (138). The position of the carbon-carbon stretching mode of HC≡CH in $(\eta^5\text{-}C_5H_5)_2MoHC{\equiv}CH$ (1613 cm^{-1}) indicates a large reduction in the carbon-carbon bond order of HC≡CH on coordination to the molybdenum atom (139). A linear relation has been found between the ν(MC≡) mode frequencies and the Taft polar constants for the R groups in the complexes $(RC{\equiv}C)_2ML_2$ (M=Ni,Pd,Pt;$L{=}(CH_3)_3P,(C_2H_5)_3P,(C_6H_5)_3P,(C_2H_5)_3Sb$;$R{=}CH_3,CH_2F,C_6H_5,CH_2{=}CH,HC{\equiv}C,C_6H_5C{\equiv}C$) ($\nu_a(MC{\equiv})$=597 to 543 cm^{-1},ν(C≡C)=2160 to 1985 cm^{-1}) (140). Infrared data have been listed for $(\eta^5\text{-}C_5H_5)_2GdC{\equiv}CC_6H_5$ (141), $\eta^5\text{-}C_5H_5Ho(C{\equiv}CC_6H_5)_2$ (141) and $(\eta^5\text{-}C_5H_5)_3UC{\equiv}CR$ (R=H (142),C_6H_5 (143)).

The ν(C≡C) mode for [*trans*-$(RC{\equiv}CR'Pt(CH_3)L_2]PF_6$ ($R{=}R'{=}CH_3,C_2H_5,L{=}C_6H_5(CH_3)_2P;R{=}C_6H_5,R'{=}(C_6H_5)_2COH,L{=}C_6H_5(CH_3)_2P;R{=}C_6H_5,R'{=}C_6H_5,(C_6H_5)_2COH,L{=}(CH_3)_3As$) in which the alkyne group is π-bonded to the platinum atom was observed only in the Raman spectra as a very intense band at approximately 2116 to 2024 cm^{-1} as compared to a range of approximately 2314 to 2208 cm^{-1} for this mode in the free alkyne (144). Although the ν(Pt-CH_3) mode for these derivatives (556 to 542 cm^{-1}) was easily assigned, the ν(PtC≡) mode could not be observed in the Raman spectra from 600 to 160 cm^{-1}. On coordination, a decrease of approximately 446 to 412 cm^{-1} has been observed in the alkyne ν(C≡C) mode frequency of the complexes HC≡CRPt[P(C_6H_5)$_3$]$_2$ (R=CF_3,C_6H_5,C(=O)OH,CH_2OCH_3, (CH_3)$_2$COH,CH_2OH). The trend in the ν(C≡C) mode frequencies for this series indicates that the platinum-alkyne bond stability decreases for a given R group in the order CF_3 > C_6H_5 > alkyl, although this conclusion is based on a small number of compounds and a small variation in the ν(C≡C) mode frequency (145). The ν(C≡C) mode has been assigned for the π-bonded alkyne complexes [(C_6H_5)$_3$P]$_2$PtHC≡CR (R=C_6H_5,C_6H_{10}OH,CH_3(C_2H_5)COH, (CH_3)$_n$H$_{2-n}$COH (n=0 to 2)) (1720 to 1680 cm^{-1}) (146). It has also been suggested in this study that a medium intensity infrared band at approximately 560 cm^{-1} may be due to one of the ν(PtC≡) modes of these complexes. The ν(OH) mode frequency of di(hydroxyalkyl)acetylenes is lowered by 140 to 110 cm^{-1} upon coordination to a platinum(II) atom. This shift has been interpreted by one group as an indication of intramolecular hydrogen-bonding between the OH groups and the *cis*-chlorine atoms of the complex (2.15). More recently, however, another group

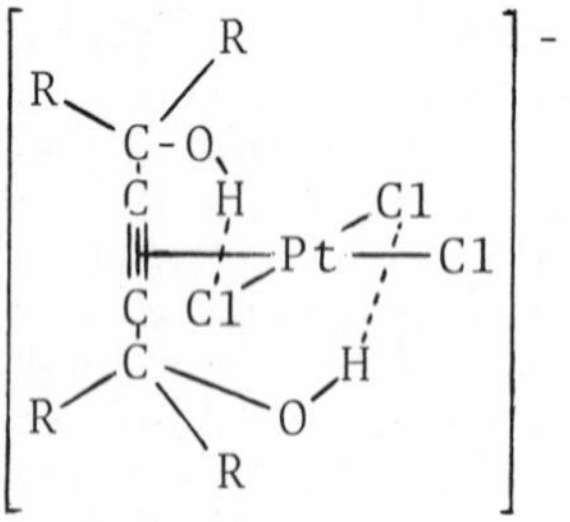

(2.15)

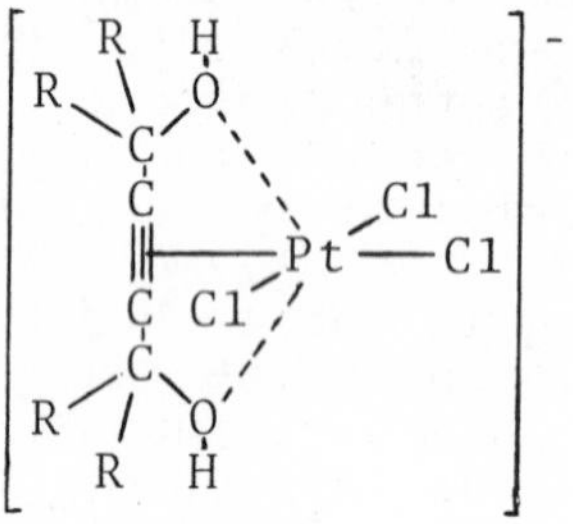

(2.16)

has attributed it to an interaction between the OH group and the platinum atom (2.16), which would also result in a decrease in the ν(OH) mode frequency (148).

The ν(C≡C) mode has been assigned for several other π-type-transition-metal complexes (147-158). The extent to which the alkyne ν(C≡C) mode frequency is lowered on coordination has been used in discussions of the structures of these derivatives. These structures have been represented in terms of two extreme valence bond pictures (2.17 and 2.18). In structure 2.17, the alkyne remains essentially undistorted and the ν(C≡C)

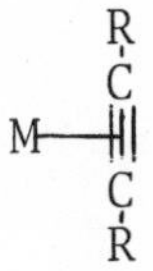

(2.17)

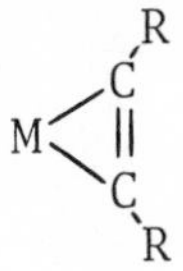

(2.18)

mode frequency is lowered by approximately 230 to 130 cm^{-1} (156). In structure 2.18, the skeleton is distorted with a reduction in the C≡C bond order and a decrease of approximately 500 cm^{-1} in the ν(C≡C) mode frequency. Complexes of the second type are stabilized with electron-withdrawing substituents on the alkyne. The structures of all π-type alkyne-transition-metal complexes lie somewhere between these two extremes. The structural picture that is most realistic for a given compound will depend on the properties of the metal and the substituents on the alkyne group. In Table 2.6, the ν(C≡C) mode assignments are given for selected alkyne derivatives. The complex (NCC≡CCN)Pt-$[P(C_6H_5)_3]_2$ has been represented in terms of 2.18 because of the position of the ν(C≡C) mode (cf. Table 2.6) and also because the C≡C bond length changes from 1.19 A in the free ligand (159) to 1.40 A in the complex (160). This complex can undergo a photochemical reaction to produce a σ-type compound, the structure of which (2.19)

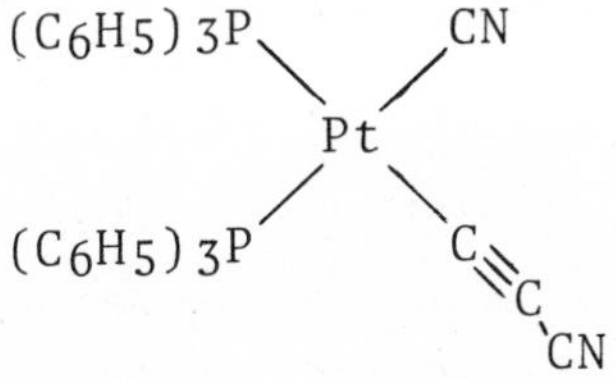

(2.19)

has been determined in a single crystal X-ray study (161). Infrared bands at 1683 (m) (ν(C≡C)), 2177 (m), 2185 (s), and 2196 (m) cm^{-1} (ν(C≡N)) in the π-bonded compound have been replaced with bands at 2070 (w), 2140 (w), and 2235 (vs) cm^{-1} in the σ-bonded complex (161).

Complete infrared assignments have been reported for $HC≡CHCo_2(CO)_6$ and the isotopically substituted analogs

TABLE 2.6
The ν(C≡C) Mode Assignments (cm^{-1}) for π-Bonded Alkyne-Metal Compounds

R=	$CF_3C{\equiv}CCF_3$			NCC≡CCN			$CH_3OC(=O)C{\equiv}CC(=O)OCH_3$		
Compound[a]	ν(C≡C)	Δν(C≡C)	Refs.	ν(C≡C)	Δν(C≡C)	Refs.	ν(C≡C)	Δν(C≡C)	Refs.
$C_5H_5Mn(CO)_2R$	1919	381	149						
$L_2Rh(Cl)R$	1917	383	150						
$L_2Rh(CO)(Cl)R$				1775	344	151			
L_2PdR	1811	489	149,150	1751	368	151	1830	305	150
$(C_5H_5)_2VR$	1800	500	153				1821	329	153
L_2NiR	1790	510	154						
L_2PtR	1775	525	149,154	1683	436	151	1782	362	151,152
$L_2Ir(CO)(Cl)R$	1773	527	151,155	1725	394	151	1770	380	156

[a] $L = (C_6H_5)_3P$.

$HC{\equiv}CDCo_2(CO)_6$, $DC{\equiv}CDCo_2(CO)_6$, and $H^{13}C{\equiv}CHCo_2(CO)_6$ (162). The fact that the ν(C≡C) mode frequency of the coordinated HC≡CH (1402 cm^{-1}) is much lower than that of free HC≡CH indicates that the HC≡CH molecule is highly distorted in the complex (2.20). It has been proposed that

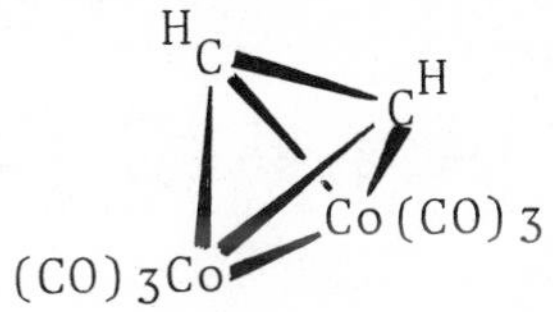

(2.20)

the infrared spectrum of the coordinated HC≡CH is very similar to that of HC≡CH in an excited electronic state. The ν(CoC≡) modes have been assigned at 605 and 551 cm^{-1}. Using these data, an approximate normal coordinate analysis has been reported for the $HC{\equiv}CHCo_2$ unit of the $HC{\equiv}CHCo_2(CO)_6$ molecule (163). The force constants obtained for the HC≡CH molecule in the complex were shown to be nearly equal to the corresponding force constants obtained for the HC≡CH molecule in the ^1A_u excited electronic state.

D. ALLYLIC COMPOUNDS

Allyl derivatives of the main group elements are found with either an ionic structure and symmetric, anionic allylic group, or a *monohapto*, covalent structure in which the allyl group has a localized C=C bond. These two structures can be differentiated by the fact that the delocalized, ionic structure has no bands from 1650 to 1600 cm^{-1}, which are characteristic of the ν(C=C) mode.

In an infrared study of C_3H_5M ($M={}^6Li, {}^7Li, Na$) at 25°C and -180°C, and a Raman study of solid $C_3H_5{}^7Li$ at 25°C, the allyl and metal-allyl skeletal modes have been assigned (164). These data suggest strong, ionic character in these compounds with the ν_a(CCC) and ν(MC) modes assigned at 1570 to 1510 cm^{-1} and 438 to 318 cm^{-1}, respectively, for $M={}^6Li$ and ^7Li, and 1541 to 1512 cm^{-1} and 252 to 175 cm^{-1}, respectively for M=Na. In addition, the infrared spectral data (1700 to 200 cm^{-1}) listed in this study suggest some ionic character for $C_3H_5{}^6Li$ and $C_3H_5{}^7Li$ in diethyl ether and tetrahydrofuran solutions. Earlier, similar conclusions were reached in an infrared study of allyl derivatives of lithium, sodium, and potassium (165,166); the ν_a(CCC) mode of C_3H_5Li in tetrahydrofuran, diethyl ether, and as a Nujol mull was assigned to a strong intensity infrared band at 1540 to 1525 cm^{-1} (165), while analogous infrared bands of allyl-, 2-methylallyl-, and cyclohexenylsodium were assigned at 1535, 1520, and 1560 cm^{-1}, respectively (166). Because of the high energy barrier to rotation in unsymmetrically substituted allyl anions, two rotational isomers (2.21 and 2.22) are structural possibilities. The two strong

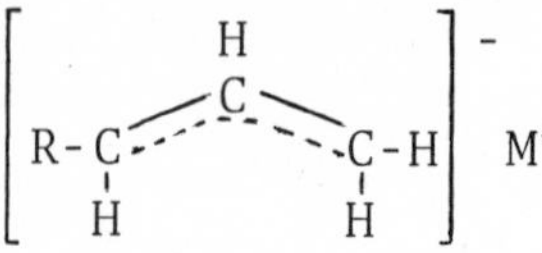

syn isomer

(2.21)

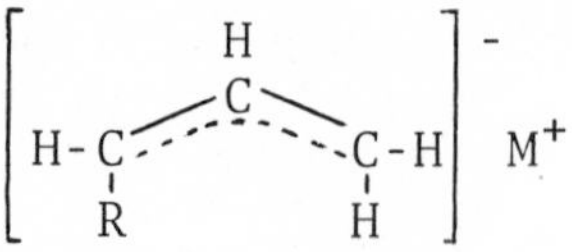

anti isomer

(2.22)

intensity infrared bands observed for penylsodium at 1560 and 1525 cm^{-1} have been attributed to the presence of both isomers with the 1560 cm^{-1} band assigned to the *syn* isomer (2.21) and the 1525 cm^{-1} band to the *anti* isomer (2.22). The fact that cyclohexenylsodium can exist only in a form geometrically equivalent to 2.22 and exhibits a band at 1522 cm^{-1} supports these assignments (167).

Complete assignments have been reported of the allyl modes and magnesium-allyl skeletal modes in the infrared spectra at 25°C and -180°C, and the Raman spectra at 25°C of solid $(C_3H_5)_2Mg$ and C_3H_5MgX (X=Cl,Br) (164). These data are consistent with strong ionic character for these compounds in the solid state. The ν_a(CCC) and ν(MgC) modes have been assigned at 1450 to 1510 cm^{-1} and 432 to 298 cm^{-1}, respectively. The magnesium-halide stretching mode has been assigned at 255 cm^{-1} for C_3H_5MgCl and 214 cm^{-1} for C_3H_5MgBr from the infrared data at 25°C. Infrared data for C_3H_5MgCl in solution (tetrahydrofuran and diethyl ether) have been reported to be consistent with some ionic character (164), although nmr studies of allylmagnesium derivatives indicate (167) the presence of rapidly equilibrating mixtures of the classical *monohapto* or σ-bonded structures (2.23 and 2.24) in solution. An infrared band at 1580 cm^{-1} has been observed for the product obtained on reacting magnesium with allylchloride in diethyl ether or in dibutyl ether; the corresponding products found on using allylbromide or allyliodide gave a band at 1588 cm^{-1} (168). Addition of tetrahydrofuran

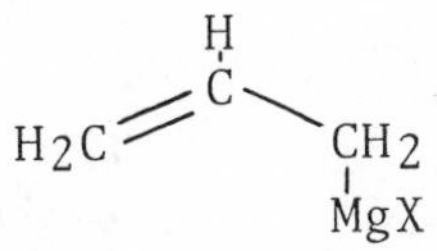

(2.23)

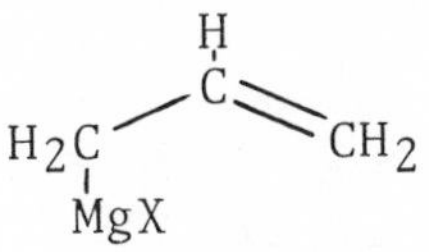

(2.24)

to the bromide or iodide caused this band to shift to 1570 and 1560 cm^{-1}, respectively, while addition of dioxane to the bromide lowered the frequency of this band to 1577 cm^{-1}. As the electronegativity of the main group element increases, the ionic character of the allyl derivative decreases to produce an increase in the higher-frequency carbon-carbon stretching mode. Therefore, infrared bands have been observed at 1575 cm^{-1} in $(C_3H_5)_2Mg$, 1610 cm^{-1} in $(C_3H_5)_2Zn$, 1630 cm^{-1} in $(C_3H_5)_3B$, and 1650 to 1640 cm^{-1} in the allylhalides (165,169,170). An ionic solid-state structure has been proposed for $(C_3H_5)_2Zn$ while the infrared spectra in benzene and tetrahydrofuran solutions show bands at 1613 and 1622 cm^{-1}, respectively, which are characteristic of a localized C=C bond (171). Allyl derivatives of mercury have a covalent *monohapto*-type structure. Complete infrared and Raman assignments have been made for $(\eta^1\text{-}C_3H_5)_2Hg$ (172,173) and $\eta^1\text{-}C_3H_5HgX$ (X=Cl (172,174-176),Br (172, 174,176),I (174,175)). A comparison of the ν(C=C) mode assignments for these mercury derivatives (1626 to 1617 cm^{-1}) with similar data for allylhalides (1650 to 1640 cm^{-1}) as well as with data obtained in an extended Hückel molecular orbital calculation for $\eta^1\text{-}C_3H_5HgBr$ suggestes that there is π-electron delocalization from the allyl group to the mercury atom (177). Infrared data have also been reported for RHgX (R=substituted allyl group,X=Cl,Br) (175). In the series $\eta^1\text{-}C_3H_5Tl(CH_3)X$ (X=Cl,CH_3CO_2,$C_2H_5CO_2$,*i*-$C_3H_7CO_2$,$(CH_3)_2NCSS$,tropolinate) the ν(C=C) mode has been assigned to a strong intensity infrared band from 1626 to 1621 cm^{-1} while the ν_a(TlC) and ν_s(TlC) modes have been assigned from approximately 528 to 507 cm^{-1} and 476 to 470 cm^{-1}, respectively (178).

Infrared data have been reported for $(\eta^1\text{-}C_3H_5)_4M$ (M=Si,Ge,Sn) (179). In a much more detailed infrared and Raman study of $(\eta^1\text{-}C_3H_5)_4Si$ and $(\eta^1\text{-}C_3H_5)_4Sn$ (the infrared and Raman spectra of $(\eta^1\text{-}C_3H_5)_4Si$ are illustrated in Figure 2.8), it has been concluded that there is no evidence from the vibrational data for an interaction between the empty *d*-orbitals on the central atom and the π-electrons of the allyl groups (180). Trends

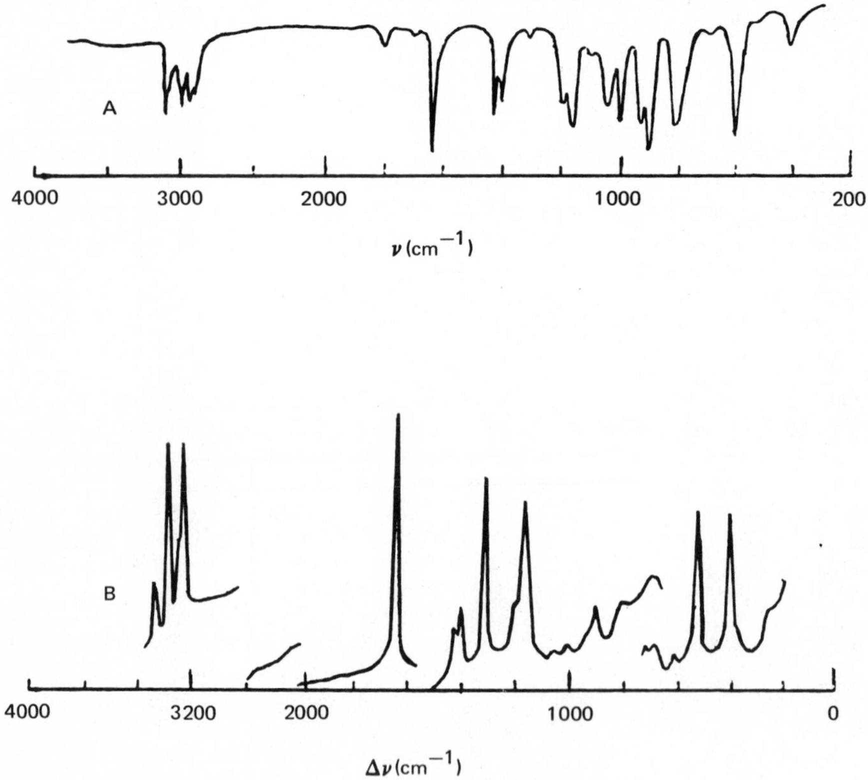

Figure 2.8. Liquid-phase (A) infrared spectrum and (B) Raman spectrum of $(CH_2{=}CHCH_2)_4Si$ (180).

observed in the infrared (181,182) and Raman (3,183) data for $R_3M(CH_2)_nCH{=}CH_2$ (R=alkyl group,Cl,Br;M=C,Si,Ge, Sn;n=0 to 2), however, have been interpreted as suggesting that this type of interaction is important. The same interpretation is possible for data reported for $(C_2H_5)_3Sn(CH_2)_nCH{=}CHC_6H_5$ (n=1,ν(C=C)=1634 cm^{-1};n=2, ν(C=C)=1648 cm^{-1}) and $(C_6H_5)_3Sn(CH_2)_nCH{=}CHC_6H_5$ (n=1, ν(C=C)=1635 cm^{-1};n=2,ν(C=C)=1650 cm^{-1}) (184). Although the intensity of the ν(C=C) mode in σ-type vinyl derivatives decreases from silicon to germanium, in the corresponding allyl derivatives the intensity remains approximately constant (2). Infrared assignments have

have been presented for η^1-$C_3H_5SnR_3$ (R=C_2H_5,*n*-C_4H_9,C_6H_5, *p*-$CH_3C_6H_4$), 2-methylallylSnR_3 (R=C_2H_5,*n*-C_4H_9,C_6H_5,*p*-CH_3-C_6H_4) (η^1-C_3H_5)$_2SnR_2$ (R=C_2H_5,*n*-C_4H_9), and (2-methylallyl)$_2$Sn(*n*-C_4H_9)$_2$ (185). Infrared and molecular weight data for (η^1-C_3H_5)$_3$SnR (R=CH_3CO_2,CH_2ClCO_2) show them to be polymeric in the solid state with five-coordinate tin atoms, a trigonal bipyramidal skeletal structure, and bridging carboxylate groups, while in solution depolymerization occurs to produce monomeric structures (186). In contrast to the (alkyl)$_2$Sn(carboxylate)$_2$ derivatives, which are monomeric, the (η^1-C_3H_5)$_2SnR_2$ (R=CH_2ClCO_2, $CHCl_2CO_2$) derivatives are dimeric in solution. These allyl derivatives show an infrared band at approximately 1700 cm^{-1} that is characteristic of an ester like carboxylate group and bands at 1610 to 1600 cm^{-1} and 1380 to 1370 cm^{-1} that are characteristic of chelating carboxylate groups (186). For the distannoxanes [(η^1-C_3H_5)$_2$-Sn(O_2CR)]$_2$O, the ν_a(SnOSn) mode has been tentatively assigned to strong intensity infrared bands at 625, 610, and 615 cm^{-1} and the ν_s(SnOSn) mode to medium intensity bands at 505, 515, and 510 cm^{-1} for R=CH_2Cl, $CHCl_2$, and CCl_3, respectively. Infrared bands from 495 to 470 cm^{-1} in all of the above allyl-tin carboxylates have been assigned to the ν(SnC) mode (186). In Table 2.7, vibrational assignments are given for several η^1-C_3H_5 derivatives of the main group elements.

Allyl derivatives of the transition elements have been found with either a *monohapto* structure (2.25), a

M-$CH_2CH=CH_2$

(2.25)

(2.26)

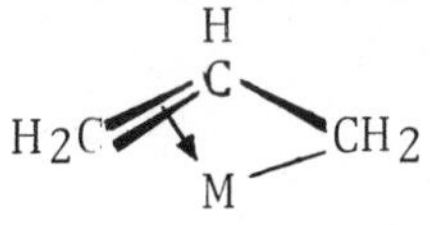

(2.27)

trihapto structure (2.26), or an intermediate structure (2.27) (187,188) in which there are both σ- and π-type interactions. Vibrational assignments have only been reported for compounds with *monohapto* and *trihapto* structures. The vibrational spectra of η^1-C_3H_5 derivatives of the transition elements are similar to those reported for the corresponding derivatives of the main group elements (cf. Figure 2.8) and show a characteristic, strong intensity infrared and Raman band at approximately 1600 cm^{-1} due to the ν(C=C) mode. The vibrational spectra of η^3-C_3H_5 derivatives of the transition elements show no bands in the 1600 cm^{-1} region. The

TABLE 2.7
Vibrational Assignments (cm^{-1}) for η^1-Allyl Derivatives of the Main Group Elements

Mode	$(C_3H_5)_4Si$[a]	$(C_3H_5)_4Sn$[a]	$(C_3H_5)_2Hg$[b]	C_3H_5HgCl[c]	C_3H_5HgBr[c]	C_3H_5HgI[d]
$\nu_a(=CH_2)$	3083	3080	3071	3078	3078	3077
$\nu(CH)$	2998	2996	3016	3032	3020	3020
$\nu_s(=CH_2)$	2977	2968	2987	3000	3005	2996
$\nu_a(CH_2)$	2908	2953	2962	2975	2970	2968
$\nu_s(CH_2)$	2886	2916	2906	2924	2918	2930
$\nu(C=C)$	1631	1624	1617	1626	1624	1622
$\delta_s(CH_2)$	1421	1419	1421	1431	1432	1427
$\delta_s(=CH_2)$	1395	1392	1391	1400	1395	1393
$\rho_r(CH)$	1298	1295	1293	1300	1296	1295
$\rho_w(CH_2)$	1191	1188	1098	1119	1104	1103
$\rho_r(CH_2)$	1153	1094	1188	1187	1190	1188
$\rho_t(CH_2)$	1040	1026	1030	1046	1045	1035
$\rho_t(=CH_2)$	992	989	987	985	989	989
$\nu(CC)$	932	928	931	936	935	935
$\rho_w(=CH_2)$	898	883	876	901	900	899
$\rho_r(CH_2)$	809	749	756	773	772	764
$\rho_w(CH)$	597	670	676	684	685	683
$\nu(MC)$	707 597 526	487 475	495 475	511	500	486
$\nu(MX)$				292	195	112
$\delta(C=CC)$	678 413	670 389	387	384	383	385
$\delta(MCC)$	397	198	236 230	232	235	225

[a]Reference 180.
[b]References 172 and 173.
[c]References 174 and 176.
[d]Reference 174.

vibrational spectra of these derivatives do, however, show three characteristic bands from 1510 to 1375 cm^{-1} that appear with strong to moderate infrared intensity and moderate to weak Raman intensity. Although it is agreed that these bands arise from the ν_a(CCC) (A″), $\delta_a(CH_2)$ (A″), and $\delta_s(CH_2)$ (A′) modes, there has been disagreement with respect to the specific assignments. This is observed in Table 2.8 in which the assignments are summarized for the allyl modes of η^3-allyl derivatives of transition elements. The infrared spectrum (1600 to 600 cm^{-1}) of η^3-$C_3H_5Fe(CO)_2NO$ and its deuterated analog are illustrated in Figure 2.9 (190). Relatively complete vibrational data have also been reported for (η^3-2-methylallyl)$_2$Ni (193) and (η^3-1-methylallyl-PdX)$_2$ (X=Cl,Br) (194). Also, a complete normal coordinate analysis has been reported of the η^3-allyl group as an isolated molecular fragment at the level of the harmonic oscillator approximation (196).

Partial infrared assignments made principally in the 1650 to 1450 cm^{-1} region have been reported for several additional allyl and alkyl substituted allyl derivatives of the transition elements. These assignments have aided in determining whether η^1- and/or η^3-allyl groups are present. Several of these assignments are summarized in Table 2.9. From a comparison of the data in Table 2.9 for the titanium complexes (198) with previously discussed data and conclusions for related substituted allylsodium derivatives (166) it has been concluded that the methyl substituents at the number one and three allylic carbon atoms of the titanium complexes occupy the *syn* positions. A single crystal X-ray study of (η^5-C_5H_5)$_2$Ti(1,2-dimethylallyl) has confirmed the presence of a η^3-1,2-dimethylallyl group and confirmed that the methyl group at the number one carbon atom is in the *syn* position (198). Assignments for the ν(CH) (3017 cm^{-1}) and $\nu(CH_2)$ (3078,3038, and 2908 cm^{-1}) modes of $C_3H_5VCl_3$ (211) are similar to those found for η^3-allyl complexes. Several *trihapto* complexes of the type $RMo(CO)_2L_2X$ (R=allyl,substituted allyl group;L=$(C_6H_5)_3P$,CH_3CN,pyridine, or other Lewis base;X=halide, NCS) have been characterized (212). They all exhibit infrared bands at approximately 1460 and 1030 cm^{-1}, which have been assigned to the ν_a(CCC) and ν_s(CCC) modes, respectively. The nmr spectra and the absence of infrared bands in the 1650 to 1600 cm^{-1} region have been given as evidence for *trihapto* allylic groups in 2-$RC_3H_4RhL_2X$ (R=H,CH_3,L=$(C_6H_5)_3P$,$(C_6H_5)_3As$,$(C_6H_5)_3Sb$, X=Cl;R=H,CH_3,L=$(C_6H_5)_3P$,X=Br), [($(C_6H_5)_3P)_2PtR$]X

TABLE 2.8

Vibrational Assignments (cm^{-1}) for the Allyl Modes of η^3-Allyl Derivatives of the Transition Elements

Mode	Sym.	Compound[a] RMn-$(CO)_4$[b]	RFe-$(CO)_2NO$[c]	$RCo(CO)_3$		R_2Ni[f]	R_2Pd[f]	R_3Rh[f]	R_3Ir[f]	$(RPdCl)_2$	
$\nu(CH_2)$	A''	3078	3082	3078[d]	3087[e]	3067	3090	3069	3063		3079[h]
$\nu(CH)$	A'	3025	3048	3056	3023	2993		3028	3016		3048
$\nu(CH_2)$	A'	2973	3016	3014	2971	2993		3007			3008
$\nu(CH_2)$	A''	2964	2968	2955		2955		2961	3001		
$\nu(CH_2)$	A'	2948	2932	2922	2940	2955		2961			2955
$\delta_a(CH_2)$	A''	1503	1387	1385	1487	1499	1501	1493	1465	1461[g]	1488
$\delta_s(CH_2)$	A'	1462	1466	1469	1473	1460	1485	1466	1455	1383	1456
$\nu_a(CCC)$	A''	1397	1492	1484	1389	1387		1391	1387	1492	1380
$\pi(CH)$	A'	1214	1229	1224	1228	1215	1221	1221	1220	920	1225
							1213				

δ(CH)	A''	1155	1202	1186	1189	1160	1137 1125	1205	1188	1199	1197
ρ_s(CCC)	A'	1017	966	950	1020	1010	1009	1015	1018	1024	1019
$\rho_t(CH_2)$	A'	1007	1018	1017	1020	990	991	1015	1018	998	993
$\rho_w(CH_2)$	A'	920	926	927	951	905	878	931	967	974	954
$\rho_w(CH_2)$	A''	883	916	932	934	862	863	914	926	943	937
$\rho_t(CH_2)$	A''	980	752	772	927	915	930 910	989	1018	763	
$\rho_r(CH_2)$	A''	788	722	735	805	802	800	895	898	1230	766
$\rho_r(CH_2)$	A'	774	778	812	775	755	752	871	800	767	755
δ(CCC)	A'	521	561	555	525	500	493	557	604 580	510	510

[a] $R=\eta^3-C_3H_5$.
[b] Reference 189.
[c] Reference 190.
[d] Reference 191.
[e] Reference 192.
[f] Reference 193.
[g] Reference 194.
[h] Reference 195.

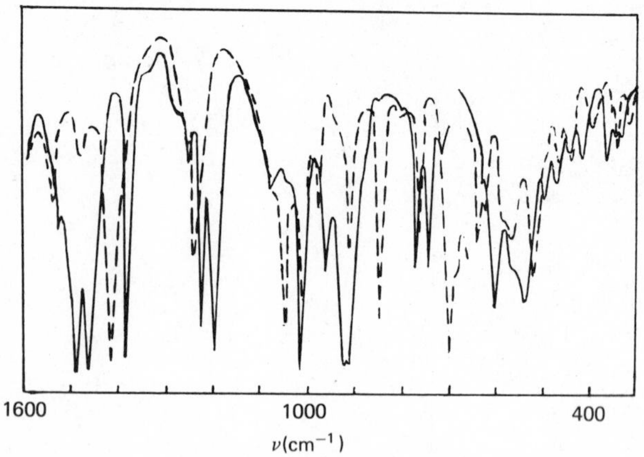

Figure 2.9. Infrared spectra of liquid η^3*-*$C_3H_5Fe(CO)_2NO$ *(——) and* η^3*-*$C_3D_5Fe(CO)_2NO$ *(---) (190).*

(R=allyl,X=Cl;R=1,1-dimethylallyl,X=Cl) (213), and C_3H_5-$Rh[P(OR)_3]_4^{2+}$ and $C_3H_5Rh(Cl)[P(OR)_3]_3^+$ (R=CH_3,C_2H_5,*i*-C_3H_7) (214). A structure with bridging carboxylate ligands has been proposed for $C_6H_8Pd_2(O_2CR)_2$ (R=CH_3,C_2H_5,$(CH_3)_2CH$, C_6H_5,$C_6H_5CH_2$,CF_3,CCl_3) (2.28) while a structure with

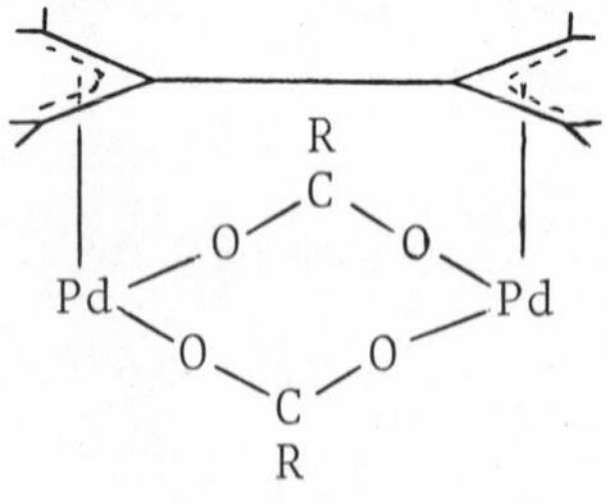

(2.28)

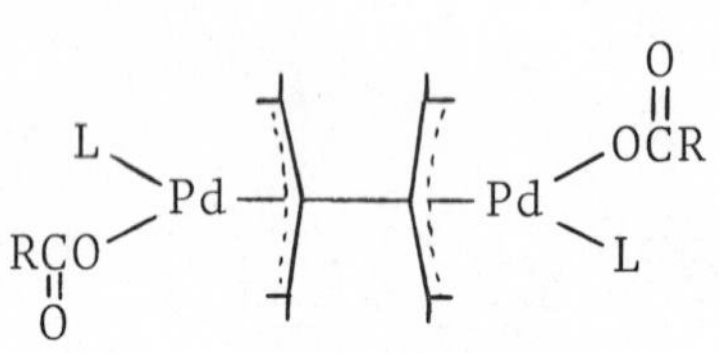

(2.29)

terminal, monodentate carboxylates has been proposed for $C_6H_8PdL_2(O_2CR)_2$ (R=CH_3,CF_3,CCl_3;L=pyridine,$(C_6H_5)_3P$ (2.29) (215). The presence of different structures is indicated by the fact that for the compounds with the bridging carboxylates, the difference between the ν_a(CO) and ν_s(CO) mode frequencies is approximately 185 to 150 cm^{-1}, while for the compounds with terminal carboxylates this difference is approximately 375 to 240 cm^{-1}.

TABLE 2.9
Bands (cm^{-1}) Characteristic of η^1- and η^3-Allyl and Substituted Allyl Groups in Transition-Metal Complexes

Compound	*Monohapto* Group	*Trihapto* Group	Refs.
$(C_5H_5)_2Sc(\eta^3\text{-}C_3H_5)$		1475	107
$\eta^1\text{-}C_3H_5Ti[N(C_2H_5)_2]_3$	1602		197
η^1-2-methylallylTi$[N(C_2H_5)_2]_3$	1607		197
η^1-3-methylallylTi$[N(C_2H_5)_2]_3$	1623		197
$(C_5H_5)_2Ti(\eta^1\text{-}C_3H_5)CH_3$	1597		169
$(C_5H_5)_2Ti(\eta^1$-2-methylallyl$)CH_3$	1603		169
$(C_5H_5)_2Ti(\eta^3\text{-}C_3H_5)$		1509	198
$(C_5H_5)_2Ti(\eta^3$-1-methylallyl$)$		1533	198
$(C_5H_5)_2Ti(\eta^3$-2-methylallyl$)$		1480	198
$(C_5H_5)_2Ti(\eta^3$-1,1-dimethylallyl$)$		1558	198
$(C_5H_5)_2Ti(\eta^3$-1,2-dimethylallyl$)$		1499	198
$(C_5H_5)_2Ti(\eta^3$-1,3-dimethylallyl$)$		1546	198
$(C_5H_5)_2Ti(\eta^3$-1,2,3-trimethylallyl$)$		1499	198
$(C_5H_5)_2Ti(\eta^3$-3-ethyl-1-methylallyl$)$		1539	198
$(C_5H_5)_2Zr(\eta^1\text{-}C_3H_5)Cl$	1598		169
$(C_5H_5)_2Zr(\eta^1$-2-methylallyl$)$	1606		169
$(C_5H_5)_2Zr(\eta^1\text{-}C_3H_5)(\eta^3\text{-}C_3H_5)$	1589	1533	169
$(C_5H_5)_2Zr(\eta^1$-2-methylallyl$)(\eta^3$-2-methylallyl$)$	1603	1520	169
$(C_5H_5)_2Zr(\eta^1\text{-}C_3H_5)(\eta^3\text{-}C_3H_5)Cl$	1599 1589	1533	169
$(\eta^3\text{-}C_3H_5)_4Zr$		1515	169
$(C_5H_5)_2V(\eta^1\text{-}C_3H_5)$	1588		169
$(C_5H_5)_2V(\eta^1$-2-methylallyl$)$	1598		169
$(\eta^3\text{-}C_3H_5)_3V$		1520-1505	169
$C_5H_5W(\eta^1\text{-}C_3H_5)(CO)_3$	1609		199
$C_5H_5W(\eta^3\text{-}C_3H_5)(CO)_2$	1476		199

TABLE 2.9 (continued)

Compound	*Monohapto* Group	*Trihapto* Group	Refs.
$[(C_2H_5)_4N][M_2X_3(\eta^3\text{-}C_3H_5)_2(CO)_4]$ (M=Mo,W; X= C_3H_5Cl,C_3H_5Br)		1500-1400	200
$\eta^1\text{-}C_3H_5Mn(CO)_5$	1617		201,202
$\eta^3\text{-}C_3H_5Fe(CO)_2Cl$		1468	203
$\eta^3\text{-}C_3H_5Fe(CO)_2Br$		1470	203
$\eta^3\text{-}C_3H_5Fe(CO)_2I$		1464	203
η^3-2-methylallylFe$(CO)_3Br$		1460	203
η^3-2-methylallylFe$(CO)_3I$		1460	203
η^3-2-bromoallylFe$(CO)_3Br$		1462	203
$\eta^3\text{-}C_3H_5Ru(CO)_3Cl$		1462	204
$\eta^3\text{-}C_3H_5Ru(CO)_3Br$		1467	204
$\eta^3\text{-}C_3H_5Ru(CO)_3I$		1462	204
η^3-2-methylallylRu$(CO)_3Cl$		1481	204
η^3-1-methylallylCo$(CO)_3$		1460	205
η^3-1-ethylallylCo$(CO)_3$		1451	205
η^3-1,1,2-trimethylallylCo$(CO)_3$		1462	205
η^3-1,2-dimethylallylCo$(CO)_3$		1450	205
$[(\eta^3\text{-1-methylallyl})_2CoL]X$ (L= Bypyr,Phen; X= ClO_4,PF_6)		1520	206
$(\eta^3\text{-}C_3H_5NiBr)_2$		1455	207
$(\eta^3\text{-}C_3H_5NiI)_2$		1449	208
$(\eta^3\text{-}C_3H_5NiC_5H_5)_2$		1449	208
$(C_5H_5)_2Nb(\eta^1\text{-}C_3H_5)CS_2$	1605		209
$(C_5H_5)_2Nb(\eta^1\text{-}C_3H_5)(S_2CR)I$	1610		209
$(C_5H_5)_2Nb(\eta^3\text{-}C_3H_5)$		1500	210

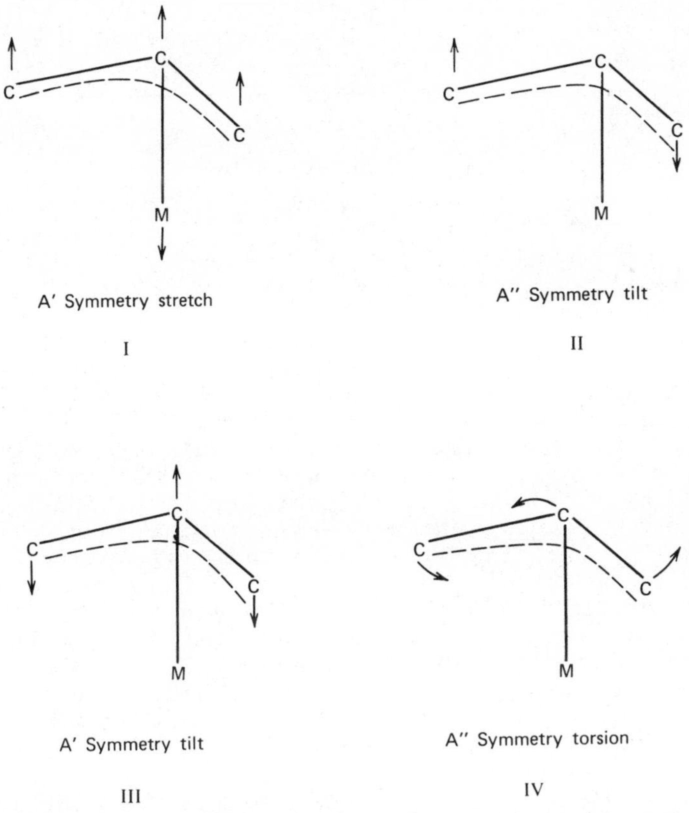

Figure 2.10. Skeletal modes of the η^3*-*C_3H_5M *unit.*

The lanthanide complexes $(\eta^5\text{-}C_5H_5)_2M$ (M=Sm,Er,Ho) show a strong intensity infrared band at 1533 cm^{-1} that has been assigned to the delocalized C-C stretching mode of a η^3-allyl ligand (216), while $(\eta^5\text{-}C_5H_5)_3ThC_3H_5$ (217) and $(\eta^5\text{-}C_5H_5)_3UC_3H_5$ (218) show a weak intensity band at 1650 cm^{-1} (Raman) and 1630 cm^{-1} (infrared), respectively, arising from the ν(C=C) mode of a η^1-allyl ligand.

The four skeletal modes associated with the η^3-C_3H_5M unit are illustrated in Figure 2.10. They can be described as one stretching (I), two tilting (II and III), and one torsional (IV) mode, or as three stretching (I to

TABLE 2.10
Skeletal Modes of $(\eta^3\text{-}C_3H_5)_2M$ Compounds

Mode Description[a]	C_{2v} Irred. Rep.	C_{2v} Activity	C_{2h} Irred. Rep.	C_{2h} Activity
M-R stretch	A_1	R(p),IR	A_g	R(p)
	A_2	R(dp)	B_u	IR
M-R tilt	A_1	R(p),IR	A_g	R(p)
	A_2	R(dp)	A_u	IR
	B_1	R(dp),IR	B_g	R(dp)
	B_2	R(dp),IR	B_u	IR
R-M-R def.	A_1	R(p),IR	A_g	R(p)
M-R torsion	A_2	R(dp)	A_u	IR
	B_1	R(p),IR	A_g	R(p)

[a] $R=\eta^3\text{-}C_3H_5$.

III) and one torsional mode (IV). The description and activity of the skeletal modes for $(\eta^3\text{-}C_3H_5)_2M$ and $(\eta^3\text{-}C_3H_5)_3M$ compounds are summarized in Tables 2.10 and 2.11, respectively, while the assignments made for the metal-allyl skeletal stretching and tilting modes for several η^3-allyl derivatives of the transition elements are summarized in Table 2.12. To help confirm the skeletal mode assignments for $(\eta^3\text{-}C_3H_5PdCl)_2$, this complex was prepared using ^{104}Pd and ^{110}Pd isotopes (194). The palladium-allyl modes at 402.0, 379.0, and 366.4 cm^{-1} and the ν(PdCl) modes at 255.0 and 245.0 cm^{-1} in the ^{104}Pd derivative were shifted to 399.0, 376.0, 252.8, and 243.6 cm^{-1}, respectively, in the corresponding ^{110}Pd derivative. The assignments made for the metal-allyl modes in $(\eta^3$-2-methylallylCo$(CO)_3$ (219) were complicated by the presence of in-plane and out-of-plane (with respect to the allyl group) H_3CCC bending modes in the 400 cm^{-1} region. Skeletal mode assignments have also been given for $\eta^3\text{-}XC_3H_4Co(CO)_3$ (X=1-Cl, 2-Cl,1-CH_3) (219).

E. ALLENIC COMPOUNDS

Uncomplexed allene, $H_2C=C=CH_2$, shows a characteristic infrared band at 1970 cm^{-1}, due to the ν_a(C=C=C) mode (220). A product obtained on reacting allene and *n*-butyllithium has been characterized as allenyllithium, C_3H_3Li (221). In a solution of tetrahydrofuran and

TABLE 2.11
Skeletal Modes of $(\eta^3\text{-}C_3H_5)_3M$ Compounds

Mode Description[a]	D_{3h} Irred. Rep.	D_{3h} Activity	C_{3v} Irred. Rep.	C_{3v} Activity	C_{2v} Irred. Rep.	C_{2v} Activity	C_{2h} Irred. Rep.	C_{2h} Activity
M-R stretch	A_1'	R(p)	A_1	R(p),IR	$2A_1$	R p),IR	2A'	R(p),IR
	E'	R(dp),IR	E	R(dp),IR	B_2	R(dp),IR	A"	R(dp),IR
R_3M i.p. def.	E'	R(dp),IR	A_1	R(p),IR	A_1	R(p),IR	2A'	R(p),IR
			E	R(dp),IR	B_1	R(dp),IR	A"	R(dp),IR
					B_2	R(dp),IR		
R_3M o.o.p. def.	A"	IR	A_1	R(p),IR	A_1	R(p),IR	3A'	R(p),IR
M-R tilt	A'	inactive	A_2	inactive	A_2	R(dp)	3A"	R(dp),IR
	A"	IR	2E	R(dp),IR	$2B_1$	R(dp),IR		
	E'	R(dp),IR			$2B_2$	R(dp),IR		
	E"	R(dp)						

[a] $R=\eta^3\text{-}C_3H_5$.

TABLE 2.12
Skeletal Mode Assignments (cm-1) for η3-Allyl Complexes of the Transition Metals

Compound	M-Allyl Tilt	M-Allyl Stretch	ν(MX)	Refs.
η^3-$C_3H_5Mn(CO)_4$	411(dp) 385(p)	327(p)	478(p) 454 430(p)	189
η^3-$C_3H_5Fe(CO)_2NO$	417 335	365	627[a] 446[b] 348[b]	190
η^3-$C_3H_5Co(CO)_3$	415 340	376	456 364	191
	419(p) 375	362(p)	482(p) 455(dp)	192
$(\eta^3$-$C_3H_5)_2Ni$	406(p) 398 385(dp) 380	350 324(p)		193
$(\eta^3$-$C_3H_5)_2Pd$	468 402	375 329		193
$(\eta^3$-$C_3H_5)_2Rh$	541 535 395(p)	357(dp) 326(p)		193
$(\eta^3$-$C_3H_5)_2Ir$	570 413(p)	377 354(p) 320		193
$(\eta^3$-$C_3H_5PdCl)_2$	398(p) 381	369(p)	254 243	194,195
$(\eta^3$-$C_3H_5PdBr)_2$	403(p) 372	363(p)	183 174	194,195

[a]ν(FeN) mode.
[b]ν(Fe-CO) mode.

hexane, allenyllithium exhibits a strong intensity infrared band at 1894 cm^{-1}, while solid allenyllithium gives rise to a strong intensity band at 1894 cm^{-1}; these bands have been assigned to the ν_a(C=C=C) mode (221). In addition, solid allenyllithium shows weaker intensity infrared bands at 2336, 2222, and 1795 cm^{-1}. Although $CH_3C{\equiv}CH$ shows no infrared bands between 2100 and 1600 cm^{-1}, after the addition of $CH_3C{\equiv}CH$ to four equivalents of *n*-butyllithium in hexane, the solution develops three bands at 1870, 1770, and 1675 cm^{-1} (222). None of these bands could be attributed to $CH_3C{\equiv}CLi$ since this compound is insoluble in hexane and shows a band at 2050 cm^{-1}. Therefore, these bands have been assigned to the ν_a(C=C=C) mode of the allenyl derivatives $C_3H_2Li_2$, C_3HLi_3, and C_3Li_4, respectively (222). Similarly, a strong intensity infrared band has been observed at 1880 cm^{-1} for allenyl Grignard (223,224).

Themmolysis or pyrolysis of $[(CH_3)_3Si]_2CHC{\equiv}CSi(CH_3)_3$ (ν(C≡C)=2180 cm^{-1}) resulted in almost complete conversion to 1,1,3-tris(trimethylsilyl)allene, $[(CH_3)_3Si]_2C{=}C{=}C(H)Si(CH_3)_3$, which showed a strong intensity infrared band at 1910 cm^{-1}, which has been assigned to the ν_a(C=C=C) mode, and a weak intensity band at 3160 cm^{-1}, which was assigned to the allenic ν(CH) mode (225). In the same study, the ν_a(C=C=C) mode of $[(CH_3)_3Si]_2C{=}C{=}C[Si(CH_3)_3]_2$ was assigned to a strong intensity infrared band at 1890 cm^{-1}. Infrared data consisting mainly of assignments for the ν_a(C=C=C) mode from 1935 to 1878 cm^{-1} have also been reported for several other allene-silicon derivatives (222,226-229). Heating $(C_6H_5)_2(CH_3)PCHC{\equiv}CH$ (ν(C≡C)=2257 cm^{-1}) at approximately 130°C resulted in the production of a small amount of $(C_6H_5)_2(CH_3)PCH{=}C{=}CH_2$ (ν_a(C=C=C)= 1912 cm^{-1}) (230).

Protonation of η^5-$C_5H_5Fe(CO)_2(CH_2C{\equiv}CH)$ has produced a η^3-allene derivative (231). This study included a listing of infrared bands observed above 700 cm^{-1} for both η^5-$C_5H_5Fe(CO)_2(CH_2C{\equiv}CH)$ and $[\eta^5$-$C_5H_5Fe(CO)_2(\eta^3$-$H_2C{=}C{=}CH_2)]SbCl_6$. The only assignments presented were for the ν(CO) modes of both compounds and the ν(C≡C) mode (2141 cm^{-1}) of the alkyne derivative

F. CONJUGATED AND NONCONJUGATED DIOLEFIN AND RELATED COMPOUNDS

1. Butadiene Complexes

Free butadiene, C_4H_6, has a symmetric *trans* configuration of C_{2h} symmetry at room temperature (232). A comparison of the infrared spectrum of complexed butadiene in $K_2[C_4H_6(PtCl_3)_2]$ (233) with that of free butadiene (234) has been interpreted as indicating that rather than taking on the *cis* configuration as had previously been suggested (2.30) (49,235), the complexed

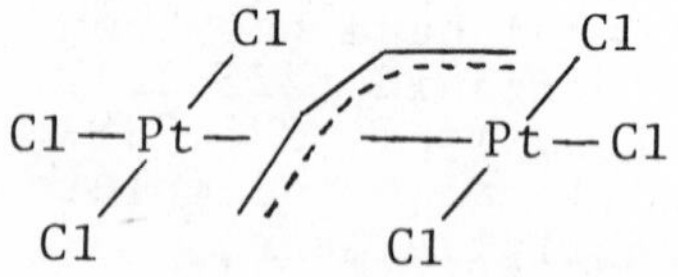

(2.30)

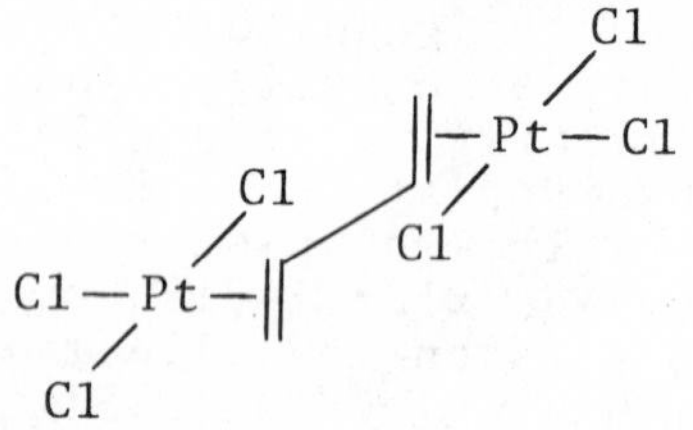

(2.31)

butadiene molecule retains the *trans* configuration (2.31) found for the free molecule (234). The infrared spectrum of $(C_4H_6PdCl_2)_2$ has been interpreted (235) in terms of a butadiene-bridged structure (2.32).

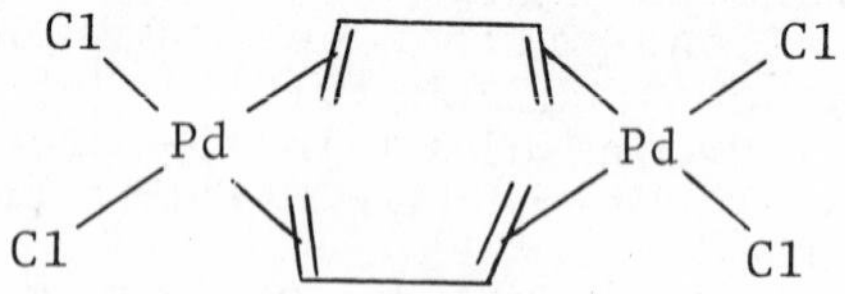

(2.32)

The butadiene molecule in $C_4H_6Fe(CO)_3$ has been shown to have a *cis* planar structure (236). A comparison of the infrared spectra of $K_2[C_4H_6(PtCl_3)_2]$ and $C_4H_6Fe(CO)_3$ (233), which are both illustrated in Figure 2.11, shows the spectrum of $K_2[C_4H_6(PtCl_3)_2]$ to be simpler, especially in the 1500 to 700 cm^{-1} region, than that of $C_4H_6Fe(CO)_3$ (234). This is to be expected since the butadiene molecule has a center of symmetry in the platinum complex and obeys the mutual exclusion rule while the butadiene molecule has no center of symmetry in the iron complex. A complete assignment has

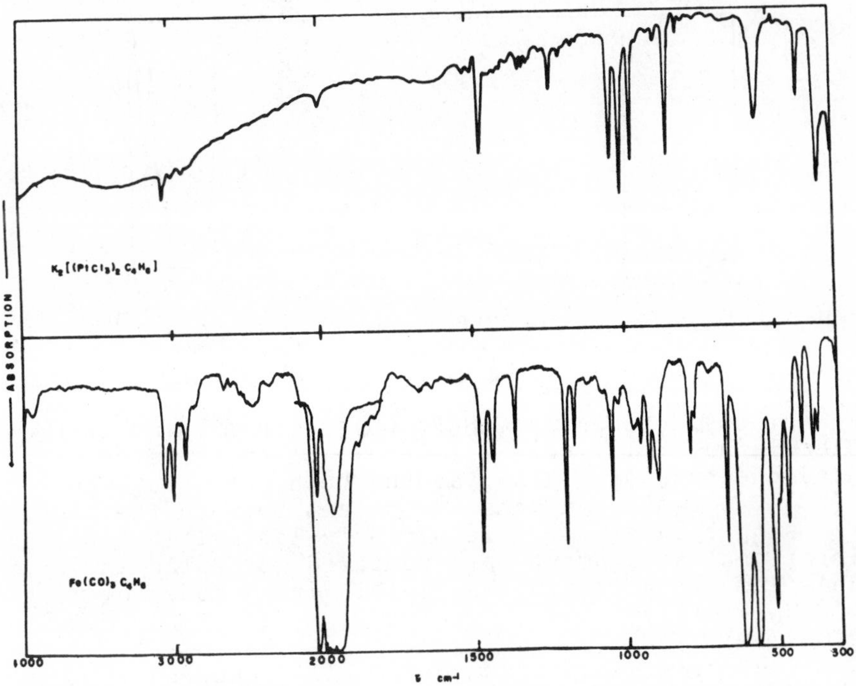

Figure 2.11. Infrared spectra of $K_2[C_4H_6(PtCl_3)_2]$ *and* $C_4H_6Fe(CO)_3$ *(233).*

been made of the infrared and Raman spectra of liquid $C_4H_6Fe(CO)_3$ (237). The Raman spectrum of $C_4H_6Fe(CO)_3$ is illustrated in Figure 1.12 (237). It was cautioned in this study that extensive vibrational mixing is present. A thorough infrared and Raman study of $C_4H_6Fe(CO)_3$ in the ν(CO) mode region in different phases has lead to the conclusion that the solid-state spectra can not be explained using a simple factor group analysis; adequate explanation requires a detailed consideration of the crystal structure (238). An analysis has also been reported of the infrared and Raman spectra of $(C_4H_6)_2FeCO$ in the solid state and solution (CS_2 and benzene) (239). From the assignments presented for $(C_4H_6)_2FeCO$, it has been concluded that coupling between the modes of the two butadiene molecules is rather weak and that the replacement of

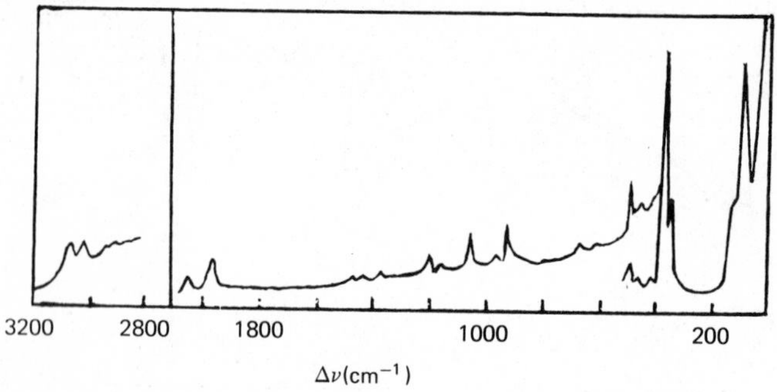

Figure 2.12. Liquid-phase Raman spectrum of $C_4H_6Fe(CO)_3$ (237).

two carbonyl groups in $C_4H_6Fe(CO)_3$ with a butadiene molecule has relatively little effect on the iron-butadiene bonding. Also, as has been noted for C_4H_6-$Fe(CO)_3$, extensive mixing is present between the modes of a single butadiene molecule. In Table 2.13, the vibrational assignments are compared for butadiene (234) and the butadiene and metal-ligand modes of $K_2[C_4H_6(PtCl_3)_2]$ (233), $C_4H_6Fe(CO)_3$ (237), and $(C_4H_6)_2$-FeCO (239).

The infrared spectrum originally reported for $C_4H_6(CuCl)_2$ has been observed to be similar to those reported for platinum-butadiene complexes although it is also more complex, with several bands split into doublets (235). Using more recent infrared data, it has been proposed that the copper atoms are π-bonded to both C=C bonds of the butadiene group since the infrared active ν(C=C) mode at 1600 cm^{-1} for free butadiene was replaced by infrared bands at 1565 and 1500 cm^{-1} in the complex (240). The ν(CuCl) modes were assigned at 263 and 180 cm^{-1}.

The infrared spectra of the π-type complexes $[RCo(CO)_2]_2$ (R=butadiene,2-methylbuta-1,3-diene,2,3-dimethylbuta-1,3-diene,1,2,3,4-tetramethylbuta-1,3-diene) indicate the presence of a planar $Co(CO)_2Co$ bridge (241). In solution, the *trans*- (2.33) or *cis*-bridged (2.34) isomers, or a mixture of both isomers, were proposed as possible (241). Solid-state infrared data (2100 to 1700 cm^{-1}) for $RFeCo(CO)_4R'$ ($R=C_5H_5,CH_3C_5H_4$, indenyl;R'=2,3-dimethylbuta-1,3-diene) show the presence of *cis* or *trans* carbonyl-bridged structures similar

TABLE 2.13

Assignments (cm^{-1}) for 1,3-Butadiene and Various Transition-Metal Complexes

Mode	1,3-Butadiene[a]	$K_2[C_4H_6(PtCl_3)_2]$[b]	$C_4H_6Fe(CO)_3$[c]	$(C_4H_6)_2FeCO$[d]
$\nu(=CH_2)$	3102	3070	3067	3063
	3101	3060	3012	3042
	3014	2985	2950	3002
	2985	2945		
$\nu(=CH)$	3056	3010	3012	3002
	3014	2995	2929	2934
$\nu(C=C)$	1643	1521	1477	1481
	1599	1471	1439	1435
$\delta(=CH_2)$	1442	1402	1499	1474
	1385	1340	1370	1373
$\delta(=CH)$	1285	1238	1174	1189
	1279	1310	1060	1054
$\nu(CC)$	1205	1205	1205	1220
$\rho_t(=CH_2)$	1013	849	1048	1021
	967	810	954	940
$\rho_r(=CH_2)$	987	959	968	969
	890	888	926	920
$\rho_w(=CH_2)$	910	1039	926	920
	909	1008	896	875
π(skeletal)	686 520	553	791 669	766 651
$\delta(C=CC)$	513 301	305	493 417	481 453
$\nu(M-C_4H_6)$		415	351	394 297
$Fe-C_4H_6$ tilt			363	334
$\delta(Pt-C_4H_6)$		210		
$\delta(Pt-C_4H_6)$		121		

[a]Reference 234.
[b]Reference 233.
[c]Reference 237.
[d]Reference 238.

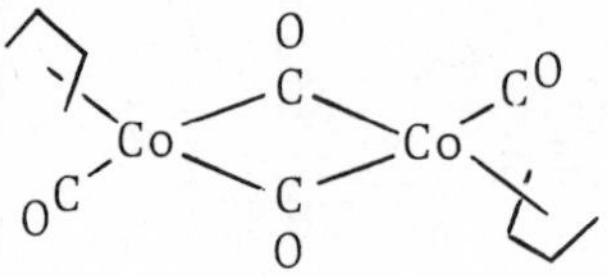

trans

(2.33)

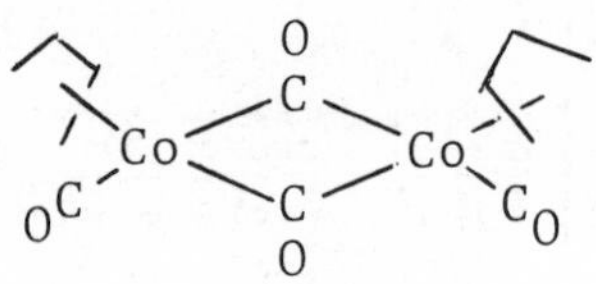

cis

(2.34)

to those proposed for similar cobalt complexes (242). In solution, however, an equilibrium of both isomers is present (242). Bands in the 1485 to 1460 cm^{-1} region, characteristic of conjugated, coordinated olefins, have been reported for $(C_5H_5)(C_4H_6)V(CO)_2$ (243) and RR'RhCl (R=R'=1,3-butadiene,2,3-dimethylbuta-1,3-diene; R=2,3-dimethylbuta-1,3-diene,R'=cyclooctene,*p*-toluidene) (244). Infrared data have also been listed for (2,3-dimethylbuta-1,3-diene)$Os_2(CO)_6$ (245), while the ν(CO) modes have been assigned for $C_4H_6Co(CO)_4$ (246). From infrared data of CCl_4 solutions, complexed ν(C=C) modes have been assigned at 1485 and 1440 cm^{-1} for $(C_4H_6)_3W$ and 1437 cm^{-1} for $(C_4H_6)_3Mo$ (247).

2. Pentadiene Complexes

Two possible structures (2.35 and 2.36) have been proposed for the 1,4-pentadiene complex $(C_5H_7)_2Cr$ with in-

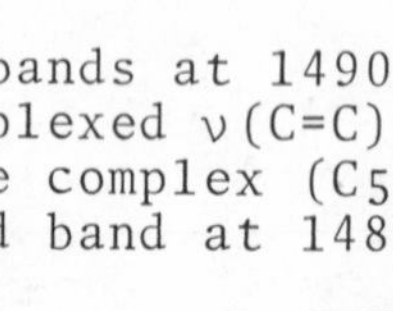

(2.35)

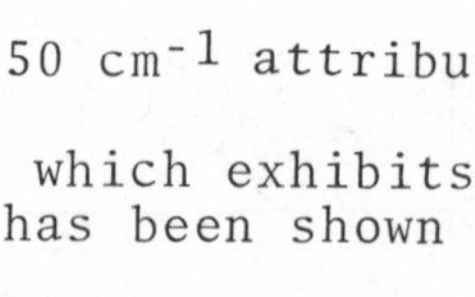

(2.36)

frared bands at 1490, 1485, and 1450 cm^{-1} attributed to the complexed ν(C=C) modes (248).

The complex $(C_5H_7)_2Ni_2$ (2.37) which exhibits an infrared band at 1480 cm^{-1} (249), has been shown in an

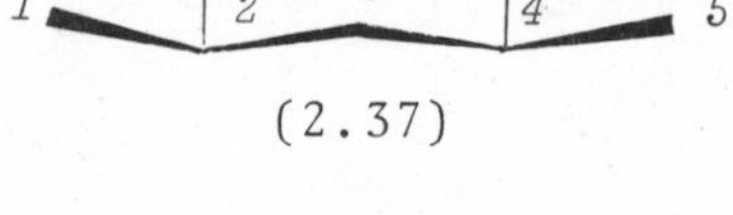

(2.37)

X-ray study to have a structure in which the terminal carbon-carbon bonds (1.40 A) of the asymmetric π-allyl ligand show more double-bond character than the C_2-C_3 and C_3-C_4 bonds (1.44 A) (250).

3. Hexadiene Complexes

Using infrared and pmr data, it was originally proposed that the 1,5-hexadiene ligand in $C_6H_{10}MX_2$ (M=Pd,X=Cl; M=Pt,X=Cl,I) is chelating and has a *gauche* configuration, while in the complexes $K_2[C_6H_{10}(PtCl_3)_2]$ and $C_6H_{10}(CuCl)_2$ it is nonchelating but bridging and retains the *trans* configuration found in the free ligand (251). More recently, however, pmr and single crystal X-ray studies have extablished that the 1,5-hexadiene molecule has a *cis* configuration in $C_6H_{10}PdCl_2$ (2.38)

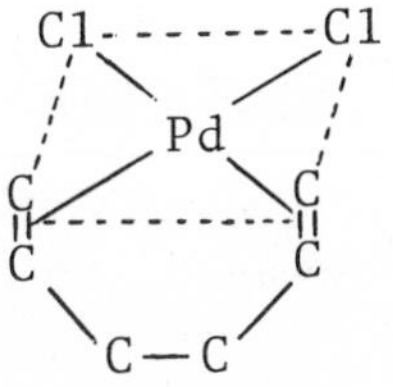

(2.38)

(252). On the basis of this study, and the similarity of the infrared and Raman spectra of $C_6H_{10}PdCl_2$ to the infrared and in many cases Raman spectra of $C_6H_{10}MX_2$ (M=Pd,X=Br;M=Pt,X=Cl,Br,I) and $(C_6H_{10}RhCl)_2$, it has been concluded that the hexadiene ligand in all of these complexes has a *cis* configuration (253). The Raman spectra of these complexes show a strong intensity band at approximately 1240 cm^{-1} similar to that found for π-type ethylene complexes. It has also been suggested that bands in the 500 to 340 cm^{-1} and 350 to 250 cm^{-1} regions may arise from the ν(MC) (M=Pd,Pt) and ν(RhC) modes, respectively (253). In alcoholic solvents, at ambient temperature, *cis,trans*-1,5-cyclodecadiene apparently undergoes a metal-catalyzed isomerization to produce *cis*-1,2-divinylcyclohexane complexes of palladium and platinum (2.39). Although the infrared spectra of both of these complexes have been reported, analysis was not attempted because of the complexity of these spectra (254). It was noted, however, that the 1645 cm^{-1} band of free 1,2-divinylcyclohexane was lowered by 150 to 140 cm^{-1} on complexation.

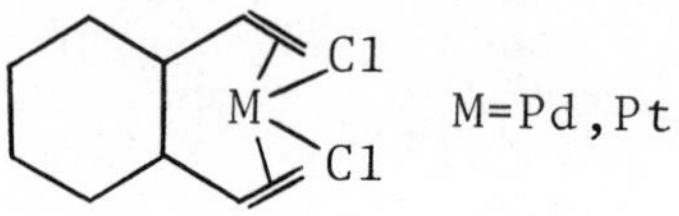

(2.39)

Since the infrared active ν(C=C) mode of free 1,5-hexadiene (1643 cm^{-1}) is lowered by 60 to 45 cm^{-1} in $[(C_6H_{10})_3Ag_2](ClO_4)_2$, it has been suggested (255) that each silver(I) ion is coordinated to three C=C bonds (2.40).

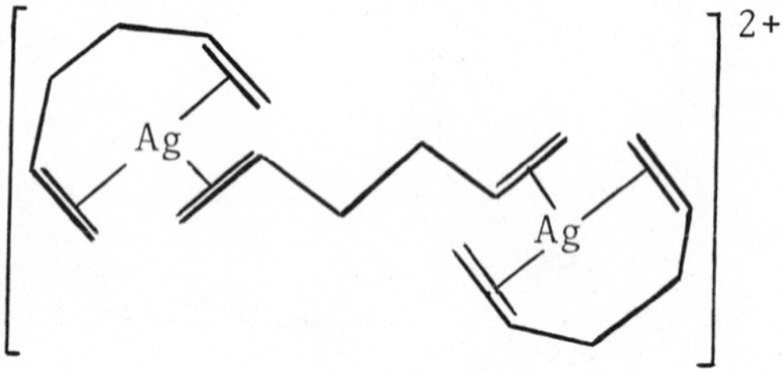

(2.40)

4. Miscellaneous Olefin Complexes

It has been proposed that the silver(I) ion in the 1,7-octadiene and 1,9-decadiene complexes $C_8H_{14}AgClO_4$ and $C_{10}H_{14}AgClO_4$, respectively, is coordinated to both C=C bonds in each diolefin molecule, which results in a decrease in the ν(C=C) mode frequency from 1645 cm^{-1} in the free ligands to 1595 to 1583 cm^{-1} in the complexes (255). In Figure 2.13, the infrared spectra (2000 to 400 cm^{-1}) are illustrated for free 1,9-decadiene and $C_{10}H_{18}AgClO_4$ (255).

A structure in which the tungsten atom is π-bonded to both the C=C bond and the carbonyl group is indicated for $(C_4H_6O)_3W$ (2.41) by the absence of the carbonyl

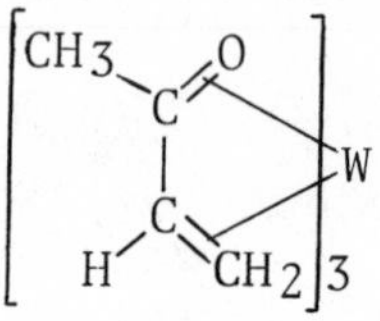

(2.41)

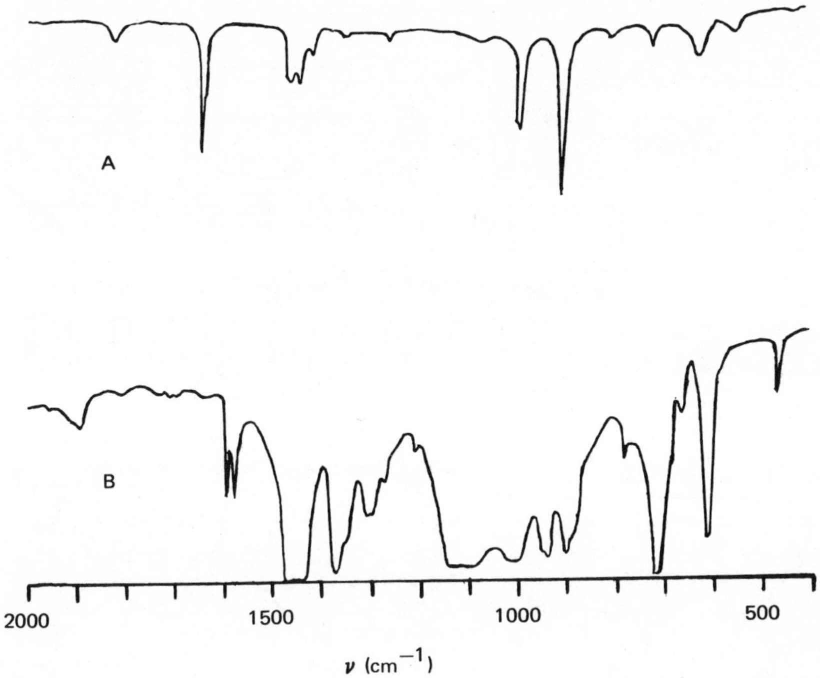

Figure 2.13. Infrared spectra if (A) pure 1,9-decadiene and of (B) $C_{10}H_{18}AgClO_4$ as a Nujol mull (255).

stretching bands in the ketonic carbonyl region (256). A similar structure was proposed earlier for the cinnamaldehyde complex $(C_6H_5CH{=}CHCHO)Fe(CO)_3$, for which partial infrared data have been reported (257). Infrared bands have been listed for $(C_{12}H_{13}O)Mn(CO)_3$ (2.42)

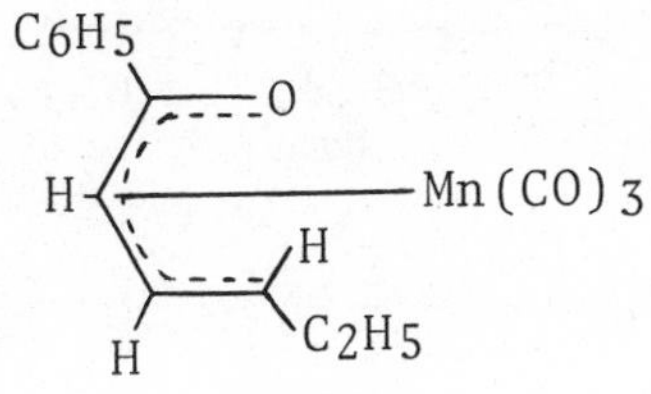

(2.42)

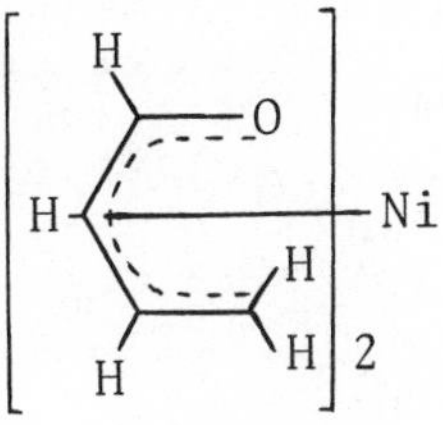

(2.43)

(258). A π-type structure has also been proposed (259, 260) for the acrolein complex $(C_4H_5O)_2Ni$ (2.43) with the carbonyl stretching, ν(C=C), and ν_a(Ni-ligand) modes assigned (259) at 1518, 1605, and 451 cm^{-1}, respectively.

It has been proposed that the geometry of ligand in the dibenzylideneacetone complexes $(C_6H_5CH{=}CHC({=}O){-}CH{=}CHC_6H_5)_2M$ (M=Pd,Pt) (2.44) prevents effective over-

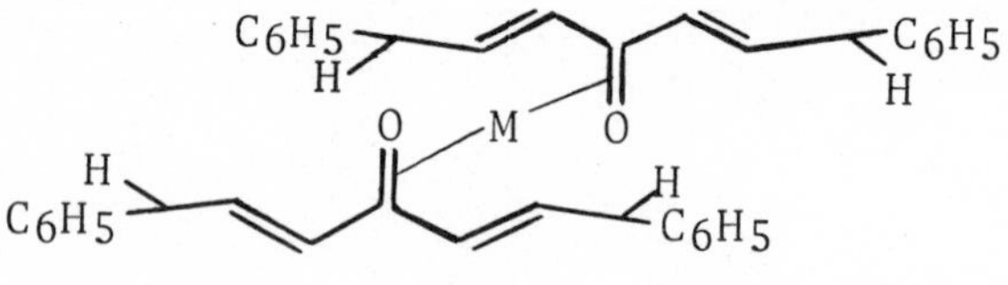

(2.44)

lap of the C=C bonds and the metal orbitals, and that the metal atom therefore forms π-bonds only with the ketone carbonyl group from each ligand molecule (261). The basis for this conclusion is that the intensity and frequency of the strongly infrared active ν(C=C) mode in the free ligand (1621 cm^{-1}) remain relatively constant in the palladium (1613 cm^{-1}) and platinum (1624 cm^{-1}) complexes. Also, the intensity of the carbonyl stretching mode found for the free ligand (1652 cm^{-1}) decreases in the complexes with the appearance of a medium intensity band, which is absent in the free ligand, in the palladium (1544 cm^{-1}) and platinum (1527 cm^{-1}) complexes. This new band has been attributed to the complexed ketone carbonyl group (261).

The structure proposed for the acrylonitrile complex $(NCCH{=}CH_2)_2Ni$ (259,260) is similar to that proposed for the corresponding acrolein complex (2.43) (259,260). The infrared bands have been listed for both free acetonitrile and the complex (259). Although coordination in the acetonitrile complexes $(CH_3CN)_2M(CO)_3X$ (M=Mn,Re;X=Cl,Br) is through the lone pair of electrons on the nitrogen atoms as indicated by the characteristic increase in the ν(C≡N) mode frequency in the complexes relative to the ν(C≡N) mode frequency of the free ligand, a π-type structure has been proposed for the dinitrile complexes (2.45) since the ν(C≡N)

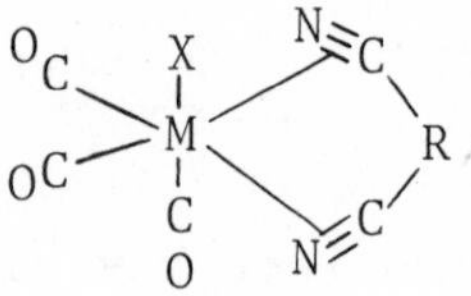

R= $(CH_2)_n$ (n=1 to 3), 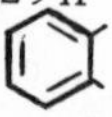

M=Mn,Re;X=Cl,Br.

(2.45)

mode frequency of the dinitriles is lower by 230 to 185 cm^{-1} in the complexes than in the free dinitriles (262).

The infrared and Raman spectra have been assigned in the vapor (263), liquid (264), and solid (263) phases for trimethylenemethaneiron tricarbonyl, $[(CH_2)_3C]Fe(CO)_3$, and its perdeutero derivative (263). The $\nu(CC_3)$ modes of this complex in the vapor phase have been assigned at 1348 (E) and 916 (A_1) cm^{-1}, while the Fe-CC_3 stretching and tilting modes were assigned at 369 and 350 cm^{-1}, respectively (263). Also, on the basis of a comparison of the $\nu(CO)$ mode assignments presented for $C_4H_6Fe(CO)_3$ (2051, 1998, and 1970 cm^{-1}) and $[(CH_2)_3C]Fe(CO)_3$ (2061 and 1994 cm^{-1}), it has been suggested that the electron-donating power of butadiene is greater than that of trimethylenemethane (264). Two normal coordinate analyses have been reported for $[(CH_2)_3C]Fe(CO)_3$ using a valence force field (263,265). The K(CC) force constant obtained in the first study is approximately that expected for a C-C bond (265). In the more recent study, however, it was concluded that this bond has a significant amount of double-bond character (263).

Infrared data in the $\nu(CO)$ mode region have been reported for phenylmethylenediiron octacarbonyl, $[(C_6H_5)(H)C]Fe_2(CO)_8$ (266).

On the basis of an intense infrared band at 1800 cm^{-1}, it has been postulated that C_5Li_4 has mainly the pentatetraene structure, $Li_2C{=}C{=}C{=}C{=}CLi_2$ (226).

Infrared and Raman assignments have been made for several vinylacetylene silicohydrocarbons with the $CH_2{=}CHC{\equiv}CSi$ unit (267). Infrared and Raman bands have been listed for $CH_2{=}C(CH_3)C{\equiv}CH$ and $CH_2{=}C(CH_3)C{\equiv}CCu$ (93). Infrared data have also been presented for silicon and germanium derivatives with $C{=}C{=}CC{\equiv}C$ and $C{\equiv}CC{\equiv}C$ structural features (267) as well as for $C_6H_5C{\equiv}CC{\equiv}CCu(PR_3)_n$ (n=1 to 3, R=CH_3, C_2H_5, n-C_3H_7) (88).

REFERENCES

1. G. Davidson, *Spectrochim. Acta*, *27A*, 1161 (1971).
2. N. A. Chumaevskii, *Opt. Spectrosk.*, *13*, 68 (1962).
3. V. F. Mironov, Yu. P. Egorov, and A. D. Petrov, *Izv. Akad. Nauk SSSR, Otd. Khim. Nauk*, 1400 (1959).
4. T. Shimanouchi, *J. Chem. Phys.*, *26*, 594 (1957).
5. E. A. V. Ebsworth, *Organometallic Compounds of the Group IV Elements*, Dekker, New York, 1968, Vol. 1, part 1, p. 69.

6. D. Seyferth, *Prog. Inorg. Chem., 3,* 129 (1962).
7. B. J. Hewitt, A. K. Holliday, and R. J. Puddephatt, *J. Chem. Soc., Dalton Trans.,* 801 (1973).
8. J. A. Waters and G. A. Mortimer, *J. Organometal. Chem., 22,* 417 (1970).
9. H. D. Kaesz and F. G. A. Stone, *Spectrochim. Acta, 15,* 360 (1959).
10. J. Mink and Y. A. Pentin, *Acta Chem. Acad. Sci. Hung., 66,* 277 (1970).
11. A. K. Holliday, W. Reade, K. R. Seddon, and I. A. Steer, *J. Organometal. Chem., 67,* 1 (1974).
12. J. D. Odom, L. W. Hall, S. Riethmiller, and J. R. Durig, *Inorg. Chem., 13,* 170 (1974).
13. G. Masetti and G. Zerbi, *Spectrochim. Acta, 26A,* 1891 (1970).
14. U. Kunze, E. Lindner, and J. Koola, *J. Organometal. Chem., 57,* 319 (1973).
15. J. Mink and Y. A. Pentin, *Acta Chim. Acad. Sci. Hung., 67,* 435 (1971).
16. J. R. Durig, R. O. Carter, and J. D. Odom, *Inorg. Chem., 13,* 701 (1974).
17. J. R. Durig, L. W. Hall, R. O. Carter, C. J. Wurrey, V. F. Kalasinsky, and J. D. Odom, *J. Phys. Chem., 80,* 1188 (1976).
18. S. G. Frankiss, *Spectrochim. Acta, 22,* 295 (1966).
19. J. R. Durig and K. L. Hellams, *J. Mol. Struct., 6,* 315 (1970).
20. K. Licht, P. Koehler, and H. Kriegsmann, *Z. Anorg. Allg. Chem., 415,* 31 (1975).
21. J. R. Durig and J. B. Turner, *Spectrochim. Acta, 27A,* 1623 (1971).
22. J. R. Durig and J. B. Turner, *Spectrochim. Acta, 27A,* 395 (1971).
23. E. L. Lines and L. F. Centofanti, *Inorg. Chem., 13,* 1517 (1974).
24. F. E. Brinckman and F. G. A. Stone, *J. Amer. Chem. Soc., 82,* 6218 (1960).
25. D. S. Matteson, *J. Amer. Chem. Soc., 82,* 4228 (1960).
26. T. D. Coyle, S. L. Stafford, and F. G. A. Stone, *J. Chem. Soc.,* 3103 (1961).
27. J. P. Oliver and L. G. Stevens, *J. Inorg. Nucl. Chem., 24,* 953 (1962).
28. F. Glockling, M. A. Lyle, and S. R. Stobart, *J. Chem. Soc., Dalton Trans.,* 2537 (1974).
29. J. Knižek, M. Horak, and V. Chvalovsky, *Collect. Czech. Chem. Commun., 28,* 3079 (1963).
30. V. Peruzzo, G. Tagliavini, and R. E. Hester, *J. Organometal. Chem., 56,* 185 (1973).

31. M. C. Henry and J. G. Noltes, *J. Amer. Chem. Soc.*, *82*, 555 (1960).
32. G. S. Biserni and V. Peruzzo, *J. Organometal. Chem.*, *67*, 265 (1974).
33. V. Peruzzo, G. Plazzogna, and G. Tagliavini, *J. Organometal. Chem.*, *24*, 347 (1970).
34. R. E. Hester and D. Mascord, *J. Organometal. Chem.*, *51*, 181 (1973).
35. P. G. Harrison, *Inorg. Nucl. Chem. Lett.*, *8*, 555 (1972).
36. D. Seyferth and L. G. Vaughan, *J. Organometal. Chem.*, *1*, 138 (1963).
37. C. S. Kraihanzel and M. L. Losee, *J. Organometal. Chem.*, *10*, 427 (1967).
38. W. A. McAllister and L. T. Farias, *Inorg. Chem.*, *8*, 2806 (1969).
39. A. N. Nesmeyanov, M. I. Rybinskaya, L. V. Rybin, V. S. Kaganovich, and P. V. Petrovskii, *J. Organometal. Chem.*, *31*, 257 (1971).
40. B. F. G. Johnson, J. Lewis, J. D. Jones, and K. A. Taylor, *J. Chem. Soc., Dalton Trans.*, 34 (1974).
41. B. L. Dyatkin, L. G. Zhuravkova, B. I. Martynov, E. I. Mysov, S. R. Sterlin, and I. L. Knunyants, *J. Organometal. Chem.*, *31*, C15 (1971).
42. S. L. Stafford and F. G. A. Stone, *J. Amer. Chem. Soc.*, *82*, 6238 (1960).
43. S. L. Stafford and F. G. A. Stone, *Spectrochim. Acta*, *17*, 412 (1961).
44. A. N. Nesmeyanov, A. E. Borisov, N. V. Novikova, and E. I. Fedin, *J. Organometal. Chem.*, *15*, 279 (1968).
45. H. Murata, *J. Chem. Phys.*, *21*, 181 (1953).
46. J. Dunogues, R. Calas, J. Malzac, N. Duffaut, C. Biran, and P. Lapouyade, *J. Organometal. Chem.*, *27*, C1 (1971).
47. R. E. Bichler, M. R. Booth, and H. C. Clark, *Inorg. Nucl. Chem. Lett.*, *3*, 71 (1967).
48. M. J. Grogan and K. Nakamoto, *J. Amer. Chem. Soc.*, *88*, 5454 (1966).
49. J. Chatt and L. A. Duncanson, *J. Chem. Soc.*, 2939 (1953).
50. H. B. Jonassen and J. E. Field, *J. Amer. Chem. Soc.*, *79*, 1275 (1957).
51. D. B. Powell and N. Sheppard, *Spectrochim. Acta*, *13*, 69 (1958).
52. J. Pradilla-Sorzano and J. P. Fackler, Jr., *J. Mol. Spectrosc.*, *22*, 80 (1967).
53. M. J. Grogan and K. Nakamoto, *J. Amer. Chem. Soc.*, *90*, 918 (1968).

54. J. A. Kieft and K. Nakamoto, *J. Inorg. Nucl. Chem.*, *30*, 3103 (1968).
55. A. A. Baboushkin, L. A. Gribov, and A. D. Gelman, *Dokl. Akad. Nauk SSSR*, 461 (1958).
56. D. B. Powell and N. Sheppard, *J. Chem. Soc.*, 2519 (1960).
57. D. M. Adams and J. Chatt, *Chem. Ind.*, 149 (1960).
58. D. M. Adams and J. Chatt, *J. Chem. Soc.*, 2821 (1962).
59. J. Hiraishi, *Spectrochim. Acta*, *25A*, 749 (1969).
60. D. P. Powell, J. G. V. Scott, and N. Sheppard, *Spectrochim. Acta*, *28A*, 327 (1972).
61. D. C. Andrews and G. Davidson, *J. Organometal. Chem.*, *35*, 161 (1972).
62. J. Hiraishi, D. Finseth, and F. A. Miller, *Spectrochim. Acta*, *25A*, 1657 (1969).
63. H. P. Fritz and D. Sellmann, *J. Organometal. Chem.*, *6*, 558 (1966).
64. J. Hubert, P. C. Kong, F. D. Rochon, and T. Theophanides, *Can. J. Chem.*, *50*, 1596 (1972).
65. M. A. Bennett, R. J. H. Clark, and D. L. Miller, *Inorg. Chem.*, *6*, 1647 (1967).
66. G. T. Behnke and K. Nakamoto, *Inorg. Chem.*, *7*, 2030 (1968).
67. L. E. Nance and H. G. Frye, *J. Inorg. Nucl. Chem.*, *38*, 637 (1976).
68. P. E. Garrou and G. E. Hartwell, *J. Organometal. Chem.*, *71*, 443 (1974).
69. M. A. Meester, D. J. Stufkens, and K. Vrieze, *Inorg. Chim. Acta*, *14*, 25 (1975).
70. H. Huber, D. McIntosh, and G. A. Ozin, *J. Organometal. Chem.*, *112*, C50 (1976).
71. K. Fischer, K. Jonas, and G. Wilke, *Angew. Chem., Inter. Ed. Engl.*, *12*, 565 (1973).
72. R. M. Atkins, R. MacKenzie, P. L. Timms, and T. W. Turney, *Chem. Commun.*, 764 (1975).
73. E. O. Fischer and P. Kuzel, *Z. Naturforsch.*, *16b*, 475 (1961).
74. H. P. Kögler and E. O. Fischer, *Z. Naturforsch.*, *15b*, 676 (1960).
75. R. Cramer, *Inorg. Chem.*, *1*, 722 (1962).
76. H. J. Taufen, M. J. Murray, and F. F. Cleveland, *J. Amer. Chem. Soc.*, *63*, 3500 (1941).
77. J. D. Prentice, A. Lesiunas, and N. Sheppard, *Chem. Commun.*, 76 (1976).
78. R. Fields, G. L. Godwein, and R. N. Haszeldine, *J. Organometal. Chem.*, *26*, C70 (1971).
79. H. L. M. Van Gaal and A. Van der Ent, *Inorg. Chim. Acta*, *7*, 653 (1973).

80. D. C. Andrews, G. Davidson, and D. A. Duce, *J. Organometal. Chem.*, *101*, 113 (1975).
81. L. Schäfer, J. D. Ewbank, S. J. Cyvin, and J. Brunvoll, *J. Mol. Struct.*, *14*, 185 (1972).
82. J. W. Fitch, D. P. Flores, and J. E. George, *J. Organometal. Chem.*, *29*, 263 (1971).
83. J. W. Fitch and H. E. Herbold, *Inorg. Chem.*, *9*, 1926 (1970).
84. D. M. Roundhill, *J. Inorg. Nucl. Chem.*, *33*, 3367 (1971).
85. R. E. Sacher, B. C. Pant, F. A. Miller, and F. R. Brown, *Spectrochim. Acta*, *28A*, 1361 (1972).
86. R. Nast and J. Gremm, *Z. Anorg. Allg. Chem.*, *325*, 62 (1963).
87. M. A. Coles and F. A. Hart, *J. Organometal. Chem.*, *32*, 279 (1971).
88. G. E. Coates and C. Parkin, *J. Inorg. Nucl. Chem.*, *22*, 59 (1961).
89. D. Blake, G. Calvin, and G. E. Coates, *Proc. Chem. Soc.*, 396 (1959).
90. R. Nast and H. Schindel, *Z. Anorg. Allg. Chem.*, *326*, 201 (1963).
91. M. F. Shostakovskii, L. A. Polyakova, L. V. Vasil'eva, and A. I. Polyakov, *Zh. Org. Khim.*, *2*, 1899 (1966).
92. I. A. Garbuzova, V. T. Aleksanyan, L. A. Leites, I. R. Gol'ding, and A. M. Sladkov, *J. Organometal. Chem.*, *54*, 341 (1973).
93. V. T. Aleksanyan, I. A. Garbuzova, I. R. Gol'ding, and A. M. Sladkov, *Spectrochim. Acta*, *31A*, 517 (1975).
94. A. G. Meister, *J. Chem. Phys.*, *16*, 950 (1948).
95. G. W. King and S. So, *J. Mol. Spectrosc.*, *36*, 468 (1970).
96. R. W. R. Corfield and H. M. M. Shearer, *Acta Crystallogr.*, *21*, 957 (1966).
97. A. Camus and N. Marsich, *J. Organometal. Chem.*, *21*, 249 (1970).
98. P. W. R. Corfield and H. M. M. Shearer, *Acta Crystallogr.*, *20*, 502 (1966).
99. E. A. Jeffery and T. Mole, *J. Organometal. Chem.*, *11*, 393 (1968).
100. J. M. Burke, J. J. Ritter, and W. J. Lafferty, *Spectrochim. Acta*, *30A*, 993 (1974).
101. P. R. Reed, Jr. and R. W. Lovejoy, *J. Chem. Phys.*, *56*, 183 (1972).
102. M. F. Lappert and B. Prokai, *J. Organometal. Chem.*, *1*, 384 (1964).
103. A. D. Petrov, L. L. Shchukovskaya, and Yu. P. Egorov, *Dokl. Akad. Nauk SSSR*, *93*, 293 (1953).

104. J. C. Masson and P. Cadiot, *Bull. Soc. Chim. Fr.*, *32*, 3518 (1965).
105. R. Nast and K. Käb, *J. Organometal. Chem.*, *6*, 456 (1966).
106. H. Kurosawa, M. Tanaka, and R. Okawara, *J. Organometal. Chem.*, *12*, 241 (1968).
107. R. S. P. Coutts and P. C. Wailes, *J. Organometal. Chem.*, *25*, 117 (1970).
108. F. A. Hart, A. G. Massey, and M. S. Saran, *J. Organometal. Chem.*, *21*, 147 (1970).
109. R. E. Sacher, D. H. Lemmon, and F. A. Miller, *Spectrochim. Acta*, *23A*, 1169 (1967).
110. D. I. MacLean and R. E. Sacher, *J. Organometal. Chem.*, *74*, 197 (1974).
111. F. A. Miller and D. H. Lemmon, *Spectrochim. Acta*, *23A*, 1099 (1967).
112. H. Buchert and W. Zeil *Z. Phys. Chem. (Frankfurt)*, *29*, 317 (1961).
113. R. E. Sacher, W. Davidsohm, and F. A. Miller, *Spectrochim. Acta*, *26A*, 1011 (1970).
114. H. Buchert, B. Hass, and W. Zeil, *Spectrochim. Acta*, *19*, 379 (1963).
115. W. Steingross and W. Zeil, *J. Organometal. Chem.*, *6*, 464 (1966).
116. R. B. Reeves, R. E. Wilde, and D. W. Robinson, *J. Chem. Phys.*, *40*, 125 (1964).
117. E. A. V. Ebsworth, S. G. Frankiss, and W. J. Jones, *J. Mol. Spectrosc.*, *13*, 9 (1964).
118. R. W. Lovejoy and D. R. Baker, *J. Chem. Phys.*, *46*, 658 (1967).
119. H. Buchert and W. Zeil, *Spectrochim. Acta*, *18*, 1043 (1962).
120. W. Steingross and W. Zeil, *J. Organometal. Chem.*, *6*, 109 (1966).
121. E. L. Lines and L. F. Centofanti, *Inorg. Chem.*, *12*, 598 (1973).
122. D. H. Lemmon and J. A. Jackson, *Spectrochim. Acta*, *29A*, 1899 (1973).
123. H. F. Reiff and B. C. Pant, *J. Organometal. Chem.*, *17*, 165 (1969).
124. H. Fast and H. L. Welsh, *J. Mol. Spectrosc.*, *41*, 203 (1972).
125. W. R. Cullen and M. C. Waldman, *Inorg. Nucl. Chem. Lett.*, *6*, 205 (1969).
126. J. Parker and J. A. Ladd, *Trans. Faraday Soc.*, *66*, 1907 (1970).
127. V. Devarajan and S. J. Cyvin, *Aust. J. Chem.*, *25*, 1387 (1972).

128. R. C. Lord, D. W. Mayo, H. E. Opitz, and J. S. Peake, *Spectrochim. Acta, 12,* 147 (1958).
129. H. Kriegsmann and H. Beyer, *Z. Anorg. Allg. Chem., 311,* 180 (1961).
130. F. A. Cotton and T. J. Marks, *J. Amer. Chem. Soc., 91,* 7281 (1969).
131. P. J. Moehs and W. E. Dahidsohn, *J. Organometal. Chem., 20,* 57 (1969).
132. H. Köpf and M. Schmidt, *J. Organometal. Chem., 10,* 383 (1967).
133. A. D. Jenkins, M. F. Lapert, and R. C. Srivastava, *J. Organometal. Chem., 23,* 165 (1970).
134. R. Nast and K. Fock, *Chem. Ber., 109,* 455 (1976).
135. K. Oguro, M. Wada, and R. Okawara, *Chem. Commun.,* 899 (1975).
136. K. Yasufuku and H. Yamazaki, *Bull. Chem. Soc. Japan, 45,* 2664 (1972).
137. O. M. Abu Salah and M. I. Bruce, *Aust. J. Chem., 29,* 73 (1976).
138. M. L. H. Green and T. Mole, *J. Organometal. Chem., 12,* 404 (1968).
139. K. L. T. Wong, J. L. Thomas, and H. H. Brintzinger, *J. Amer. Chem. Soc., 96,* 3694 (1974).
140. H. Masai, K. Sonogashira, and N. Hagihara, *J. Organometal. Chem., 26,* 271 (1971).
141. N. M. Ely and M. Tsutsui, *Inorg. Chem., 14,* 2681 (1975).
142. M. Tsutsui, N. Ely, and A. Gebala, *Inorg. Chem., 14,* 78 (1975).
143. A. E. Gebala and M. Tsutsui, *J. Amer. Chem. Soc., 95,* 91 (1973).
144. M. H. Chisholm and H. C. Clark, *Inorg. Chem., 10,* 2557 (1971).
145. P. B. Tripathy and D. M. Roundhill, *J. Organometal. Chem., 24,* 247 (1970).
146. A. Furlani, P. Carusi, and M. V. Russo, *J. Organometal. Chem., 67,* 315 (1974).
147. J. Chatt, R. G. Guy, L. A. Duncanson, and D. T. Thompson, *J. Chem. Soc.,* 5170 (1963).
148. A. D. Allen and T. Theophanides, *Can. J. Chem., 44* 2703 (1966).
149. J. L. Boston, S. O. Grim, and G. Wilkinson, *J. Chem. Soc.,* 3468 (1963).
150. M. J. Mays and G. Wilkinson, *J. Chem. Soc.,* 6629 (1965).
151. G. L. McClure and W. H. Baddley, *J. Organometal. Chem., 27,* 155 (1971).
152. E. O. Greaves and P. M. Maitlis, *J. Organometal. Chem., 6,* 155 (1971).

153. R. Tsumura and N. Hagihara, *Bull. Chem. Soc. Japan, 38,* 1901 (1965).
154. E. O. Greaves, C. J. Lock, and P. M. Maitlis, *Can. J. Chem., 46,* 3879 (1968).
155. G. W. Parshall and F. N. Jones, *J. Amer. Chem. Soc., 87,* 5356 (1965).
156. J. P. Collman and J. W. Kang, *J. Amer. Chem. Soc., 89,* 844 (1967).
157. J. Chatt, G. A. Rowe, and A. A. Williams, *Proc. Chem. Soc.,* 208 (1957).
158. J. Chatt, R. G. Guy, and L. A. Duncanson, *J. Chem. Soc.,* 827 (1961).
159. R. B. Hannan and R. L. Collins, *Acta Crystallogr., 6,* 350 (1953).
160. G. L. McClure and W. H. Baddley, *J. Organometal. Chem., 25,* 261 (1970).
161. W. H. Baddley, C. Panattoni, G. Bandoli, D. A. Clement, and U. Belluco, *J. Amer. Chem. Soc., 93,* 5590 (1971).
162. Y. Iwashita, F. Tamura, and A. Nakamura, *Inorg. Chem., 8,* 1179 (1969).
163. Y. Iwashita, *Inorg. Chem., 9,* 1178 (1969).
164. C. Sourisseau and B. Pasquier, *Spectrochim. Acta, 31A,* 287 (1975).
165. P. West, J. I. Purmort, and S. V. McKinley, *J. Amer. Chem. Soc., 90,* 797 (1968).
166. E. J. Lanpher, *J. Amer. Chem. Soc., 79,* 5578 (1957).
167. G. M. Whitsides, J. E. Nordlander, and J. D. Roberts, *J. Amer. Chem. Soc., 84,* 2010 (1962); *Discus. Faraday Soc., 35,* 185 (1962).
168. C. Prévost and B. Gross, *C. R. Acad. Sci., Paris, 252,* 1023 (1961).
169. H. A. Martin, P. J. Lemaire, and F. Jellinek, *J. Organometal. Chem., 14,* 149 (1968).
170. G. Wilke, B. Bogdanović, P. Hardt, P. Heimbach, W. Keim, M. Kröner, W. Oberkirch, K. Tanaka, E. Steinrücke, D. Walter, and H. Zimmerman, *Angew. Chem., Inter. Ed. Engl., 5,* 151 (1966).
171. K. H. Thiele and P. Zdunneck, *J. Organometal. Chem., 4,* 10 (1965).
172. J. Mink and Yu. A. Pentin, *J. Organometal. Chem., 23,* 293 (1970).
173. C. Sourisseau and B. Pasquier, *J. Organometal. Chem., 39,* 65 (1972).
174. C. Sourisseau and B. Pasquier, *J. Organometal. Chem., 51,* 51 (1972).
175. A. N. Nesmeyanov, A. Z. Rubezhov, L. A. Leites, and S. P. Gubin, *J. Organometal. Chem., 12,* 187 (1968).

176. J. Mink and Yu. A. Pentin, *Acta Chim Acad. Sci. Hung.*, *70*, 41 (1971).
177. R. D. Bach and P. A. Scherr, *J. Amer. Chem. Soc.*, *94*, 220 (1972).
178. T. Abe, H. Kurosawa, and R. Okawara, *J. Organometal. Chem.*, *25*, 353 (1970).
179. M. Fishwick and M. G. H. Wallbridge, *J. Organometal. Chem.*, *25*, 69 (1970).
180. G. Davidson, P. G. Harrison, and E. M. Riley, *Spectrochim. Acta*, *29A*, 1265 (1973).
181. V. F. Mironov and N. A. Chumaevskii, *Dokl. Akad. Nauk SSSR*, *146*, 1117 (1962).
182. I. V. Obreimov and N. A. Chumaevskii, *Zh. Struckt. Khim.*, *5*, 137 (1964).
183. Yu. P. Egorov, L. A. Leites, I. D. Kravtsova, and V. F. Mironov, *Izvest. Akad. Nauk SSSR*, 114 (1963).
184. R. M. G. Roberts, *J. Organometal. Chem.*, *18*, 307 (1969).
185. W. T. Schwartz, Jr. and H. W. Post, *J. Organometal. Chem.*, *2*, 357 (1964).
186. V. Peruzzo, G. Plazzogna, and G. Tagliavini, *J. Organometal. Chem.*, *40*, 121 (1972).
187. J. Powell, S. D. Robinson, and B. L. Shaw, *Chem. Commun.*, 78 (1965).
188. R. Mason and D. R. Russell, *Chem. Commun.*, 26 (1966).
189. G. Davidson and D. C. Andrews, *J. Chem. Soc., Dalton Trans.*, 126 (1972).
190. G. Paliani, A. Poletti, G. Cardaci, S. M. Murgia, and R. Cataliotti, *J. Organometal. Chem.*, *60*, 157 (1973).
191. G. Paliani, S. M. Murgia, G. Cardaci, and R. Cataliotti, *J. Organometal. Chem.*, *63*, 407 (1973).
192. D. C. Andrews and G. Davidson, *J. Chem. Soc., Dalton Trans.*, 1381 (1972).
193. D. C. Andrews and G. Davidson, *J. Organometal. Chem.*, *55*, 383 (1973).
194. K. Shobatake and K. Nakamoto, *J. Amer. Chem. Soc.*, *92*, 3339 (1970).
195. D. M. Adams and A. Squire, *J. Chem. Soc. A*, 1808 (1970).
196. I. R. Beattie, J. W. Emsley, J. C. Lindon, and R. M. Sabine, *J. Chem. Soc., Dalton Trans.*, 1264 (1975).
197. H. J. Neese and H. Bürger, *J. Organometal. Chem.*, *32*, 213 (1971).
198. R. B. Helmholdt, F. Jellinek, H. A. Martin, and A. Vos, *Rec. Trav. Chim. Pays-Bas*, *86*, 1263 (1967).
199. M. L. H. Green and A. W. Steer, *J. Organometal. Chem.*, *1*, 230 (1964).
200. H. D. Murdoch, *J. Organometal. Chem.*, *4*, 119 (1965).

201. H. D. Kaesz and R. B. King, *Z. Naturforsch., 15b*, 682 (1960).
202. H. L. Clarke and N. J. Fitzpatrick, *J. Organometal. Chem., 40*, 379 (1972).
203. H. D. Murdoch and E. Weiss, *Helv. Chim. Acta, 45*, 1927 (1962).
204. G. Sbrana, G. Braca, F. Piacenti, and P. Pino, *J. Organometal. Chem., 13*, 240 (1968).
205. J. R. Bertrand, H. B. Jonassen, and D. W. Moore, *Inorg. Chem., 2*, 601 (1963).
206. G. Mestroni and A. Camus, *Inorg. Nucl. Chem. Lett., 5*, 215 (1969).
207. E. O. Fischer and G. Bürger, *Z. Naturforsch., 16b*, 77 (1961).
208. E. O. Fischer and G. Bürger, *Chem. Ber., 94*, 2409 (1961).
209. G. W. A. Fowles, L. S. Pu, and D. A. Rice, *J. Organometal. Chem., 54*, C17 (1973).
210. F. W. Siegert and H. J. De Liefde Meijer, *J. Organometal. Chem., 23*, 177 (1970).
211. K. H. Thiele and S. Wagner, *J. Organometal. Chem., 20*, P25 (1969).
212. H. Tom Dieck and H. Friedel, *J. Organometal. Chem., 14*, 375 (1968).
213. H. C. Volger and K. Vrieze, *J. Organometal. Chem., 9*, 527 (1967).
214. L. M. Haines, *J. Organometal. Chem., 25*, C85 (1970).
215. R. P. Huges and J. Powell, *J. Organometal. Chem., 54*, 345 (1973).
216. M. Tsutsui and N. Ely, *J. Amer. Chem. Soc., 97*, 3551 (1975).
217. T. J. Marks and W. A. Wachter, *J. Amer. Chem. Soc., 98*, 703 (1976).
218. T. J. Marks, A. M. Seyam, and J. R. Kolb, *J. Amer. Chem. Soc., 95*, 5529 (1973).
219. H. L. Clarke and N. J. Fitzpatrick, *J. Organometal. Chem., 43*, 405 (1972).
220. T. L. Jacobs and S. Singer, *J. Org. Chem., 17*, 475 (1952).
221. F. Jaffe, *J. Organometal. Chem., 23*, C85 (1970).
222. R. West and P. C. Jones, *J. Amer. Chem. Soc., 91*, 6156 (1969).
223. T. L. Jacobs and T. L. Moore, *Abst. 141st National ACS Meeting*, Washington, D. C., p. 9 (1962).
224. C. Prévost, M. Gaudemar, L. Miginiac, F. B. Gaudemar, and M. Andrac, *Bull. Soc. Chim. Fr., 26*, 679 (1959).

225. R. West, P. A. Carney, and I. C. Mineo, *J. Amer. Chem. Soc.*, *87*, 3788 (1965).
226. T. L. Chwang and R. West, *J. Amer. Chem. Soc.*, *95*, 3324 (1973).
227. D. Ballard and H. Gilman, *J. Organometal. Chem.*, *14*, 87 (1968).
228. R. Mantione and Y. Leroux, *J. Organometal. Chem.*, *31*, 5 (1971).
229. Y. Leroux and R. Mantione, *J. Organometal. Chem.*, *30*, 295 (1971).
230. M. Crawford and V. R. Supanekar, *J. Chem. Soc. C*, 1832 (1970).
231. J. K. P. Ariyaratne and M. L. H. Green, *J. Organometal. Chem.*, *1*, 90 (1963).
232. R. S. Rasmussen, D. D. Tunnicliff, and R. R. Brattain, *J. Chem. Phys.*, *11*, 432 (1943).
233. M. J. Grogan and K. Nakamoto, *Inorg. Chim. Acta*, *1*, 228 (1967).
234. R. K. Harris, *Spectrochim. Acta*, *20*, 1129 (1964).
235. P. J. Hendra and D. B. Powell, *Spectrochim. Acta*, *18*, 1195 (1962).
236. O. S. Mills and G. Robinson, *Proc. Chem. Soc.*, 421 (1960).
237. G. Davidson, *Inorg. Chim. Acta*, *3*, 596 (1969).
238. D. A. Duddell, S. F. A. Kettle, and B. T. Kontnik-Matecka, *Spectrochim. Acta*, *28A*, 1571 (1972).
239. G. Davidson and D. A. Duce, *J. Organometal. Chem.*, *44*, 365 (1972).
240. B. W. Cook, R. G. J. Miller, and P. F. Todd, *J. Organometal. Chem.*, *19*, 421 (1969).
241. P. McArdle and A. R. Manning, *J. Chem. Soc. A*, 2123 (1970).
242. A. R. Manning, *J. Chem. Soc.*, *Dalton Trans.*, 821 (1972).
243. E. O. Fischer, H. P. Kögler, and P. Kuzel, *Chem. Ber.*, *93*, 3006 (1960).
244. L. Porri and A. Lionetti, *J. Organometal. Chem.*, *6*, 422 (1966).
245. E. O. Fischer, K. Bittler, and H. P. Fritz, *Z. Naturforsch.*, *18b*, 83 (1963).
246. E. Koerner von Gustorf, O. Jaenicke, and O. E. Polansky, *Angew. Chem.*, *Inter. Ed. Engl.*, *11*, 532 (1972).
247. P. S. Skell, E. M. Van Dam and M. P. Silvon, *J. Amer. Chem. Soc.*, *96*, 626 (1974).
248. U. Giannini, E. Pellino, and M. P. Lachi, *J. Organometal. Chem.*, *12*, 551 (1968).
249. R. Rienacker and H. Yoshiura, *Angew. Chem.*, *Inter. Ed. Engl.*, *8*, 677 (1969).
250. C. Krüger, *Angew. Chem.*, *Inter. Ed. Engl.*, *8*, 678 (1969).

251. P. J. Hendra and D. B. Powell, *Spectrochim. Acta, 17,* 909 (1961).
252. I. A. Zakharova, G. A. Kukina, T. S. Kuli-Zade, I. I. Moiseev, G. Yu. Pek, and M. A. Pori-Koshits, *Zh. Neorg. Kim., 11,* 2543 (1966).
253. I. A. Zakharova, L. A. Leites, and V. T. Aleksanyan, *J. Organometal. Chem., 72,* 283 (1974).
254. J. C. Trebellas, J. R. Olechowski, and H. B. Jonassen, *J. Organometal. Chem., 6,* 412 (1966).
255. G. Bressan, R. Broggi, M. P. Lachi, and A. L. Segre, *J. Organometal. Chem., 9,* 355 (1967).
256. R. B. King and A. Fronzaglia, *Inorg. Chem., 5,* 1837 (1966).
257. K. Stark, J. E. Lancaster, H. D. Murdoch, and Weiss, *Z. Naturforsch., 19b,* 284 (1964).
258. M. Green and R. I. Hancock, *J. Chem. Soc. A,* 109 (1968).
259. H. P. Fritz and G. N. Schrauser, *Chem. Ber., 94,* 642 (1961).
260. G. N. Schrauser, *J. Amer. Chem. Soc., 82,* 1008 (1960).
261. K. Moseley and P. M. Maitlis, *Chem. Commun.,* 982 (1971).
262. M. F. Farona and K. F. Kraus, *Inorg. Chem., 9,* 1700 (1970).
263. D. H. Finseth, C. Sourisseau, and F. A. Miller, *J. Phys. Chem., 80,* 1248 (1976).
264. D. C. Andrews and G. Davidson, *J. Organometal. Chem., 43,* 393 (1972).
265. D. C. Andrews, G. Davidson, and D. A. Duce, *J. Organometal. Chem., 97,* 95 (1975).
266. E. O. Fischer, W. Kiener, and R. D. Fischer, *J. Organometal. Chem., 16,* P60 (1969).
267. M. F. Shostakovskii, N. I. Shergina, N. V. Komarov, and Yu. V. Maroshin, *Izvest. Akad. Nauk SSSR,* 1606 (1964).

III

Cyclic, Unsaturated Organometallic Derivatives

A. THREE-CARBON RINGS

Although the cyclopropenium cation, $C_3H_3^+$, is expected to act as a three-electron donor in forming η^3-cyclopropenium complexes with transition metals, no such complexes have been reported. Compounds with a trisubstituted cyclopropenium ring, however, have been reported. Spectroscopic (including infrared) data support an ionic structure for $[R_3C_3^+]_2M_2Cl_6^{2-}$ (R=$(CH_3)_2N$,n-C_3H_7; M=Pd,Pt) (1). Assignments in the ν(CO) mode region have been reported for nickel carbonyl complexes with alkyl-substituted η^3-cyclopropenium rings (2) and $[\eta^3$-$(C_6H_5)_3$-$C_3]Ni(1,7$-$B_9H_9CHPCH_3)$ (4).

Infrared bands at 1850 and 1630 cm^{-1} for diphenylcyclopropenone, $(C_6H_5)_2C_3$=O, arise from the strongly coupled ν(C=O) and ν(C=C) modes with the 1630 cm^{-1} band predominantly ν(C=O) in character (5.6). Magnetic measurements and electronic and infrared spectra for $[(C_6H_5)_2C_3{=}O]_2MX_2$ (M=Zn,X=Cl,Br,I;M=Co,Cu,X=Cl,Br;M=Ru, Pd,X=Cl;M=Ni,X=Br), $[(C_6H_5)_2C_3{=}O]_2Pt_nCl_4$ (n=1,2), $[(C_6H_5)_2C_3{=}O]_2CuCl_2 \cdot 2H_2O$, $[(C_6H_5)_2C_3{=}O]_3RuCl_3$, and $[(C_6H_5)_2C_3{=}O]_6M(ClO_4)_2$ (M=Ni,Co) have been used to characterize the structures of these compounds (7). Since the 1630 cm^{-1} band of free diphenylcyclopropenone shifted to approximately 1590 cm^{-1} in all of the complexes mentioned above while the position of the 1850 cm^{-1} band remained relatively constant, it was concluded that the metal atom coordinates to the oxygen atom of the ligand and not to the C=C bond. This study (7) also included assignments of the metal-halogen stretching modes. The same structural conclusions have been reached using infrared data for $[(C_6H_5)_2C_3{=}O]PtCl_2R$ ($R{=}C_2H_4$, *cis*-C_4H_8, CO, $(C_4H_9)_3P$) and $[(C_6H_5)_2C_3{=}O]Co$-$[Co(CO)_4]_2$ (8).

B. FOUR-CARBON RINGS

Infrared data and some assignments have been reported for the perfluorocyclobutene derivatives $C_4F_5M(CO)_5$ (M=Mn, ν(C=C)=1623 cm^{-1}; M=Re, ν(C=C)=1633 cm^{-1}) in which it has been concluded that the metal atoms are σ-bonded to one carbon atom of the C=C bond (9), and a tetrakis-(trifluoromethyl)cyclobutene derivative (3.1) in which

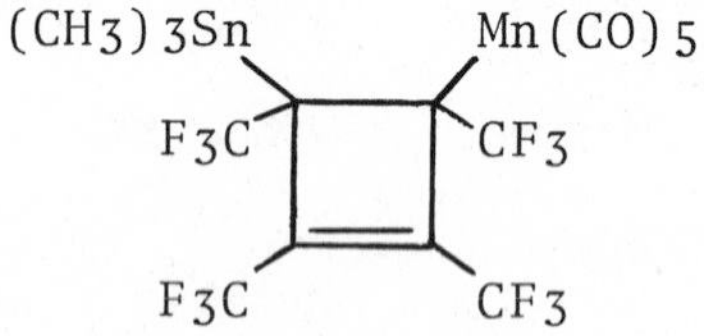

(3.1)

it has been proposed that the tin and manganese atoms are σ-bonded to the *3*- and *4*-carbon atoms (10).

When the cyclobutadiene ring, C_4H_4, is bonded in a *tetrahapto* manner to a metal atom, it has an extensively delocalized electronic structure. The infrared and Raman assignments reported for η^4-$C_4H_4Fe(CO)_3$ are summarized in Table 3.1 (11). The high frequency of the ν_s(CC) or "ring breathing" mode of the C_4 ring in η^4-$C_4H_4Fe(CO)_3$ (1234 cm^{-1}) relative to that of the analogous mode in complexes with a η^5-cyclopentadienyl (1100 cm^{-1}) or η^6-benzene (1000 cm^{-1}) ring has been attributed to the high strain in the small C_4 ring. A normal coordinate analysis of η^4-$C_4H_4Fe(CO)_3$ shows that many of the normal modes involve more than one type of internal coordinate (12). Since this is especially true in the 650 to 350 cm^{-1} region, the assignment of bands in this region must be considered only approximate. The infrared spectrum has been reported for $C_4H_4AgNO_3$, although the structure of this compound has not been fully characterized (13). Other vibrational data for cyclobutadiene derivatives have only included assignments for the ν(CO) modes of $C_4H_4M(CO)_4$ (M=Mo, ν(CO)= 2064, 1994, and 1950 cm^{-1}; M=W, ν(CO)=2068, 1983, and 1950 cm^{-1}) (14) and $C_4H_4Ru(CO)_3$ (ν(CO)=2050 and 1975 cm^{-1}) (14).

TABLE 3.1
Vibrational Assignments (cm^{-1}) for Liquid η^4-$C_4H_4Fe(CO)_3$[a]

Assignment	Symmetry	Raman Spectrum	Infrared Spectrum
ν(CH)	A_1		
	B_1	3134 w,dp	3130 w
	E		
ν(CO)	A_1	2051 m,p	2045 vs
	E	1979 s,dp	1965 vs
ν(CC)	A_1	1235 s,p	1232 s
	B_2	?1330 m,dp	1325 s
	E	1330 m,dp	1325 s
δ(ring)	B_1	957 sh	
π(ring)	B_1	506 m,dp	
δ(CH)	A_2	964 m,dp	
	B_2	964 m,dp	
	E		975 m
π(CH)	A_1	825 m,dp	822 m
	B_1	774 w	775 w
	E	940 m,dp	937 s
ν(Fe-C_4H_4)	A_1	399 s,p	396 m
Fe-C_4H_4 tilt	E	472 m,dp	470 s

[a]Reference 11.

An assignment of the infrared spectrum of the tetramethylcyclobutadiene complex $[(CH_3)_4C_4]NiCl_2$ (15) is not in complete agreement with the analogous assignments reported for $C_4H_4Fe(CO)_3$ (11). The infrared spectrum and partial assignments have been reported for $[(CH_3)_4C_4]Co(CO)_2I$ (16). Methylation of the cyclobutadiene ring in the series $[(CH_3)_{4-n}C_4H_n]Fe(CO)_3$ (n=0,2,4) results in a decrease in the ν(CO) mode frequencies, indicating increased electron donation to the iron atom (17). The opposite trend has been observed when the electron-withdrawing chlorine atom was used in the complex $(C_4H_3Cl)Fe(CO)_3$ (18).

The tetraphenylcyclobutadiene derivatives $[(C_6H_5)_4C_4]MX_2$ (M=Ni,X=Cl (15),Br (19);M=Pd,X=Cl (15,19), Br (20),I (20)) all show an infrared band from 1380 to 1360 cm^{-1} that is characteristic of the η^4-tetraphenylcyclobutadiene ring. The analogous infrared band for $C_4H_4Fe(CO)_3$ (1325 cm^{-1}) was assigned to the ν(CC) mode of E symmetry or to the ν(CC) modes of both E and B_2 symmetry, which are accidentally degenerate (cf. Table

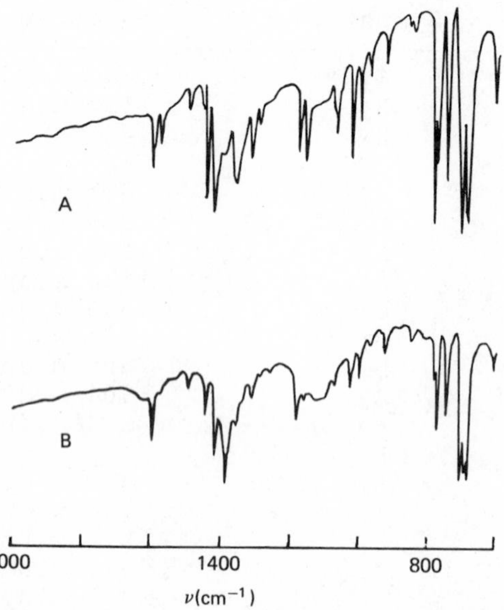

Figure 3.1. Infrared spectra of (A) $[[(C_6H_5)_4C_4]PtCl_2]_n$ *and (B)* $[[(C_6H_5)_4C_4]PtI_2]_2$ *in KBr pellets (21).*

3.1) (11). The similarity of the infrared spectra illustrated in Figure 3.1 for $[[(C_6H_5)_4C_4]PtCl_2]_n$ and $[[(C_6H_5)_4C_4]PtI_2]_2$ to those of the related palladium and nickel complexes has led to the conclusion that these platinum complexes also have structures with η^4-tetraphenylcyclobutadiene ligands (21). Heating $[[(C_6H_5)_4C_4]PtCl_2]_n$ (as well as the corresponding bromide or iodide salts) in methanol or ethanol with potassium acetate has been explained in terms of the reaction

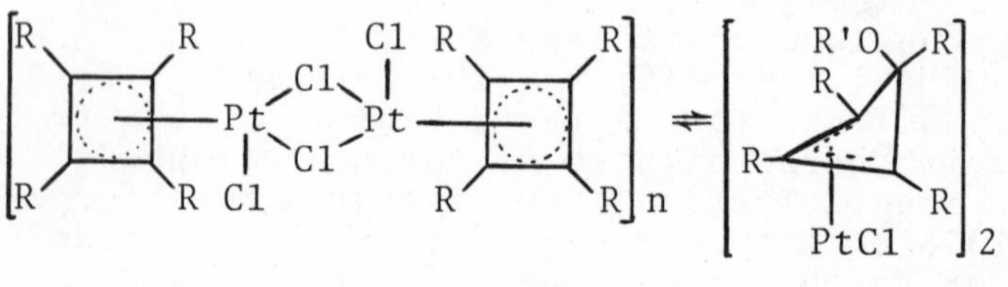

$R = C_6H_5$
$R' = CH_3, C_2H_5$

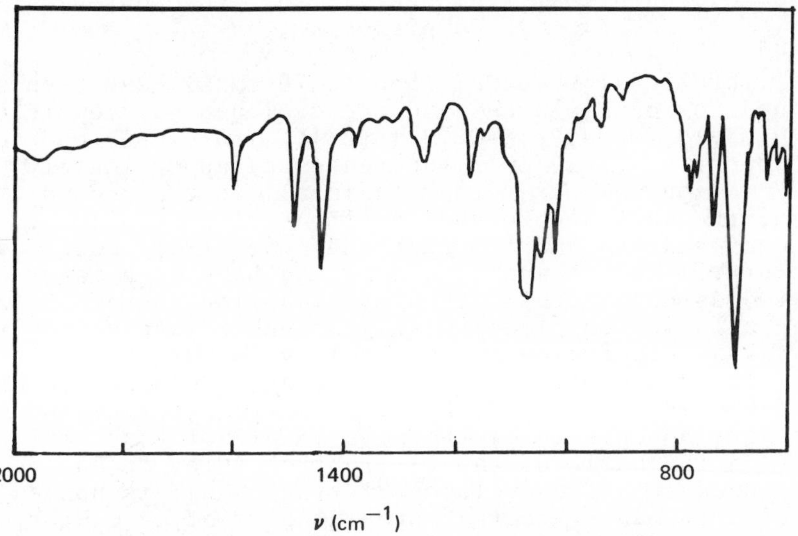

Figure 3.2. Infrared spectrum of $[[(C_6H_5)_4C_4OCH_3]PtCl]_2$ *in a KBr pellet (21).*

where the product was identified from elementary analyses, molecular weight measurements, and nmr and infrared data as dimeric chloro(1-*exo*-methoxy-1,2,3,4-tetraphenylcyclobutenyl)platinum, $[[(C_6H_5)_4C_4OCH_3]PtCl]_2$, or chloro(1-*exo*-ethoxy-1,2,3,4-tetraphenylcyclobutenyl)-platinum, $[[(C_6H_5)_4C_4OC_2H_5]PtCl]_2$ (21). In the infrared spectrum of the methoxy derivative, illustrated in Figure 3.2, the 1380 cm^{-1} band of the $[[(C_6H_5)_4C_4]PtCl_2]_n$ starting material has been replaced by a strong intensity methoxy band at 1075 cm^{-1} (21). Additional changes are also observed in the 800 to 600 cm^{-1} region.

The ν(C≡O) and ν(C=O) mode assignments for $[(CH_3)_3C_4{=}O]Co(CO)_3$ (22) are consistent with structure 3.2.

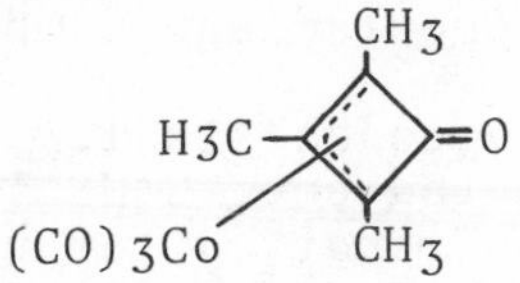

(3.2)

C. FIVE-CARBON RINGS

1. Cyclopentene, Cyclopentyl, and Cyclopenteneone Complexes

The vibrational spectra (1500 to 70 cm^{-1}) have been assigned for cyclopentene and the cyclopentene complexes $(C_5H_8MX_2)_2$ (M=Pd,Pt;X=Cl,Br) (23). The $\nu(C=C)$ mode of cyclopentene (1619 cm^{-1}) decreases by approximately 200 cm^{-1} on complex formation in all four compounds. Also, the shifts observed for the cyclopentene modes on complexation and the fact that one $\nu(PdC)$ mode and two $\nu(PtC)$ modes have been assigned have been interpreted as consistent with a model analogous to 2.3 for the palladium complexes with a strong σ interaction and a weak π interaction and a model analogous to 2.4 for the platinum complexes with a strong π interaction and a weak σ interaction. The total $\sigma + \pi$ interaction, however, appears to be the same in all of the complexes.

On the basis of the 18-electron rule, it has been concluded that the cyclopentyl ring, C_5H_7, is bonded in a *trihapto* manner to the chromium atom of η^5-$C_5H_5Cr(CO)_2$-(η^3-C_5H_7). Partial infrared data have been reported for this compound, including the assignment of a band at 2865 cm^{-1} to a $\nu(CH_2)$ mode of the η^3-cyclopentyl ring (24). Infrared data have also been listed for η^5-CH_3-$C_5H_4Ni(\eta^3$-$C_5H_7)$ (25), and the tetrakis(trifluoromethyl)-cyclopenteneone complex $[(CF_3)_4C_4{=}O]Rh(\eta^5$-$C_5H_5)$ (3.3) (26).

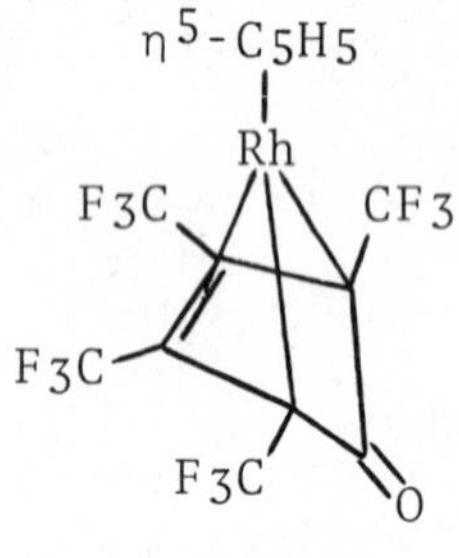

(3.3)

2. 1,3-Cyclopentadiene Complexes

Although detailed vibrational studies have been reported for 1,3-cyclopentadiene, C_5H_6 (27-30), no complete studies have been reported for π-type complexes of 1,3-cyclopentadiene. The complex $C_5H_6Fe(CO)_3$ has been characterized in terms of a π-type structure (3.4)

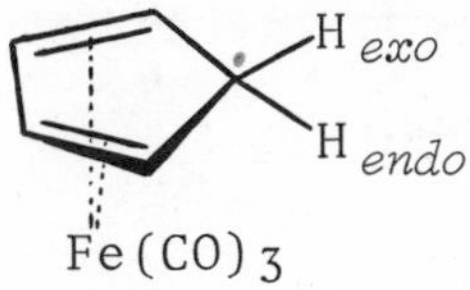

(3.4)

with the $\nu(CH_{exo})$ and $\nu(CH_{endo})$ modes assigned at approximately 2945 and 2750 cm^{-1}, respectively, for $C_5H_6MC_5H_5$ (M=Co,Rh) (32). In a more recent study, however, which included infrared data for *exo*-1-$RC_5H_5Fe(CO)_2[P(C_6H_5)_3]$ ($R=C_6H_5,C_6F_5$), it was concluded that these assignments should be reversed with the $\nu(CH_{exo})$ mode appearing at a lower frequency than the $\nu(CH_{endo})$ mode (33). The basis for this conclusion is that while metal-cyclopentadiene complexes show the two $\nu(CH_2)$ bands at approximately 2950 and 2750 cm^{-1}, replacement of the *exo*-hydrogen with a phenyl or pentafluorophenyl group results in the disappearance of the 2750 cm^{-1} band but not the 2950 cm^{-1} band. Additional infrared data and some assignments have also been listed for $C_5H_6Fe(CO)_2$-$[P(C_6H_5)_3]$ (34), $C_5H_6FeC_5H_5$ (35), $C_5H_6RhC_5H_5$ (36), $C_5H_6IrC_5H_5$ (37), $(C_5H_6)_2Ni$ (38), and C_5H_6Pd(1,3-cyclohexadiene) (39). Infrared and nmr data have been used to characterize diphenylfulvene-metal complexes of cobalt, rhodium, and iridium (3.5) with a band at 1600

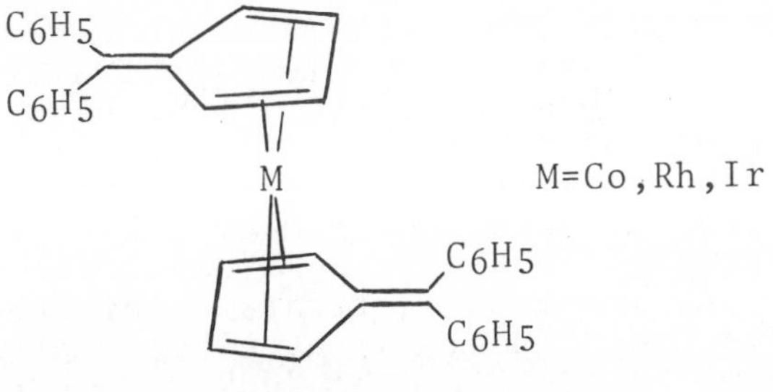

(3.5)

cm^{-1} assigned to the $\nu(C{=}C)$ mode of the uncomplexed C=C bond (40,41).

The solid-state infrared spectra of several (η-disubstituted-cyclopentadienone)(η^5-cyclopentadienyl)Co complexes (3.6) all show a broad ketonic $\nu(CO)$ band near 1500 cm^{-1} (42). The position of this band is somewhat sensitive to the state of the sample. Therefore, for the 1,3-di(t-butyl) substituted derivatives,

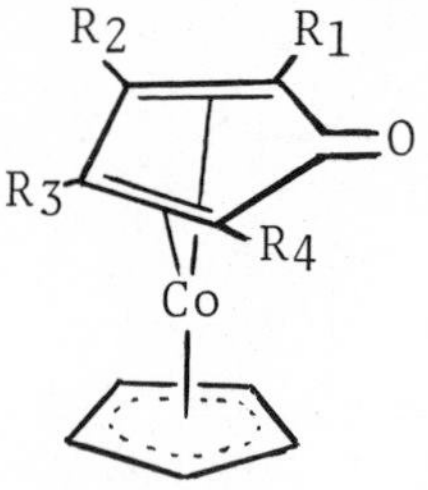

(3.6)

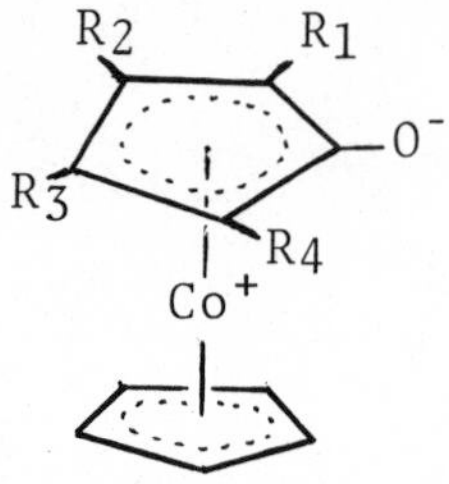

(3.7)

the ν(CO) mode appears at 1563 cm^{-1} in the solid state, 1565 cm^{-1} in $CHCl_3$, and 1599 cm^{-1} in CCl_4 (42). The relatively low frequency of the ν(CO) mode in these complexes has been explained by assuming a mesomeric structure (3.6 and 3.7) (42,43). Infrared data have also been reported for η-(2,5-diperfluorophenyl-cyclopentadienone)$Fe(CO)_3$ (44), η-tetraphenylcyclopentadienone$Ru(CO)_3$ (45), and η-cyclopentadieneone$Os(CO)_3$ (45).

3. Anionic Cyclopentadienides

The vibrational spectra of cyclopentadienide complexes are relatively simple because of the high D_{5h} symmetry of the $C_5H_5^-$ ring. Therefore, four bands are infrared active and seven bands are Raman active. The normal modes of the cyclopentadienide anion and their vibrational activity are illustrated in Figure 3.3.

The solid-state infrared spectra of the group Ia derivatives C_5H_5M (M=K,Rb,Cs) have been interpreted in terms of an ionic structure (36,47,48), as have the ether solution infrared and Raman spectra of C_5H_5K (47). The infrared spectrum of a KBr pellet of C_5H_5K is illustrated in Figure 3.5 (36). The Nujol mull infrared spectra of the derivatives of the lighter group Ia metals C_5H_5M (M=Li,Na) have been interpreted in terms of a "centrally σ-bonded" structure, that is a structure in which there is a weak covalent interaction between the metal cation and delocalized cyclopentadienide anion (48). The vibrational spectra of centrally σ-bonded complexes are discussed in Sec. C.4 of this chapter. The infrared, uv, and nmr spectra of 1M tetrahydrofuran solutions of C_5H_5M (M=Li,Na) have been reported (49). These data are consistent with the presence of metal-cyclopentadienide ion pairs in which the metal ion lies on or near the C_5 axis of the delocalized cyclopentadieneide ring.

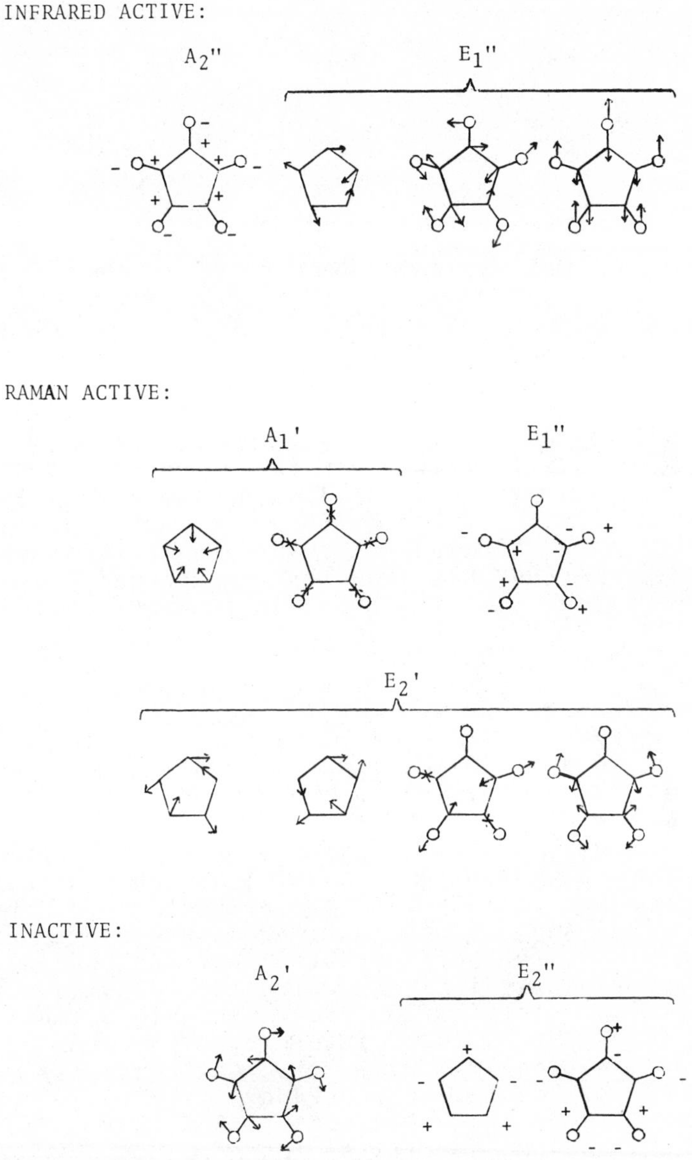

Figure 3.3. Normal modes of vibration for the cyclopentadienide anion and the vibrational activity of these modes.

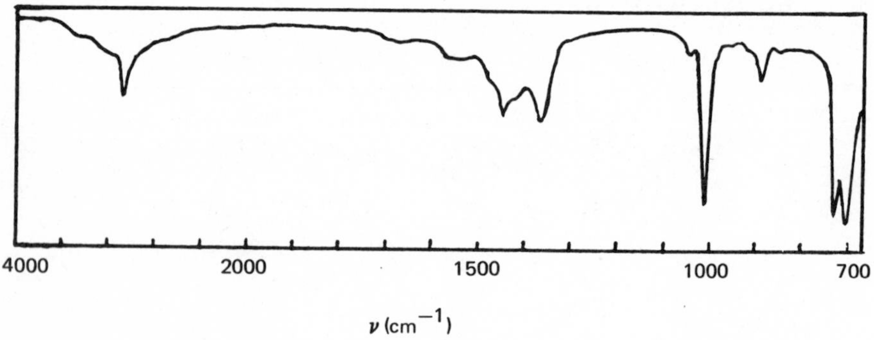

Figure 3.4. Infrared spectrum of C_5H_5K as a KBr pellet (36).

The Nujol mull infrared spectra of the heavier group IIa derivatives $(C_5H_5)_2M$ (M=Ca,Sr,Ba) had originally been interpreted as indicating the presence of an essentially ionic structure (50). It was also reported in another infrared study of $(C_5H_5)_2Ca$ that the spectrum is similar to that expected for centrally σ-bonded complexes (51). In the same study it was reported that a comparison of the infrared spectrum of solid $(C_5H_5)_2Ca$ with spectra of $(C_5H_5)_2CaL_2$ (L=Py,L_2=Bipyr,1,2-dimethoxyethane,N,N,N',N'-tetramethylenediamine) and $(C_5H_5)_2CaS_n$ (n=0.5 to 2,S=tetrahydrofuran,Py) suggests that the metal-ring interaction is more highly ionic in the adducts than in uncomplexed $(C_5H_5)_2Ca$ (51). More recently, however, an X-ray crystallographic study has shown that the structure of solid $(C_5H_5)_2Ca$ is more complex than had been assumed in the infrared studies (52). This X-ray study shows that each calcium atom has a coordination number of four with two η^5-C_5H_5 rings, one η^3-C_5H_5 ring, and one η^1-C_5H_5 ring in the coordination sphere. It was further suggested that the bonding, while predominantly ionic in character, does have a small covalent contribution (52). The vibrational spectrum of $(C_5H_5)_2Mg$ has been interpreted in terms of both ionic (53) and weakly covalent (54) or centrally σ-bonded (48) structures in the vapor state (53), benzene solution (54), or the solid state (48,50,54). The infrared, uv, and nmr spectra of $(C_5H_5)_2Mg$ in tetrahydrofuran indicate the presence of ion triplets, while C_5H_5MgX (X=Cl,Br) are probably ion pairs in the same solvent (51).

An ionic structure has been proposed for the group IIIa derivative C_5H_5Tl using chemical evidence and an analysis of the infrared spectrum of the vapor in the ν(CH) mode region (53). In a relatively complete infrared and Raman study of crystalline C_5H_5Tl (55), the spectra have also been interpreted in terms of predominantly ionic bonding with D_{5h} symmetry for the cyclopentadienide ring. Bands appearing below 400 cm^{-1} were assigned to lattice modes with no evidence for the presence of metal-ring vibrations since these bands shifted to higher frequencies on cooling the sample to liquid-nitrogen temperature, as expected for lattice modes. It was also noted in this study, however, that there is some evidence for a weak covalent interaction and that it might be more appropriate to classify C_5H_5Tl as a centrally σ-bonded complex, as had been done in an earlier infrared study (48).

The physical and chemical properties of $(C_5H_5)_2Mn$ indicate the presence of an ionic structure (56). This conclusion is also supported by the simplicity of the infrared spectrum in the cyclopentadienyl region (48,54, 56) and the absence of metal-ring modes in the low-frequency region (48).

The chemical properties and infrared spectra of the lanthanide complexes $(C_5H_5)_2Eu$ (57,58) and $(C_5H_5)_2Yb$ (58) are similar to those of the ionic dicyclopentadienides of the group IIa elements. The infrared spectra have also been reported for $(C_5H_5)_3M$ (M=Ho,Tm,Tb,Lu) which, though highly ionic, are believed to possess partial covalent bonding (58). The infrared spectra of the cyclohexylisonitrile (CNC_6H_{11}) complexes $(C_5H_5)_3M$-(CNC_6H_{11}) (M=Y,Nb,Tb,Ho,Yb) are similar to those of the uncomplexed tricyclopentadienyl lanthanide complexes (59,60). The ν(C≡N) mode of cyclohexylisonitrile is approximately 70 cm^{-1} higher in the complex than for the free ligand. This indicates that the cyclohexylisonitrile molecule is acting mainly as a σ-donor but not as a π-acceptor and is consistent with the facts that lanthanide ions are poor π-donors and that the cyclopentadienyl compounds are highly ionic (60). In Table 3.2, the assignments are compared for the infrared active fundamental modes of several ionic cyclopentadienide complexes.

4. Covalent Cyclopentadienyls

Various limiting structures have been found for covalent cyclopentadienyl compounds, namely *trihapto* or π-allylic

TABLE 3.2
Assignments (cm^{-1}) of the Infrared Active Modes for Ionic Cyclopentadienide Complexes

Compound	ν(CH)	ν(CC)	δ(CH)	π(CH)	Refs.
C_5H_5K	3048 m	1455 m	1009 s	702 vs	47
C_5H_5Rb	3030 m	1501 vw	1011 s	696 vs	47
C_5H_5Cs	3021 s	1494 m	1008 s	668 s	47
$(C_5H_5)_2Sr$	3077 m	1435 m	1008 s	739 vs,br	58
$(C_5H_5)_2Ba$	3065 m	1435 w	1009 s	736 vs	50
$(C_5H_5)_2Mn$	3063 m	1428 m	1004 s	758 s	54
$(C_5H_5)_2Eu$	3077 m	1435 m	1007 s	739 vs,br	58
$(C_5H_5)_2Yb$	3086 m	1433 m	1006 s	752 vs,br	58

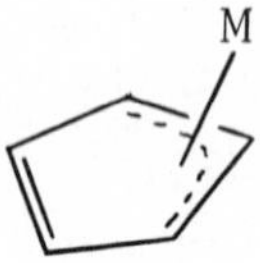

(3.8)

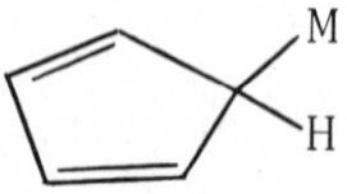

(3.9)

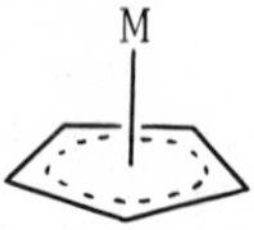

(3.10)

(3.8), *monohapto* or σ (3.9), and *pentahapto* or π (3.10). Bridged structures with μ-(η^1,η^1-cyclopentadienyl), μ-(η^1,η^5-cyclopentadienyl), or μ-(η^5,η^5-cyclopentadienyl) rings have also been found. In compounds with more than one cyclopentadienyl ring, each ring may be present with any one of these structures. It should also be noted that these are limiting structures that may or may not adequately describe the structure of some metal-cyclopentadienyl complexes. Therefore, an X-ray crystallographic study of $(C_5H_5)_3Ti$ shows that while two of the rings are *pentahapto* in nature, the third is bonded to the titanium atom through only two carbon atoms (61). Also, an X-ray crystallographic study of $(C_5H_5)_3MoNO$ has shown the presence of one *monohapto* ring while the other two rings have structures that lie between those of the *trihapto* and *pentahapto* extremes (62). The vibrational spectra of each cyclopentadienyl structural extreme is expected to exhibit specific characteristic features. These features can be useful in determining the overall structure of a covalent metal-cyclopentadienyl complex.

a. η^3-Cyclopentadienyl Rings

Vibrational data for compounds with η^3-cyclopentadienyl rings are not available mainly because such compounds are almost totally unknown. An X-ray crystallographic study has shown that the C-C bond lengths of the cyclopentadienyl rings in bis(cyclopentadienyl)-2,2'-η^3-allyl-bis(nickel), $C_5H_5Ni(C_3H_4\text{-}C_3H_4)NiC_5H_5$, are consistent with a *trihapto* structure (63). No vibrational data, however, have been reported for this compound. Infrared data have been reported for $(C_5H_5)_2Mo(CO)X$ ($X=I,CH_3$); it has been proposed that in each of these compounds there is one η^3- and one η^5-cyclopentadienyl ring (64). The experimental data for both compounds, however, are also consistent with the structure found for solid $(C_5H_5)_3MoNO$ in which, rather than one *trihapto* and one *pentahapto* ring being present, both rings adopt structures between these two extremes. Therefore, more data are needed before structural conclusions can be drawn for $(C_5H_5)_2Mo(NO)X$ ($X=I,CH_3$). A single crystal X-ray study of $(C_5H_5)_2Ca$ indicates that one of the four cyclopentadienyl rings in the coordination sphere of each calcium atom is *trihapto* in nature (52). A monomeric structure with a η^3-cyclopentadienyl ring has been proposed for gaseous $C_5H_5Al(CH_3)_2$ as the result of an electron diffraction study (65). This study contrasts with an earlier benzene solution nmr study in which a structure with a η^1-cyclopentadienyl ring was proposed (66) and a solid-state vibrational study in which a structure with a η^5-cyclopentadienyl ring was proposed for the solid state and also for the vapor state and solution (67).

b. η^1-Cyclopentadienyl Rings

Several similarities are expected between the vibrational spectra of cyclopentadiene and the η^1-cyclopentadienyl ring because of their analogous structures (68). The vibrational spectra of η^1-cyclopentadienyl-metal complexes, however, are relatively complex because of the low C_s symmetry of a η^1-C_5H_5M group relative to the C_{2v} symmetry of cyclopentadiene. Therefore, all 27 modes (15A' + 12A") of the η^1-C_5H_5M unit are both infrared and Raman active. The most complete vibrational analysis for *monohapto* complexes has been reported for $(C_5H_5)_2Hg$ and C_5H_5HgX ($X=Cl,Br,I$) (68,69). The infrared spectrum of $(C_5H_5)_2Hg$ is illustrated in Figure 3.5. In addition to the richness of the spectra, *monohapto*-type

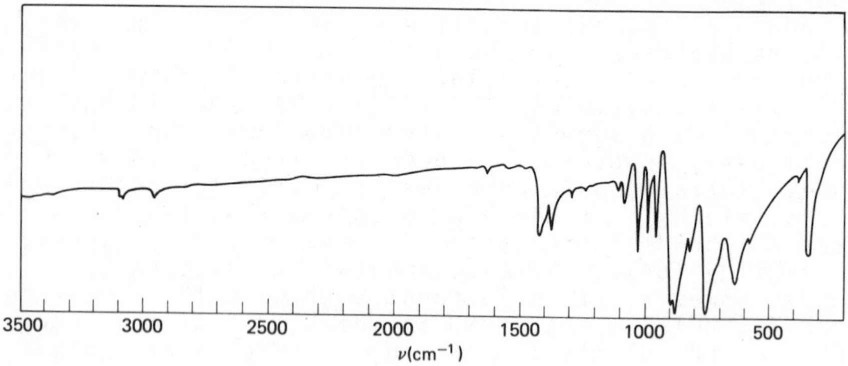

Figure 3.5. Infrared spectrum of (η^1-C_5H_5)$_2$Hg as a KBr pellet.

complexes have several additional characteristic features. Among these is the complexity of the ν(CH) mode region. For the η^1-C_5H_5M unit, all five ν(CH) modes are vibrationally active. The aliphatic ν(CH) modes are expected from approximately 3000 to 2800 cm^{-1}, while the olefinic ν(CH_2) modes are expected from approximately 3100 to 3000 cm^{-1}. Another feature is a relatively weak intensity infrared band at approximately 1600 to 1500 cm^{-1}, which has been assigned to a ν(C=C) mode. In the infrared spectrum of $(C_5H_5)_2Hg$, this band appears at 1530 cm^{-1}. The third and most easily identified feature of a *monohapto* ring is the strong intensity infrared band from approximately 760 to 690 cm^{-1}, which has been assigned to a π(CH) mode.

If more than one cyclopentadienyl ring is bonded to a metal atom, there is the theoretical possibility of inter ring coupling which would aid in determining the relative orientation of the rings. This possibility has been discussed for $(C_5H_5)_2Hg$ (68), which can be pictured in terms of *transoid* (3.11) or *cisoid* (3.12) structural extremes. The highest symmetry of the former structure (C_{2h}) would give rise to 26 infrared active and 31 Raman active vibrations, while the highest symmetry of the latter structure (C_{2v}) would give rise to 47 infrared active and 57 Raman active vibrations. The infrared spectrum of $(C_5H_5)_2Hg$ is essentially identical to the spectra of C_5H_5HgX (X=Cl,Br,I) (68). These results indicate a lack of significant coupling between the two

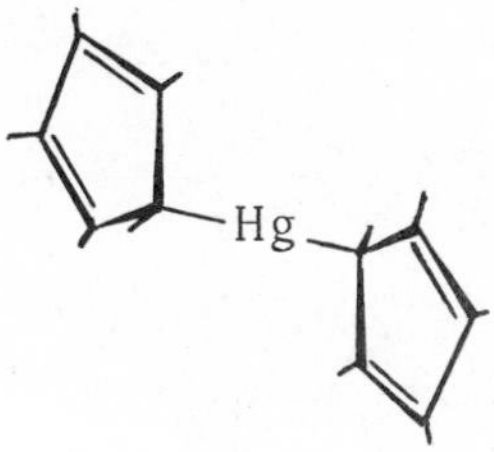

(3.11)

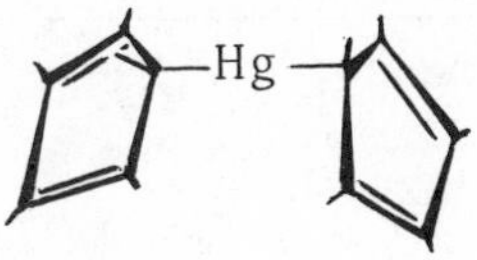

(3.12)

rings in $(C_5H_5)_2Hg$ and the ineffectiveness of vibrational spectroscopy in determining the relative orientation of two or more η^1-cyclopentadienyl rings in a complex. The metal-ring skeletal modes of η^1-cyclopentadienyl complexes have been observed in the region below 500 cm^{-1}, with the metal-ring stretching modes appearing at lower frequencies than the analogous modes in metal-alkyl complexes

c. η^5-Cyclopentadienyl Rings

Compounds with a covalent η^5-cyclopentadienyl ring have been further subdivided by one investigator (48) into centrally σ-bonded, and "genuine" or centrally π-bonded structures. The former group includes principally main group and f-block elements as well as some d-block elements in all of which the metal-ring interaction is relatively weak. The latter group includes mainly d-block elements with relatively strong metal-ring interactions. The basis for this division was that the centrally σ-bonded derivatives, except for those of relatively light metals, showed no bands above 250 cm^{-1}, which could be attributed to metal-ring skeletal modes. It has been noted, however, (48) that this division is artificial and that a strict classification of all *pentahapto* compounds into one of these two subgroups is not always possible.

The relatively high symmetry (C_{5v}) of the η^5-cyclopentadienyl ring results in a relatively simple vibrational spectrum with 15 Raman active bands ($4A_1 + 5E_1 + 6E_2$), nine infrared active bands ($4A_1 + 5E_1$), and one inactive mode (A_2). In Figure 3.6, the infrared spectrum is illustrated for the typical *pentahapto* compound $(C_5H_5)_2Fe$. Among the characteristic spectral features of a η^5-cyclopentadienyl ring is a sharp band of variable infrared intensity but strong Raman intensity at

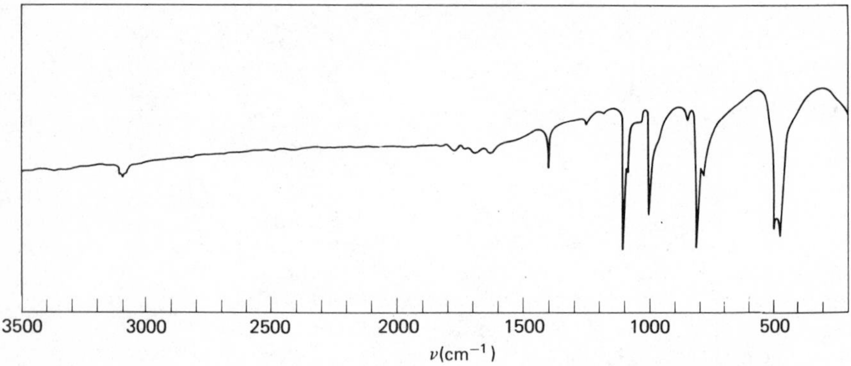

Figure 3.6. Infrared spectrum of (η^5-C_5H_5)$_2$Fe as a KBr pellet.

approximately 1100 cm^{-1}. The assignment of this band, which is absent in the infrared spectrum of ionic derivatives, has been somewhat controversial. It was assigned to the ν_s(CC) or "ring breathing" mode in a complete vibrational study of $(C_5H_5)_2Fe$ (70), as compared to 983 cm^{-1} for the same mode in the cyclopentadienide anion (47). This assignment for $(C_5H_5)_2Fe$ was later questioned with the 1100 cm^{-1} band assigned to the π(CH) mode of A_{1g} symmetry and the ring breathing mode assigned at 1390 cm^{-1} (48). More recently, however, the original assignment of the 1100 cm^{-1} band to the ring breathing mode has been supported by a vibrational study of $(C_5H_5)_2Fe$ and $(C_5H_5)_2Co^+$ (71), and this assignment will be used in the following discussions. It has also been observed that the intensity of the 1100 cm^{-1} band in the infrared spectrum decreases as the degree of ionic character in the metal-cyclopentadienyl interaction increases until it becomes infrared inactive for the cyclopentadienide anion (48). In addition to this intensity change, significant frequency changes have been observed for two infrared active cyclopentadienyl modes as the degree of ionic character in the metal-cyclpentadienyl interaction increases. One such band is the infrared active ν_a(CC) mode which is observed at 1410 cm^{-1} in $(C_5H_5)_2Fe$ but at a noticeably higher frequency in ionic cyclopentadienide derivatives (cf. Table 3.2). The π(CH) mode also shows a noticeable frequency change and appears at 800 cm^{-1} in the infrared spectrum of $(C_5H_5)_2Fe$ but at lower frequencies in the ionic derivatives (cf. Table 3.2).

Dicyclopentadienyl compounds with parallel rings are found with either a staggered (3.13) or an eclipsed

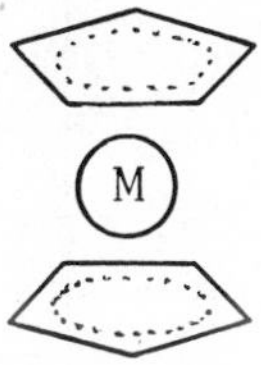

(3.13)

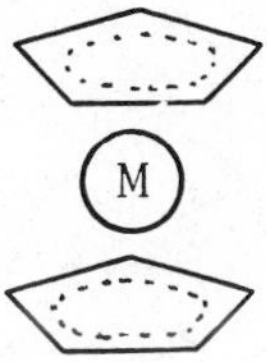

(3.14)

(3.14) configuration. Since the staggered configuration has a center of symmetry while the eclipsed configuration does not, different vibrational spectra might be anticipated for both of these structures. As seen in Table 3.3, however, the vibrational spectra expected for $(\eta^5\text{-}C_5H_5)_2M$ compounds with staggered (D_{5d}) and eclipsed (D_{5h}) configurations are the same. Also, in a vibrational study of $(C_5H_5)_2M$ (M=Fe,Ru), which have staggered and eclipsed solid-state structures, respectively, it has been observed (70) that the spectra are not only very similar but that the frequency differences between the in-phase and out-of-phase combinations of a given mode for each ring are very small. A possible explanation for this is that little vibrational coupling is found between the rings and therefore the spectra expected for the cyclopentadienyl rings can be interpreted in terms of the monocyclopentadienyl model $\eta^5\text{-}C_5H_5M$.

Metal-cyclopentadienyl skeletal modes have been observed from approximately 500 to 100 cm^{-1}. The two skeletal modes expected for $\eta^5\text{-}C_5H_5M$ complexes are both infrared and Raman active and are illustrated in Figure 3.7. The six skeletal modes expected for $(\eta^5\text{-}C_5H_5)_2M$ complexes with parallel rings are also illustrated in Figure 3.7. The lowering of the symmetry from D_{5d} or D_{5h} in $(\eta^5\text{-}C_5H_5)_2M$ complexes with parallel rings to a maximum of C_{2v} in the corresponding complexes with nonparallel rings removes any degeneracy associated with the skeletal modes and therefore increases the predicted complexity of the spectrum due to the metal-ring skeletal modes. This is noted in Table 3.4, in which the expected skeletal modes are compared for $(\eta^5\text{-}C_5H_5)_2M$ derivatives as the symmetry is changed from D_{5h} to C_{2v}.

TABLE 3.3
Description of the Vibrational Modes for $(\eta^5\text{-}C_5H_5)_2M$ Complexes with Staggered (D_{5d}) and Eclipsed (D_{5h}) Rings[a]

D_{5h}	D_{5d}	Mode Number	Mode Description	Mode Number	D_{5d}	D_{5h}
A_1' (R)	A_{1g} (R)	1	ν(CH)	8	A_{2u} (IR)	A_2'' (IR)
		2	π(CH)	9		
		3	Ring breathing	10		
		4	Metal-ring stretching	11		
A_2' (ia)	A_{2g} (ia)	7	δ(CH)	5	A_{1u} (ia)	A_1'' (ia)
			Torsion	6		
E_1'' (R)	E_{1g} (R)	12	ν(CH)	17	E_{1u} (IR)	E_1' (IR)
		13	δ(CH)	18		
		14	π(CH)	19		
		15	ν(CC)	20		
		16	Metal-ring tilt	21		
			Ring-metal-ring defor.	22		
E_2' (R)	E_{2g} (R)	23	ν(CH)	29	E_{2u} (ia)	E_2'' (ia)
		24	δ(CH)	30		
		25	π(CH)	31		
		26	ν(CC)	32		
		27	δ(CC)	33		
		28	π(CC)	34		

[a]The numbering of the modes is taken from Ref. 70.

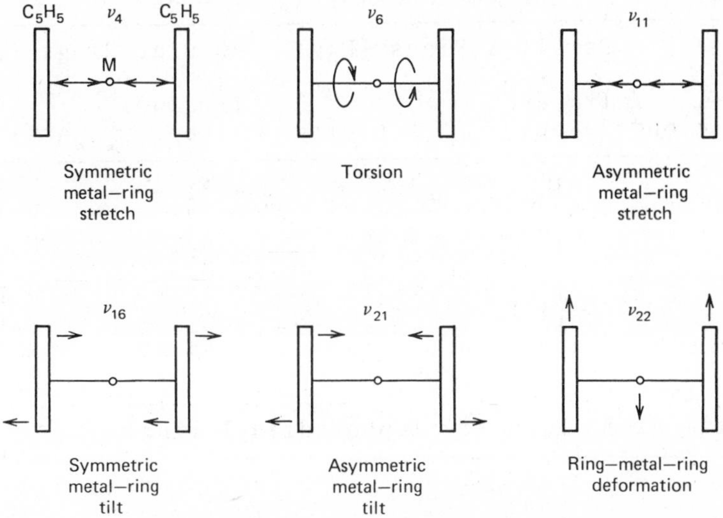

Figure 3.7. Skeletal mode vibrations for (η^5-C_5H_5)$_2$*M derivatives.*

d. *Bridging Cyclopentadienyl Rings*

Very limited vibrational data are available for compounds with bridging cyclopentadienyl rings. The infrared spectrum of $(C_5H_5)_3In$ was originally interpreted in terms of a *monohapto* structure (50,72). More recently, a crystal study has shown the structure to consist of infinite polymeric chains as illustrated in Figure 3.8, in which each indium atom is bonded to two μ-(η^1,η^1-cyclopentadienyl) rings (73). It was noted, however, in one of the infrared studies (72) that the spectrum appeared to be more complex than observed for other compounds that are known to have purely η^1-cyclopentadienyl rings. It has been suggested (73) that these additional bands arise from the bridging cyclopentadienyl rings, although no analysis has been presented of the data. A structure similar to that found for $(C_5H_5)_3In$ has also been found in a single crystal X-ray study of $(C_5H_5)_3Sc$ (74), although no vibrational data have been reported for $(C_5H_5)_3Sc$.

There has been interest in compounds in which the cyclopentadienyl ring functions simultaneously in a *monohapto* and *pentahapto* fashion in bridging two metal

TABLE 3.4
Skeletal Modes and Activities for $(\eta^5\text{-}C_5H_5)_2M$ Complexes with Parallel (D_{5h}) and Angular (C_{2v}) Rings

	Parallel Rings (D_{5h})		Angular Rings (C_{2v})	
Description[a]	Irreduc. Rep.	Activity	Irreduc. Rep.	Activity
ν(MR)	A_1'	R	A_1	R,IR
	A_2''	IR	B_2	R,IR
M-R tilt	E_1''	R	A_2	R
			B_2	R,IR
	E_1'	IR	A_1	R,IR
δ(RMR)	E_1'	IR	A_1	R,IR
Torsion	A_1''	ia	A_2	R
			B_1	R,IR

[a]M, metal atom; R, η^5-cyclopentadienyl ring.

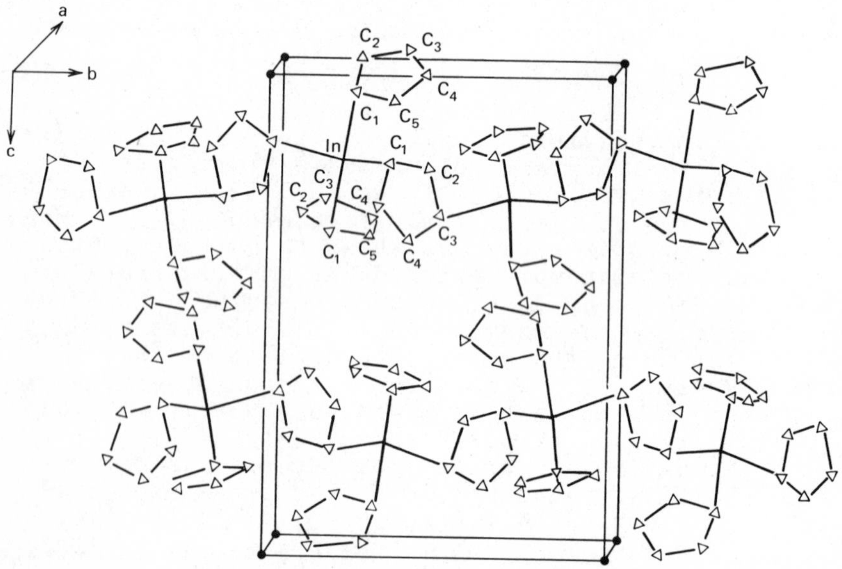

Figure 3.8. The molecular structure of $(C_5H_5)_3In$ viewed down the a axis (73).

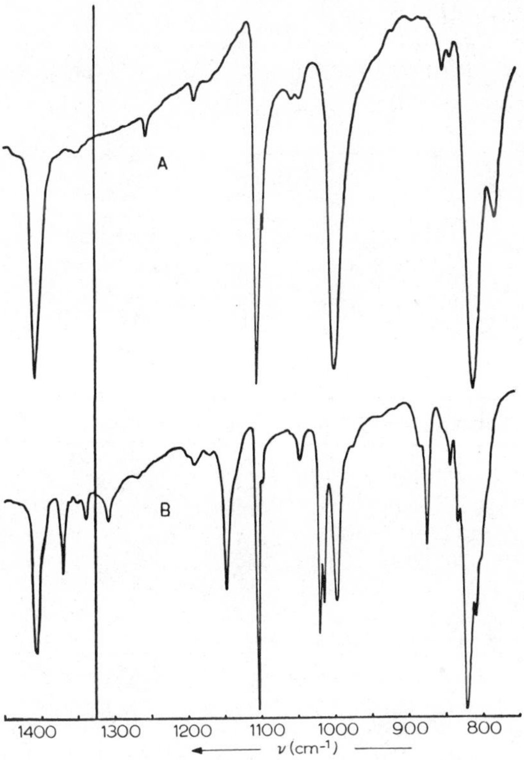

Figure 3.9. The hexachloro-1,3-butadiene (1450-1325 cm^{-1}) and Nujol mull (1325-750 cm^{-1}) infrared spectra of (A) (η^5-C_5H_5)$_2$Fe and (B) (η^5-C_5H_5)Fe(C_5H_4HgCl) (78).

atoms (75-77). These compounds have mainly been characterized using single crystal X-ray studies and, to a lesser extent, nmr spectroscopy with little vibrational data reported. The infrared spectrum of η^5-C_5H_5Fe-(C_5H_4HgCl) in which the C_5H_4 ring is bonded to the iron atom in a *pentahapto* fashion and to the mercury atom in a *monohapto* fashion has been assigned and compared with that of (η^5-C_5H_5)$_2$Fe (78). The similarity of the spectra in the ν(CH) mode and 1800 to 1600 cm^{-1} regions shows these portions of the spectra to be of little use in distinguishing between these two complexes. The 1450 to 750 cm^{-1} region illustrated in Figure 3.9, however,

TABLE 3.5
Assignments (cm^{-1}) of the q- and r-Type X-Sensitive Modes for Various Substituted Cyclopentadienyl and Phenyl Compounds

	Substituent	Frequency	
Mode	-X	C_5H_4X[a]	C_6H_5X[b]
q	-Cl	1170	1087
	-Br	1152	1070
	-I	1142	1060
	-HgCl	1148	1067
r	-Cl	884	701
	-Br	872	669
	-I	865	654
	-HgCl	873	665

[a]Assignments for the C_5H_4X rings of $(\eta^5\text{-}C_5H_4X)_2Fe$ (X=Cl,Br,I) are from Ref. 73; those for the C_5H_4HgCl ring in $\eta^5\text{-}C_5H_5Fe(C_5H_4HgCl)$ are from Ref. 78.
[b]Assignments for C_6H_5X (X=Cl,Br,I) are from Ref. 80; those for C_6H_5HgCl are from Ref. 81.

contains more dramatic and useful differences. Several bands appear in the spectrum of $\eta^5\text{-}C_5H_5Fe(C_5H_4HgCl)$ that are absent in the spectrum of $(\eta^5\text{-}C_5H_5)_2Fe$. The most prominent of these are the relatively strong intensity bands at 1148 and 873 cm^{-1}. Similar bands have been assigned in the infrared spectra of the 1,1'-dihalo ferrocenes to modes analogous to the X-sensitive or metal-sensitive q- and r-modes found in monosubstituted benzenes (80). The two X-sensitive modes arise from the A_1 symmetry ring breathing and $\nu(CX)$ modes which both interact strongly. Further support for the assignments given for these bands comes from a comparison of the assignments made for the q- and r-modes in $(\eta^5\text{-}C_5H_4X)_2Fe$ (X=Cl,Br,I) (79) and the C_5H_4Hg ring in $\eta^5\text{-}C_5H_5Fe(C_5H_4HgCl)$ (78) with the corresponding assignments made for C_6H_5X (X=Cl,Br,I) (80) and C_6H_5HgCl (81). These assignments are summarized in Table 3.5. It appears that the q- and r- modes are the most characteristic in ferrocene derivatives and should serve as a diagnostic test for the presence of the C_5H_4 ring in analogous complexes . Other differences between the spectra of $\eta^5\text{-}C_5H_5Fe(C_5H_4HgCl)$ and $(\eta^5\text{-}C_5H_5)_2Fe$ include the presence of two strong intensity bands at 1020 and 1013 cm^{-1}, which have been assigned to the C-H bending modes of η^5-$C_5H_5Fe(C_5H_4HgCl)$, and several additional bands of much weaker intensity that are present in the spectrum of η^5-$C_5H_5Fe(C_5H_4HgCl)$ but absent in that of $(\eta^5$-$C_5H_5)_2$-Fe. The antisymmetric iron-ring stretching and tilting modes appear at essentially the same frequency in both η^5-$C_5H_5Fe(C_5H_4HgCl)$ and $(\eta^5$-$C_5H_5)_2Fe$. Infrared data have also been listed for tris(η^5-cyclopentadienyl)-(ferrocenyl)uranium (3.15) and 1,1'-bis[tris(η^5-cyclopentadienyl)uranium]ferrocene (3.16) (82)

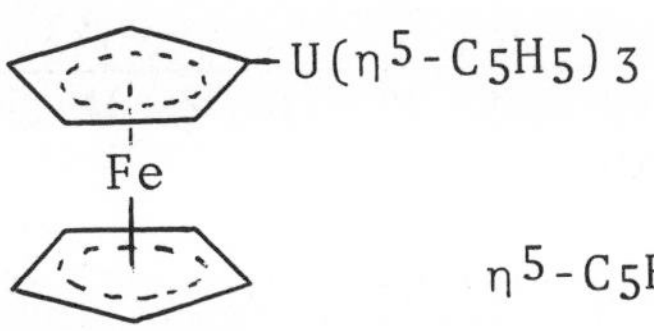

(3.15)

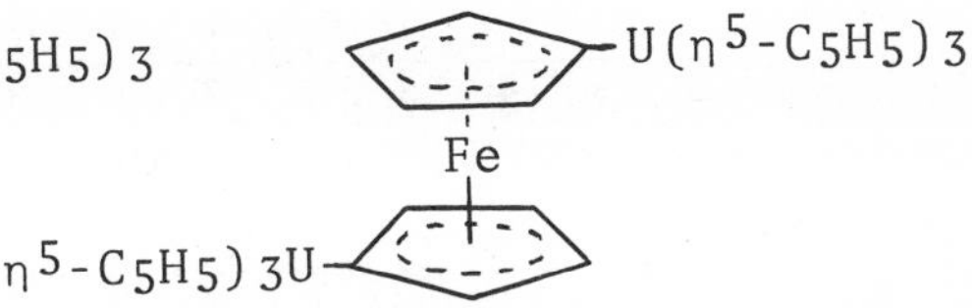

(3.16)

Polymeric cyclopentadienyl compounds with μ-(η^5,η^5-cyclopentadienyl) rings are also known. Such a structure (3.17) has been found in single crystal X-ray

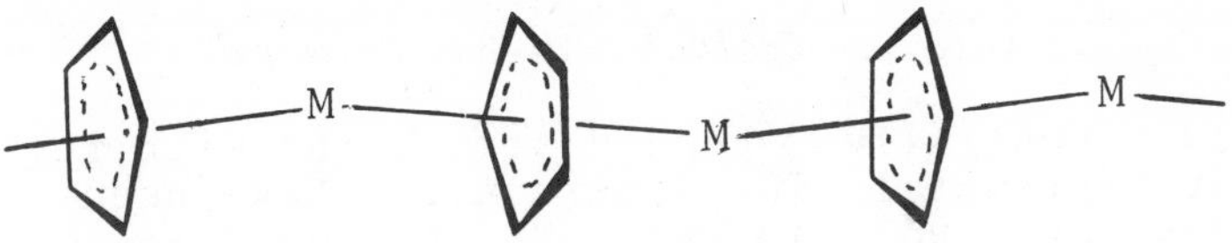

(3.17)

studies of C_5H_5M (M=In,Tl) (83). Because of the high symmetry of the μ-(η^5,η^5-cyclopentadienyl) ring, there are no characteristic features of the vibrational spectrum that make it possible to distinguish it from the η^5-cyclpentadienyl ring. Therefore, the infrared spectrum of solid C_5H_5In is very similar to the spectra observed for complexes with a centrally σ-bonded cyclopentadienyl ring (72). Also, a complete infrared and Raman study of C_5H_5Tl has been interpreted in terms of predominantly ionic bonding and a symmetry of D_{5h} for the cyclopentadienyl rings (55). Solid $(C_5H_5)_2Pb$ can exist in either a monoclinic or orthorhombic form (84). A crystal study of the orthorhombic modification shows the presence of a polymeric structure with each lead atom bonded to two μ-(η^5,η^5-cyclopentadienyl) rings and

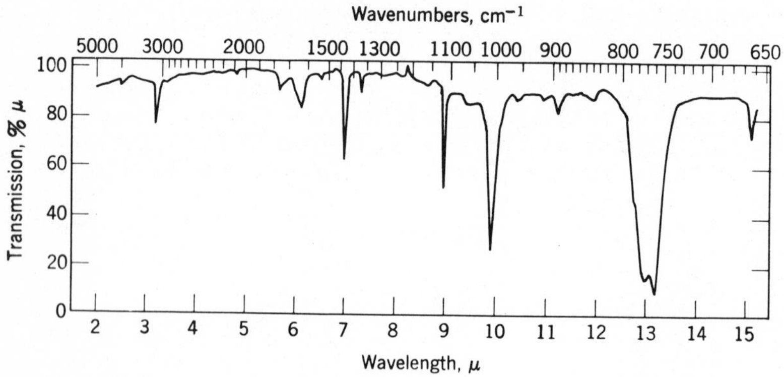

Figure 3.10. Infrared spectrum of $(C_5H_5)_2Pb$ *(86).*

one η^5-cyclopentadienyl ring (85). The infrared spectrum of solid $(C_5H_5)_2Pb$, illustrated in Figure 3.10 (86), is essentially identical to that of $(C_5H_5)_2Fe$; this again illustrates the fact that the bridging and terminal η^5-cyclopentadienyl rings in solid $(C_5H_5)_2Pb$ can not be distinguished using vibrational spectroscopy.

e. *Data for Cyclopentadienyl Compounds*

Group Ia. Centrally σ-bonded structures with relatively weak covalent metal-ring interactions have been proposed for C_5H_5M (M=Li,Na) (47,48). The infrared assignments reported for these two compounds are summarized in Table 3.6. The frequencies are similar to those found for the corresponding modes in the ionic derivatives (cf. Table 3.2). Spectroscopic data for 1M tetrahydrofuran solutions of C_5H_5M (M=Li,Na), however, are consistent with the presence of metal-cyclopentadienide ion pairs (49). Ionic structures have been found for the remaining cyclopentadienyl-group Ia derivatives (47,48).

Group IIa. A pure *monohapto* structure was originally proposed for $(C_5H_5)_2Be$ on the basis of infrared (86) and dipole moment (88) data. An electron diffraction study of $(C_5H_5)_2Be$ vapor, however, has been interpreted in terms of a staggered structure of C_{5v} symmetry (3.18) in which the beryllium atom is not equidistant from the two parallel η^5-cyclopentadienyl rings (89,90). Infrared data for $(C_5H_5)_2Be$ in the polycrystalline phase and

TABLE 3.6
Solid-State Infrared Assignments (cm^{-1}) for Covalent Cyclopentadienyl Complexes of the Group Ia and Group IIa Elements

Mode	Mode Number[a]	C_5H_5Li[b]	C_5H_5Na[b]	$(C_5H_5)_2Be$[c]	$(C_5H_5)_2Mg$[d]	$(C_5H_5)_2Ca$[e]
ν(CH)	1	3048 m	3048 m	3077 m	3063 m	3078 m
ν(CH)	2	2906 w	2907 w		2913	2693 vw
ν(CC)	3	1426 m	1422 w	1470 m 1430 s	1428 m	1437 vw
ν(CC)	4	1110 w	1144 w	1122 ms 1102 m	1108 m	1122 vw
δ(CH)	5	1003 s	998 s	1015 vs	1004	1009 s
π(CH)	6			962 vs 868 vs 828 vs	886[f]	
π(CH)	7	746 vs	712 vs	768 736 vs	779 s	751 vs
ν(M-C_5H_5)	8	538 vs	315 w	888 vs 416 w	526 m 191[g] ms	284[h] m
Tilt	9			487 vw 472 vw	440 m 229[g] w	284[h] m

[a]The numbering of the modes is taken from Ref. 48.
[b]References 47 and 48.
[c]Reference 95.
[d]References 36, 48, and 54.
[e]Reference 51.
[f]Observed in the crystal spectrum using a diamond cell.
[g]Data taken from the Raman spectrum.
[h]Not assigned to either a calcium-ring stretching or tilting mode.

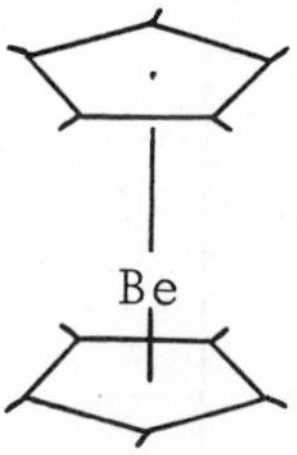

(3.18)

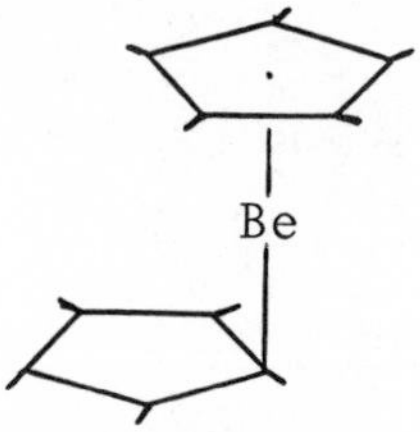

(3.19)

solution (benzene and cyclohexane) was later reported to be consistent with this structure (91). This interpretation of the condensed-phase infrared data has been questioned in single crystal X-ray studies of $(C_5H_5)_2Be$ at both -120°C (92) and room temperature (93). The crystal study at -120°C showed the presence of a "slip" sandwich structure (3.19) in which the two symmetric rings are staggered relative to each other, and the beryllium atom is bonded in a *pentahapto* manner to one ring and in a *monohapto* manner to the other. At room temperature, a slip structure was also found. It differs from the low-temperature structure in that the rings are neither staggered nor eclipsed relative to each other, but have orientations somewhere between these two extremes. A dynamic aspect of the structure of $(C_5H_5)_2Be$ includes the fact that the rings are undergoing rotation with respect to each other and that the beryllium atom moves back and forth from two alternate positions. It has been further observed (93) that the slip structure in which the dynamic effect is occurring with a high enough frequency would also be consistent with solution nmr data, which indicates that the hydrogen atoms are equidistant, and also with the structure of C_{5v} symmetry proposed in the electron-diffraction study. The presence of a structural difference between the vapor and condensed phases of $(C_5H_5)_2Be$ has received support from a comparison of the solid-state X-ray data with the vapor-state electron-diffraction data (94). It is also supported by a recent infrared study of $(C_5H_5)_2Be$ in which it was noted that the vapor-state spectrum is simpler than the solid-state spectrum (95). The infrared spectra of gaseous and solid $(C_5H_5)_2Be$ are compared in Figure 3.11, while the assignments made from the solid-state spectrum are included in Table 3.6. A complete infrared and Raman study has been reported of liquid and solid C_5H_5BeCl, $C_5H_5BeBH_4$, and $C_5H_5BeBD_4$ (96). These data are consis-

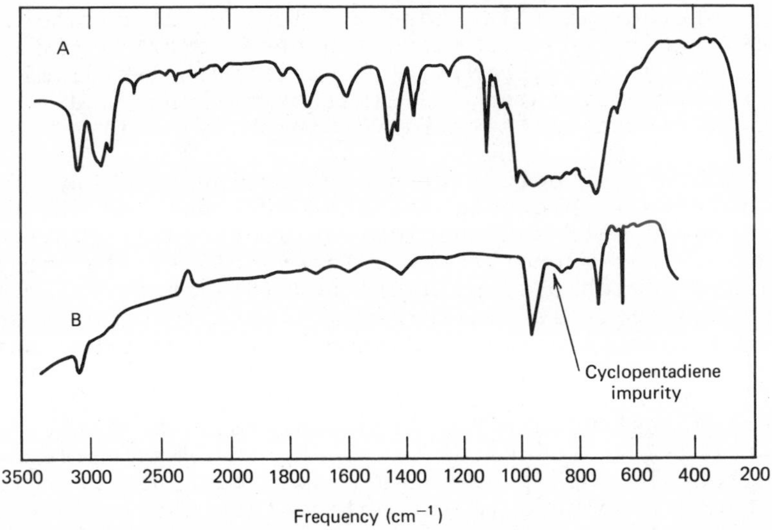

Figure 3.11. Infrared spectrum of (A) solid $(C_5H_5)_2Be$ *and (B)* $(C_5H_5)_2Be$ *vapor (95).*

tent with the presence of a η^5-cyclopentadienyl ring and a double hydrogen bridged borohydride ligand for $C_5H_5BeBH_4$.

The infrared spectrum of $(C_5H_5)_2Mg$ as a solid (KBr pellet) (36,48) and the infrared and Raman spectra in benzene (54) indicate the presence of a very weak covalent or a centrally σ-bonded magnesium-ring interaction. The assignment of the fundamentals of $(C_5H_5)_2Mg$ made in these studies is given in Table 3.6. The infrared spectra of tetrahydrofuran solutions of $(C_5H_5)_2Mg$ and C_5H_5MgX (X=Cl,Br) have been interpreted in terms of magnesium-cyclopentadienide ion triplets for $(C_5H_5)_2Mg$ and ion pairs for the two halides (49).

The infrared spectrum of $(C_5H_5)_2Ca$ was originally interpreted as consistent with an ionic structure (48). More recently it has been observed (51) that the infrared spectrum of $(C_5H_5)_2Ca$ (cf. Table 3.6) is similar to that of $(C_5H_5)_2Mg$, which, as noted above, has been described in terms of a centrally σ-bonded structure. A single crystal X-ray study of solid $(C_5H_5)_2Ca$, however, has shown that the calcium atom is four-coordinate with two η^5-, one η^3-, and one η^1-cyclopentadienyl ring in the coordination sphere (52). It was also concluded in this study (52) that the simplicity of the infrared data in solution indicates that $(C_5H_5)_2Ca$ is primarily held together in this phase by electrostatic interactions. Infrared data have also been reported for $(C_5H_5)_2CaL_2$ (L=Py,L_2=N,N,N',N'-tetramethylenediamine, Bipyr,tetrahydrofuran,1,2-dimethoxymethane) and $(C_5H_5)_2Ca(Phen)L_n$ (L=tetrahydrofuran,Py) (51).

Group Ib. A structure with a η^1-cyclopentadienyl ring was originally proposed for $C_5H_5CuP(C_2H_5)_3$ on the basis of infrared data and chemical evidence (97). This structure was supported by a later solid-state infrared (48) and variable temperature nmr (98) study. A more recent and detailed infrared (mull and toluene-d_8 solution) investigation of $C_5H_5CuPR_3$ ($R=C_2H_5,C_4H_9$) has led to the conclusion that the cyclopentadienyl ring is *pentahapto* and not *monohapto* in nature (69). The *pentahapto* structure has been confirmed in single crystal X-ray studies of $C_5H_5CuPR_3$ ($R=C_6H_5$ (99),C_2H_5 (100)). Although a *pentahapto* structure has been found for these compounds in the solid state and in common organic solvents, the nmr spectrum of $C_5H_5CuP(C_2H_5)_3$ using SO_2 as the solvent showed the presence of a η^1-cyclopentadienyl ring (98). It has been suggested (101) that this observation may be due to the formation of a SO_2 adduct of the type η^1-$C_5H_5Cu[P(C_2H_5)_3](SO_2)_n$ (n=1,2), as has been proposed (102) in an nmr study of $C_5H_5Tl(CH_3)_2$ in which SO_2 was used as the solvent.

The infrared spectrum of $C_5H_5AuP(C_2H_5)_3$ has been reported to be consistent with the presence of a η^1-cyclopentadienyl ring, although it was also noted that a more complex spectrum would have been predicted for the 1400 to 600 cm^{-1} region (103). Infrared data for $C_5H_5Au(CH_3)_2[P(C_2H_5)_3]$ also suggest the presence of a η^1-cyclopentadienyl ring (104).

Group IIb. The infrared spectra of $(C_5H_5)_2Zn$, C_5H_5Zn-C_2H_5, and $C_5H_5CdC_2H_5$ have been interpreted in terms of structures with a η^5-cyclopentadienyl ring (48).

A pure *monohapto* structure was originally proposed (97) for $(C_5H_5)_2Hg$ on the basis of its infrared spectrum (cf. Figure 3.5) and chemical evidence. The *monohapto* structure was, however, questioned; it being suggested that the infared spectrum was too complex for a pure *monohapto* type structure and that the uv spectrum cannot be accounted for on the basis of this structure (105). Detailed vibrational studies for $(C_5H_5)_2Hg$ and C_5H_5HgX

(X=Cl,Br,I) in the solid state and solution ($CHCl_3$ and CS_2) have now been shown to be consistent with the originally proposed *monohapto* structure (68,69). In Table 3.7 a comparison is made of the infrared and Raman data reported for cyclopentadiene, $(C_5H_5)_2Hg$, and C_5H_5HgX (X=Cl,Br,I). The skeletal mode assignments for cyclopentadienyl-mercury derivatives are given in Table 3.8. The low frequency of the ν(HgC) mode in these cyclopentadienyl-mercury derivatives relative to the frequency of the ν(HgC) mode in alkyl-mercury derivatives (approximately 575 to 485 cm^{-1}) has been attributed to the large mass of the cyclopentadienyl ring. Since the ring is more rigid than noncyclic alkyl groups it is more likely to act as a unit in any metal-ring vibration than an alkyl group.

Group IIIb. The infrared data listed for $(C_5H_5)_3Al$ are reported to be consistent with a *monohapto* structure (48). The absence of a η^1-cyclopentadienyl ring and bridging methyl groups had originally been suggested on the basis of infrared and Raman data reported for C_5H_5-$Al(CH_3)_2$ (67). It was further concluded that these data, together with the low solubility, low volatility, and high melting point of this compound, are consistent with a polymeric structure with μ-(η^5,η^5-cyclopentadienyl) rings similar to that found for C_5H_5M (M=In,Tl) (83) and $(C_5H_5)_2Pb$ (85). More recently, electron-diffraction data for $C_5H_5Al(CH_3)_2$ vapor have shown the presence of a η^3-cyclopentadienyl ring (65). It was therefore suggested in the electron-diffraction study that the structural model for solid $C_5H_5Al(CH_3)_2$ should be modified to include μ-(η^1,η^1-cyclopentadienyl) rings, as has been found for solid $(C_5H_5)_3In$ (73), and that as this polymeric structure is broken down on vaporization, the cyclopentadienyl ring of monomeric $C_5H_5Al(CH_3)_2$ becomes *trihapto* in nature. The infrared and Raman spectra have been assigned for $C_5H_5Al(C_2H_5)_2$ (ν(Al-C_5H_5)=340 cm^{-1} (106). The infrared spectra of $C_5H_5AlCl_2\cdot 3ROH$ (R=CH_3, C_2H_5,*n*-C_3H_7,*i*-C_3H_7,*n*-C_4H_9,*i*-C_4H_9,*i*-C_5H_{11}) suggest the presence of a η^5-cyclopentadienyl ring and that the alcoholate groups are bonded to the aluminum atom by σ bonds (107).

In contrast to the corresponding aluminum derivative, a centrally σ-type *pentahapto* structure has been proposed for $(C_5H_5)_3Ga$ on the basis of its infrared spectrum (48). The infrared and Raman spectra have also been assigned for $C_5H_5Ga(C_2H_5)_2$ (ν(Ga-C_5H_5)=280 cm^{-1}) (106).

TABLE 3.7
Comparison of Vibrational Data (cm^{-1}) for Cyclopentadiene and η^1-Cyclopentadienyl Derivatives of Mercury[a]

Cyclopentadiene Raman	Cyclopentadiene Infrared	$(C_5H_5)_2Hg$ Infrared	C_5H_5HgCl Infrared	C_5H_5HgBr Infrared	C_5H_5HgI Infrared
	3105 ms	3090 s	3105 m	3105 s	3105 s
3091 s		3088 s	3098 m	3090 s	3075 s
3075 sh	3075 m	3076 sh	3072 s	3063 s	3060 sh
	3043 m	3040 sh	3040 w	3030 sh	3040 sh
2985 w	2986 vvw	2970 w	2950 w	2960 w	2960 w
	1829 vvw		1830 w	1825 w	1821 w
	1816 mw	1808 w	1810 w	1810 w	1808 w
	1625 m	1630 w		1631 m	1635 w
1500 vs	1500 w	1530 w	1545 w[b]	1540 w[b]	1540 w[b]
1441 w		1427 m	1461 m[b]	1460 m[b]	1457 m[b]
1378 m		1383 m	1379 m	1378 m	1379 m
1366 ms	1356 s				
	1292 mw	1296 m	1290 m	1289 m	1293 m
	1239 m	1234 m	1228 m	1224 w	1230 w
1107 s	1106 w	1109 m	1108 w	1110 vw	1110 vw
1093 sh	1088 m	1084 m	1089 m	1086 m	1086 m
		1026 m	1026 m	1019 m	1022 m
995 mw	993 w	988 w	990 m	989 s	988 s
958 w	959 s	957 m	962 m	958 m	952 m
914 m	915 m	907 vs	941 s	938 vs	925 vs
893 sh	891 vs	885 vs	902 s	901 s	
			885 w	887 w	
803 w	805 ms	822 m	819 m	818 m	820 s
		748 vs	756 vs	753 vs	750 vs
700 vw		720 w	720 w	718 w	718 w
666 vw	664 vs	646 s	653 s	650 s	643 s
		575 w	565 w	564 w	566 w
349 w		352 ms	370 vw[c]		

[a]Data for liquid cyclopentadiene from Ref. 28. Data were recorded in CS_2 and $CHCl_3$ for $(C_5H_5)_2Hg$ and CS_2 for C_5H_5HgX (X=Cl,Br,I) unless otherwise indicated and taken from Ref. 68.
[b]Taken from the perfluorocarbon mull spectrum.
[c]Taken from the Nujol mull spectrum.

Although C_5H_5In vapor has a monomeric structure with approximate C_{5v} symmetry (108), solid C_5H_5In is polymeric (3.17) with μ-(η^5,η^5-cyclopentadienyl) rings (83). The infrared spectrum of solid $(C_5H_5)_3In$ is consistent with this structure (72). A $CHCl_3$ solution of

TABLE 3.8
Skeletal Mode Assignments (cm^{-1}) for η^1-Cyclopentadienyl Derivatives of Mercury[a]

Compound	Phase	ν(HgC)	ν(HgX)	δ(HgCC)	δ(CHgC) or δ(CHgX)
C_5H_5HgCl	Solid[b]	345 s	295 s	200 m	90 w
	Solution[c]	359 s	315 s	197 m	108 w
C_5H_5HgBr	Solid	340 s	214 s	200 m	64 w
		333 s			
	Solution	342 s	232 s	220 m	53 w
C_5H_5HgI	Solid	331 s	175 s	200 m	63 w
	Solution	336 s	186 s	220 m	51 w
$(C_5H_5)_2Hg$	Solid	345 s		200 m	100 w
	Solution	348 s		200 m	100 w

[a]Reference 68.
[b]From the Nujol mull spectra.
[c]From the CS_2 solution spectra.

C_5H_5In reacts with gaseous BX_3 (X=F,Cl,Br,CH_3) to produce the solid complexes $C_5H_5In{\cdot}BX_3$ (109). In contrast to the polymeric, *pentahapto* structure found for C_5H_5In, the vibrational spectra of the $C_5H_5In{\cdot}BX_3$ adducts have been reported to be consistent with a monomeric *monohapto* structure. The change in the mode of indium-ring bonding on complexation of C_5H_5In has been attributed to the large stabilization energy achieved on complex formation, which is sufficient to promote the cyclopentadienyl ring from the *pentahapto* structure to the higher-energy *monohapto* structure (109).

The infrared and Raman spectra of solid $(C_5H_5)_3In$ are more complex than the infrared spectrum of C_5H_5In (72). This is consistent with the polymeric structure found for solid $(C_5H_5)_3In$ (cf. Figure 3.8) (73), in which there are two η^5-cyclopentadienyl rings and one μ-(η^1,η^1-cyclopentadienyl) ring. Although no attempt has been made to assign the vibrational data for $(C_5H_5)_3In$, infrared bands at 339 (s) and 316 (m) cm^{-1}, and Raman bands at 338 (m) and 322 (s) cm^{-1} are in the region expected for the ν(In-C_5H_5) modes. Low-frequency infrared and Raman bands reported for $(C_5H_5)_3InL$ (L=$(C_6H_5)_3P$,Bipyr,Phen) are from 10 to 40 cm^{-1} lower in frequency than the corresponding bands in uncomplexed $(C_5H_5)_3In$ (73). The infrared spectra of $C_5H_5InI_2$, $C_5H_5InI_3^-$, and $C_5H_5InX_2L$ (X=F,L=Bipyr;X=Cl,L=Phen) are consistent with a η^1-cyclopentadienyl structure with the ν(In-C_5H_5) mode assigned from 329 to 315 cm^{-1} (110). The physical properties and infrared data for solid $C_5H_5In(CH_3)_2$ suggest a polymeric structure similar to that proposed for $C_5H_5Al(CH_3)_2$ (111). Infrared data and assignments have also been reported for C_5H_5In-$(C_2H_5)_2$ (106).

While C_5H_5Tl has a monomeric structure of C_{5v} symmetry in the vapor state (53,108,112), it has a polymeric solid-state structure (83). The infrared spectrum of C_5H_5Tl vapor in the ν(CH) mode region has been interpreted in terms of ionic bonding (53). A more recent and detailed infrared and Raman study of polycrystalline C_5H_5Tl has been reported (55). The vibrational spectrum is relatively simple with few coincidences between the corresponding infrared and Raman bands and is therefore consistent with the presence of cyclopentadienyl rings with D_{5h} symmetry. No bands were observed that could be assigned to thallium-ring skeletal modes. It was noted, however, that in a highly ionic complex such as C_5H_5Tl, any skeletal modes would be incorporated into the lattice modes that involve coordinated motion of the groups making up the polymeric chain. Several bands observed from 389 to 66 cm^{-1} were assigned to lattice modes. Since their positions were temperature dependent, it was concluded that although the bonding in C_5H_5Tl is predominently ionic, there is some evidence for weak covalent bonding. Therefore, it might be more appropriate to classify the bonding as centrally σ in nature. In Table 3.9, the infrared and Raman data and assignments are given for solid C_5H_5Tl. Infrared data have also been listed for $(C_5H_5)_3Tl$ (113), $(C_5H_5)_2TlH$ (ν(TlH)=2105 cm^{-1}) (114), $(C_5H_5)_2TlCl$ (115), and $(C_5H_5)_2TlX$ (X=N_3,NCO,NCS,CN) (116). The carboxylate-bridged structure proposed for η^1-$C_5H_5Tl(CH_3)X$ (X=CH_3CO_2,$C_2H_5CO_2$,*i*-$C_3H_7CO_2$,tropolinate, 4-isopropyltropolinate) (3.20) in both the solid state

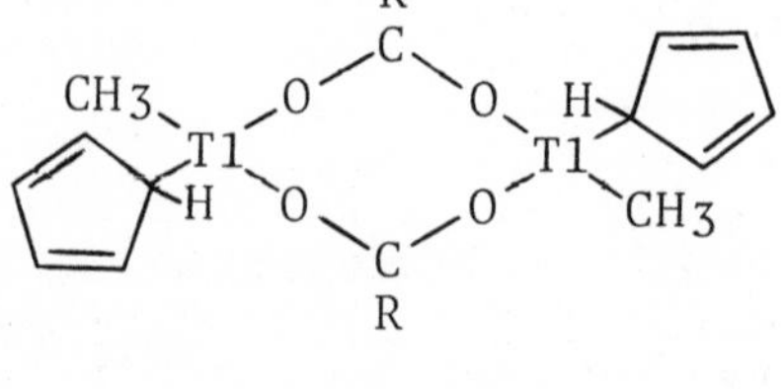

(3.20)

TABLE 3.9
Vibrational Assignments (cm^{-1}) for Solid C_5H_5Tl[a]

Mode	Activity[b]	Raman	Infrared
ν(CH)	A_1'	3092 s	
ν(CH)	E_1'		3070 w
ν(CH)	E_2'	3065 m	
			2920 w
			1710 vw
			1600 w
		1497 vw	
ν(CC)	E_1'		1425 w
ν(CC)	E_2'	1370 m,sh	
		1342 s	
			1300 w
δ(CH)	E_2'	1208 s	
δ(CH)	A_2'		1158 w
Ring breathing	A_1'	1118 vs	1120 vw
π(CH)	E_1''	1058 m	
δ(CH)	E_1'		1000 s
π(CH)	E_2''	861 w,sh	
		841 m	
δ(CC)	E_2'	762 s,sh	
		751 s	755 m,sh
π(CH)	A_2''	727 s	
			555 vw,br
π(CC)	E_2''	445 w	
		418 w	
		388 w	389 vw
			377 vw
		360 w	362 vw
			266 vw
Lattice modes		188 m	175 m
			114 s
			66 vs

[a]Reference 55.
[b]A D_{5h} symmetry was assumed for the cyclopentadienyl ring.

and solution (117) is supported by the presence of bands in the olefinic ν(CH) mode region as well as bands characteristic of bridging carboxylate groups. The $\nu(Tl\text{-}C_5H_5)$ and $\nu(Tl\text{-}CH_3)$ modes have been assigned from 318 to 315 cm^{-1} and 513 to 500 cm^{-1}, respectively.

Group IVb. Although $(C_5H_5)_2Si$ is unknown, $(C_5H_5)_2M$ (M=Ge,Sn,Pb) have all been prepared. A monomeric, angular-sandwich structure has been found in an electron-diffraction study of gaseous $(C_5H_5)_2Sn$ and $(C_5H_5)_2Pb$ with angles of $45\pm15^{\circ}$ and approximately 55°, respectively, between the planes of the η^5-cyclopentadienyl rings (118). This study corroborated similar structures suggested earlier on the basis of solid-state and solution infrared (36,86 119-121) and dipole moment (122) data. An analogous structure has been suggested as likely for $(C_5H_5)_2Ge$ (123). Solid $(C_5H_5)_2Ge$ has been found to polymerize completely within 3 hours (123), while $(C_5H_5)_2Sn$ becomes 90% polymerized in 5 days at room temperature (124). The nature of these polymeric structures, however, has not been investigated. A single crystal X-ray study of the orthorhombic form of $(C_5H_5)_2Pb$ has shown it to have a polymeric structure consisting of both η^5-cyclopentadienyl and μ-(η^5,η^5-cyclopentadienyl) rings (84). Infrared data for $(C_5H_5)_2Ge$ in benzene-d_6 and methylene chloride solutions are included in Table 3.10. It has been observed in this study that the frequencies fall between the ranges expected for centrally σ- and centrally π-bonded compounds, pointing out the lack of a clearcut division between these two subgroups. Six prominent bands, one of which was possibly split into a doublet were also observed at $-100^{\circ}C$ in the Raman spectrum of $(C_5H_5)_2Ge$. Although these bands appear in the region expected for the skeletal modes, no assignments were attempted. It was further observed that these Raman data are consistent with either an angular structure or the polymeric structure found for solid $(C_5H_5)_2Pb$. In a detailed infrared and Raman study of $(C_5H_5)_2Sn$ in both the solid state and CCl_4 solution, it has been concluded that the angular vapor-state structure is preserved in the solid state and solution (125). The cyclopentadienyl ring mode assignments made in this study are summarized in Table 3.10. No bands have been observed that could be assigned to the tin-ring tilting modes. Raman bands at 240 (s) and 170 (w-m) cm^{-1} have, however, been assigned to the ν_a(Sn-C_5H_5) and ν_s(Sn-C_5H_5) modes, respectively. The former band is also observed in the infrared spectrum (121,125). Cyclopentadienyl ring assignments reported for $(C_5H_5)_2Pb$ are also included in Table 3.10. No skeletal mode assignments have been reported for $(C_5H_5)_2Pb$. Although infrared data indicate that the indium-cyclopentadienyl bond changed from the *pentahapto* to the *monohapto* type when C_5H_5In formed the adduct $C_5H_5In{\cdot}BF_3$ (109), similar

TABLE 3.10
Assignments of the Cyclopentadienyl Fundamental Modes (cm^{-1}) for $(C_5H_5)_2Ge$, $(C_5H_5)_2Sn$, and $(C_5H_5)_2Pb$

		$(C_5H_5)_2Ge$[a]	$(C_5H_5)_2Sn$			$(C_5H_5)_2Pb$[b]
Assignment	Symmetry[c]	Infrared	Raman[d]	Infrared	Infrared	Infrared
ν(CH)	A_1		3101 s		3100 m,sh[e]	
ν(CH)	E_1, E_2	3088 m		3061 w[f]	3090 m	3021 w
ν(CC)	E_2				1534 vw,br	
ν(CC)	E_1	1429 m	1431 s	1424 m	1429 vs	1416 m
δ(CH)	A_2				1263 vw	1263 w
δ(CH)	E_2				1161 vvw	1164 w
Ring breathing	A_1	1112 ms	1118 s	1112 m	1114 vs	1112 m
π(CH)	E_2		1063 w	1059 w	1062 w	1057 w
δ(CH)	E_1	1005 ms	1005 w	1003 s	1011 vs	1005 s
δ(CC)	E_2		886 w	890 w	890 vw,sh	898 w
π(CH)	E_1	811 ms	796 w	792 s		756 s
π(CH)	A_1	759 s	738 w	751 s		740 s
π(CC)	E_2			544 vw		

[a]Data from Ref. 123 for benzene-d_6 and methylene chloride solutions.
[b]Data from Ref. 121 for KBr pellet and Nujol mull of the sample.
[c]Taken from Ref. 125 assuming a symmetry of C_{5v} for the cyclopentadienyl rings.
[d]Data and assignments from Ref. 125 for a solid sample.
[e]Data and assignments from Ref. 125 for a CCl_4 solution.
[f]Data from Ref. 121 for KBr pellet of the sample.

data indicate that the cyclopentadienyl rings are *pentahapto* in both $(C_5H_5)_2Sn$ and $(C_5H_5)_2Sn \cdot BF_3$ (126).

While η^5-cyclopentadienyl rings are found in the divalent derivatives of the group IVb elements, η^1-cyclopentadienyl rings are found in the tetravalent derivatives of these elements. Chemical evidence and infrared and uv data for $C_5H_5Si(CH_3)_3$ are consistent with the presence of a η^1-cyclopentadienyl ring (97). More recently, a *monohapto* structure has been proposed for $C_5H_5M(CH_3)_3$ (M=Ge,Sn) on the basis of infrared data (127). The infrared spectra above 3000 cm^{-1} for $C_5H_5M(CH_3)_3$ (M=Si,Ge,Sn) provided a clearcut means of concluding that the cyclopentadienyl ring is *monohapto* in nature (127). All five ν(CH) modes for the η^1-cyclopentadienyl ring are infrared active with the olefinic modes clearly observed and easily distinguished from the $\nu(CH_3)$ modes, which appear below 3000 cm^{-1}. For $C_5H_5Si(CH_3)_3$, the olefinic cyclopentadienyl ν(CH) modes have been assigned at 3121, 3094, and 3075 cm^{-1}, while for the germanium and tin analogs these modes have been assigned at 3110, 3090, and 3070 cm^{-1}, and 3065, 3054, 3042, and 3014 cm^{-1}, respectively (127). The infrared spectrum of $C_5H_5Sn(CH_3)_3$ is illustrated in Figure 3.12 (127). The infrared spectra of cyclopentadienylchlorosilanes have been interpreted as consistent with the presence of three different isomers (3.21 to 3.23)

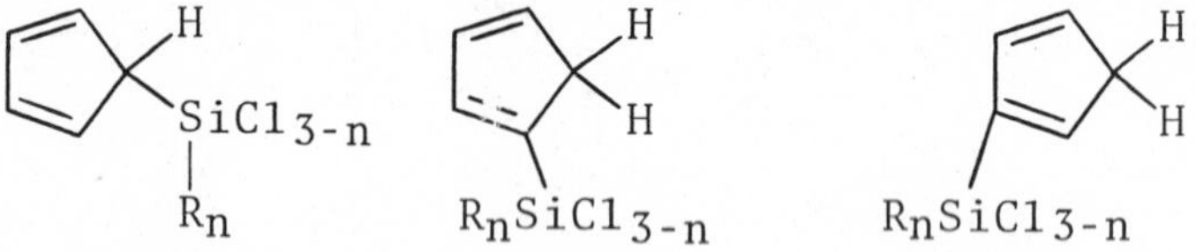

(3.21) (3.22) (3.23)

(128). These derivatives showed bands in the 2950 to 2860 cm^{-1} and 1379 to 1375 cm^{-1} regions that have been assigned to the stretching and deformation modes, respectively, of the methylene cyclopentadienyl group. Vibrational assignments have been made for $C_5H_5GeH_3$ ($\nu(Ge\text{-}C_5H_5)$=369 cm^{-1}) (129), while infrared data have been listed for $(C_5H_5)_2Si(CH_3)_2$ (48), $(C_5H_5)_4Sn$ (48), and $(C_5H_5)_2Pb(C_2H_5)_2$ (130).

Group Vb. Infrared and nmr data for $(C_5H_5)_3M$ (M=As,Sb) and a yellow and black form of $(C_5H_5)_2Bi$ indicate the presence of different dypes of structures for these compounds (131). There is a trend from the *monohapto*

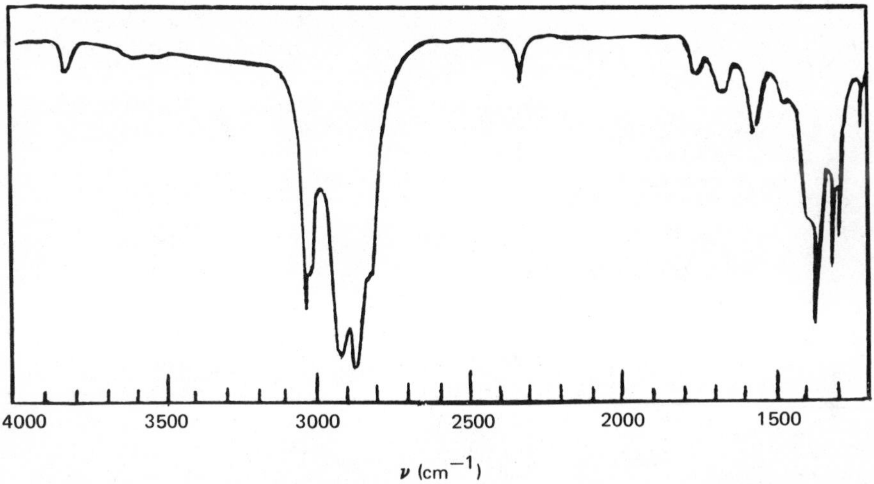

Figure 3.12. Infrared spectrum of $C_5H_5Sn(CH_3)_3$ (4000-1200 cm^{-1}) (127).

structure for $(C_5H_5)_3As$ to the *pentahapto* structure for the black form of $(C_5H_5)_3Bi$. Intermediate structures in which there is a rapid interconversion of *monohapto* and *pentahapto* rings have been proposed for $(C_5H_5)_3Sb$ and the yellow form of $(C_5H_5)_3Bi$. The infrared spectrum of the black form of $(C_5H_5)_3Bi$ is very simple, as expected for a compound with highly symmetric η^5-cyclopentadienyl rings. The infrared spectrum of $(C_5H_5)_3As$ is much more complex, again as expected for a compound with η^1-cyclopentadienyl rings. The infrared spectra of $(C_5H_5)_3Sb$ and the yellow form of $(C_5H_5)_3Bi$, however, show a complexity that is consistent with the presence of both *monohapto* and *pentahapto* rings.

The complexity of the infrared spectra of solid $C_5H_5M(CH_3)_2$ (M=As (111),Sb (132,133),Bi (132)) are consistent with the presence of a η^1-cyclopentadienyl ring in all three compounds. Infrared data have also been listed for $C_5H_5SbCl_2$ and other monocyclopentadienylstilbines (133).

Group IIIa. Although no vibrational data have been reported for $(C_5H_5)_3Sc$, a single crystal X-ray study shows this compound to be polymeric with each scandium atom bonded to two η^5-cyclopentadienyl rings and two μ-(η^1,η^1-cyclopentadienyl) rings to give a coordination number of four (74). A dimeric structure (3.24) has

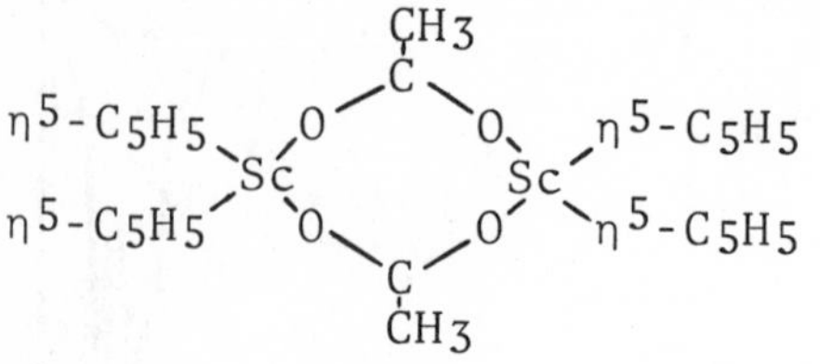

(3.24)

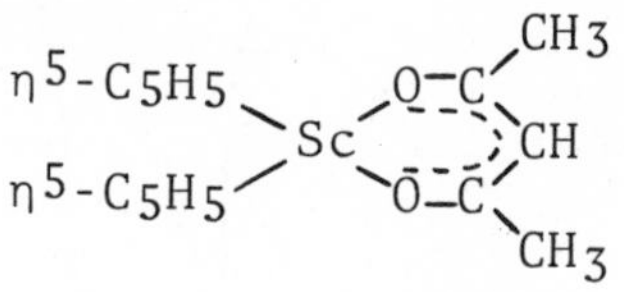

(3.25)

been proposed for $(C_5H_5)_2ScO_2CCH_3$ on the basis of molecular weight data in benzene and a separation of 135 cm^{-1} in the ν(CO) mode frequencies (134). A monomeric structure (3.25) is likely for $(C_5H_5)_2$ScAcac with the ν(CO) modes assigned to infrared bands at 1515 and 1370 cm^{-1} and a band at 1575 cm^{-1} assigned to a ν(C$\overset{...}{-}$C) mode (134).

Group IVa. The similarity of the infrared spectra of compounds with the empirical formula $(C_5H_5)_2M$ (M=Ti (135),Zr (136),Hf (137)) to those of $(\eta^5\text{-}C_5H_5)_2M$ (M=Fe, Ru,Ni) was proposed (137) to support the presence of η^5-cyclopentadienyl rings. It was also observed, however, that the titanium derivative is dimeric in benzene (135) and that although $(\eta^5\text{-}C_5H_5)_2M$ (M=Fe,Ru,Ni) all show five infrared bands in the 1700 cm^{-1} region, these bands were apparently shifted to lower frequencies in the infrared spectra of the titanium, zirconium, and hafnium analogs (i.e., 1302, 1262, and 1232 cm^{-1} for the titanium derivative) (135-137). It was subsequently proposed (138) that the infrared and nmr spectra of dimeric $(C_5H_5)_2Ti$ were too complex for a pure *pentahapto* structure and that in solution the most probable structure consists of a dimer with a Ti-Ti bond and both η^5- and η^1-cyclopentadienyl rings. More recently, mass spectral and infrared data have been presented which indicate that rather than formulating the titanium derivative as $[(C_5H_5)_2Ti]_2$, it is better represented as $[(C_5H_5)(C_5H_4)TiH]_2$.(139). It was therefore proposed that the strong intensity band at 1232 cm^{-1} in the infrared spectrum of $[(C_5H_5)(C_5H_4)TiH]_2$ (illus-

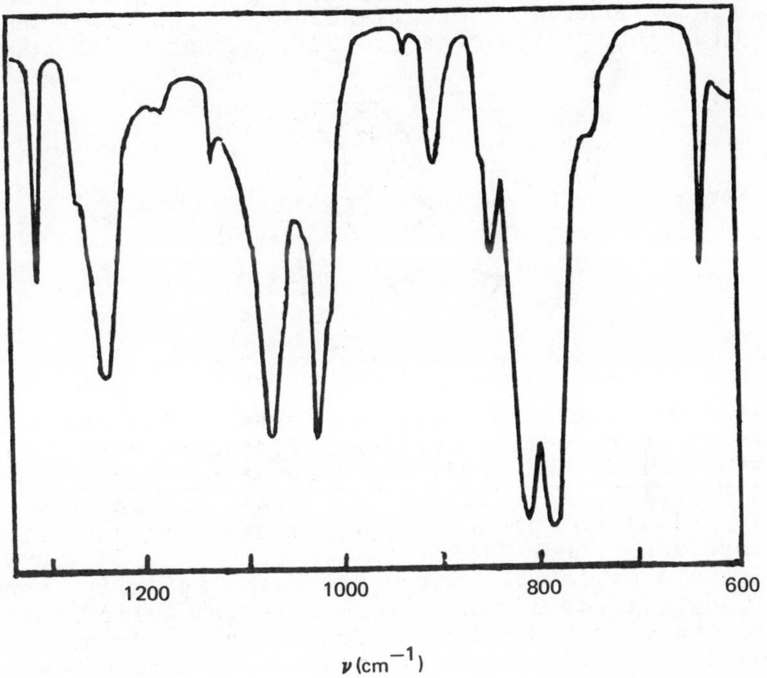

Figure 3.13. Infrared spectrum (1325-600 cm^{-1}) of $(C_{10}H_9TiH)_2$ in a Nujol mull (137).

trated in Figure 3.13) does not arise from a combination band as had originally been suggested (137) but from the ν_a(TiH) mode of the two bridging hydride ligands (139). Although two structures (3.26 and 3.27)

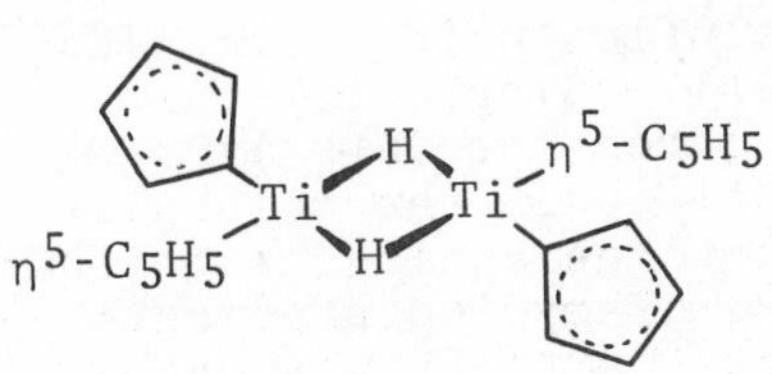

(3.26)

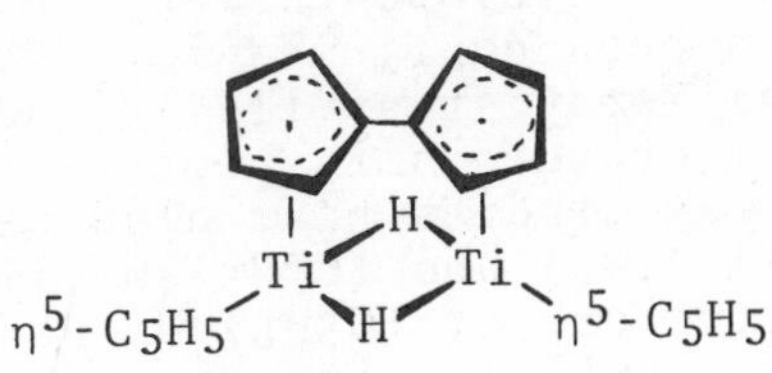

(3.27)

were suggested as possibilities, the data could not be used to distinguish between them (139). In addition to the two structures mentioned above, a third has been proposed (76) in which the C_5H_4 ring is bonded in a *pentahapto* fashion to one titanium atom and in a *monohapto* fashion to the other (3.28), as has been found in

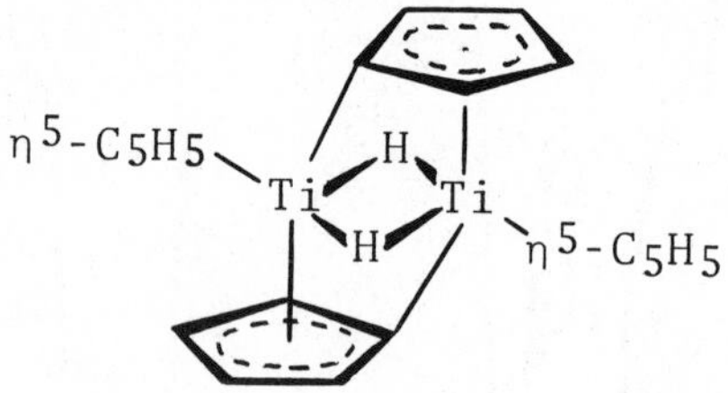

(3.28)

single crystal X-ray studies of several other compounds (75-77). Both chemical evidence (140) and a carbon-13 nmr study (141), however, have been reported as supporting a structure for "titanocene" that contains a μ-(η^5,η^5-fulvalene) group (3.27), as had been suggested in a previous study (139).

A pure *pentahapto* structure was originally proposed for $(C_5H_5)_3Ti$ (48,142). A more detailed solid-state infrared and tetrahydrofuran solution nmr study of this compound was interpreted in terms of a structure with one η^1- and two η^5-cyclopentadienyl rings (143). A single crystal X-ray study of $(C_5H_5)_3Ti$ has now shown that although in the solid state there are two η^5-cyclopentadienyl rings, the third is bonded to the titanium atom through only two adjacent carbon atoms (61). No studies have been reported for $(C_5H_5)_3M$ (M=Zr,Hf).

The infrared spectrum of solid $(C_5H_5)_4Ti$ has been interpreted in terms of a structure with two η^1- and two η^5-cyclopentadienyl rings (143). This structure has been confirmed for the solid state in a single crystal X-ray study (144), and in solution using variable temperature nmr spectroscopy (101). Single crystal X-ray studies have also been reported for $(C_5H_5)_4Zr$ (145,146) and $(C_5H_5)_4Hf$ (147). Although the molecular structure of $(C_5H_5)_4Hf$ was found to be analogous to that of $(C_5H_5)_4Ti$ with two η^1- and two η^5-cyclopentadienyl rings, the molecular structure of $(C_5H_5)_4Zr$ was characterized as containing one η^1- and three η^5-cyclopentadienyl rings. The structural difference between $(C_5H_5)_4Zr$ and $(C_5H_5)_4Hf$ is also indicated by the report that their infrared spectra differ (148). It has more recently been suggested (144), however, that the structure proposed for $(C_5H_5)_4Zr$ (145,146) is not justified by the reported data or by generally accepted bonding considerations. Although there appears to be one η^1-cyclpentadienyl ring, only one of the other three rings is probably *pentahapto* in nature, the other two having structures analogous to the severely tilted or *polyhapto* type found (62) for two of the rings in $(C_5H_5)_3MoNO$. This structure would give zirconium an effective atomic number of 18, which is much more reasonable than the value of 20 obtained with three η^5-cyclopentadienyl rings

In addition to the proposal that titanocene has a hydride-bridged structure (139) and that $[(C_5H_5)_2TiH]_2$ occurs as a reaction intermediate (149), $(C_5H_5)_2ZrH_2$ is the only other cyclopentadienyl-group IVa hydride to be characterized (150). The relatively simple infrared spectrum exhibited by this compound shows the presence of η^5-cyclopentadienyl rings (150). Although a polymeric structure has been proposed for $(C_5H_5)_2ZrH_2$ with two hydride ligands bridging neighboring zirconium atoms, the degree of association could not be determined (150). Solid-state infrared bands at 1520 and 1300 cm^{-1} have been assigned to the bridging $\nu(ZrH)$ modes; these bands shift to 1100 and 960 cm^{-1}, respectively, for the corresponding deuteride (151). Assignments in the same regions have been made for the $\nu(ZrH)$ and $\nu(ZrD)$ modes of $(C_5H_5)_2Zr(H)Cl$, $(C_5H_5)_2Zr(H)CH_3$, and $[(C_5H_5)_2ZrH]_2O\cdot(C_5H_5)_2ZrH_2$ and the analogous deuterides for which associated structures with bridging hydride ligands have also been proposed (151).

The BH_4^- group can interact with metals to give one of four possible structures. These are illustrated in Table 3.11 together with frequency ranges observed for the infrared active modes arising from each structural type (152). These structures and data must be considered in interpreting the infrared spectra of $(C_5H_5)_2TiBH_4$ (152-155), $(C_5H_5)_2M(BH_4)_2$ (M=Zr (152,155),Hf (155)), the corresponding BD_4^- complexes (154,155), and $(C_5H_5)_2Zr(H)BH_4$ (150,152). The infrared spectra of $(C_5H_5)_2TiBH_4$, $(C_5H_5)_2Zr(BH_4)_2$, and $(C_5H_5)_2Zr(H)BH_4$ are compared with those of $(C_5H_5)_2MCl_2$ (M=Ti,Zr) and $Al(BH_4)_3$ (152). Although it is agreed that the infrared spectra show the presence of η^5-cyclopentadienyl rings in all of the BH_4^- derivatives mentioned above and that the for the zirconium and hafnium complexes the BH_4^- ligand has the bidentate structure illustrated in Table 3.11, there is less general agreement concerning the assignment of the

TABLE 3.11
Possible Structures for the BH_4^- Group in Metal Complexes and the Characteristic Infrared Spectral Features and Frequency Ranges (cm^{-1}) Associated with Each[a]

Structure	Approximate Frequency	Mode[b]	Symmetry	Comments
$M^+BH_4^-$ Ionic	2300-2200	ν(BH)	T_2	Strong, broad
	1150-1050	BH_2 def.	T_2	Strong, broad
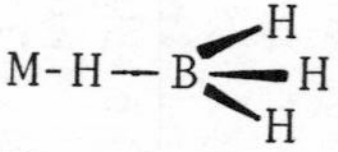 Monodentate	2450-2300	ν(BH)	A_1,E	Strong, probably doublet
	ca. 2000	ν(BH')	A_1	Strong
	1150-1000	BH_3 def.	A_1,E	Strong band, possibly with a weaker one at slightly higher frequency
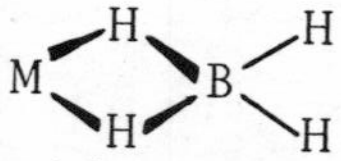 Bidentate	2600-2400	ν(BH)	A_1,B_1	Strong doublet; 80-40 cm^{-1} splitting
	2150-1950	ν(BH')	A_1,B_2	Strong band, possible shoulder or second band
	1500-1300	Bridge st.	A_1	Strong, broad
	1200-1100	BH_2 def.	B_2	Strong
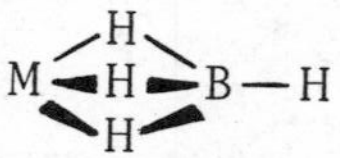 Tridentate	2600-2450	ν(BH)	A_1	Strong singlet
	2200-2100	ν(BH')	A_1,E	Doublet; 80-50 cm^{-1} splitting
	1250-1150	Bridge def.	E	Strong

[a]Reference 152.
[b]ν(BH) = terminal stretching mode; ν(BH') = bridging stretching mode.

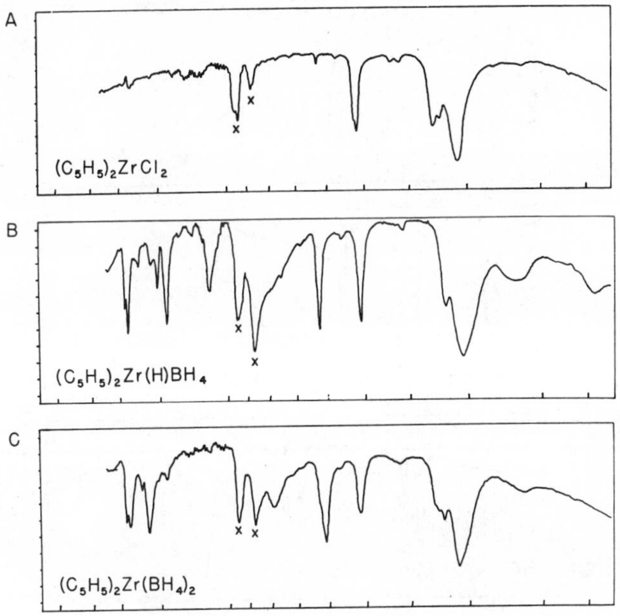

Figure 3.14. Infrared spectra of (A) (η^5-C_5H_5)$_2$$ZrCl_2$ in Nujol, (B) (η^5-C_5H_5)$_2$Zr(H)BH_4 in Nujol, and (C) (η^5-C_5H_5)$_2$Zr$(BH_4)_2$ in Nujol (sharp bands marked x are due to Nujol) (152).

spectra arising from the hydride ligands and the structure of $(C_5H_5)_2TiBH_4$. In the most recent study, it has been concluded that the BH_4^- ligand is bidentate as found for the zirconium and hafnium compounds, and that infrared bands at 1945 and 1320 cm^{-1} arise from the A_1 ν(BH') and A_1 bridge stretching mode, respectively (152). The corresponding modes of $(C_5H_5)_2Zr(BH_4)_2$ were assigned at 2145 and 1295 cm^{-1}, respectively (152). In the same study, an infrared band of $(C_5H_5)_2Zr(H)BH_4$ at 1945 cm^{-1}, which had previously been assigned to the terminal ν(ZrH) mode (150), has been reassigned to the A_1 ν(BH') mode of the bidentate BH_4^- group with the ν(ZrH) mode assigned at 1620 (mull) and 1595 (C_6H_6 solution) cm^{-1} (152). Although $(C_5H_5)_2Zr(H)BH_4$ is monomeric, a polymeric structure has been proposed for $(C_5H_5)_2Zr(H)AlH_4$ (151). While a Zr-H-Zr structure (3.29) has been proposed as most likely, an adduct of the zirconim hydride

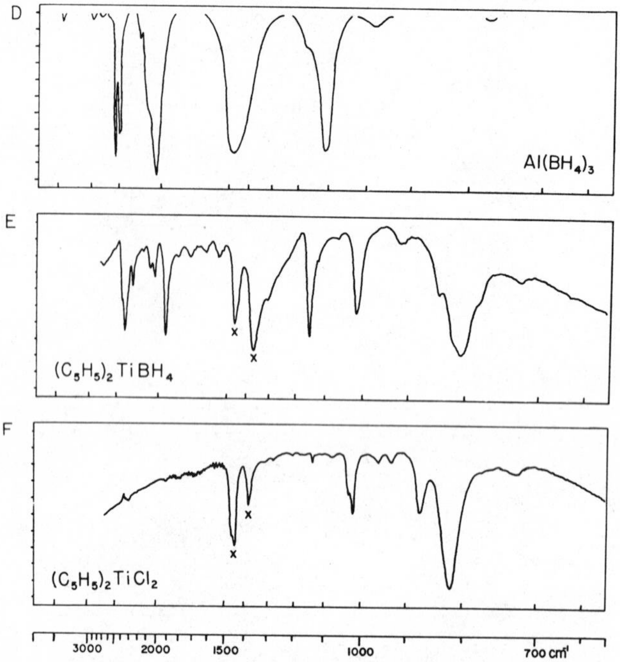

Figure 3.14 (continued). Infrared spectra of (D) $Al(BH_4)_3$ vapor, (E) $(\eta^5\text{-}C_5H_5)_2TiBH_4$ in Nujol, and (F) $(\eta^5\text{-}C_5H_5)_2TiCl_2$ in Nujol (sharp bands marked x are due to Nujol) (152).

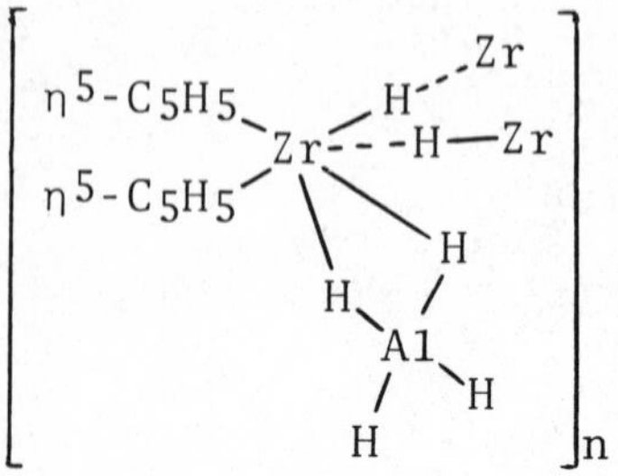

(3.29)

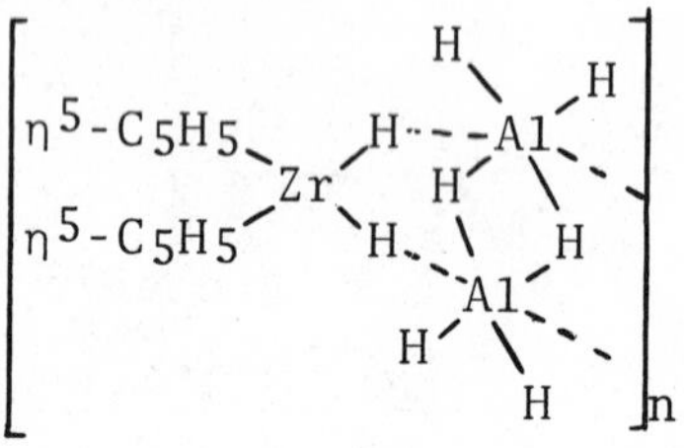

(3.30)

and the aluminum hydride (3.30) could not be ruled out (151). A strong intensity infrared band at 1425 cm^{-1} that shifted to 1055 cm^{-1} on deuteration has been attributed to a Zr-H-Zr mode, while bands at 1790 and

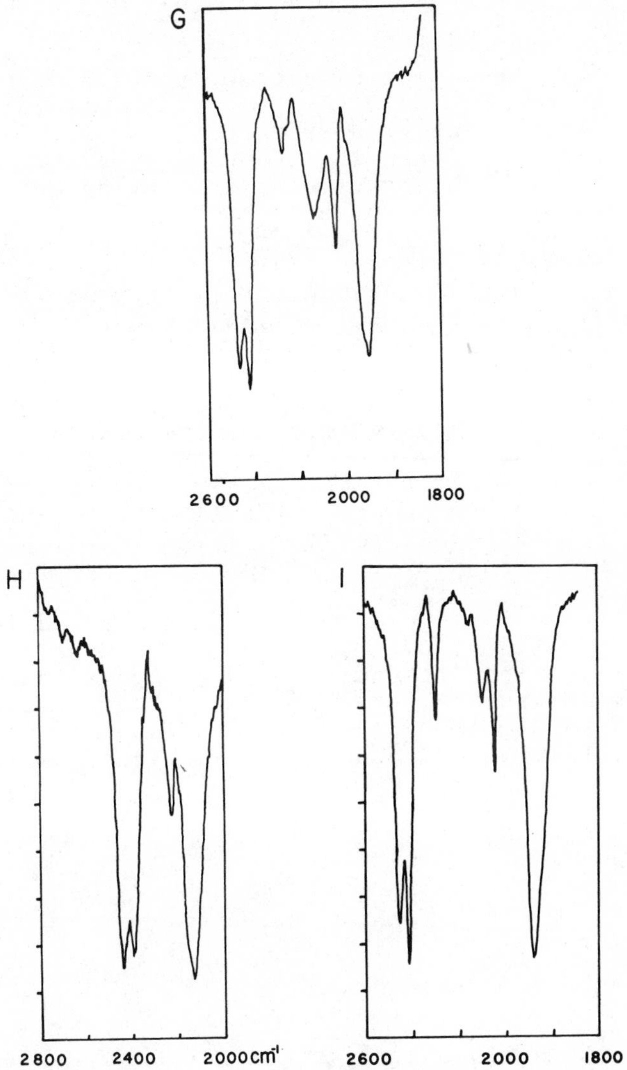

Figure 3.14 (continued). Infrared spectra in the B-H stretching region of (G) (η^5-C_5H_5)$_2$Zr(H)BH_4 in benzene, (H) (η^5-C_5H_5)$_2$Zr(BH_4)$_2$ in benzene, and (I) (η^5-C_5H_5)$_2$TiBH_4 in benzene (152).

1700 cm^{-1}, and 1310 and 1260 cm^{-1} were assigned to AlH_4^- and AlD_4^- modes, respectively (151).

Green, crystalline $(C_5H_5)_2TiCl$ is dimeric in benzene (156) and has an infrared spectrum (KBr pellet) consistent with the presence of η^5-cyclopentadienyl rings (138). Treatment of titanocene with dry HCl gas produces a red compound, also formulated as $(C_5H_5)_2TiCl$ but which has an infrared spectrum (KBr pellet) of greater complexity than that found for the green form; this having been interpreted as indicating the presence of both η^1- and η^5-cyclopentadienyl rings (138). Mass spectral, chemical, and further infrared evidence have now been reported which suggest that the red form, like the green form, is dimeric but more accurately formulated as $[(C_5H_5)(C_5H_4)TiCl]_2$ (139). Therefore, the same structural possibilities discussed earlier for titanocene (3.26 to 3.28) also exist for this compound. The complexity of the infrared spectrum for the red form can be explained in terms of the presence of both a η^5-cyclopentadienyl ring and a C_5H_4 ring. Reaction of zirconocene and $(C_5H_5)_2ZrCl_2$ gives two products: a red soluble benzene adduct, $[(C_5H_5)_2ZrCl]_2 \cdot C_6H_6$, and an olive-green, insoluble polymer, $[(C_5H_5)_2ZrCl]_n$ (157). The greater complexity of the infrared spectrum of the green form has led to the proposal that while only η^5-cyclopentadineyl rings are present in the red form, the green form contains both η^5- and η^1-cyclopentadienyl rings (157). The presence of C_5H_4 rings in the green zirconium derivative, however, can not be ruled out. The infrared spectrum has also been reported for a white crystalline compound now established as the oxygen-bridged species $[(C_5H_5)_2ZrCl]_2O$ (3.31) (158,159)

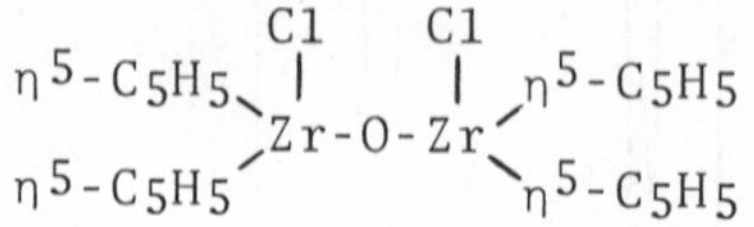

(3.31)

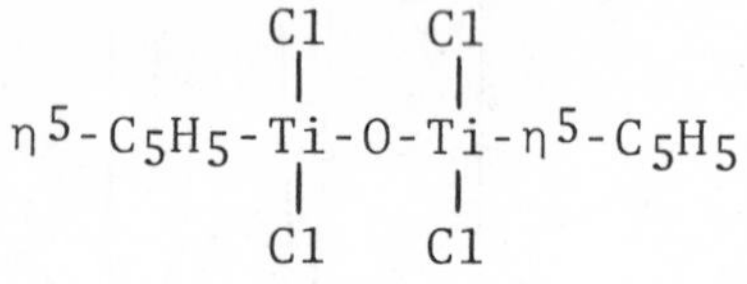

(3.32)

with the strong intensity infrared band originally reported at 775 to 750 cm^{-1} (158) assigned to a $\nu(ZrO)$ mode (159). The analogous titanium derivative has strong intensity infrared bands at 795 and 720 cm^{-1} (160,161) while the oxygen-bridged compound $(C_5H_5TiCl_2)_2O$ (3.32) shows a strong intensity band at 760 cm^{-1} (160). The infrared spectra of solid $C_5H_5TiX_2$ (X=Cl,Br,

I) show the presence of η^5-cyclopentadienyl rings (162). Although a halogen-bridged dimeric structure seems likely, more evidence is needed before a structure can be proposed. In tetrahydrofuran, the $C_5H_5TiX_2$ derivatives form a complex (3.33) as indicated by the shift

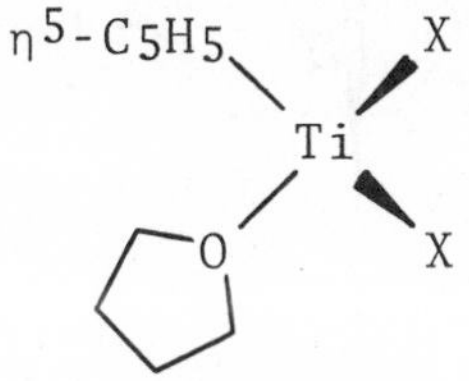

(3.33)

of the $\nu_a(CO)$ mode from 1070 cm^{-1} for free tetrahydrofuran to 1035 cm^{-1} in the presence of the $C_5H_5TiX_2$ derivatives (162).

The vibrational data reported for $(C_5H_5)_2MX_2$ (M=Ti,Zr,Hf;X=F,Cl,Br,I) (163-167) are consistent with the presence of η^5-cyclopentadienyl rings and approximate tetrahedral skeletons as have been found for $(C_5H_5)_2TiCl_2$ (168,169) and $(C_5H_5)_2ZrX_2$ (X=F,I) (170). The infrared spectra of $(C_5H_5)_2TiCl_2$ and $(C_5D_5)_2TiCl_2$ are illustrated in Figure 3.15 (171). Although the vibrational spectra above 650 cm^{-1} are relatively simple, the metal-skeletal region below 650 cm^{-1} is much more complex, as illustrated by the infrared spectra of $(C_5H_5)_2TiX_2$ (X=F,Cl,Br,I) in Figure 3.16 (165). This results from the angular orientation of the two rings, which gives rise to seven infrared active and nine Raman active metal-ring skeletal modes (cf. Table 3.4) in addition to the metal-halide skeletal modes. A weak intensity infrared band at 595 cm^{-1} was originally assigned to a titanium-ring tilting mode for $(C_5H_5)_2TiCl_2$ (48). Since this band appears at the same frequency in all of the group IVa derivatives $(C_5H_5)_2MX_2$, it has been reassigned to a cyclopentadienyl ring mode (162). The $\nu(MF)$ modes of $(C_5H_5)_2MF_2$ (M=Ti,Zr,Hf) are easily assigned to bands from 564 to 526 cm^{-1}, while the $\nu(TiCl)$ modes of $(C_5H_5)_2TiCl_2$ have been assigned at 400 cm^{-1} (164-166). Conflicting assignments have been given for several of the other skeletal modes (164-166). The difficulty in making specific assignments of these modes is due to their appearance in a relatively narrow frequency range and the existence of significant coupling. In general, however, the metal-ring stretching

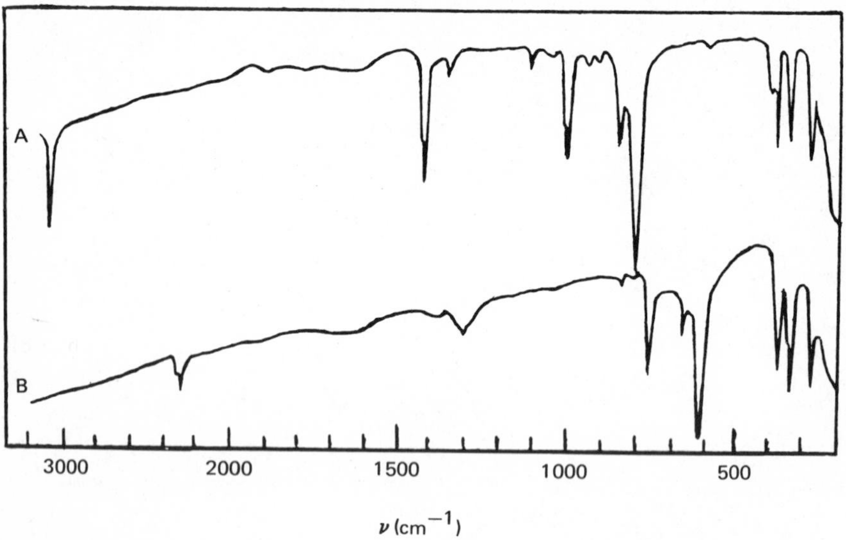

Figure 3.15. Infrared spectra of (A) (η^5-C_5H_5)$_2$$TiCl_2$ and (B) (η^5-C_5D_5)$_2$$TiCl_2$ (171).

and tilting modes of the dihalides have been assigned from approximately 430 to 245 cm^{-1} (165,166). Although the red form of (C_5H_5)$_2$$TiCl_2$ and the white form of (C_5H_5)$_2$$ZrCl_2$ are monomeric with η^5-cyclopentadienyl rings, green, polymeric forms of both the titanium (138) and zirconium (157) derivatives have also been reported. These studies have included solid-state infrared data which indicate that their structures are more complex than those found for the monomeric, *pentahapto* analogs. Infrared data listed for η^5-$C_5H_5MX_3$ (167) are similar to those reported for the corresponding (η^5-C_5H_5)$_2$$MX_2$ (M=Ti,X=Cl;M=Zr,X=Cl,Br,I) complexes. A strong intensity infrared band at 770 cm^{-1}, previously reported as characteristic of monocyclopentadienyltitanium trihalides (172), has been observed to be absent from the infrared spectrum of purified $C_5H_5TiCl_3$ (167). This band is, however, very similar to an infrared band at 760 cm^{-1} which has been assigned to a ν(TiO) mode of ($C_5H_5TiCl_2$)$_2$O, and therefore most likely is due to an oxide impurity (160). Infrared and Raman assignments have been reported for $C_5H_5TiX_3$ (X=Cl (165,166),Br (166)) in the skeletal mode

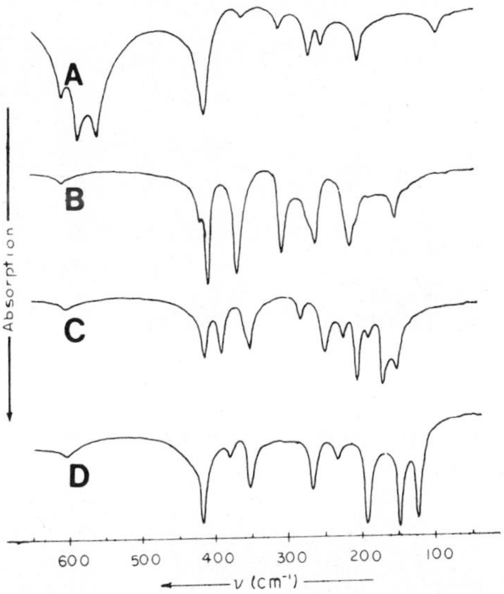

Figure 3.16. Infrared spectra (650 to 50 cm^{-1}) of Nujol mulls of (A) (η^5-C_5H_5)$_2$$TiF_2$, (B) ($\eta^5$-$C_5H_5$)$_2$$TiCl_2$, (C) ($\eta^5$-$C_5H_5$)$_2$$TiBr_2$, and (D) ($\eta^5$-$C_5H_5$)$_2$$TiI_2$ (171).

mode region below 650 cm^{-1}. The infrared spectra of several compounds of the type (C_5H_5)$_2$Ti(alkyl)Cl show bands at 465 to 459 cm^{-1}, 417 to 410 cm^{-1}, and 395 to 375 cm^{-1} that have been assigned to ν(Ti-alkyl), titanium-ring, and ν(TiCl) modes, respectively (173). The skeletal modes have also been assigned for (C_5H_5)$_2$M-(CH_3)$_2$ (M=Ti,Zr,Hf) and C_5H_5Ti(CH_3)$_3$ with the ν(M-CH_3) modes assigned from 532 to 462 cm^{-1} (166).

While infrared, uv-visible, and mass spectral data indicate that (C_5H_5)$_2$$MX_2$ (M=Ti,Zr,Hf,X=NCS;M=Ti,X=NCSe) and (C_5H_5)$_2$TiNCO have nitrogen-bonded *iso*-structures, oxygen- and not nitrogen-bonded structures have been proposed for (C_5H_5)$_2$M(OCN)$_2$ (M=Ti,Zr,Hf) (174).

The frequency separation of the ν(CO) modes in cyclopentadienyl-titanium carboxylates has been offered as a means of distinguishing between the different possible structures (175). This is illustrated in Figure 3.17 (175) by the infrared spectra of (C_5H_5)$_2$TiO_2CCH_3, C_5H_5Ti(O_2CCH_3)$_2$, and (C_5H_5)$_2$Ti(O_2CCH_3)$_2$ for which

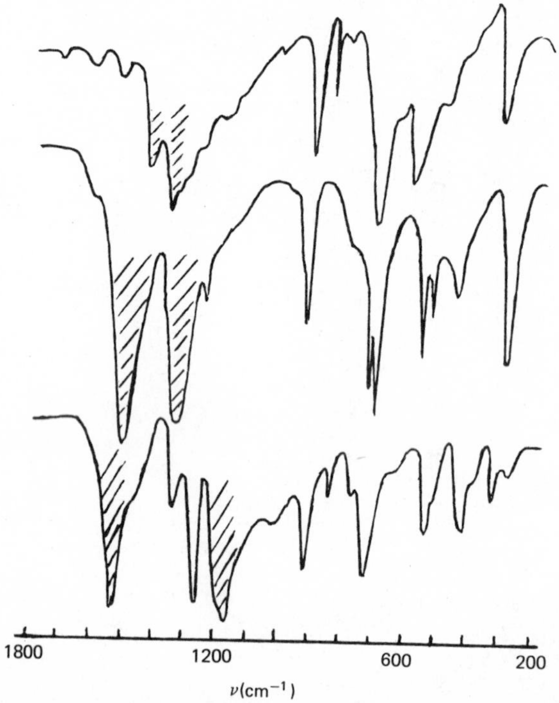

Figure 3.17. Infrared spectra of (η^5-$C_5H_5)_2TiO_2CCH_3$ *(top),* η^5-$C_5H_5Ti(O_2CCH_3)_2$ *(middle), and* (η^5-$C_5H_5)_2Ti(O_2CCH_3)_2$ *(bottom) (175). The shaded bands arise from the C-O stretching modes.*

structures with chelating (3.34), dimeric bridging (3.35), and terminal (3.36) acetate groups, respectively, have been proposed (175,176). Therefore, the frequency separation of the ν(CO) modes in $(C_5H_5)_2TiO_2CR$ (R=H,CH_3, $CH_3(CH_2)_8$,$CH_3(CH_2)_{16}$,C_6H_5) and $[(C_5H_5)_2TiO_2C]_2R$ (R= CH_2CH_2,CH=CH (*cis* and *trans*)) (115 to 60 cm^{-1}) has been proposed to be characteristic of a chelated carboxylate structure (176). For the $C_5H_5Ti(O_2CR)_2$ (R=CH_3,C_2H_5, *n*-C_3H_7,C_6H_5) derivatives that are presumed to have a dimeric, carboxylate bridged structure, a frequency separation of approximately 175 to 166 cm^{-1} has been reported (156). The largest frequency separation of over 300 cm^{-1} has been observed for $(C_5H_5)_2Ti(O_2CCH_3)_2$ (156) and other analogous dicyclopentadienyltitanium(IV)

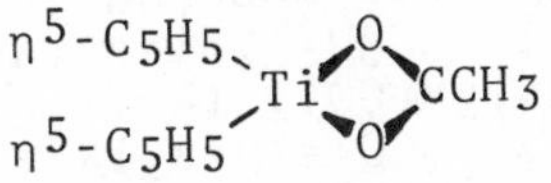

(3.34)

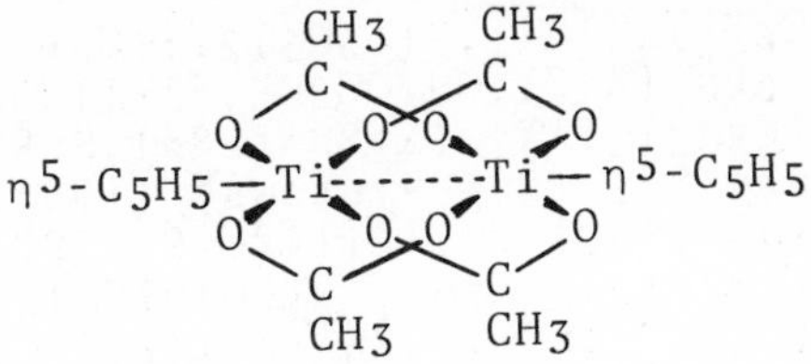

(3.35)

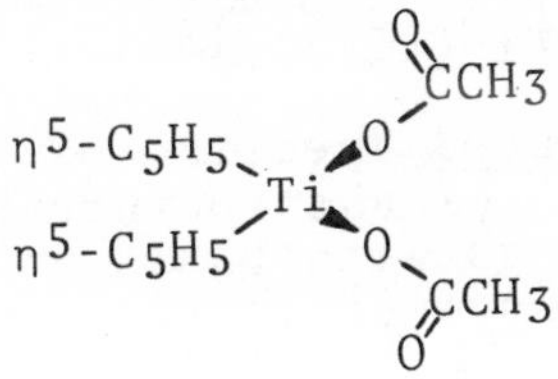

(3.36)

dicarboxylates (177,178) for which a terminal carboxylate structure has been proposed. The infrared spectra have been reported for the thiocarboxylate derivatives $(C_5H_5)_2TiSOCR$ (R=H,CH_3,C_2H_5,$CH_3(CH_2)_{16}$,C_6H_5) (179). A strong to medium intensity infrared band from 1435 to 1390 cm^{-1} has been assigned to the SCO group; its low frequency has been interpreted as indicating that both the oxygen and sulfur atoms are coordinated to the titanium atom. A monomeric structure with a chelated thiocarboxylate ligand has been proposed (179), as has been found for the analogous carboxylate compounds (3.34) (176). The infrared spectra of $(C_5H_5)_2TiAcac$ (180) and $[(C_5H_5)_2TiAcac]X$ (X=ClO_4,CF_3SO_3) (181) have been shown to be consistent with a chelated acetylacetonate ligand. Infrared assignments have also been reported for the monomeric dithiocarbamates $C_5H_5Ti(S_2CNR_2)X_2$ (R=CH_3,C_2H_5,*n*-C_4H_9, -$(CH_2)_5$-;X=Cl,Br), which contain five coordinate titanium (182).

Spectroscopic, magnetic, and conductivity data for $C_5H_5TiCl_2L$ (L=Bipyr,*o*-phenylenediamine) and $C_5H_5TiCl_2L_2$ (L=Py,α-picolylamine) have been interpreted in terms of either six-coordinate, dimeric structures or five-coordinate, monomeric structures (183). Although the infrared spectra were recorded in the 600 to 250 cm^{-1} region, few assignments were attempted, because of their complexity

Infrared data have also been recorded for the following η^5-cyclopentadienyl derivatives: $(C_5H_5)_2TiR$ (R=C_6H_5;*o*-,*m*-,*p*-$CH_3C_6H_4$;2,6-$(CH_3)_2C_6H_3$;2,4,6-$(CH_3)_3C_6H_2$;

C_6F_5) (184), $(C_5H_5)_2Ti(C_6F_5)_2$ (185,186), $(C_5H_5)_2Ti(X)$-C_6F_5 (X=Cl,OH,OC_2H_5,F) (185), $(C_5H_5)_2Ti(C{\equiv}CC_6H_6)_2$ (187), $(C_5H_5)_2TiX_2$ (X=SR (188),SeR (189)), $(C_5H_5)_2TiX_2$ (X=S_5 and Se_5 ring systems) (190), $(C_5H_5)_2M(CO)_2$ (M=Ti, ν(CO)=1975,1897 cm^{-1} (142,191,192);M=Zr, ν(CO)=1978,1888 cm^{-1} (192-194);M=Hf, ν(CO)=1961,1866 cm^{-1} (192,194)), $(C_5H_5)_2ZrCl_{2-n}R_n$ (n=1,2;R=C_4H_7,C_3H_5) (195), $(C_5H_5)_4Zr_2$-$(C_3H_5)Cl$ (195), $(C_5H_5)_2Zr(ON(CH_3)NO)X$ (X=CH_3,Cl (196), $(C_5H_5)_2ZrX_2$ (X=SC_6H_5,SeC_6H_5 plus other sulfur ligands) (197), and $(C_5H_5)_2Zr(C_6F_5)_2$ (186).

Group Va. The infrared spectrum of $(C_5H_5)_2V$ has been interpreted in terms of a structure with parallel η^5-cyclopentadienyl rings with bands at 422 and 379 cm^{-1} assigned to the antisymmetric vanadium-ring tilting and stretching modes, respectively (48,120,198). More complex structures have been proposed for the analogous niobium and tantalum complexes. Solid-state and solution spectroscopic evidence show that niobocene is best formulated as the dimer $[(C_5H_5)(C_5H_4)NbH]_2$ with η^5-cyclopentadienyl and μ-(η^1,η^5-C_5H_4) rings (3.37) (199).

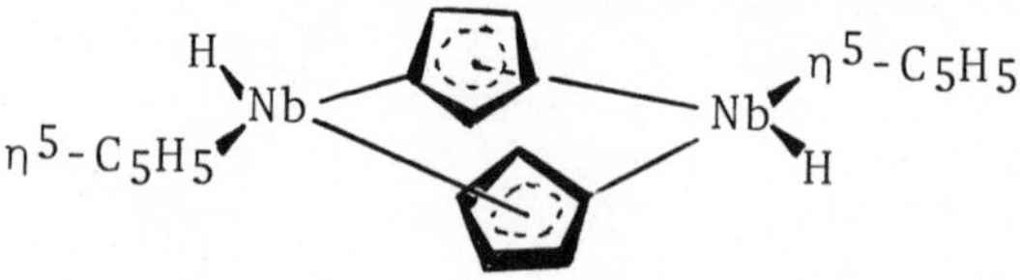

(3.37)

The Nb-H bond gives rise to an infrared band at 1680 cm^{-1}. This structure has been confirmed in a single crystal X-ray study of the niobocene dimer (75,200). A similar structure has been proposed for the corresponding tantalum derivative, $[(C_5H_5)(C_5H_4)TaH]_2$, since it is isomorphous with and has the same spectral properties as the niobocene dimer (199,200).

Infrared and pmr data for $(C_5H_5)_3V$ have been interpreted to indicate that there are one η^1-and two η^5-cyclopentadienyl rings (201). The infrared spectrum of $(C_5H_5)_3Nb$ has been reported to be similar to that of $(C_5H_5)_3Ti$ (202), which at the time was thought to contain one η^1- and two η^5-cyclopentadienyl rings (143). Since the structure of $(C_5H_5)_3Ti$ has now been shown to consist of two η^5-cyclopentadienyl rings and one that is bonded to the titanium atom through only two adjacent carbon atoms (61), more data are needed before the structures of $(C_5H_5)_3V$ and $(C_5H_5)_3Nb$ can be fully characterized.

Although attempts to prepare $(C_5H_5)_4V$ have reportedly led to reduction of the metal (143), the analogous niobium compound has been characterized (203,204) and its infrared spectrum interpreted as consistent with the presence of two η^5- and two η^1-cyclopentadienyl rings (143).

Infrared bands at 1710 and 1735 cm^{-1} have been assigned to the ν(MH) modes of $(C_5H_5)_2NbH_3$ (199) and $(C_5H_5)_2TaH_3$ (205), respectively. The ν(MH) modes have also been assigned in the infrared spectra of $(C_5H_5)_2Nb$-(H)CO (1695 cm^{-1}), $(C_5H_5)_2Nb(H)P(CH_3)_3$ (1635 cm^{-1}), $(C_5H_5)_2Nb(H)P(C_2H_5)_3$ (1650 cm^{-1}), $(C_5H_5)_2Nb(H)C_2H_4$ (1735 cm^{-1}), $(C_5H_5)_2Ta(H)CO$ (1750 cm^{-1}), and $(C_5H_5)_2Ta$-(H)$P(C_2H_5)_3$ (1705 cm^{-1}) for which structures with nonparallel *pentahapto* rings have been proposed (199).

The infrared spectra of $(C_5H_5)_2VCl_2$ (198,206), $C_5H_5VCl_3$ (198), $(C_5H_5)_2NbCl_2$ (202), and $(C_5H_5)_2TaI_2$ (207) have all been characterized in terms of a tetrahedral skeleton and η^5-cyclopentadienyl rings. The vanadium-ring tilting and stretching modes of $(C_5H_5)_2VCl_2$ and $C_5H_5VCl_3$ have been assigned to infrared bands at 400 and 354 cm^{-1}, and 447 and 320 cm^{-1}, respectively (198). The infrared spectrum of $(C_5H_5)_2NbI_3$ has been interpreted as showing the *pentahapto* rings to be parallel or nearly parallel (207). The basis for this was the observation of a strong intensity infrared band at 800 cm^{-1} rather than a medium intensity doublet that is reported to be more characteristic of compounds with angular η^5-cyclopentadienyl rings. Also, the 1000 cm^{-1} band has been reported to be relatively weak in intensity. This interpretation of the infrared data for $(C_5H_5)_2NbI_3$, lowever, is far from conclusive and more information is required before a structure can be proposed for this compound. The infrared spectra have also been reported for $(C_5H_5)_2MBr_3$ (M=Nb,Ta) (206).

Infrared, uv-visible, and mass spectral data for $(C_5H_5)_2VX_2$ (X=NCS,NCO) indicate that both derivatives have the nitrogen-bonded *iso*-structure (174).

Infrared data have been reported for $(C_5H_5)_2V$-$(O_2CCH_3)_2$ (ν_a(CO)=1610 cm^{-1}) (208) and $(C_5H_5)_2V$-$(O_2CCF_3)_2$ (ν_a(CO)=1720 cm^{-1}) (209), which are both dimeric compounds. A single crystal X-ray study of dimeric $(C_5H_5)_2V(O_2CCF_3)_2$ has shown that the vanadium atoms are bridged by both trifluoroacetate ligands (209) to give a structure analogous to that proposed for $[C_5H_5Ti(O_2CCH_3)_2]_2$ (3.35) (176). A similar structure has been presumed for $(C_5H_5)_2V(O_2CCH_3)_2$. The similarity of the vibrational spectrum of $(C_5H_5)_2VAcac^+$ to that of the analogous titanium compound indicates that both have the same structure (181). The synthesis, and infrared, electronic, and esr spectra have been discussed for the xanthate complexes $[(C_5H_5)_2V(S_2COR)]BF_4$ (R=CH_3,C_2H_5, i-C_3H_7,C_4H_9,C_6H_{11}) for which a monomeric structure with chelating xanthate ligands (3.38) has been proposed

$$\left[\begin{array}{c} \eta^5\text{-}C_5H_5 \\ \eta^5\text{-}C_5H_5 \end{array} \diagdown V \diagup \begin{array}{c} S \\ S \end{array} \diagdown C\text{-}OR \right]^+ \quad BF_4^-$$

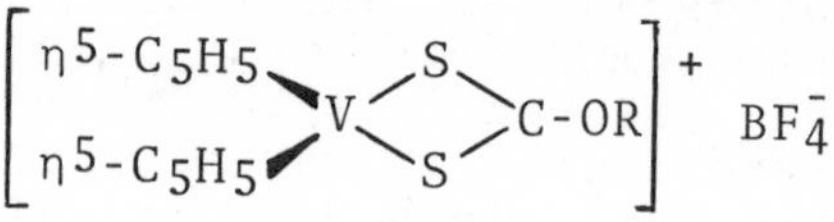

(3.38)

(210), as well as for the chelate complexes formed between the $(\eta^5\text{-}C_5H_5)_2V(IV)$ unit and the maleonitriledithiolate, N-cyanodithiocarbimate, O-ethyl thioacetothioacetate, dithiobiuret, and dimethyldithioarsinate anions (211).

A complete analysis has been made of the infrared (CS_2 and CCl_4 solutions, mull, and KBr pellet) and Raman (cyclohexane solution and solid state) spectra of η^5-$C_5H_5V(CO)_4$ (212). The assignments given for the ν(CO) and skeletal modes of this complex are summarized in Table 3.12. On protonation of $C_5H_5M(CO)_4$ (M=V,Nb) and their phosphine derivatives, the infrared active ν(CO) mode frequencies increase by 150 to 100 cm^{-1} (213). Infrared assignments in the ν(CO) mode region have also been made for other carbonyl complexes of vanadium (214-216) and niobium (199,217), as well as for $(C_5H_5)_2$-Ta(H)CO (1885 cm^{-1}).

Additional infrared data have been listed for the following η^5-cyclopentadienyl derivatives: $(C_5H_5)_3VO$ (218), $(C_5H_5)_2VR$ (R=η^5-allyl,2-methylallyl,2-butenyl, C_6F_5) (201), $(C_5H_5)_2NbR$ (R=allyl,C_6H_5) (202), $(C_5H_5)_2$-Nb(η^1-allyl)CS_2 (219), and $[(C_5H_5)_2Nb(\eta^1$-allyl)(C(=S)-SR)]I$ (R=CH_3,C_2H_5,n-C_3H_7,n-C_4H_9) (219).

Group VIa. The infrared spectra of both $(C_5H_5)_2Cr$ (48,120,220) and $[(C_5H_5)_2Cr]X$ (X=Br,I) (48,220) have been interpreted in terms of a η^5-cyclopentadienyl structure. For the former compound, the asymmetric chromium-ring tilting and stretching modes have been assigned at 464 and 335 cm^{-1}, respectively (48,198), while for the latter compounds these modes have been assigned at 500 and 461 cm^{-1}, respectively (48). The infrared spectrum of $(C_5H_5)_4Mo$ has been reported to be consistent with the presence of one η^1- and three η^5-cyclopentadienyl rings (221).

TABLE 3.12
Vibrational Assignments (cm^{-1}) for the Noncyclopentadienyl Modes of η^5-$C_5H_5V(CO)_4$[a]

Mode[b]	Sym.	Infrared (CS_2 soln.)	Raman (solid)
ν(CO)	A_1	2030 s	2015 m
	B_1		1940 s
	E	1920 vs,br	1927 s, 1897 s
δ(VCO)	A_1		623 w
	E		610 vw
	E	595 vs	
	B_1,B_2		584 m
ν(V-CO)	E	495 vs	
	B_1		456 w
	A_1	425 m[c]	432 s
Tilt	E		349 m
ν(VR)	A_1	325 w[c]	329 vs
$RV(CO)_4$ twist	A_2		240 w
δ(CVC)	A_1,E		122 m,br
	B_1		84 w
δ($RV(CO)_4$)	E		99 m

[a]Reference 212.
[b]R=η^5-C_5H_5.
[c]Sample in KBr pellet.

Infrared data have been reported for $(\eta^5\text{-}C_5H_5)_2MoH_2$ (ν(MoH)=1845 cm^{-1}) (48,205,222,223), $[(\eta^5\text{-}C_5H_5)_2WH_3]X$ (ν(WH)=1901 cm^{-1}) (48,205,222-224), $[(\eta^5\text{-}C_5H_5)_2MoH_3]X$ and $[(\eta^5\text{-}C_5H_5)_2WH_3]X$ (X=Cl (48,222),PF_6 (225)) (the ν(MoH) and ν(WH) modes have been assigned at 1915 and 1943 cm^{-1}, respectively, for the PF_6 salts) (225). Although a structure with an approximately tetrahedral skeleton and angular *pentahapto* rings has been found in a single crystal X-ray study of $(C_5H_5)_2MoH_2$ (226) and is therefore presumed for $(C_5H_5)_2WH_2$, the structures of $[(C_5H_5)_2MH_3]X$ (M=Mo,W;X=Cl,PF_6) have still to be fully characterized. Structures in which the two η^5-cyclopentadienyl rings are either parallel or tilted with respect to each other have both been proposed as possible (48,226).

The infrared active cyclopentadienyl bands have been listed for $C_5H_5CrCl_3$ (227). A single crystal X-ray study has been reported for $[LiC_5H_5CrCl_3\cdot 2THF]$-.Dioxane (228). This study also included an assignment of the infrared spectrum taken at both 300°K and 100°K.

Infrared data have been listed for $(C_5H_5)_2MoX_2$ (X=Cl (ν(MoCl)=293,262 cm^{-1}),Br,I), $(C_5H_5)_2WX_2$ (X=Cl (ν(WCl)= 283,266 cm^{-1}),Br,I), $[(C_5H_5)_2MoCl_2]X$ ($X=Cl_{1-2}$ (ν(MoCl)= 334,311 cm^{-1}),PF_6), $[(C_5H_5)_2MoBr_2]X$ ($X=HBr_2$ (ν(MoBr)= 212,204 cm^{-1}),PF_6), $[(C_5H_5)_2WCl_2]X$ ($X=HCl_2$ (ν(WCl)=311, 296 cm^{-1}),PF_6 (ν(WCl) 307,300 cm^{-1})), and $[(C_5H_5)_2WBr_2]X$ ($X=HBr_2,PF_6$) (229), as well as for several cyclopentadienyl-molybdenum and -tungsten complexes of thiols and pseudohalides (230). The cyclopentadienyl rings for these halides, pseudohalides, and thios are *pentahapto* in nature.

Several studies have appeared on η^5-cyclopentadienyl-group VIa complexes that also contain either CO ligands or CO and other ligands such as hydrides, halides, nitrosyls, amines, phosphines, etc. The infrared assignments for the noncyclopentadienyl modes of these complexes are summarized in Table 3.13. A detailed discussion has also appeared of the infrared active modes of $[C_5H_5M(CO)_3]_2$ (M=Mo,W), $C_5H_5M(CO)_3X$ (M= Mo,X=Cl,I;M=W,X=I), $C_5H_5Mo(CO)_2NO$, $[C_5D_5Mo(CO)_3]_2$, and $C_5D_5Mo(CO)_2NO$ (240). A single crystal X-ray study has shown solid $[C_5H_5Mo(CO)_3]_2$ to have a centrosymmetric *trans* structure (3.39) (241). A detailed solid-state

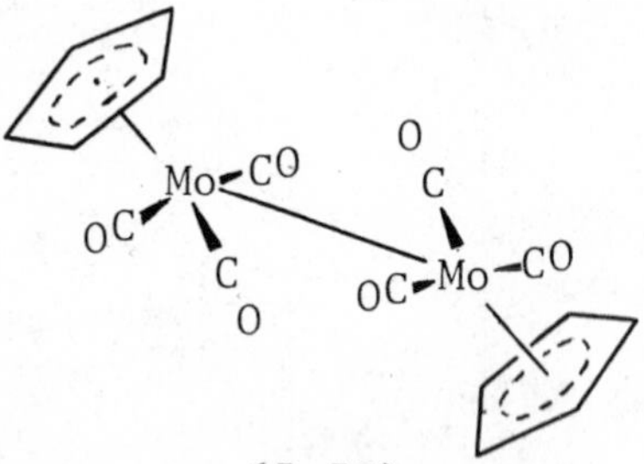

(3.39)

and solution infrared study of $[C_5H_5M(CO)_3]_2$ (M=Cr,Mo, W) in the ν(CO) mode region has been reported to be consistent with this structure (232). The infrared spectrum (KBr pellet) of $[C_5H_5Cr(CO)_3]_2$ (2050 to 1750 cm^{-1}) is illustrated in Figure 3.18 (232). In another infrared study in the ν(CO) mode region and nmr study in various solvents, it has been concluded that while the $[C_5H_5Mo(CO)_3]_2$ molecules have the *trans* structure in solvents with low dielectric constants, as the dielectric constant of the solvent increases, an increasing proportion of these molecules rearrange to the *gauche* structure (242). Fairly complex changes in the ν(CO) mode region have also been observed on the addi-

TABLE 3.13
Assignment of the Noncyclopentadienyl Modes (cm^{-1}) for Carbonyl and Substituted Carbonyl Derivatives of η^5-Cyclopentadienyl-Group VIa Complexes

Compound[a]	ν(CO)	ν(MH), ν(NO), ν(M-halide), or ν(SO)	δ(MCO)	ν(M-CO)	M-R Tilt	ν(MR)	Refs.
$Na[RCr(CO)_3]$	1880 1695						48
$[RCr(CO)_4]BF_4$	2105 2028					431	48
$[RMo(CO)_4]PF_6$	2128 2041						48
$[RW(CO)_4]PF_6$	2128 2028						48
$R_2W(CO)_2$	1955 1872						231
$[RCr(CO)_3]_2$	1951 1923 1910					371	48,220,232
$[RMo(CO)_3]_2$	1963 1919 1911		588 549 503	479 412	370 351	336	48,232-234
$[RW(CO)_3]_2$	1958 1893		572 553 495 457	480 438	348	327	48,234,235
$[RMo(CO)_3]_2H^+$	2074 2054 1988						235
$[RW(CO)_3]_2H^+$	2028 1961						235

TABLE 3.13 (continued)

Compound[a]	ν(CO)	ν(MH), ν(NO), ν(M-halide), or ν(SO)	δ(MCO)	ν(M-CO)	M-R Tilt	ν(MR)	Refs.
$[(RW(CO)_3)_2I]BF_4$	2058 1984 1938						232
$RCr(CO)_3H$	2020 1938	1828					48,220
$RCr(CO)_3D$	2020 1934	1299					48
$RMo(CO)_3H$	2028 1934	1809					48,225
$RMo(CO)_3D$	2024 1942	1285					48,225
$RW(CO)_3H$	2024 1927	1828				347	48,225,236
$RW(CO)_3D$	2016 1916	1306					48,225
$RMo(CO)_3Cl$	2062 1992 1965	280	563 524 470	429 411		359	48,234,237
$RMo(CO)_3I$	2042 1974 1960	177	567 526 474	437 420	380	360	234
$RW(CO)_3Cl$	2045 1961 1927					352	48,237
$RW(CO)_3Br$	2028 1961 1927					351	48

$RW(CO)_3I$	2041 1953 1905		564 519 468	433 402	348	48,234
$RCr(CO)_2NO$	2018 1947	1704				238
$RMo(CO)_2NO$	2016 1936	1674	608 445[b] 584 434[b] 552	433 402	324	234,238
$RW(CO)_2NO$	2006 1923	1665				238
$RCr(CO)(NO)$- $P(C_6H_5)_3$	1921	1654				238
$RMo(CO)(NO)$- $P(C_6H_5)_3$	1906	1617				238
$RW(CO)(NO)$- $P(C_6H_5)_3$	1899	1605				238
$[RMo(CO)_3NH_3]$- $B(C_6H_5)_4$	2070 2008 1980					237
$[RW(CO)_3NH_3]$- $B(C_6H_5)_4$	2058 1976 1953					237
$RMo(CO)_3$- (SO_2CH_3)	2058 1975	1190 1051				239
$RMo(CO)_3$- $(SO_2C_2H_5)$	2056 1985	1171 1038				239
$RMo(CO)_3$- $(SO_2CH_2C_6H_5)$	2056 1996 1973	1191 1041				239

[a] $R=\eta^5-C_5H_5$.
[b] $\delta(MoNO)$ mode.

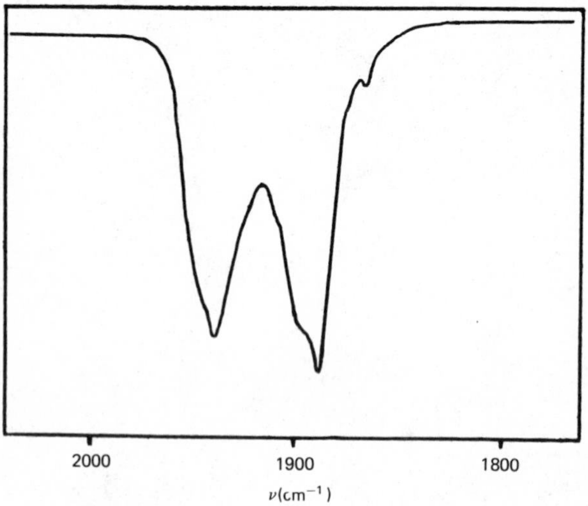

Figure 3.18. Infrared spectrum of [η^5-$C_5H_5Cr(CO)_3]_2$ *(KBr pellet) in the* ν*(CO) mode region (232).*

tion of trialkylaluminum complexes to heptane solutions of [$C_5H_5Mo(CO)_3]_2$ (243). To explain these data, the possibility of a structural change has been suggested in which two of the CO ligands bridge the molybdenum atoms and a trialkylaluminum molecule coordinates to the oxygen atom of each of these bridging CO ligands.

The conformation of [$(C_5H_5M(CO)_3)_2X]^n$ (M=Cr,Mo,W) varies with the nature of the bridging ligand, X. For cationic complexes where X=H or I, three infrared active ν(CO) modes have often been assigned (cf. Table 3.13); this is consistent with the presence of the centrosymmetric structure. Dipole moment and infrared data in the ν(CO) mode region for the neutral complexes where X=Hg, however, have been explained in terms of a skew configuration and a linear M-Hg-M skeleton. For neutral complexes where X=YR_2 (Y=Ge,Sn,Pb;R=CH_3,C_6H_5, etc.), the expected bent skeleton is found (232). The complex nature of the infrared spectrum of [$C_5H_5Cr(CO)_3]_2Hg$ (KI pellet) in the ν(CO) mode region (2050 to 1750 cm^{-1}) is illustrated in Figure 3.19 (232). In a detailed analysis of the solid-state and methylene

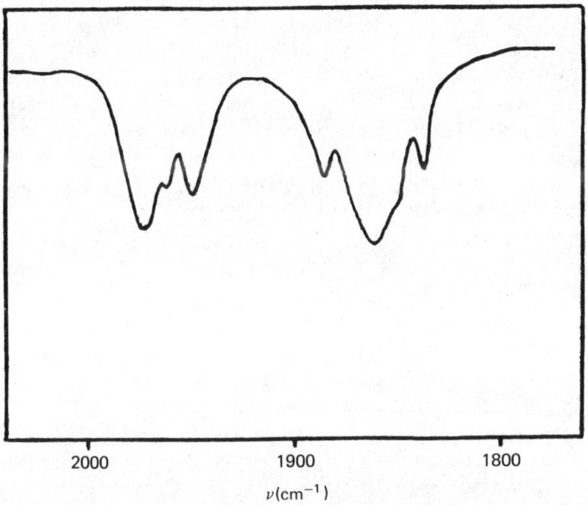

Figure 3.19. Infrared spectrum of $[\eta^5\text{-}C_5H_5Cr(CO)_3]_2Hg$ *(KI pellet) in the* $\nu(CO)$ *mode region (232).*

chloride solution infrared and Raman spectra of $C_5H_5W(CO)_3CH_3$ (244), the $\nu(W\text{-}CH_3)$, tungsten-ring tilt, and $\nu(W\text{-}C_5H_5)$ modes were assigned to medium intensity Raman bands at 430, 365, and 333 cm^{-1}, respectively. The infrared and Raman spectra (550 to 40 cm^{-1}) have been recorded and (MM') mode assignments made for $C_5H_5M(CO)_3[Ge(CH_3)_3]$ (M=Cr,ν(CrGe)=191 cm^{-1};M=Mo,ν(MoGe)=180 cm^{-1}; M=W,ν(WGe)=171 cm^{-1}) and $C_5H_5M(CO)_3[Sn(CH_3)_3]$ (M=Cr, ν(CrSn)=183 cm^{-1};M=Mo,ν(MoSn)=168 cm^{-1};M=W,ν(WSn)=167 cm^{-1} (245).

In an attempt to observe the Raman spectrum of $[C_5H_5Cr(NO)_2]_2$ using a He-Ne source, the sample exploded violently and immediately on entering the sample compartment (246).

Group VIIa. An ionic structure has been proposed for $(C_5H_5)_2Mn$ (48,54,56). The only other pure cyclopentadienyl derivative of a group VIIa element for which vibrational data have been presented is $[(C_5H_5)_2Tc]_2$ (48).

Infrared data have been listed and some assignments made for the hydrides $(C_5H_5)_2MH$ (M=Tc,ν(TcH)=1984 cm^{-1}

(247);M=Re,ν(ReH)=2016 cm^{-1} (36,222,224,247)) and $[(C_5H_5)_2MH_2]PF_6$ (M=Tc,ν(TcH)=1984 cm^{-1};M=Re,ν(ReH)= 2058 cm^{-1}) (247) and the halides $[(C_5H_5)_2ReX_2]X$ (X=Cl, ν(ReCl)=314,295 cm^{-1},Br,I) (229).

Complete infrared and Raman assignments have been reported for η^5-$C_5H_5Mn(CO)_3$ (233,234,240,248,249), η^5-$C_5D_5Mn(CO)_3$ (233,240), and η^5-$C_5H_5Re(CO)_3$ (250). More limited infrared data and assignments have been reported for η^5-$C_5H_5Tc(CO)_3$ (48) and η^5-$C_5H_5Mn(CO)_2$ (which is an unstable product of the pyrolysis of η^5-$C_5H_5Mn(CO)_3$) (251). Infrared assignments in the ν(CS) mode region have been reported for several η^5-cyclopentadienylmanganese thiocarbonyl and carbon disulfide complexes (253). The absolute integrated infrared intensities of the ν(CO) and ν(CS) modes have been compared for η^5-$C_5H_5Mn(CO)_2CS$ (254). The complete infrared and Raman spectra have also been assigned for η^5-$C_5H_5Mn(CO)(CS)_2$ in the solid state and η^5-C_5H_5Mn-$(CO)_2CS$ in the vapor, solution, and solid phases (255). The noncyclopentadienyl mode assignments of CO derivatives and several derivatives that contain CO and other ligands are summarized in Table 3.14. Protonation of η^5-$C_5H_5Re(CO)_3$ and η^5-$C_5H_5Re(CO)_3P(C_6H_5)_3$ in an acidic medium results in a shift of the ν(CO) modes to higher frequencies (213).

Group VIIIa. The first detailed infrared and Raman study of η^5-cyclopentadienyl complexes was reported for $(C_5H_5)_2Fe$, $(C_5D_5)_2Fe$, and $(C_5H_5)_2Ru$ (70). This study also included the infrared spectrum of $(C_5H_5)_2Ni$. Although several reinvestigations have been made of the vibrational spectra of $(C_5H_5)_2Fe$ (48,71,256-266) and $(C_5H_5)_2Ru$ (265-268), the main features of the original study are still valid. The data for $(C_5H_5)_2Fe$ have included both single crystal (256,258,262,265), matrix isolation (259,264), and solid-state, low-temperature (262-265) vibrational studies. Single crystal and low-temperature vibrational studies have also been reported for $(C_5H_5)_2Ru$ (265,267,268). The infrared spectrum of $(C_5H_5)_2Fe$ in a nitrogen matrix at 20°K is illustrated in Figure 3.20 (259). An interesting feature of this spectrum is the resolution of the ν_a(Fe-C_5H_5) mode at 480 cm^{-1} into two components, the most intense of which arises from the ^{56}Fe isotope (natural abundance, 91.7%), while the much weaker intensity sattelite is due to the ^{54}Fe isotope (natural abundance, 5.82%). This analysis is supported by an infrared study of ^{57}Fe- and ^{54}Fe-substituted $(C_5H_5)_2Fe$ (260), in which the antisymmetric

TABLE 3.14
Assignment of the Noncyclopentadienyl Modes (cm^{-1}) for Carbonyl and Substituted Carbonyl Derivatives of η^5-Cyclopentadienyl-Group VIIa Compounds

Compound[a]	ν(CO)	δ(MCO)	ν(M-CO)	M-R Tilt	ν(MR)	Refs.
$RMn(CO)_2$	2026 1938					251
$RMn(CO)_3$	2041 1965	661 631 533	495 488	372	347	233,234, 248,249
$RTc(CO)_3$	2040 1949				437	48
$RRe(CO)_3$	2026 1931	617 507	515 500	349	327	250
$RMn(CO)_2L$[b]		663- 529	540- 481			234
$RMn(CO)_2CS$	2025 1979 1279[c]	645[d] 607[d] 516[d] 473[d]	458 437 364[e]	382	339	254
$RMn(CO)(CS)_2$	1997 1989 1980 1298[c] 1220[c]	625[d] 590[d] 570[d] 499[d] 478[d]	429 404[e] 359[e]	350	321	254
$RRe(CO)_2N_2$	1970 1915 2141[f]					255

[a]$R=\eta^5$-C_5H_5.
[b]$L=(C_6H_5)_3P, (C_6H_5)_3As, (n$-$C_4H_9)_3P$.
[c]Assigned to a ν(CS) mode.
[d]In the region expected for both the δ(MCO) and δ(MCS) modes.
[e]Assigned to a ν(M-CS) mode.
[f]Assigned to the $\nu(N_2)$ mode.

iron-ring tilting and stretching modes, which were assigned at 497.0 and 482.0 cm^{-1}, respectively, for the ^{54}Fe-substituted molecule were shifted by 6.9 and 8.0 cm^{-1}, respectively, to lower frequencies for the ^{57}Fe-substituted molecule. The infrared and Raman assignments for $(C_5H_5)_2M$ (M=Fe,Ru,Os) and $(C_5D_5)_2Fe$, and the infrared assignments for $(C_5H_5)_2M$ (M=Co,Ni) are summarized in Table 3.15. Infrared data have been listed

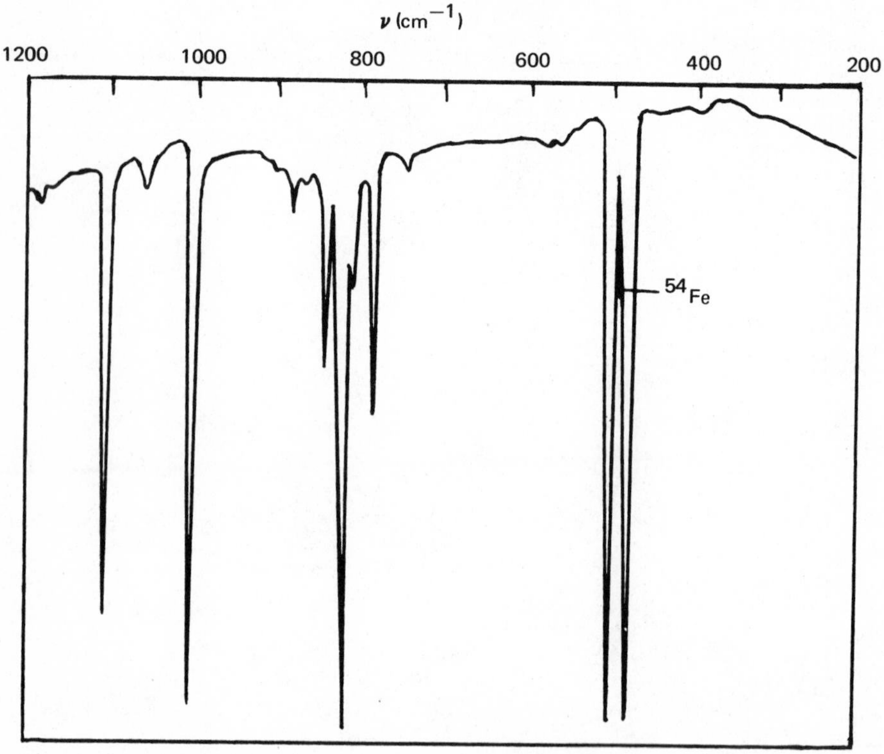

Figure 3.20. Infrared spectrum (1200-200 cm^{-1}) of (η^5-C_5H_5)$_2$Fe isolated in a nitrogen matrix at 20°K (259).

for $(C_5H_5)_2Pt$ for which a dimeric structure with both η^1- and η^5-cyclopentadienyl rings has been proposed (269).

A normal coordinate analysis has been reported of the cyclopentadienyl rings of $(C_5H_5)_2Fe$ (70). An approximate calculation of the $(C_5H_5)_2Fe$ iron-ring stretching and bending modes has been carried out assuming the rings to be single atoms (48,70). More recently, normal coordinate analyses have been reported for the complete $(C_5H_5)_2Fe$ (270-272) and $(C_5H_5)_2Ru$ (275) molecules. In one of these complete studies, a value of 1.4 dynes/cm was calculated for the K(FeC) force constant (271).

TABLE 3.15
Vibrational Assignments (cm^{-1}) for η^5-Cyclopentadienyl-Group VIIIa Derivatives

Mode Desc. and Number[b]		R_2Fe h_{10}[c]	R_2Fe d_{10}[d]	R_2Ru[c]	R_2Os[f]	R_2Co[f]	R_2Ni[g]	R_2Fe^+ h_{10}[h]	R_2Fe^+ d_{10}[i]	R_2Co^{+}[j]	R_2Ir^{+}[k]	RNi^{+}[l]
		Compound[a]										
ν(CH)	1	3110	2335	3112						3122		
	8				3061	3041	3075	3108-3100	2370-2360	3094	3077	3120
	12	3086	2318	3084						3097		
	17	3077	2354	3076	3061	3041	3075	3108-3100	2370-2360	3094	3077	
	23	3100	2318	3104						3102		
	29	3085										
ν(CC)	3	1102	1056	1100						1109		
	10	1110	1044	1095	1098	1101	1110	1116-1110	1060-1057	1113	1106	
	15	1410	1300	1410						1421		
	20	1410	1316	1410	1400	1412	1430	1421-1412	1304	1419	1409	1410
	26	1356	1268 1255	1360						1361		
	32	1351	1360									
δ(CH)	5	1255	998 996	1255								
	7		987	1250								
	13	998	751	1000						1010		
	18	1005	770	1005	998	995	1000	1017-1001	782-780	1010	1009	1002
	24	1191	1060 1047	1205						1188		

TABLE 3.15 (continued)

Mode Desc.	and Number[b]	Compound[a]										
		R_2Fe h_{10}[c]	R_2Fe d_{10}[d]	R_2Ru[c]	R_2Os[f]	R_2Co[f]	R_2Ni[g]	R_2Fe^+ h_{10}[h]	R_2Fe^+ d_{10}[i]	R_2Co^+[j]	R_2Ir^+[k]	RNi^+[l]
	30	1189	1063 1057	1188								
π(CH)	2	814	619	818						859		
	9	820	637	808	823	778	773	805-779	628	860	818	
	14	844	631	842						843		
	19	855	677	834	831	828	839	860-841	782-780	895	862	872
	25	1058	861	1050								
δ(CC)	27	897	708	900						947		
	33	885	743	868								
	28	597	552	603						585		
	34	569	552	600								
Tilt	16	389	361	400	415					386		
	21	492	480	450[e]	428	464	355	501-490	467	495		
ν(MR)	4	309	295	333	356					315		
	11	478	451	380[e]	353	355	355	423-405	385	455		
δ(RMR)	22	179	175-161	170	160					172		
Torsion	6			134								

[a]$R=\eta^5\text{-}C_5H_5$.

[b]The description of each mode is summarized in Table 3.3 and taken from Ref. 70.

[c]Reference 265.
[d]References 70 and 279.
[e]In Ref. 33, ν_{21} and ν_{11} were assigned at 380 and 450 cm^{-1}, respectively.
[f]References 36, 48, 198, and 266.
[g]References 48 and 70.
[h]Data and assignments from Ref. 275 for the following salts: $FeCl_4^-$, $FeBr_4^-$, HgI_4^{2-}, $Cr(NH_3)_2(SCN)_4^-$, MoO_4^-, $C_6H_2O_7N_3^-$, BF_4^-, I_5^-, and $GaCl_4^-$.
[i]Data and assignments from Ref. 279 for the following salts: $FeCl_4^-$, HgI_3^-, and $Cr(NH_3)_2(SCN)_4^-$.
[j]Data and assignments from Ref. 71 for the Cl^- and Br^- salts; data for the $CoCl_4^-$ salt from Ref. 280.
[k]Reference 48.
[l]Data for the BF_4^- salt from Ref. 281.

Data have been reported for cationic η^5-cyclopentadienyl complexes of iron (48,198,270-279), cobalt (71, 280), nickel (281), and iridium (48). The vibrational assignments made for several of these derivatives are included in Table 3.15. Several conflicting reports have appeared concerning the structure of the salts of $(C_5H_5)_2Fe^+$, particularly the $FeCl_4^-$ salt (274-276,278, 279). A dimeric structure with bridging chloride ligands (3.40) was originally proposed for $(C_5H_5)_2Fe_2Cl_4$

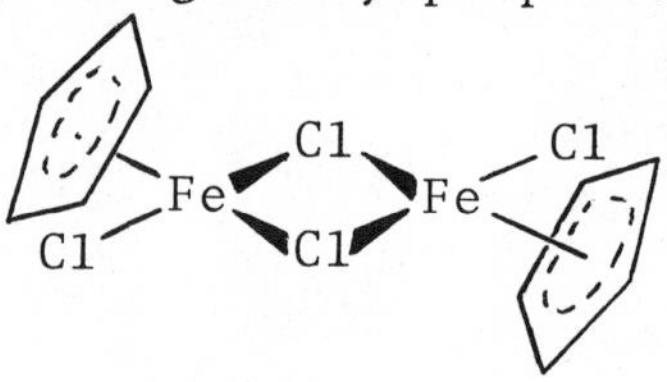

(3.40)

(274). At the same time, however, another group (275, 279) formulated the compound as $[(C_5H_5)_2Fe]FeCl_4$ with parallel η^5-cyclopentadienyl rings. Although other groups have agreed with the formulation $[(C_5H_5)_2Fe]$-$FeCl_4$, they have concluded that the η^5-cyclopentadienyl rings are not parallel but angular and that there is some degree of interaction between the $(C_5H_5)_2Fe^+$ and $FeCl_4^-$ ions (276,278). Oxidation of $(C_5H_5)_2Ru$ with iodine or bromine gives $[(C_5H_5)_2RuI]I_3$ and $[(C_5H_5)_2RuBr]$-Br_3, respectively (282). A single crystal X-ray study has shown that the η^5-cyclopentadienyl rings in $[(C_5H_5)_2RuI]I_3$ are eclipsed with each ring tilted back by 16° to accomodate the iodine ligand (282). The similarity of the infrared spectrum of $[(C_5H_5)_2RuBr]Br_3$ to that of the iodide indicates that both compounds have the same structure (282). The shift of the $\nu_a(Ru\text{-}C_5H_5)$ mode from 450 cm^{-1} in $(C_5H_5)_2Ru$ to 419 cm^{-1} in $[(C_5H_5)_2RuI]I_3$ indicates a weakening of the ruthenium-ring bond in the latter complex. It was further suggested that infrared bands at 225 and 185 cm^{-1} may be due to the $\nu(RuBr)$ and $\nu(RuI)$ modes, respectively.

Infrared data have been listed for $[(C_5H_5)_3Ni_2]$-BF_4 (3.41) (283). The triple-decker structure proposed for this compound has been confirmed in a single crystal X-ray study (284).

The tetramer $(C_5H_5CoH)_4$ has been proposed to have a structure in which the cobalt atoms of the η^5-C_5H_5Co units are connected by the four μ_3-hydrido bridges (285). In addition to bands characteristic of η^5-cyclopentadienyl rings, the infrared spectrum also

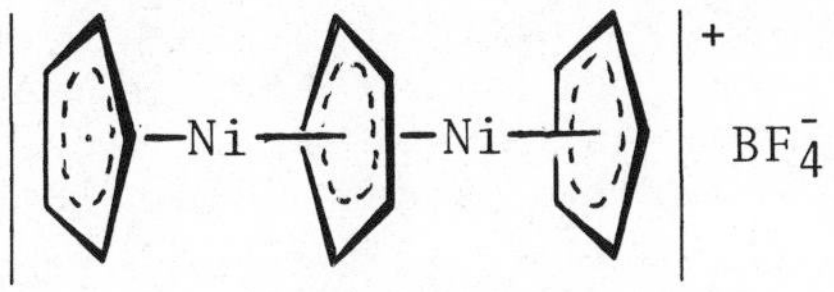

(3.41)

exhibits a band at 950 cm^{-1} that has been assigned to bridging ν(CoH) modes. A shoulder at 890 cm^{-1} and an additional band at 1052 cm^{-1} might also arise from the ν(CoH) modes. A slightly distorted tetrahedron of nickel atoms with η^5-cyclopentadienyl rings at each corner and μ_3-hydrido bridges has also been found for $(C_5H_5Ni)_4H_3$ (286). Although infrared bands were easily assigned to the cyclopentadienyl modes, no bands characteristic of the ν(NiH) modes could be observed.

Several studies have appeared concerning the structures of $[C_5H_5M(CO)_2]_2$ (M=Fe,Ru,Os). These compounds can be pictured in terms of either one of two CO bridged structures (i.e., the *trans* (3.42) or *cis* (3.43) forms) or three structures with nonbridging CO

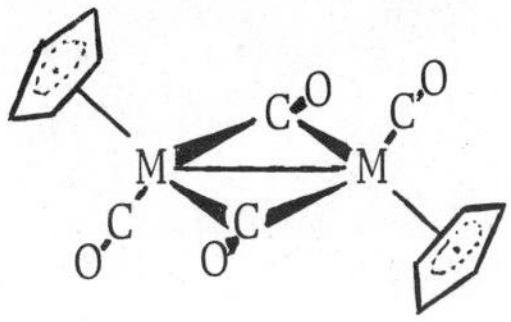

(3.42)

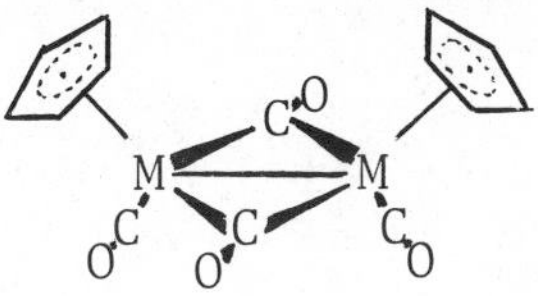

(3.43)

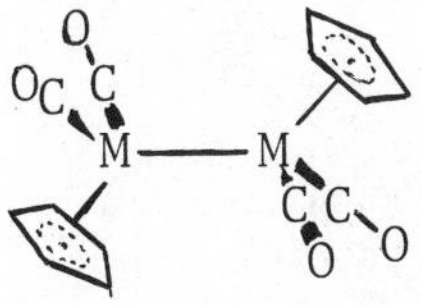

(3.44)

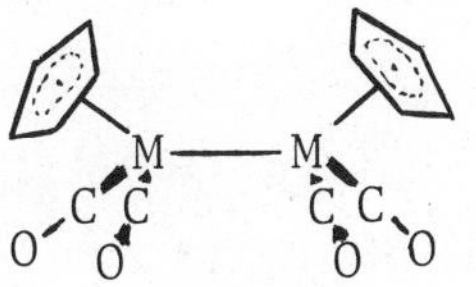

(3.45)

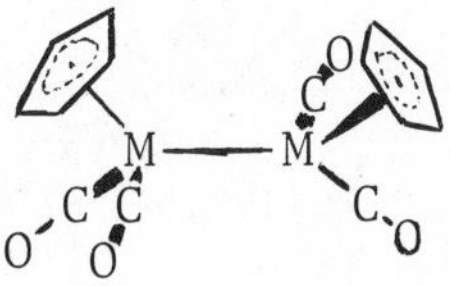

(3.46)

ligands (i.e., the *trans* (3.44), *cis* (3.45), or staggered (3.46) forms). A single crystal X-ray study has shown $[C_5H_5Fe(CO)_2]_2$ to have the CO-bridged *trans* structure (3.42) in the solid state (287,288). Crystallization of $[C_5H_5Fe(CO)_2]_2$ from polar solvents at low-temperatures, however, produced the CO-bridged *cis* structure (3.43) in the solid state (289,290). A single crystal X-ray study has shown solid $[C_5H_5Ru(CO)_2]_2$ to have the CO-bridged *trans* structure (291). Solid-state (KBr pellet) infrared assignments have been made for the ν(CO) modes of $[C_5H_5Fe(CO)_2]_2$ (terminal: 1958, 1936 cm^{-1}; bridging: 1773, 1761 cm^{-1}) (292). A nonbridged structure has been proposed for solid $[C_5H_5Os(CO)_2]_2$ as indicated by the presence of ν(CO) bands in the terminal (1972 and 1927 cm^{-1}) but not in the bridging region (292).

The solution structures of $[C_5H_5M(CO)_2]_2$ (M=Fe,Ru, Os) have proved more difficult to characterize. Detailed solution infrared studies of $[C_5H_5M(CO)_2]_2$ (M=Fe,Ru) in the ν(CO) fundamental mode (292,293) and overtone (293) regions originally led to the proposal that both compounds are present as equilibrium mixtures of the *cis* bridged (3.43) and nonbridged (3.45) isomers (292,293). The two-structure proposal was soon questioned on the basis of an analysis of the solvent dependence of the infrared spectra of $[C_5H_5Fe(CO)_2]_2$ (294) and $[C_5H_5Ru(CO)_2]_2$ (295) in the ν(CO) mode region. On the basis of this analysis, a three-structure (3.42 ⇌ 3.45 ⇌ 3.43) and four-structure (3.42 ⇌ 3.44 ⇌ 3.45 ⇌ 3.43) model were proposed for the iron and ruthenium derivatives, respectively. A more recent study has been reported of the 100-MHz nmr and the infrared spectra (2100 to 1700 cm^{-1}) of $[C_5H_5M(CO)_2]_2$ (M=Fe,Ru) as a function of changes in the nonpolar solvent and temperature (296). It was concluded that a four-structure equilibrium in which the *trans* and *cis* CO-bridged structures (3.42 and 3.43) interconvert via the *trans* nonbridged (3.44) and a polar nonbridged (probably 3.46) structures is consistent with these data for the ruthenium and, presumably, the iron derivative. The presence of both bridged and nonbridged $[C_5H_5Ru(CO)_2]_2$ isomers in solution is also supported by low-frequency Raman data. Therefore, in the solid state, where only the CO-bridged isomer is found, one ν(RuRu) mode was observed at 217 cm^{-1} (297). In solution, however, two polarized bands were observed at 221 and 180 cm^{-1}, and assigned to the ν(RuRu) mode of the bridged and nonbridged isomers, respectively (297). The ν(FeFe) mode could not be detected for $[C_5H_5Fe(CO)_2]_2$ (297). Although $[C_5H_5Os(CO)_2]_2$ is found with only a nonbridged structure in solution (292), it is not known which isomer(s) is(are) present.

The structure of $(C_5H_5FeCO)_4$ consists of a regular tetrahedron of iron atoms with η^5-cyclopentadienyl rings at each corner and μ_3-carbonyl ligands (298). The infrared active ν(CO) modes of the μ_3-carbonyl ligands have been assigned at 1649 cm^{-1} (243). Also, in a $CHCl_3$ solution Raman study, a band at 214 cm^{-1}, the intensity of which has been enhanced by a resonance effect, was assigned to the ν_s(FeFe) mode (299). Solid $(C_5H_5CoCO)_3$ is found with a triangle of cobalt atoms (each of which is bonded to a η^5-cyclopentadienyl ring), a triply bridging CO ligand, and two semibridging CO ligands (300). An infrared band at 1673 cm^{-1} has been assigned to the ν(CO) mode of the μ_3-carbonyl ligand while bands at 1833 and 1775 cm^{-1} were assigned to the ν_s(CO) and ν_a(CO) modes of the semibridging CO ligands (300). The solution-state infrared spectrum of $(C_5H_5CoCO)_3$ differs from the solid-state spectrum in that it exhibits bands at 1959, 1811, and 1753 cm^{-1} which are indicative of a structure with a triangle of cobalt atoms, one terminal CO ligand on cobalt atom one, and two CO ligands which bridge cobalt atoms two and three (300). An analogous structure has been found for solid $(C_5H_5RhCO)_3$ (301) which exhibits infrared bands at 1973, 1827, 1794, and 1744 cm^{-1} in the solid state (300,301). The low frequency of the ν(CO) mode in $(C_5H_5)_3Ni_3(CO)_2$ (1761 cm^{-1}) arises because one μ_3-carbonyl ligand lies above and one below the triangular plane formed by the nickel atoms (243). The solution-state infrared spectra in the ν(CO) mode region have been investigated for $(C_5H_5NiCO)_2$ and $(C_5H_5)_2FeNi(CO)_3$ (302). For $(C_5H_5NiCO)_2$, a structure has been suggested in which both CO ligands bridge the nickel atoms and the *pentahapto* rings are on the same side of the $Ni(CO)_2Ni$ bridge (3.47). The bridging ν(CO) modes

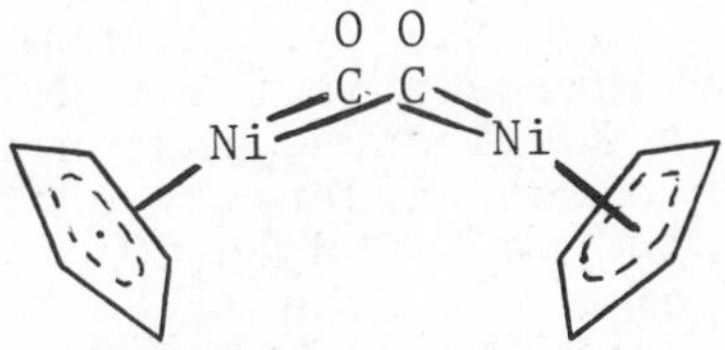

(3.47)

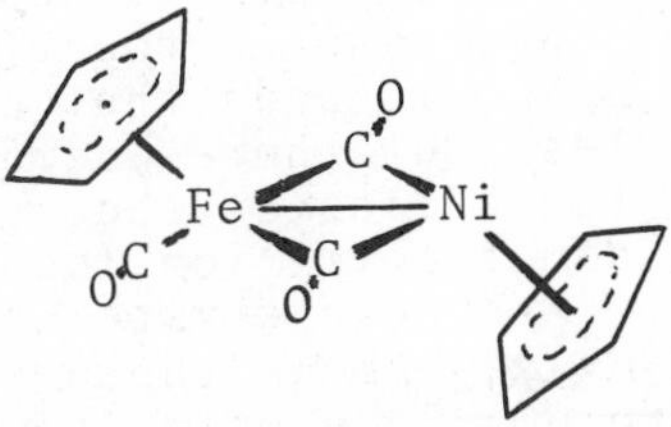

(3.48)

have been assigned at approximately 1885 and 1845 cm^{-1}. A structure with one terminal, iron-bonded and two bridging CO ligands has been proposed for $(C_5H_5)_2FeNi(CO)_3$ (3.48). The terminal ν(CO) mode appears at approximately 2000 cm^{-1} in various solvents while the bridging ν(CO) modes have been assigned at approximately 1845 and 1815 cm^{-1}. The infrared spectrum of $C_5H_5Fe(CO)_2Co(CO)_4$ in the ν(CO) mode region is consistent with a dicarbonyl-bridged species in the solid state while in solution one nonbridged and two bridged isomers are present (303). Similar data for $C_5H_5Ru(CO)_2Co(CO)_4$, however, show the presence of only the nonbridged form (303). Infrared data and some assignments have also been reported for η^5-$C_5H_5M(CO)_4$ (M=Co,ν(CO)=2037,1965 cm^{-1} (304);M=Rh,ν(CO)=2051,1987 cm^{-1} (305);M=Ir,ν(CO)= 2916,2037 cm^{-1}) (48,306), [η^5-$C_5H_5Rh(CO)_2]_2$ (ν(CO)=1961, 1812 cm^{-1}) (307), and (η^5-$C_5H_5PtCO)_2$ (ν(CO)=2005,1968 cm^{-1} (48,308).

The infrared ν(CO) mode region of $[C_5H_5M(CO)_2]_2$ (M=Fe,Ru) and $(C_5H_5NiCO)_2$ have been observed to increase in complexity on dissolving in hydrolytic solvents such as alcohols and/or phenols (309). This has been attributed to the presence of not only the bridged and/or nonbridged isomers found in nonprotonic solvents, such as cyclohexane, but also bridged isomers in which the oxygen atoms of the bridging CO ligands are hydrogen-bonded to the solvent. Similarly, the solution infrared spectra of $[C_5H_5M(CO)_2]_2$ (M=Fe,Ru), $(C_5H_5NiCO)_2$, and $(C_5H_5)_3Ni_3(CO)_2$ in the ν(CO) mode region have been observed to change with the addition of trialkylaluminum derivatives to these solutions (243). This has been attributed to the formation of a bond between the aluminum atom of the trialkylaluminum molecules and the oxygen atom of the bridging CO ligands. Representative data are shown in Figure 3.21 to illustrate the changes observed in the ν(CO) mode region as $(i\text{-}C_4H_9)_3Al$ is added to a solution of $[C_5H_5Fe(CO)_2]_2$ to form the 1:1 and 1:2 $[C_5H_5Fe(CO)_2]_2$-$Al(i\text{-}C_4H_9)_3$ complexes. On complex formation with various CO derivatives, not only was there a decrease in the ν(CO) mode frequency for the bridging CO ligands that bond to the trialkylaluminum molecules, but there was also an increase in the ν(CO) mode frequencies of the remaining CO ligands, indicating that adduct formation lowers the electron density on the metal. A single crystal X-ray study of $[C_5H_5Fe(CO)_2]_2$-$[Al(C_2H_5)_3]_2$ shows the presence of a *cis* CO-bridged $[C_5H_5Fe(CO)_2]_2$ complex that is coordinated to two $(C_2H_5)_3Al$ acceptor molecules through the oxygen atom of each bridging CO ligand (310). In addition to trialkylaluminum compounds, group IIIb halides have also been found to act as Lewis acids in forming oxygen-bonded adducts with metal carbonyl complexes (311).

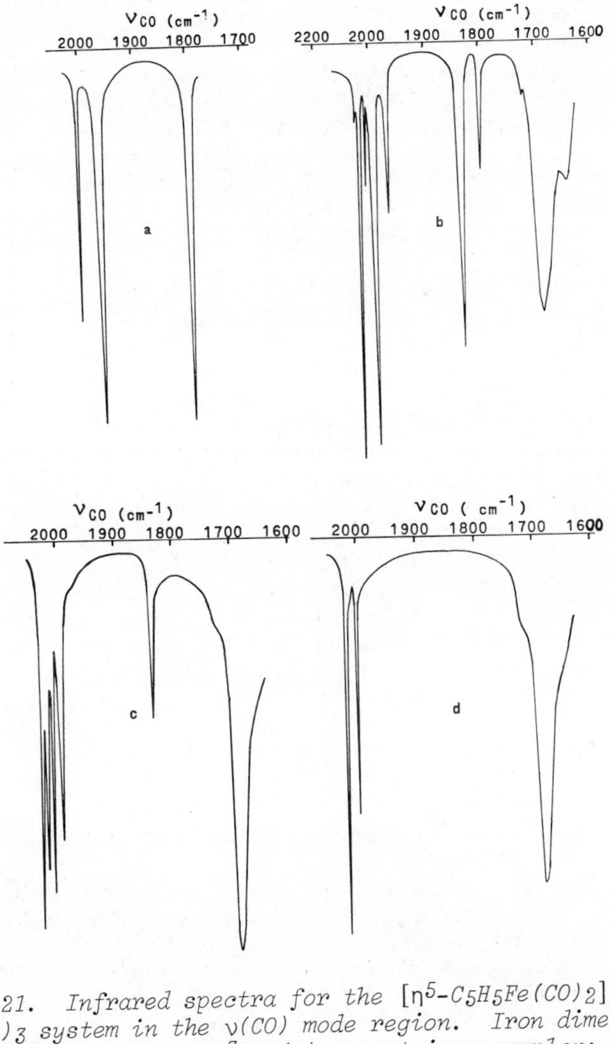

Figure 3.21. Infrared spectra for the $[\eta^5\text{-}C_5H_5Fe(CO)_2]_2$-$Al(i\text{-}C_4H_9)_3$ system in the ν(CO) mode region. Iron dimer concentration = 2.34 x 10^{-3}M: (a) parent iron complex; (b) Lewis acid added, 6.45 x 10^{-3}M, the primary species are the complex $[\eta^5\text{-}C_5H_5Fe(CO)]_2$ and its 1:1 adduct; (c) Lewis acid concentration increased to 7.48 x 10^{-3}M, the primary species are 1:1 and 1:2 adducts; (d) Lewis acid concentration increased further to 1.46 x 10^{-2}M, the primary species is the 1:2 adduct (243).

Therefore, $[C_5H_5Fe(CO)_2]_2BX_3$ (X=F,Br), $(C_5H_5FeCO)_4(MX_3)_n$ (M=B,X=F,n=1,2,4;M=B,X=Cl,Br,n=1,2;M=Al,X=Br,n=1 to 4), and $(C_5H_5)_3Ni_3(CO)_2BF_3$ all exhibit one or more very low-frequency ν(CO) mode bands characteristic of the Lewis acid coordinated CO ligand. The change in the ν_a(CO) mode frequency that is produced on coordination of a group IIIb Lewis acid to the nitrogen atom of the cyanide ligand in $C_5H_5Fe(CO)_2CN$ has been reported to provide a probe to the electron-pair acceptor strength of the acid (312). From these data, the following order of electron-pair acceptor strengths was deduced: $BH_3 < BF_3 < BCl_3 \simeq BBr_3$, $BCl_3 > GaCl_3 > AlCl_3$, and $(CH_3)_3Al > (CH_3)_3B \simeq (CH_3)_3Ga$.

The η^5-cyclopentadienyl ring assignments have been discussed for $C_5H_5Fe(CO)_2X$ (X=Cl,Br,I) (240). In another study, the ν(CO) mode and low-frequency iron-ligand assignments have been discussed for these three halides as well as for derivatives in which X=CH_3, HgCl, HgBr, HgI, $SnCl_3$, $SnBr_3$, and $(C_6H_5)_3Sn$ (313). Vibrational data have also been reported for other iron (234, 292,314-318) and ruthenium (319) derivatives in which other ligands in addition to η^5-cyclopentadienyl rings and CO ligands are bonded to the metal atom.

The reaction of $(C_5H_5)_2M$ (M=Fe,Ru) with HgX_2 (X=Cl,Br) produced derivatives with M-Hg bonds (320). Infrared data for the iron derivatives and infrared and Raman data for the ruthenium derivatives were presented and analyzed. Infrared and other spectroscopic data for the products obtained on reacting 1,1'-bis(diphenylphosphino)ferrocene with HgX_2 (X=Cl,Br,I,SCN,CN) and SnX_4 (X=Cl,Br) indicate that the mercury is bonded to the phosphorus and not to the iron atom. Ruthenocene reacts with SnX_4 (X=Cl,Br) to give compounds with Ru-Sn bonds (321). The infrared spectra have been discussed for these compounds.

Detailed infrared and Raman studies have been reported for η^5-C_5H_5NiNO and its deuterated and ^{15}NO analogs (322) and η^5-$C_5H_5Pt(CH_3)_3$ (323). The noncyclopentadienyl ring mode assignments from these studies are summarized in Table 3.16. The infrared spectra have also been reported for crystalline η^5-C_5H_5NiNO and η^5-C_5D_5NiNO using polarized light in order to establish the possible space groups for these derivatives (324). The complexes η^5-$C_5H_5M(PR_3)X$ (M=Pd,Pt;R=C_2H_5, *i*-C_3H_7,C_6H_5;X=Cl,Br,I) react with tertiary phosphines to give either the *pentahapto* cpmplexes $[\eta^5$-$C_5H_5M(PR_3)_2]^+X^-$ or unstable *monohapto* complexes η^1-C_5H_5M-

TABLE 3.16
Assignments of the Noncyclopentadienyl Modes (cm^{-1}) for C_5H_5NiNO, C_5D_5NiNO, $C_5H_5Ni^{15}NO$, and $C_5H_5Pt(CH_3)_3$[a]

Assignment[b]	RNiNO h_5	RNiNO d_5	$RNi^{15}NO$	$RPt(CH_3)_3$
ν(NO)	1809	1808	1774	
Tilt	290	275[c]	290	316
ν(MR)	322	298	321	263
ν(NiN) or $\nu(Pt\text{-}CH_3)$	649	649	642	585 561
ν(Ni-NO) or $\nu(Pt\text{-}CH_3)$	484	478	470	272 253

[a]Data and assignments taken from Ref. 322 for the nickel complexes and Ref. 323 for $C_5H_5Pt(CH_3)_3$
[b]$R=\eta^5\text{-}C_5H_5$.
[c]Deduced from combination and overtone bands.

$(PR_3)_2X$ (325). These platinum and palladium complexes have been characterized using pmr, infrared, and mass spectral data.

Lanthanides and Actinides. The chemical properties and infrared spectra indicate that $(C_5H_5)_2M$ (M=Eu (57,58), Yb (58)) have an ionic structure and that $(C_5H_5)_3M$ (M= Ho,Tm,Tb,Lu) (58) have partial covalent bonding. The infrared spectra of the cyclohexylisonitrile complexes $(C_5H_5)_2M(CNC_6H_{11})$ are similar to those of the uncomplexed $(C_5H_5)_3M$ derivatives (M=Y,Nb,Tb,Ho,Yb) with the cyclohexylisonitrile ligand acting mainly as a σ-donor and not as a π-acceptor (59,60). Infrared data have been listed for $(C_5H_5)_4Ce$ (326) as well as for the hydride (327), pseudohalide (328), alkoxide (329), carbonyl (330), thiol (331), and amide (327) derivatives of $(C_5H_5)_3Ce(IV)$.

The infrared spectra have been reported for the actinide derivatives $(\eta^5\text{-}C_5H_5)_3M$ (M=U (332),Th (333), Pu (332),Am (333)) and $(\eta^5\text{-}C_5H_5)_4M$ (M=Th (334),Pa (335),Np (336)). Although the infrared spectrum of $(C_5H_5)_4U$ had originally been interpreted as consistent with the presence of both η^1- and η^5-cyclopentadienyl rings (337,338), a single crystal X-ray study has shown that in the solid state all four cyclopentadienyl rings are *pentahapto* in nature (339). Since $(C_5H_5)_3UF$ is dimeric in benzene, it was proposed that it has a fluoride-bridged solid-state structure with an infrared band at 466 cm^{-1} assigned to a bridging ν(UF) mode (340). A structure with bridging fluoride ligands was also proposed for the adducts $(C_5H_5)_3UF \cdot M(C_5H_5)_3$ (M=Yb, U) with the bridging ν(UF) modes assigned at 432 and 423 cm^{-1}, respectively (341). In the vapor state, however, $(C_5H_5)_3UF$ is a monomer (340,342), and magnetic data were interpreted as suggesting less marked association in the solid state than in solution. A recent single crystal X-ray study has shown that solid $(C_5H_5)_3UF$ is also monomeric rather than dimeric (343). The η^5-cyclopentadienyl rings, however, are tipped toward the fluorine atom in the adjacent molecule at a distance appropriate for hydrogen bonding. A single crystal X-ray study of $(C_5H_5)_3UCl$ shows the chloride ligand and the η^5-cyclopentadienyl rings to lie at the apices of a distorted tetrahedron with the uranium atom at the center and an ionic U-Cl interaction (344). Infrared data for $(C_5H_5)_3UCl$ are consistent with this structure (152,338,345) and suggest a similar structure for $(C_5H_5)_3MCl$ (M=Th (152,346),Nb (347)).

Vibrational data for the η^5-cyclopentadienyl derivatives $(C_5H_5)_2MBH_4 \cdot THF$ (M=Sm,Er,Yb) indicate that the mode of BH_4^- coordination is sensitive to the ionic radius of the metal atom (348). Therefore, while the data are consistent with a tridentate BH_4^- ligand for M=Sm, a bidentate BH_4^- ligand is found for M=Yb (cf. Table 3.11). Vibrational data for the *pentahapto* derivatives $(C_5H_5)_2MBH_4$ (M=Sm,Er,Yb) suggest the presence of a polymeric structure with bridging BH_4^- ligands (348). The data also suggest appreciable ionic character in the bonding of the BH_4^- ligand to these trivalent lanthanide atoms (348). Infrared data for $(C_5H_5)_3UBH_4$ have previously been interpreted in terms of a bidentate, chelating BH_4^- ligand (338,346,349). A more recent infrared study of $(C_5H_5)_3MBH_4$ (M=U,Th), however, reports the data to be consistent with the assignments suggested in Table 3.11 for a tridentate BH_4^- ligand (152). This tridentate BH_4^- structure has been proposed for both compounds in the solid state and benzene solution. Vibrational data for $(C_5H_5)_3U$-H_3BR (R=C_2H_5,C_6H_5) likewise indicate that the H_3BR^- ligand is bonded to the uranium atom via a triple hydrogen bridge (350). Vibrational data for $(C_5H_5)_3U$-$NCBX_3$ (X=H,C_6H_5), however, suggest that the uranium-borate linkage is of the type UN≡CB (350). Infrared data have also been reported for the η^5-cyclopentadienyl derivatives $(C_5H_5)_2MR$ (M=Gd,Er,Yb;R=CH_3,C_6H_5) (351), $(C_5H_5)_2M(\eta^3$-allyl) (M=Sm,Er,Ho) (352),

$(C_5H_5)_2GdC{\equiv}CC_6H_5$ (353), $C_5H_5Ho(C{\equiv}CC_6H_5)_2$ (353), $(C_5H_5)_2$-$YbCH_3$ (353), $(C_5H_5)_3ThR$ (R=*i*-C_3H_7,*n*-C_4H_9,$(CH_3)_3CCH_2$, η^1-allyl,2-*trans*-2-butenyl,2-*cis*-2-butenyl) (354), $(C_5H_5)_3UR$ (R=CH_3 (355),*n*-C_4H_9 (355,356),$(CH_3)_3CCH_2$ (356),$C_6H_5CH_2$ (356),$CH_3C_6H_4CH_2$ (356),C_6H_5 (357),C_6F_5 (356),η^1-allyl (356),η^1-2-methylallyl (358)), $(C_5H_5)_3M$-OR (M=U,Th;R=CH_3,*n*-C_4H_9), $(C_5H_5)_3UX$ (X=tetrahydrofuran, cyclohexylisonitrile,$N_2C_{10}H_{14}$) (332), and $(C_5H_5)_3U$-(*p*-C_6H_4)$U(C_5H_5)_3$ (82).

f. Data for Ring-Substituted Cyclopentadienyl Compounds

Organic Substituents. Although no complete vibrational studies have been reported for complexes with ring-substituted η^1-cyclopentadienyl rings, infrared and Raman data have been listed for $(\eta^1\text{-}CH_3C_5H_4)_3In$ (72) while infrared data have been listed for η^1-$CH_3C_5H_4M(CH_3)_3$ (M=Si,Ge,Sn) (127) and $(\eta^1\text{-}CH_3C_5H_4)_nPb(CH_3)_{4-n}$ (n=1,2; R=CH_3,C_2H_5) (130).

The largest number of vibrational studies for ring-substituted η^5-cyclopentadienyl complexes have been for ferrocene derivatives with alkyl (359-366), aryl (360, 366-370), acetyl (360,361,363,368,369,371-373), or carboxylic acid (361,372,374) substituents, related ferrocenyl ketones (368,372,374,375), bridged ferrocenes (375-378), and mono- and diacetylferrocenium salts (379). The infrared active ring breathing and δ(CH) modes of ferrocene at 1110 cm^{-1} (ca. 9 μ) and 1005 cm^{-1} (ca. 10 μ), respectively, have been reported to be very useful in discussing the structures of ring-substituted ferrocenes (360). Therefore, while ferrocene derivatives in which one ring is substituted exhibit infrared bands near 9 μ and 10 μ, ferrocene derivatives in which both rings are substituted lack bands in these regions. This observation has been described as the "9-10 rule." Since substituents themselves may exhibit bands near 9 and/or 10 μ, the absence of bands in these regions may be a more reliable structural guide than their presence (380). The 9-10 rule has also been observed in studies of ring-substituted ruthenocene (372,381), osmocene (381), and mono- and dialkyl titanocene dichloride (382) derivatives and is likely for any ring-substituted η^5-cyclopentadienyl derivative.

The infrared region near 900 cm^{-1} (ca. 11 μ) has been used in distinguishing between 1,2-disubstituted ferrocenes, which show one band in this region, and 1,3-disubstituted ferrocenes, which show two bands in this region. These bands most likely arise from π(CH) ring modes (380). Therefore, a band at 894 cm^{-1}, which is characteristic of a ferrocene ring substituted with one acetyl group, is shifted to approximately 915 cm^{-1} in 1,2-disubstituted acetyl-ferrocene and appears as two bands at approximately 920 and 905 cm^{-1} in the corresponding 1,3-disubstituted analogs (380).

Several detailed vibrational studies have been reported for η^5-methylcyclopentadienyl complexes. The spectra of these complexes have been interpreted by assuming that the methyl group is a single atom and that the methylcyclopentadienyl ring has a symmetry of C_{2v}. In Table 3.17, a comparison is made of the modes expected for η^5-cyclopentadienyl and η^5-methylcyclopentadienyl rings assuming point symmetries of C_{5v} and C_{2v}, respectively, while in Tables 3.18 and 3.19, the assignments are given for the ring and nonring modes, respectively, for several η^5-methylcyclopentadienyl derivatives. Among the common features of these assignments is the shift of the 1100 (ν_3) and 1000 cm^{-1} (ν_{10}) cyclopentadienyl bands to lower frequencies (1048 to 1020 (ν_3) and 977 to 919 cm^{-1} (ν_{10a}), respectively) in the methylcyclopentadienyl derivatives as expected on the basis of the 9-10 rule. A second common feature is the assignment of the $\nu(C-CH_3)$ mode (ν_{9a}) from 1235 to 1225 cm^{-1}.

Another approach used in assigning the spectra of monosubstituted cyclopentadienyl complexes (76) has produced a reversal of the assignments for the ν_3 and ν_{10a} modes from those given in Table 3.18. Since an interaction is possible between the ring breathing and $\nu(CX)$ modes (where X is a substituent on the cyclopentadienyl ring), it is not strictly correct to assign bands to either of these modes alone. Rather, it has been suggested that it is more accurate to describe both modes as X- or mass-sensitive since intermode coupling produces a substituent-mass dependence on the positions of both bands. A similar phenomenon has been observed for monosubstituted benzenes (80). It has been concluded that the 1020 cm^{-1} band of $(CH_3C_5H_4)_2Fe$ does not appear to be mass sensitive since its position is essentially unchanged in the series $(XC_5H_4)_2Fe$ (X=Cl,Br,I) (79). Therefore, the 919 cm^{-1} band of $(CH_3C_5H_4)_2Fe$, which does not shift in frequency as the ring substituent is changed, has been assigned to the X-sensitive mode having mainly ring-breathing character with the 1025 cm^{-1} band assigned to the A_1 $\delta(CH)$ mode (ν_{10a}).

The reaction of $(CH_3C_5H_4)_2Sn$ with anhydrous $AlCl_3$ in benzene produced the adduct $(CH_3C_5H_4)_2Sn \cdot AlCl_3$ (385).

TABLE 3.17
Description of the Vibrational Modes of Cyclopentadienyl and Methylcyclopentadienyl Rings on the Basis of Local Symmetry[a]

C_5H_5 (C_{5v} "Local Symmetry")			$CH_3C_5H_4$ (C_{2v} "Local Symmetry")		
Description	Symmetry and Activity	Number	Description	Symmetry and Activity	Number
ν(CH)	A_1 (IR,R)	1	ν(CH)	A_1 (IR,R)	1
π(CH)	A_1 (IR,R)	2	π(CH)	B_2 (IR,R)	2
Ring-breathing	A_1 (IR,R)	3	Ring-breathing	A_1 (IR,R)	3
δ(CH)	A_2 (ia)	4	δ(CH)	B_1 (IR,R)	4
ν(CH)	E_1 (IR,R)	5	ν(CH)	A_1 (IR,R)	5a
			ν(CH)	B_1 (IR,R)	5b
δ(CH)	E_1 (IR,R)	6	δ(CH)	A_1 (IR,R)	6a
			δ(CH)	B_1 (IR,R)	6b
π(CH)	E_1 (IR,R)	7	π(CH)	A_2 (R)	7a
			π(C-CH_3)	B_2 (IR,R)	7b
ν(CC)	E_1 (IR,R)	8	ν(CC)	A_1 (IR,R)	8a
			ν(CC)	B_1 (IR,R)	8b
ν(CH)	E_2 (R)	9	ν(C-CH_3)	A_1 (IR,R)	9a
			ν(CH)	B_1 (IR,R)	9b
δ(CH)	E_2 (R)	10	δ(CH)	A_1 (IR,R)	10a
			δ(C-CH_3)	B_1 (IR,R)	10b
π(CH)	E_2 (R)	11	π(CH)	A_2 (R)	11a
			π(CH)	B_2 (IR,R)	11b
ν(CC)	E_2 (R)	12	ν(CC)	A_1 (IR,R)	12a
			ν(CC)	B_1 (IR,R)	12b
δ(CC)	E_2 (R)	13	δ(CC)	A_1 (IR,R)	13a
			δ(CC)	B_1 (IR,R)	13b
π(CC)	E_2 (R)	14	π(CC)	A_2 (R)	14a
			π(CC)	B_2 (IR,R)	14b

[a]Taken from Ref. 125.

TABLE 3.18
Ring Mode Assignments (cm^{-1}) for η^5-Methylcyclopentadienyl Complexes

Mode			Compound[a]					
Sym.	Description	Number	$RMn(CO)_3$[b]	RNiNO[c]	$[RMo(CO)_3]_2$[d]	R_2Fe[e]	R_2Sn[f]	R_2Pb[g]
A_1	ν(CH)	1	3120	3090	3124	3087	3100	3085
	ν(CH)	5a	3120	3090	3120	3087	3100	3085
	ν(CC)	12a	1394	1610	1390	1609	1607	1610
	ν(CC)	8a	1485	1452	1478	1440	1448	1450
	ν(C-CH_3)	9a	1234	1225	1230	1229	1229	1230
	δ(CH)	10a	1117		1108	1106	1117	1118
	Ring-breathing	3	1068	1042	1068	1025	1042	1044
	δ(CH)	6a	928	928	926	919	928	973
	δ(CC)	13a	847	849	840	851	847	848
	ν(CH_3)		2936	2911	2937	2922	2925	2925
	δ(CH_3)		1394	1385	1390	1384	1380	1378
B_1	ν(CH)	5b	3105	3106	3112	3087	3080	3085
	ν(CH)	9b	3105	3106	3107	3087	3080	3085
	ν(CC)	12b	1379	1560	1377	1558	1526	
	ν(CC)	8b	1399	1385	1401	1373	1408	1409
	δ(CH)	4	1239	1237	1242	1207	1260	1265
	δ(CH)	6b	1031	1025	1031	1115	1062	1065
	δ(CC)	13b	985	888	988	892	898 or 885	901
	δ(C-CH_3)	10b	319	344	317	328	319	
	ν(CH_3)		2907	2941	2903	2950	2965	2965
	δ(CH_3)		1459	1460	1466	1464	1458	1460
	δ(CH_3)		1048	1056	1045	1037	1026	
B_2	π(CH)	11b	1048	975	1045	975	976	750
	δ(CH)	2	837	802	828	811	767	750

	$\delta(CC)$	14b	628	624	630	632	610	612
	$\delta(C-CH_3)$	7b	219	354	214	217		
	$\nu(CH_3)$		2967	2973	2974	2966	2965	2965
	$\delta(CH_3)$		1452	1481	1461	1478	1482	1485
	$\delta(CH_3)$		1036	1056	1031	1053	1026	1028
A_2	$\delta(CH)$	11a	1064	967	1068	1002	1010	1010
	$\delta(CH)$	7a	878	819	879	833	813	811
	$\delta(CC)$	14a	608	780 or 639	630	597	561	565

[a] $R=\eta^5-CH_3C_5H_4$.
[b] References 249 and 383.
[c] Reference 384.
[d] Reference 383.
[e] Reference 359.
[f] Reference 125.
[g] Reference 385.

TABLE 3.19
Noncyclopentadienyl Mode Assignments (cm^{-1}) for η^5-Methylcyclopentadienyl Complexes

Assignment[a]	$RMn(CO)_3$[b]	RNiNO[c]	$[RMo(CO)_3]_2$[d]	R_2Fe[e]	R_2Sn[f]
ν(CO) or ν(NO)	2037 1959	1803	1954 1912 1901		
ν(MR)	326	322	347	482[g] 311[h]	237[g] 171[h]
Tilt	394	298	368	496[g] 398[h]	368[g] 295[h]
δ(RMR) or δ(RMX)		187			113
δ(MCO) or δ(MNO)	667 636 539	483	592-412		
ν(M-CO) or ν(MN)	499 491	642	592-412		
ν(MoMo)			83		

[a] $R=\eta^5-CH_3C_5H_4$.
[b] References 249 and 383.
[c] Reference 384.
[d] Reference 383.
[e] Reference 359.
[f] Reference 125.
[g] Antisymmetric mode.
[h] Symmetric mode.

It has been proposed that the angular structure of $(CH_3C_5H_4)_2Sn$ is preserved in this adduct (3.49). The

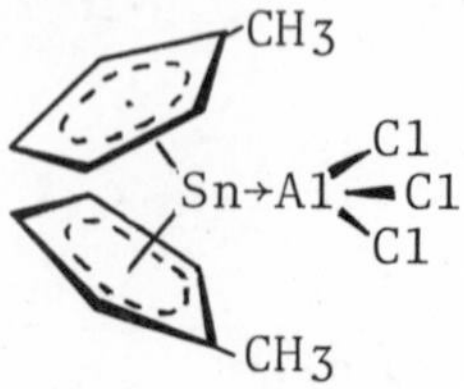

(3.49)

infrared spectrum of $(CH_3C_5H_4)_2Sn$ and its $AlCl_3$ adduct are very similar (385). The infrared spectra of $(CH_3C_5H_4)_2TiCl_2$ and $(CD_3C_5H_4)_2TiCl_2$ are illustrated in Figure 3.22 (171). Infrared data have been listed for

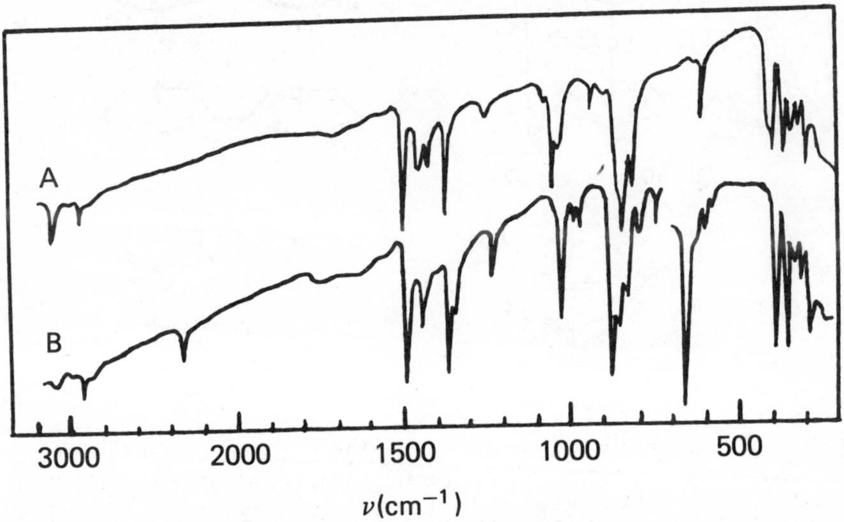

Figure 3.22. Infrared spectra of (A) (η^5-$CH_3C_5H_4)_2TiCl_2$ and (B) (η^5-$CH_3C_5D_4)_2TiCl_2$. The pmr spectrum of the latter compound shows 10% deuteration of the methyl hydrogens and full deuteration of the ring hydrogens (171).

several monosubstituted cyclopentadienyl derivatives of $C_5H_5Mn(CO)_3$ (386), while assignments of the infrared spectra in the 2100 to 1900 cm^{-1} and 700 to 225 cm^{-1} regions have been discussed for $CH_3C_5H_4M(CO)_2X$ (M=Mn,X=$(C_6H_5)_3P$,$(C_6H_5)_3As$,$(C_6H_5)_3Sb$;M=Fe,X=Cl,Br,I,$SnCl_3$,$SnBr_3$) (312). Structures have been proposed for $CH_3C_5H_4Fe$-$(CO)_2Co(CO)_4$ (303) and $(CH_3C_5H_4CoCO)_3$ (300) in the solid state and solution on the basis of infrared data in the ν(CO) mode region.

The infrared and Raman spectra have been assigned for $CH_3C(=O)C_5H_4V(CO)_4$ (212), while infrared data have been listed for $RC(=O)C_5H_4M(CO)_3$ (M=Tc,R=CH_3,C_6H_5;M=Re, R=CH_3) (387).

The infrared spectrum of [bisferrocene(Fe(II)-Fe(III))]picrate (3.50) has been reported (388). An infrared band at 489 cm^{-1} has been assigned to an iron-ring tilting mode, while bands at 445 and 399 cm^{-1} have been assigned to ν(Fe(II)-C_5H_5) and ν(Fe(III)-C_5H_5) modes, respectively. Infrared data have been listed for 1,1'biferrocene (3.51) (389), the bis(irontricarbonyl) complex of Spiro[4.4]nonatetraene (3.52) (390), and the

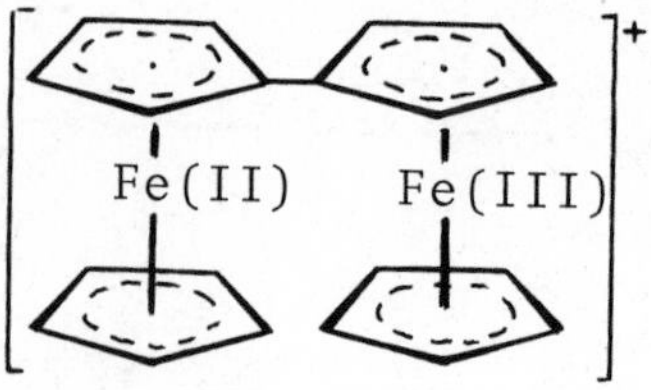

(3.50)

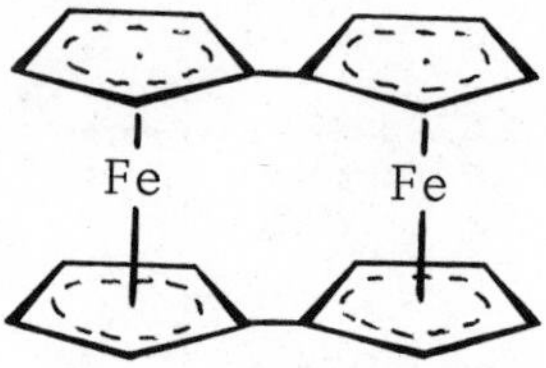

(3.51)

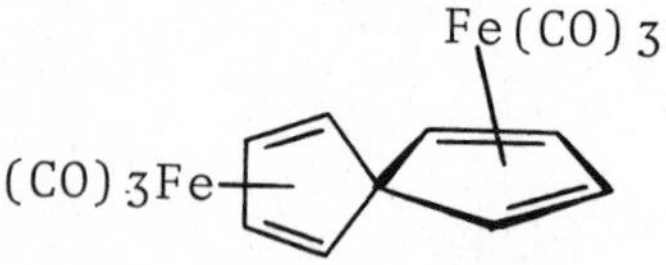

(3.52)

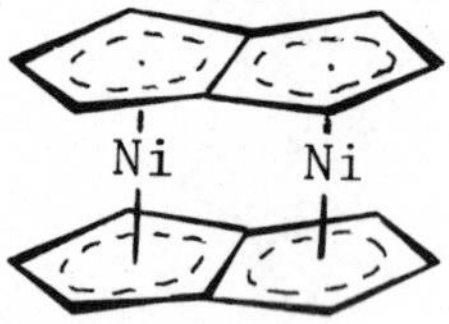

(3.53)

nickel derivative bis(pentalenyl)Ni (3.53) (391) while the infrared spectra have been illustrated for $[(t\text{-}C_4H_9)\text{-}C_5H_4]_2Ni$ and the triple-decker sandwich compound $[((t\text{-}C_4H_9)C_5H_4)_3Ni_2]BF_4$ (392).

Infrared data (2500 to 70 cm^{-1}) have been listed for the 1,2,4-tricyanocyclopentadienyl complexes $[(1,2,4\text{-}(CN)_3C_5H_2M(CO)_3]_n$ (M=Mn,Re) and the pentacyanocyclopentadienyl complexes $[((CN)_5C_5)M(CO)_3]_n$ (M=Mn,Re) as well as for $K^+[1,2,4\text{-}(CN)_3C_5H_2]^-$ and $K^+[(CN)_5C_5]^-$ (393). The increase in the ν(CN) mode frequency from approximately 2200 cm^{-1} to 2240 cm^{-1} on coordination of the 1,2,4-tricyanocyclopentadienide anion indicates that this ligand is bonded to the metal atoms not through the C_5 ring but through the nitrogen atoms of all three nitrile groups to form a polymeric structure (3.54). A

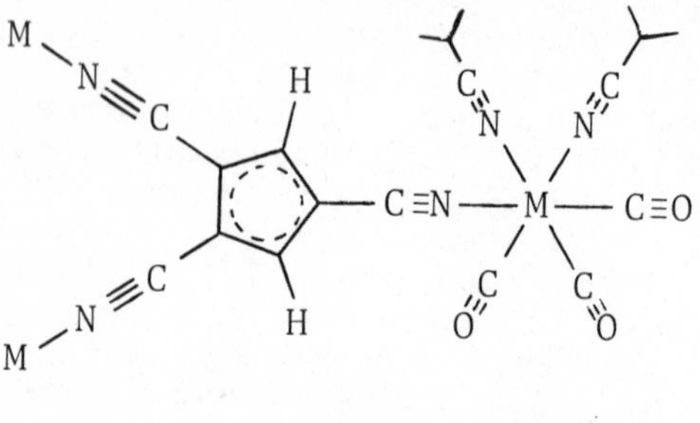

(3.54)

similar structure has been proposed for the pentacyanocyclopentadienyl complexes in which two of the nitrile groups remain nonbonded while the remaining three form the polymeric network. This is indicated by the presence of not only the infrared band at approximately 2255 cm^{-1} in the complexes, which has been attributed to the N-bonded nitrile groups, but also of a band at 2225 cm^{-1}, as compared to a similar band at 2222 cm^{-1} for $K^+[(CN)_5C_5]^-$, which has been attributed to the ν(CN) mode of the nonbonded nitrile groups. The appearance of infrared bands at 2251 and 2219 cm^{-1} in the spectrum of $[((CN)_5C_5)_2Fe{\cdot}H_2O]_n$, indicates that it also has a polymeric structure in which each ligand contains two nonbridging and three bridging nitrile groups.

Characteristic features of the infrared data listed for several η^5-pentamethylcyclopentadienyl-transition-metal complexes (364,394-398) include a strong intensity band from 1380 to 1370 cm^{-1} and weaker but more variable intensity bands at approximately 1440 and 1020 cm^{-1} (364). The appearance of ν(CO) mode bands in both the terminal (1932 cm^{-1}) and bridging (1764 cm^{-1}) regions in the cyclohexane solution spectrum of $[(CH_3)_5C_5Fe(CO)_2]_2$ indicates the presence of only the *trans* CO-bridged structure (3.42) (364). This contrasts with the fact that $[C_5H_5Fe(CO)_2]_2$ consists of a mixture of possibly four isomers in solution (296). The ν(CO) mode frequencies of η^5-pentamethylcyclopentadienyl-metal carbonyls are approximately 10 to 20 cm^{-1} lower than those of the corresponding cyclopentadienyl complexes (364,395). This is due to the greater electron-donating ability of the methyl group relative to the hydrogen atom, which results in a greater degree of metal to CO π-back bonding in pentamethylcyclopentadienyl-metal carbonyls than in the corresponding cyclopentadienyl-metal carbonyls. Both a medium intensity (2056 cm^{-1}) and a strong intensity (2023 cm^{-1}) band have been observed in the hexane solution infrared spectrum of $[(CH_3)_5C_5]_2$-TiN_2 at -65°C (397). These data have been interpreted as consistent with an equilibrium between a complex with an "edge-on" or *dihapto* N_2 ligand and a complex with the more commonly found "end-on" or *monohapto* N_2 ligand. The Nujol mull infrared spectrum of $[((CH_3)_5C_5)_2Zr]_2$-$(N_2)_3$ exhibits two strong intensity bands at 2041 and 2006 cm^{-1} and a medium intensity band at 1556 cm^{-1}; these shift to 1972, 1937, and 1515 cm^{-1}, respectively, upon substitution of doubly labeled $^{15}N_2$ (398). The two higher-frequency bands have been assigned to the $\nu(N_2)$ mode of two presumedly *monohapto* N_2 ligands, while the

1556 cm^{-1} band has been assigned to the $\nu(N_2)$ mode of a N_2 ligand, which might bridge the two zirconium atoms.

Indenyl and Fluorenyl Derivatives. Derivatives of indene (3.55) and fluorene (3.56) are included in this section

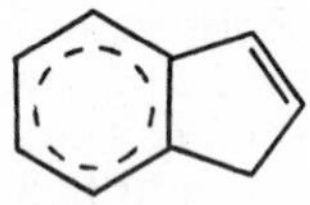

(3.55)

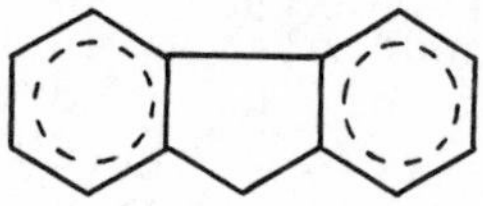

(3.56)

on substituted cyclopentadienyl derivatives since metal atoms bond mainly to the five-membered rings of these ligands rather than to the six-membered rings. Indenyl derivatives have been found with either a *monohapto* (3.57) or a *pentahapto* structure (3.58).

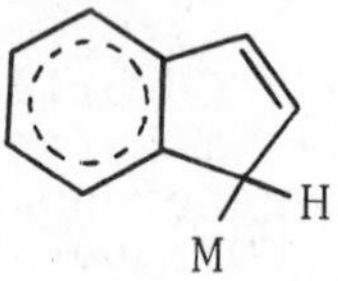

(3.57)

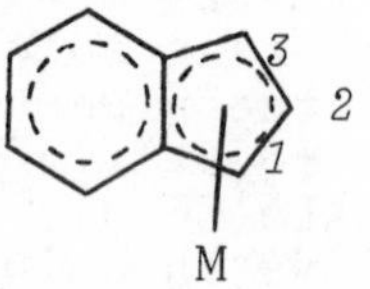

(3.58)

The most complete vibrational data reported for indenyl complexes have been reported for $(\eta^1\text{-}C_9H_7)_2Hg$ and $\eta^1\text{-}C_9H_7HgX$ (X=Cl,Br) (399,400). As illustrated by the infrared spectra of $\eta^1\text{-}C_9H_7HgCl$ and $\eta^5\text{-}C_5H_5Fe(CO)_2\text{-}(\eta^1\text{-}C_9H_7)$ in Figure 3.23, η^1-indenyl complexes show medium intensity bands at approximately 3060 to 3050 cm^{-1} that arise from the aromatic $\nu(CH)$ modes (399). Also characteristic of η^1-indenyl complexes is a band at approximately 3000 to 2800 cm^{-1} that arises from the aliphatic $\nu(C_1H)$ mode. It has been suggested that the appearance of two such bands for $\eta^1\text{-}C_9H_7HgX$ (X=Cl,Br) is due to the presence of different environments for the C_1H bond in the solid state. The $\pi(CH)$ mode region (approximately 800 to 700 cm^{-1}) is also useful in showing the presence of a η^1-indenyl ligand. This is illustrated in Figure 3.24 by the infrared spectra of $\eta^1\text{-}C_9H_7HgCl$ and its partially deuterated (at the C_1 and C_3 atoms) analogs (399). For $\eta^1\text{-}C_9H_7HgCl$, the very strong intensity band at 750 cm^{-1} has been assigned to a phenyl mode while three strong intensity bands at 788,

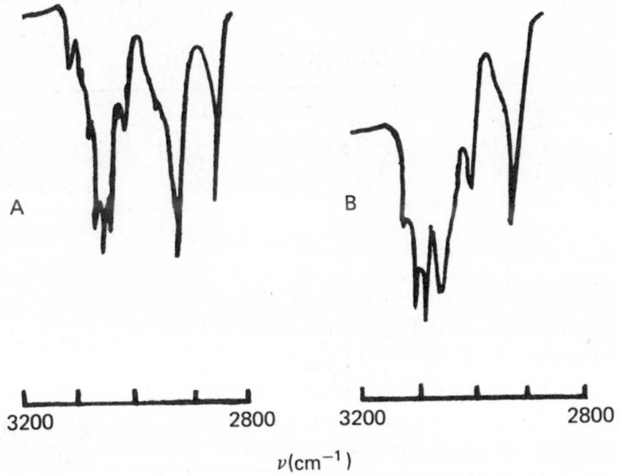

Figure 3.23. The hexachlorobutadiene mull infrared spectra of (A) η^1-C_9H_7HgCl and (B) η^5-$C_5H_5Fe(CO)_2$(η^1-C_9H_7) in the carbon-hydrogen stretching mode region (399).

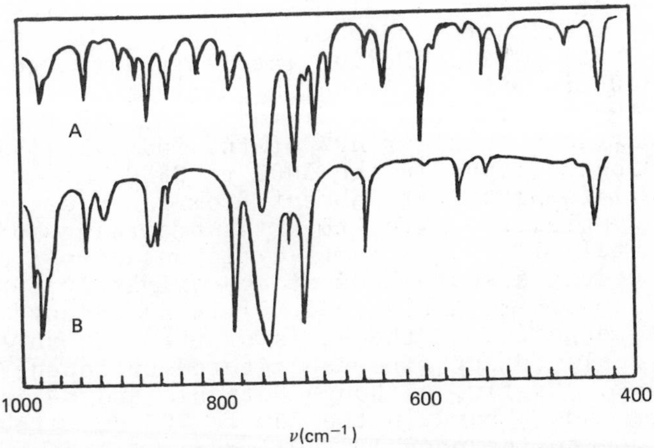

Figure 3.24. Infrared spectra (1000-400 cm^{-1}) of (A) η^1-C_9H_7HgCl and (B) the corresponding derivative which was deuterated at the C_1 and C_3 atoms (399).

TABLE 3.20
Skeletal Mode Assignments (cm^{-1}) of the Tetrahydrofuran Raman Spectra of $(\eta^1\text{-}C_9H_7)_2Hg$ and $\eta^1\text{-}C_9H_7HgX$ (X=Cl,Br)[a]

Compound	ν(HgC)	ν(HgX)	δ(CHgX) or δ(CHgC)
$(\eta^1\text{-}C_9H_7)_2Hg$	435 370		198 175
$\eta^1\text{-}C_9H_7HgCl$	380	350	207
$\eta^1\text{-}C_9H_7HgBr$	380	240	202

Reference 400.

718, and 655 cm^{-1} have been assigned to the $\pi(C_3H)$, $\pi(C_2H)$, and $\pi(C_1H)$ modes, respectively. These assignments are supported by infrared data for the partially deuterated derivative in which the 788 and 655 cm^{-1} bands disappear. The skeletal mode assignments for the indenyl-mercury derivatives using solution (tetrahydrofuran) Raman data are summarized in Table 3.20. Although a structure with η^1-indenyl ligands has been proposed for $Li[(C_9H_7)_4In][(C_2H_5)_2O]$ (72), no attempt has been made to assign the reported infrared data. A strong intensity Raman band at 352 cm^{-1}, however, has been assigned to the ν(InC) mode.

The high symmetry of a η^5-indenyl ligand leads to a simpler vibrational spectrum than found for a η^1-indenyl ligand (400). This is illustrated through a comparison of the infrared spectra of $\eta^1\text{-}C_9H_7HgCl$ (Figures 3.23 and 3.24) (399) with those of $(\eta^5\text{-}C_9H_7)_2MCl_2$ (M=Ti,Zr) (Figure 3.25) (401). Medium intensity infrared bands in the 3100 to 3000 cm^{-1} region of η^5-indenyl complexes are due to the aromatic ν(CH) modes of both the cyclopentadienyl and phenyl rings of the indenyl ligand. Infrared bands found in the 3000 to 2800 cm^{-1} region of η^1-indenyl complexes are absent from the spectra of η^5-indenyl complexes. Also, the ring breathing mode at approximately 1100 cm^{-1} in η^5-cyclopentadienyl complexes is replaced by a sharp band at approximately 1000 cm^{-1} (ca. 9.7 to 9.5 μ in Figure 3.25) in η^5-indenyl complexes as expected on the basis of the 9-10 rule, which is applicable to any ring-substituted cyclopentadienyl derivative. Lastly, although both η^1- and η^5-indenyl complexes show a band in the 750 to 725 cm^{-1} region that is due to the phenyl π(CH) mode, η^5-indenyl groups also show a strong intensity band in the 860 to 800 cm^{-1} region (ca. 12 to 11.7 μ in Figure 3.25) that arises from the π(CH) modes of the cyclopentadienyl

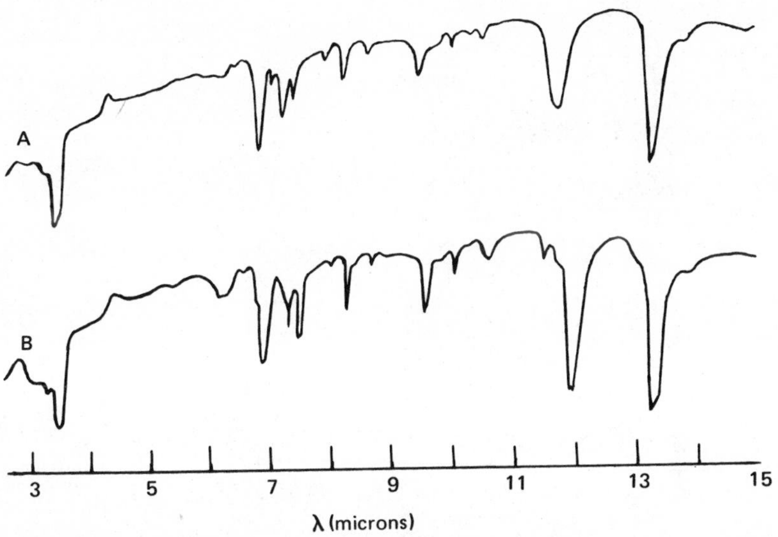

Figure 3.25. The Nujol mull infrared spectra of (A) $(\eta^5\text{-}C_9H_7)_2TiCl_2$ and (B) $(\eta^5\text{-}C_9H_7)_2ZrCl_2$ (401).

ring of the indenyl ligand (399,400,402). In contrast, η^1-indenyl complexes show three bands in the 800 to 700 cm^{-1} region due to the π(CH) modes of the indenyl cyclopentadienyl ring.

The indenyl rings are staggered in solid $(\eta^5\text{-}C_9H_7)_2$-Fe (3.59) and eclipsed in solid $(\eta^5\text{-}C_9H_7)_2Ru$ (3.60)

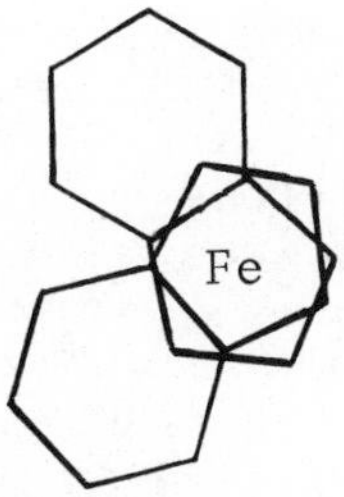

(3.59)

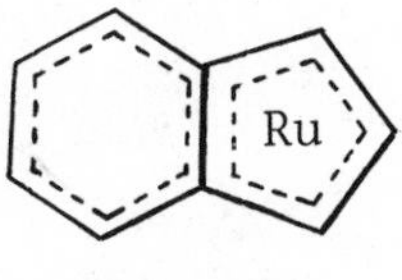

(3.60)

(403). With the exception of the A_2 modes, which are infrared active for the staggered rings but inactive for the eclipsed rings, both structures give rise to the same number of infrared active modes. The infrared spectra of $(\eta^5\text{-}C_9H_7)_2M$ (M=Fe,Ru) are very similar (400). Therefore, as was true for complexes with parallel η^5-cyclopentadienyl rings, vibrational spectroscopy cannot be used to differentiate between the staggered and eclipsed structures of $(\eta^5\text{-}C_9H_7)_2M$ complexes.

The infrared and Raman spectra have been reported and some assignments made of the indenyl modes for $(\eta^5\text{-}C_9H_7)_3MX$ (X=U,Th;X=Cl,Br) (404). Some of these assignments were for the ν(UCl), ν(ThCl), ν(UBr), and ν(ThBr) modes at 268, 268, 183, and 183 cm^{-1}, respectively.

Infrared data have also been listed for $(C_9H_7)_3Tl$ (113), $(C_9H_7)_2TlH$ (ν(TlH)=2200 cm^{-1}) (114), $(C_9H_7)_2TlCl$ (115), $(C_9H_7)_2TlX$ (X=NO_3,NCO,NCS,CN) (116), C_9H_7Ti-$(\eta^8\text{-}C_8H_8)$ (405,406), $C_9H_7Re(CO)_3$ (387) $(C_9H_7)_4Ce$ (326), $(C_9H_7)_2CeX_2$ (X=N_3,NCO,NCS,CN) (328), and $(C_9H_7)_2Ce$-$(SC_2H_5)_2$ (331) as well as for bis(tetrahydroindenyl) derivatives of iron and ruthenium (407).

The infrared spectra have been illustrated for two complexes that contain the η^5-fluorenyl ligand, namely, $C_{13}H_9Ti(\eta^8\text{-}C_8H_8)$ (405) and $(C_{13}H_9)_2ZrCl_2$ (401).

Halogen Substituents. A detailed infrared and Raman study has been made of $(\eta^5\text{-}XC_5H_4)_2Fe$ (X=Cl,Br,I) (79). The spectra of these complexes are very similar to those discussed previously for η^5-monomethylcyclopentadienyl compounds with the ring breathing (ν_3) and ν(C-halide) (ν_{9a}) modes strongly coupled. Therefore, it was not possible to assign bands to each of these modes individually. Rather, two bands, each possessing a considerable amount of ν(C-halide) character, have been observed to be mass- or X-sensitive. A similar situation has been found for monosubstituted benzenes (80) in which the higher frequency of these two X-sensitive bands has been designated as the q-mode while the lower frequency of these two modes has been designated as the r-mode. In Table 3.5, the q- and r-mode assignments are compared for $(XC_5H_4)_2Fe$ and C_6H_5X (X=Cl,Br,I).

The infrared spectra of a number of η^1-pentachlorocyclopentadienyl-mercury complexes, including $(C_5Cl_5)_2Hg$, C_5Cl_5HgX (X=Cl,Br), and the ether or double salt adducts of these derivatives, are relatively complex because of the low symmetry of the η^1-pentachlorocyclopentadienyl ring.(408,409). A single crystal X-ray study shows the presence of parallel *pentahapto* rings with an eclipsed configuration for $(C_5Cl_5)_2Ru$ (410). A similar structure is likely for $(C_5Cl_5)_2Fe$ (411). The simplicity of the

TABLE 3.21
Infrared Data (cm^{-1}) Characteristic of Organometallic Derivatives of Pentachlorocyclopentadiene

	Compound[a]				
Assignment[b]	RTl[c]	$(\eta^5\text{-}R)_2$-Fe[d]	η^5-RMn-$(CO)_3$[e]	η^1-RMn-$(CO)_5$[e]	$(\eta^1\text{-}R)_2$-Hg[f]
Ring modes	1407 m 1190 m	1350 s 1307 m	1383 s 1365 sh 1340 m 1315 sh	1565 s 1561 sh 1264 s 1191 w 1122 m 969 vw 943 vw	1576 s 1560 sh 1250 vs 1200 m 1126 s 974 w 950 m 654 m 519 vw
Allylic ν(CCl)				706 m.sh	768 w 724 s
Vinylic or Aromatic ν(CCl)	657 s	702 s	709 vs	694 vs	699 m

[a]$R=C_5Cl_5$.
[b]Description of the modes is from Ref. 409.
[c]Reference 413.
[d]Data from Ref. 412; additional bands at 509 (w), 412 (m), 378 (m), and 368 (m) cm^{-1} have not been assigned.
[e]Reference 414.
[f]Reference 408.

infrared spectrum of $(C_5Cl_5)_2Fe$ is consistent with the *pentahapto* structure (412). Although an ionic structure has been proposed for C_5Cl_5Tl (413), infrared data for this compound indicate a slight interaction between the Tl^+ and $C_5Cl_5^-$ ions, which lowers the symmetry of the pentachlorocyclopentadienide ring from D_{5h} to C_{5v}. In Table 3.21, infrared data are summarized for ionic, η^1-, and η^5-pentachlorocyclopentadienyl derivatives. A characteristic spectral feature of the *monohapto* compounds is the presence of a band at approximately 1560 cm^{-1} due to a ν(C=C) mode.

The fact that the ν(CO) mode frequencies of η^5-$C_5Cl_5Mn(CO)_3$ (ν(CO)=2048 and 1982 cm^{-1}) are on the average 28 cm^{-1} higher than those of η^5-$C_5H_5Mn(CO)_3$ has been offered as evidence that the η^5-pentachlorocyclopentadienyl ring is a poorer electron donor and/or better electron acceptor than the η^5-cyclopentadienyl ring (414). Infrared data have also been reported for $(\eta^5\text{-}C_5Cl_4Br)Mn(CO)_3$, $(\eta^1\text{-}C_5Cl_4X)Mn(CO)_5$ (X=Br,H), and $\eta^5\text{-}C_5Cl_5Rh$(1,5-cyclooctadiene) (414). The presence of two short and three long C-C bonds in the pentachlorocyclopentadienyl ring of the latter compound has been interpreted as a reflection of a considerable contribution of a bonding model where the ring is bonded to the rhodium atom via two π-bonds and one σ-bond (414).

D. FIVE-MEMBERED HETEROCYCLIC RINGS

Although cobalt complexes containing the cyclic, planar carborane $C_2B_3H_5^{4-}$ have been characterized, no vibrational data were included in this study (415). The $C_2B_3H_7^{2-}$ molecule, which may be regarded as the diprotonated derivative of $C_2B_3H_5^{4-}$, is found in $(C_2B_3H_7)Fe(CO)_3$, which is the structural and electronic analog of $\eta^5\text{-}C_5H_5Fe(CO)_3$ (416). The molecular structure determined for solid $(C_2B_3H_7)Fe(CO)_3$ (417) consists of a planar $\eta^5\text{-}C_2B_3$ ring, two B-H-B units that are directed away from the iron atom, two C-H units, and three B-H units. The infrared spectrum (CCl_4 solution) of $(C_2B_3H_7)Fe(CO)_3$ exhibits very strong intensity ν(CO) bands at 2068 and 2007 cm^{-1} as well as additional bands at 3008 (m), 2569 (vvs, terminal ν(BH) mode), 1888 (ms, bridging ν(BH) mode), 1700 (m), 1600 (m), 1350 (m), and 1050 (m) cm^{-1} (416). Several metal complexes have been reported of the pyramidal $C_2B_6H_6^{2-}$ anion (3.61) in

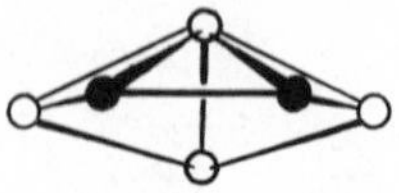

B-H = o

C-H = •

(3.61)

which the basal five-membered C_2B_3 ring contains six delocalized electrons analogous to the cyclopentadienide anion. In these complexes, the metal atom is located above the planar C_2B_3 ring to occupy one apex of a pentagonal bipyramidal cage. It has been proposed that the iron atom in $(C_2B_4H_6)Fe(CO)_3$ is equidistant from each of the atoms of the C_2B_3 ring (416). A single crystal X-ray study of $(C_2B_4H_6)GaCH_3$, however, shows that the gallium atom is closer to the boron

TABLE 3.22
Infrared Data (cm^{-1}) for η^5-Carborane Derivatives of the Main Group and Transition Elements

$(C_2B_4H_6)$-$GaCH_3$[a]	$(C_2B_4H_6)$-$InCH_3$[b]	$(C_2B_4H_6)$-$Fe(CO)_3$[c]	$(CH_3C_3B_3H_5)$-$Mn(CO)_3$[d]
		2076 s[e]	2035 vs[e]
		2017 s[e]	1956 vs[e]
3050 m	3025 m	3090 m	
2950 s	2985 s	2960 w	
2905 m	2923 s		2930 mw
2590 vs[f]	2590 vs[f]	2590 vs[f]	2590 vs[f]
	1930 m		
	1510 s		
1320 m	1346 m		
1283 s	1290 s	1270 vs	
1200 s	1155 m	1188 s	
	1060 s		
1023 vs	1022 m		
994 vs			
850 m			670 s
618 s			635 s

[a]References 418 and 419.
[b]Reference 419.
[c]Reference 416.
[d]Reference 420.
[e]ν(CO) mode.
[f]ν(BH) mode.

atoms of the C_2B_3 ring than to the carbon atoms and that the Ga-CH_3 axis is tilted 20° with respect to the perpendicular of the equatorial C_2B_3 plane (418). A similar unsymmetrical *pentahapto* structure has been proposed for $(C_2B_4H_6)InCH_3$ (418). Infrared data listed in Table 3.22 for the above mentioned $C_2B_4H_6^{2-}$ derivatives show a characteristic, strong intensity ν(BH) band at 2590 cm^{-1}. Infrared data reported for $(CH_3C_3B_3H_5)Mn(CO)_3$ (3.62) are also included in Table 3.22.

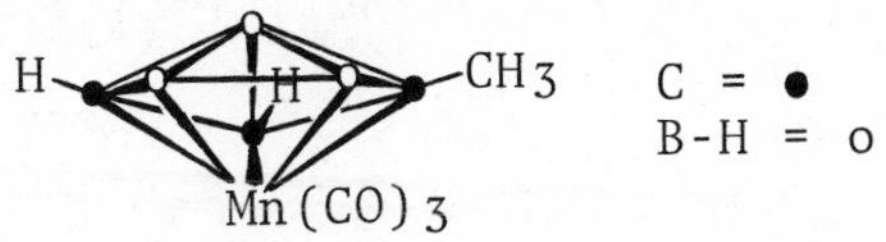

(3.62)

Because of the delocalized electronic structure of the η^5-pyrrolyl ligand in metal complexes, the local symmetry of this ligand tends toward C_{5v} rather than C_{2v}. In fact, most of the infrared bands of these pyrrolyl complexes correspond very closely in frequency to the vibrational modes of η^5-cyclopentadienyl complexes. This is illustrated in Figure 3.26 by a comparison of the infrared spectra of azaferrocene ((η^5-C_4H_4N)Fe-(η^5-C_5H_5)) and (η^5-C_5H_5)$_2$Fe (421). A complete infrared and Raman study of (η^5-C_4H_4N)Mn(CO)$_3$ (ν(CO)=2038,2053 cm^{-1};ν(Mn-ring)=347 cm^{-1};Mn-ring tilt=384,375 cm^{-1}) and (η^5-C_4D_4N)Mn(CO)$_3$ has also included a comparison of these data with those for η^5-C_5H_5Mn(CO)$_3$

The neutral pyrrole molecule, C_4H_4NH, can form π-bonded metal complexes. Infrared data and some assignments have been reported for (C_4H_4NH)Cr(CO)$_3$ and compared with the vibrational assignments of free pyrrole (423). The ν(CO) mode assignments have also been reported for chromium tricarbonyl complexes of pyrrole (which was substituted at the nitrogen and/or carbon atoms), C_4H_4S, and C_4H_4Se (423).

Detailed infrared and Raman assignments have been reported for (Maleic Anhydride)Fe(CO)$_4$ (3.63) (424,425)

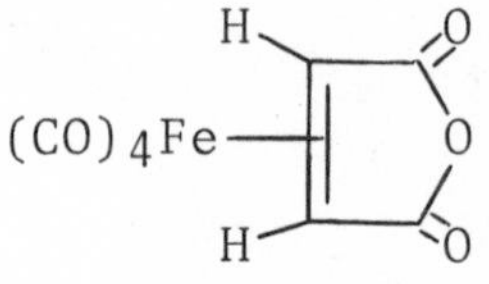

(3.63)

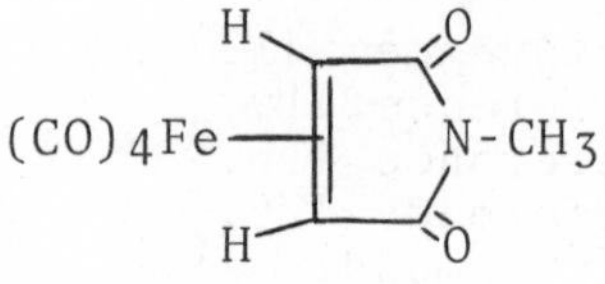

(3.64)

and (Methylmaleimide)Fe(CO)$_4$ (3.64) (426). For both compounds, there is little significant coupling between the ν(C=C) and δ(CH) modes. The shift of the ν(C=C) mode from 1595 to 1352 cm^{-1} on coordination of maleic anhydride and from 1585 to 1370 cm^{-1} on coordination of methylmaleimide is consistent with a relatively strong metal-ligand interaction in both compounds. The symmetric iron-maleic anhydride stretching mode has been assigned at 376 cm^{-1} while the corresponding mode in the methylmaleimide complex was not assigned.

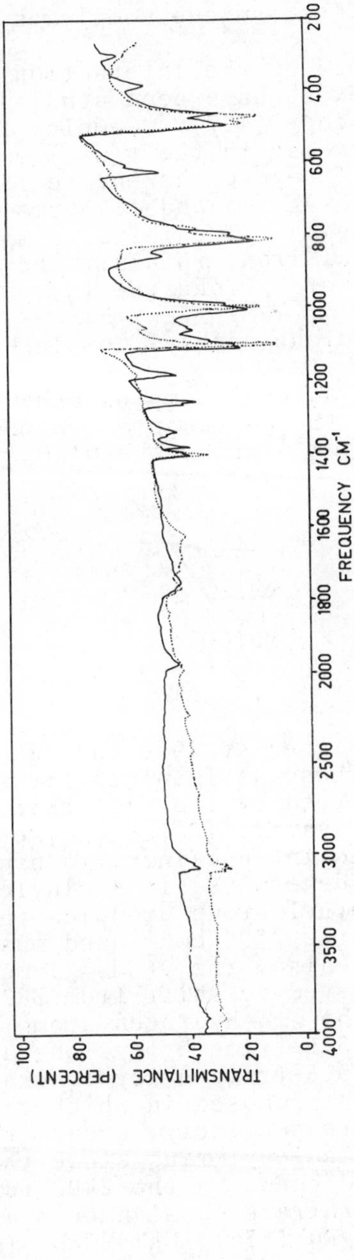

Figure 3.26. The KBr pellet infrared spectra of azaferrocene (---) and ferrocene (···) (421).

E. SIX-CARBON RINGS

1. Cyclohexene Complexes

The complexes formed by the interaction of silver(I) salts with cyclohexene have been studied using infrared and Raman spectroscopy (427). Complex formation is indicated by the decrease in the ν(C=C) mode frequency from 1653 cm^{-1} in free cyclohexene to 1582 cm^{-1} for the complex formed with $AgClO_4$ and 1575 cm^{-1} for the complex formed with $AgNO_3$. The infrared spectra have been reported for several iron, rhenium, and manganese CO complexes of perfluorocyclohexene (9).

2. Cyclohexadienyl Complexes

For the η^5-cyclohexadienyl complex $C_6H_7Mn(CO)_3$ (3.65) the ν(CH) modes of the *pentahapto* fragment appear at approximately 3050 cm^{-1} while the $\nu(CH_{endo})$ and $\nu(CH_{exo})$

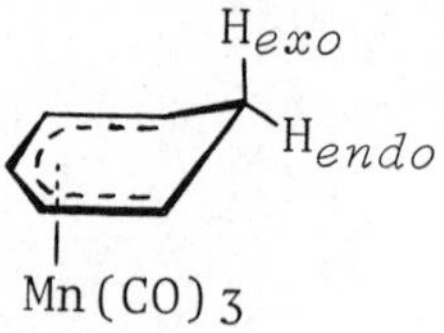

(3.65)

modes have been assigned at 2970 and 2830 cm^{-1}, respectively (428). These assignments for the *endo* and *exo* hydrogens are supported by the fact that the hexamethylcyclohexadienyl derivative $[C_6(CH_3)_6H]Re(CO)_3$, which exhibits a broad and intense infrared band at 2790 cm^{-1}, has a structure as determined in a single crystal X-ray study in which a methyl group occupies the *endo* position (429). Therefore, the 2790 cm^{-1} band must arise from the $\nu(CH_{exo})$ mode. Also, the $\nu(CH_{exo})$ mode of $C_6H_7V(CO)_3$ has been observed to shift from 2822 to 2070 cm^{-1} on deuteration of the *exo* hydrogen atom (430). For bis(6-*tert*-butyl-1,3,5-trimethylcyclohexadienyl)Fe(II) and bis(6-phenyl-1,3,5-trimethylcyclohexadienyl)Fe(II), structures have been proposed in which the *tert*-butyl and phenyl groups, respectively, occupy the *exo* position of the cyclohexadienyl ring since these complexes do not show infrared bands in the 2800 to 2700 cm^{-1} region (431,432). Infrared assignments ahve also been reported for $(C_6H_7)_2Ru$ (433), $C_6H_7MC_6H_6$ (M=Re (433),Tc (434)), and $C_6H_7CoC_5H_5^+$ (435).

3. Neutral, Cationic, and Anionic Phenyl Complexes

In assigning the vibrational spectra of organometallic phenyl derivatives, reference is often made to the assignments given for monohalobenzenes (80). Of the 30 fundamental modes of the monohalobenzene molecules, the positions of 24 are relatively independent of the nature of the halogen substituent. The frequencies of the remaining six modes (illustrated in Figure 3.27) are sensitive to the substituent and have therefore been described as mass- or X-sensitive. Of these six modes, those labeled as the q-, r-, and t-modes contain contributions from the stretching of the M-phenyl bond, while those labeled as the y-, u-, and x-modes involve M-phenyl bending. The X-sensitive modes, however, are not pure M-phenyl stretching or bending modes since they are coupled to some extent with the phenyl ring modes. It is therefore incorect to assign M-phenyl stretching and bending modes in the same sense as the corresponding M-methyl stretching and bending modes have been assigned. It has been reported that when the substituent on the phenyl ring is relatively light in mass, such as a first-row element (436,437), and/or when interactions are possible between lone-pairs of electrons on the phenyl substituent and the π-electron system of the phenyl ring (437), the major contribution of the stretching of the M-phenyl bond is to the q-mode. For heavier elements that are not found in the first row of the periodic table, however, it is the t-mode that contains the major contribution from the stretching of the M-phenyl bond (437,438).

The above considerations have mainly been with reference to one monosubstituted phenyl ring. When the substituent is a metal atom, more than one phenyl group is often found to be present. Although each ring would then be expected to give rise to a set of bands, the negligible degree of interring coupling leads to accidental degeneracy of a large number of the non-mass-sensitive phenyl modes. Therefore, the vibrational spectra arising from these modes can still be interpreted in terms of the C_6H_5M unit. Splitting or asymmetry noted for these bands in the solid state has been attributed to solid-state effects (438). The frequency ranges as derived from infrared data for the non-mass-sensitive phenyl modes of $(C_6H_5)_nM$ derivatives are summarized in Table 3.23. The same frequency ranges are found for $(C_6H_5)_mMX_n$ derivatives in which ligands other than phenyl groups are bonded to the central metal atom.

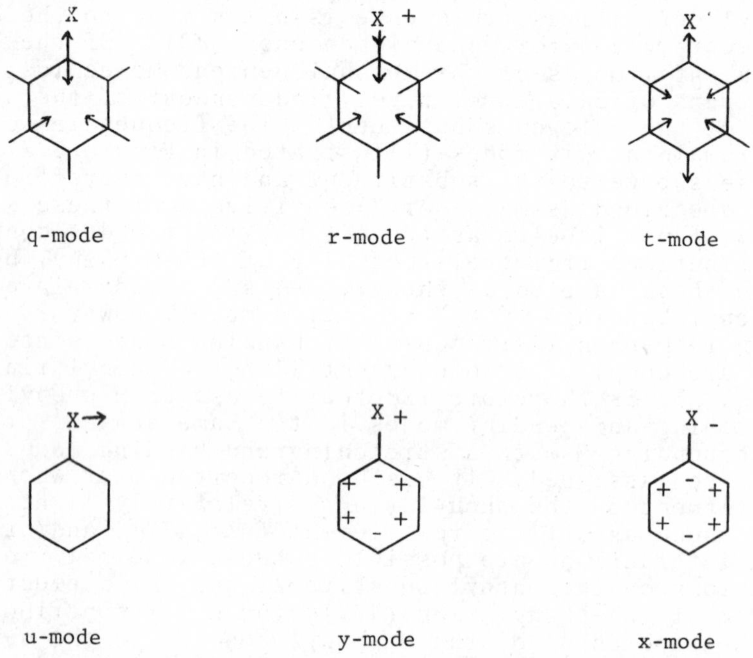

Figure 3.27. The X-sensitive modes for a monosubstituted benzene ring.

Raman data and assignments have been less frequently reported for the non-mass sensitive modes of phenyl derivatives than the corresponding infrared data. Raman data have, however, been reported for phenyl derivatives of zinc (439), mercury (440), boron (436, 441,442), silicon (443-446), germanium (446-452), tin (446,453), lead (454), phosphorus (455-459), arsenic (455,456,460,461), and antimony (461,462). Vibrational data have also been reported for a limited number of perdeuterophenyl derivatives (447,449-451,463-465). The infrared spectrum of $(C_6H_5)_4Sn$ is illustrated in Figure 3.28 (466) while the Raman spectrum of $(C_6H_5)_4Ge$ is illustrated in Figure 3.29 (447).

The six mass-sensitive modes show a greater degree of splitting than the non-mass-sensitive modes. This complexity has been attributed not only to solid-state effects (466), but more importantly to changes in coor-

TABLE 3.23
Infrared Frequency Ranges (cm^{-1}) for the Non-Mass-Sensitive Modes of Organometallic Phenyl Derivatives[a]

Mode Description		Symmetry	Frequency Range	Intensity
ν(CH)		$3A_1 + 2B_1$	3125-3000	w-m
ν(CC)	k	A_1	1600-1565	m
ν(CC)	l	B_1	1585-1560	w
ν(CC)	m	A_1	1500-1460	m-s
ν(CC)	n	B_1	1440-1415	s-vs
ν(CC)	o	B_1	1400-1320	w-m
δ(CH)	e	B_1	1280-1245	w-m
δ(CH)	a	A_1	1200-1170	w-m
δ(CH)	c	B_1	1170-1145	w-m
δ(CH)	d	B_1	1090-1040	w-vs
δ(CH)	b	A_1	1050-1015	m-s
Ring	p	A_1	1020-990	w-s
π(CH)	j	B_2	995-975	w-m
π(CH)	h	A_2	980-935	w-m
π(CH)	i	B_2	915-890	w-m
π(CH)	g	A_2	860-840	w-m
π(CH)	f	B_2	750-720	vs[b]
δ(CCC)	s	B_1	640-610	w-m
π(CC)	w	A_2	410-390	vw

[a]For a description of the modes and their symmetry, see Ref. 80.
[b]This band is sometimes found with a w-vs intensity shoulder.

dination number and symmetry about the central atom. In Table 3.24, the infrared and Raman assignments are compared for the mass-sensitive and $\delta(C_6H_5\text{-}M\text{-}C_6H_5)$ modes of neutral and cationic phenyl derivatives of the main group elements.

The Nujol mull infrared spectra have been recorded for ^6Li and ^7Li substituted C_6H_5Li (490). Infrared bands at 429 and 390 cm^{-1} in the ^6Li derivative were shifted to 421 and 378 cm^{-1} in the ^7Li derivative and therefore involve motion of the lithium atom. Phenyl derivatives of the other group Ia elements have an ionic structure, and no vibrational data have been reported for these derivatives. The infrared mull spectrum (1600 to 150 cm^{-1}) has also been recorded and assigned for the group Ib derivative C_6H_5Cu (467,491).

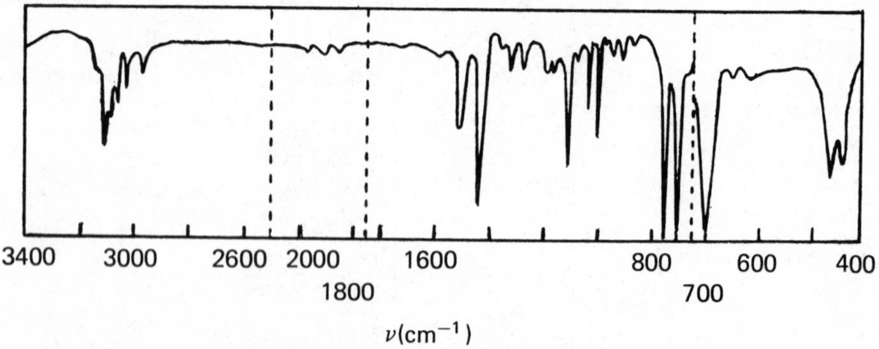

Figure 3.28. The KBr pellet infrared spectrum of $(C_6H_5)_4Sn$ (466).

Infrared data for $(C_6H_5)_2M$ (M=Zn (465),Cd (465), Hg (81,463,465,492)) and $(C_6D_5)_2M$ (M=Zn (465),Hg (491)) have all been interpreted in terms of a monomeric structure with a linear C-M-C skeleton. Also, Raman data (3500 to 200 cm^{-1}) have been listed for solid $(C_6D_5)_2Zn$ (439).

Infrared data have been reported for $(C_6H_5)_3M$ (M= B (467,468),Al (468,469),Ga (468),In (468)) and their deuterated analogs (464). Single crystal X-ray studies have shown solid $(C_6H_5)_3B$ to have a monomeric structure (493), while solid $(C_6H_5)_3Al$ has a dimeric structure with bridging phenyl groups (494). The infrared spectrum of $(C_6H_5)_3Al$, however, has been observed to be suprisingly simple in view of the presence of two different types of phenyl groups (469). Polymeric structures in which trigonal molecules are linked in chains by weak intermolecular forces have been found in single crystal X-ray studies of $(C_6H_5)_3Ga$ and $(C_6H_5)_3In$ (495). Although $(C_6H_5)_3Tl$ is unknown, several salts of $(C_6H_5)_2$-Tl^+ have been characterized using infrared data (483, 484). The infrared spectra of $M[(C_6H_5)_4B]$ show greater complexity in the 1270 to 1120 cm^{-1} and 750 to 715 cm^{-1} regions for M=Ag and Cu than for M=Tl, Na, and NH_4 (467).

Infrared data have been listed for $(C_6H_5)_3M$ (M=Sc, Y) and $Li[(C_6H_5)_4M]$ (M=La,Pr) (496). The similarities of the spectra of $(C_6H_5)_3Sc$ and $(C_6H_5)_3Al$ have led to the conclusion that $(C_6H_5)_3Sc$ is not monomeric. The nonvolatility and insolubility of this compound support this conclusion but made determination of its molecular weight impossible.

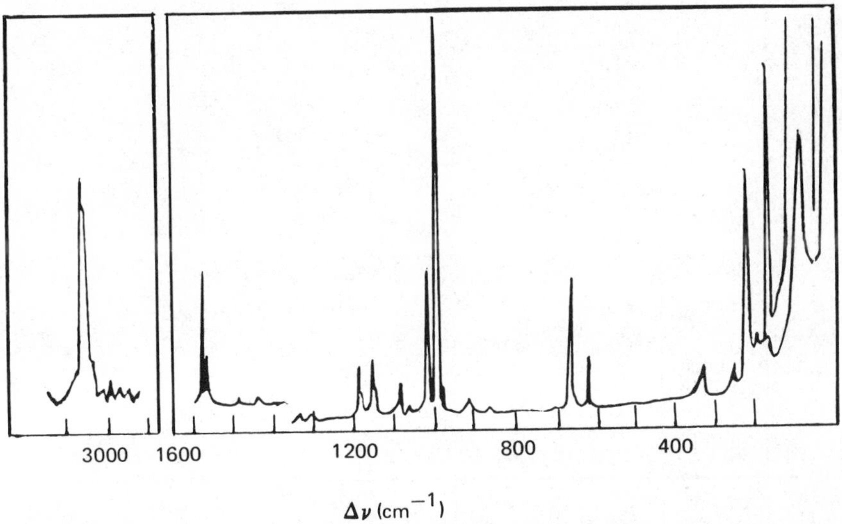

Figure 3.29. Raman spectrum of solid $(C_6H_5)_4Ge$ (447).

Several vibrational studies have appeared for $(C_6H_5)_4Si$ (443,470-472), $(C_6H_5)_4Ge$ (447,470-474), $(C_6H_5)_4Sn$ (470-472,474-476), $(C_6H_5)_4Pb$ (454,470-472) and $(C_6D_5)_4Ge$ (447), all of which have a monomeric structure with an approximate tetrahedral skeleton.

Several vibrational studies have also been reported for $(C_6H_5)_3M$ (M=P,As,Sb,Bi) (438,458,470,474 477-481). Among these has been an analysis of the low-frequency (550 to 100 cm^{-1}), solid-state, and benzene solution infrared and Raman spectra (477). Figure 3.30 illustrates the benzene solution infrared and Raman spectra of $(C_6H_5)_3M$ (M=P,As,Sb,Bi) (477). Assuming the phenyl groups to be single atoms, the triphenyl derivatives have a pyramidal skeleton of C_{3v} symmetry with two M-phenyl stretching and two M-phenyl bending modes. The splitting of the t and u modes, which are predominantly M-phenyl stretching and bending modes, respectively, into two components (cf. Table 3.24) is therefore explained (477). To aid in confirming the $(C_6H_5)_3M$ skeletal mode assignments, approximate normal coordinate calculations have been carried out assuming the phenyl groups to be single atoms (477).

Phenyl derivatives of the pentavalent group Vb elements are found with the four-coordinate, ionic struc-

TABLE 3.24
Assignments of the X-Sensitive and $\delta(C_6H_5\text{-}M\text{-}C_6H_5)$ Modes (cm^{-1}) for the Phenyl Derivatives of the Main Group Elements

	q-Mode	r-Mode		y-Mode		t-Mode		x-Mode		u-Mode			
Compd.[a]	IR	IR	Raman	IR	Raman	IR	Raman	IR	Raman	IR	Raman	δ(RMR)	Refs.
R_2Hg	1067	661		456		258				207		100	81,463
						252							
						248							
$R_3{}^{10}B$	1285	893		600		650		245[b]		408		200[b]	467,468
	1248												
$R_3{}^{11}B$	1278	883		593		635		240[b]		400		200[b]	467,468
	1235												
R_3Al[c]	1085	670		460		420		207		332		150	468,469
		643											
R_3Ga[d]	1085	665		453		315		180		245		140	468
				445						225			
R_3In[d]	1070	673		465		270		180		248		105	468
										195			
R_4Si	1108	709		519		435	236	185		261		98	443,470-
				511		239		171		223		57	472
R_4Ge	1091			481		332	327	187		232	227	98	447,470-
							222	168	172	214	216	69	473
R_4Sn	1075	616		459		268		193		225		89	470-472,
				448		212	214	152				61	474-476
R_4Pb	1061	645		450	447	223	224	147	152	181	184	101	454,470-
				440	437		214					87	472
						201	199					72	
R_3P	1089			501	501	428	423	248	252	209	212		458,470,
						398	403			190	193		474,477,
													478

R_3As	1082 1074	667		474	478	313	313	237	237	192	194	438,477- 479
R_5As			668	497		353	355	200	201		268	461
R_3Sb	1065	651		457	460	270	272	216	219	166	169	438,470, 474,477, 480
R_5Sb		662	662	475		275	288		186	227	225	461
R_3Bi	1055			448	450	237 220	237 219	207	203	157	155	470,474, 477,481
R_2Se	1082 1069	692		482 458								482
R_2Tl^+	1015	668		452 438		255 240				231 229		483-485
R_4P^+	1105	687	682	526	529	454 431	463 435	209 202 195	197	283 270	285 281 257	486,487
R_4As^+	1079	670	670		505	375 364 345	370 350		186	 245	267 261 245	438,487
R_4Sb^+	1067	665			460 446		315 291			252 239		438,487
R_4Bi^+	1052	652	655 646		445 435	242	245 214		163	230		488,489

[a] $R=C_6H_5$.
[b] Calculated value from Ref. 463.
[c] A dimeric structure has been found for this compound.
[d] A polymeric structure has been found for this compound in the solid state.

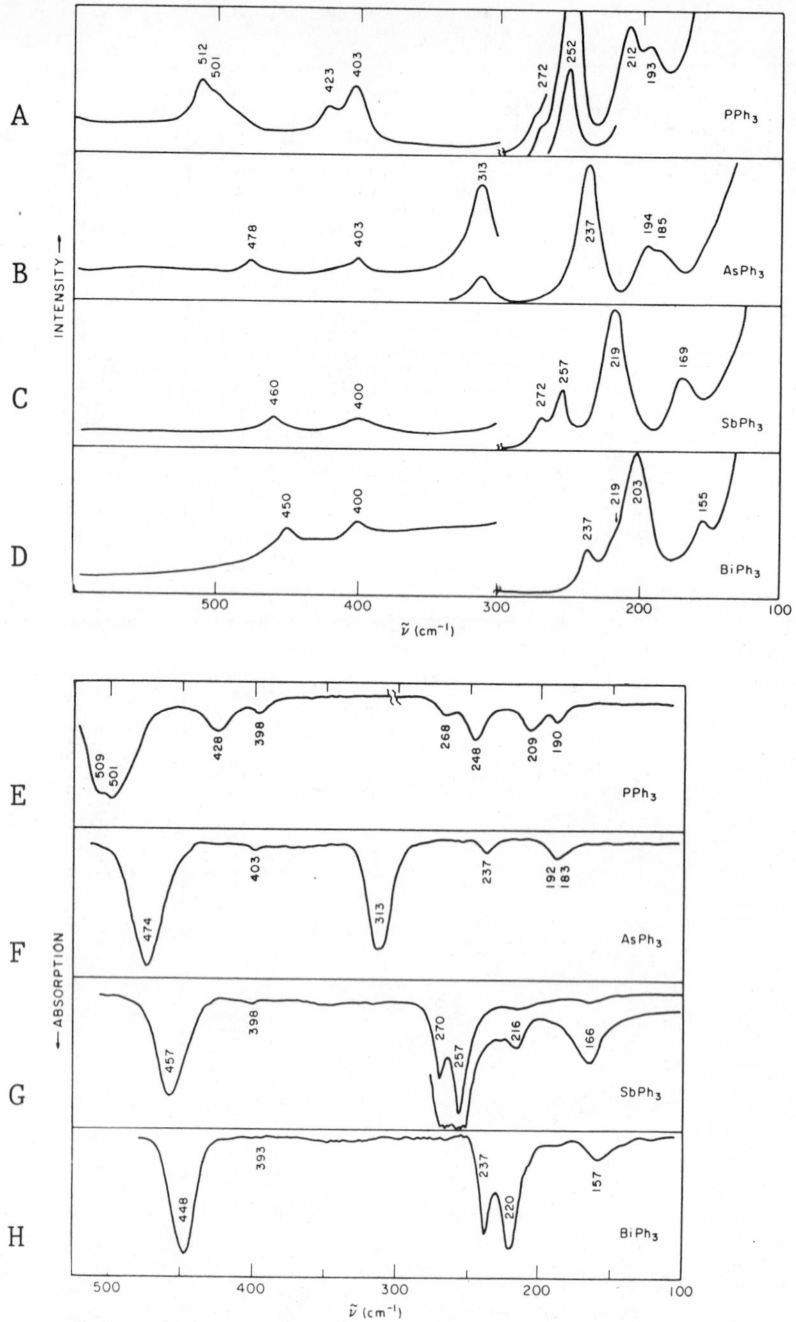

Figure 3.30. The Raman spectra (A,B,C,D) and infrared spectra (E,F,G,H) of $(C_6H_5)_3P$, $(C_6H_5)_3As$, $(C_6H_5)_3Sb$, and $(C_6H_5)_3Bi$, respectively, in benzene (477).

ture $[(C_6H_5)_4M]X$ (X=counter ion) or the five-coordinate, covalent structure $(C_6H_5)_5M$. In addition to the infrared data reported for several ionic salts of $(C_6H_5)_4M^+$ (M=P (486,487),As (438,487),Sb (438,487),Bi (488,489)), Raman studies (below 700 cm^{-1}) have also been reported to differentiate between the possibility of four-coordinate and five-coordinate structures. Therefore, while the solid-state ionic salts of $(C_6H_5)_4M^+$ (M=P,As) and $[(C_6H_5)_4Sb]ClO_4$ have a four-coordinate, tetrahedral skeleton, the derivatives $(C_6H_5)_4SbX$ (X=F,Cl,Br,OH) and $[(C_6H_5)_4Sb]_2SO_4$ appear to have a five-coordinate, trigonal bipyramidal skeleton (487). Likewise, the presence of one Raman band at 214 cm^{-1} (which is a component of the t-mode) for solid $[(C_6H_5)_4Bi]ClO_4$ has been interpreted as indicative of a tetrahedral C_4Bi skeleton, while the presence of two Raman bands at approximately 210 and 205 cm^{-1} for $(C_6H_5)_4BiX$ ($X=NO_3,NCO,NCS,CCl_3CO_2$) has been interpreted as consistent with a five-coordinate, trigonal bipyramidal skeleton (489). Although solid $(C_6H_5)_5M$ have approximate trigonal bipyramidal C_5M (M=P,As) skeletons (497), solid $(C_6H_5)_5Sb$ has an approximate square pyramidal C_5Sb skeleton (498,499). In addition to infrared data for solid $(C_6H_5)_5As$ and $(C_6H_5)_5Sb$ (438), two detailed analyses have appeared of the low-frequency, solid-state and solution (CH_2Cl_2 and CH_2Br_2 infrared and Raman spectra of both compounds (461, 500). It has been concluded in these two latter studies that both compounds retain their solid-state structures in solution, which differs from earlier suggestions (501, 502) that the unusual structure of solid $(C_6H_5)_5Sb$ is due to crystal packing factors. Infrared data have also been reported for $Li[(C_6H_5)_6Sb]$ (438).

Infrared data and some assignments have been reported for $(C_6H_5)_2M$ (M=Se (474,482),Te (503,504)) and $(C_6H_5)_4Te$ (503) as well as for $(\eta^5\text{-}C_5H_5)_2ErC_6H_5$ (351, 353) and $(\eta^5\text{-}C_5H_5)_3UC_6H_5$ (357).

4. Phenyl Compounds with Other Functional Groups

a. Hydrides

A dimeric structure with bridging hydride ligands has been proposed for $Li[(C_6H_5)_2ZnH]$ with the bridging $\nu_a(ZnH)$ mode assigned to a broad infrared band at 1650 to 1250 cm^{-1} and a ring-deformation mode assigned at 788 cm^{-1} (505). In the spectrum of $Li[(C_6H_5)_2ZnD]$, these bands shifted to 1200 to 900 cm^{-1} and 552 cm^{-1}, respectively. Both infrared and Raman data have also

TABLE 3.25
The ν(MH) and ν(MD) Mode Assignments (cm^{-1}) for $(C_6H_5)_mMH_n$ and $(C_6H_5)_mMD_n$ (M=Group IVb and Group Vb Elements) Derivatives, Respectively

Compound	ν(MH)[a]	Refs.
$C_6H_5SiH_3$	2157 (1584, 1551)	444,445
$(C_6H_5)_2SiH_2$	2147	445
$(C_6H_5)_3SiH$	2125	445
$C_6H_5GeH_3$	2072 (1983)	449
$(C_6H_5)_2GeH_2$	2053 (1474)	507
$(C_6H_5)_3GeH$	2041 (1469)	448
$C_6H_5SnH_3$	1880	509
$(C_6H_5)_2SnH_2$	1855	509
$(C_6H_5)_3SnH$	1843 (1323)	453,509,510
$C_6H_5PH_2$	2288	456,457,459,511
$(C_6H_5)_2PH$	2086	456,458
$C_6H_5AsH_2$	2089	456
$(C_6H_5)_2AsH$	2071	456
$(C_6H_5)_2SbH$	1855	512

[a]Values in parentheses refer to the ν(MD) mode assignments for the analogous deuteride derivative.

been listed and discussed for $Na[(C_6H_5)_2ZnH]$ and the di(diethyletherate) complexes of this compound as well as for $Li[(C_6H_5)_2ZnH]$ and $Li[(C_6H_5)_2ZnD]$ (439).

Additional data for phenyl-metal hydrides have been limited to compounds of the group IVb (445,448,449, 453,472,506-510) and group Vb (456-459,511,512) elements, which have monomeric structures and terminal hydride ligands. The ν(MH) and ν(MD) mode assignments for several of these derivatives are summarized in Table 3.25. The metal-hydride and -deuteride bending mode assignments included in some of these studies are less reliable than those of the corresponding stretching modes because of the possibility of coupling and/or overlap between these bending modes and the phenyl ring modes.

b. *Halides*

Group Ib. Solid-state infrared and Raman data below 400 cm^{-1} have been assigned for the tetramethylammonium and/or tetra(*n*-butyl)ammonium salts of the square planar complexes $C_6H_5AuX_3^-$ (X=Cl,Br,I) (513). The ν(AuX) mode assignments for X = Cl, Br, and I were made at approximately 290, 182, and 162 cm^{-1}, respectively, for the halide ligand *trans* to the phenyl group, while the ν_a(AuX) and ν_s(AuX) modes arising from the mutually *cis* halide ligands were assigned at approximately 359 and 340 cm^{-1}, respectively, for X=Cl, 255 and 205 cm^{-1}, respectively, for X=Br, and 183 and 153 cm^{-1}, respectively, for X=I. A chloride-bridged dimeric structure has been suggested for $C_6H_5AuCl_2$ (ν(AuCl)=368,322 cm^{-1}) (514).

Group IIb. Complete infrared data and assignments have been reported for solid C_6H_5MX (M=Zn,Cd,X=Cl,I (465); M=Hg,X=Cl,Br,I (81,463,465)). The spectra of the mercury derivatives have been interpreted in terms of covalent, monomeric structures with linear C-Hg-X skeletons. The ν(HgX) mode for X=Cl, Br, and I has been assigned to strong intensity bands at 330, 257, and 170 cm^{-1}, respectively, while the $\delta(C_6H_5\text{-}Hg\text{-}C_6H_5)$ modes appear in the 100 to 50 cm^{-1} region.

Group IIIb. Infrared (436,441,515-517) and Raman (420, 441) assignments have been reported for $(C_6H_5)_nBX_{3-n}$ (n=1,2;X=F,Cl,Br,I), all of which have a monomeric structure. When the central atom in phenyl derivatives is relatively light in mass, as is boron, the q-mode rather than the t-mode is predominantly metal-carbon stretching in character (436). The assignments made for the q-mode, the ν(B-halide) modes, and the skeletal deformation modes in the most recent vibrational studies of phenyl-boron halides are compared in Table 3.26. The spectra for the halides other than the fluorides have been reported to be consistent with structures of C_2 symmetry in which the phenyl ring(s) is (are) distorted out of the skeletal plane of the molecule (515). Important differences have been observed between the infrared spectra of the phenyl-boron fluorides and the other phenyl-boron halides. First, a larger than expected difference has been reported between the ν(BF) mode frequencies and the analogous ν(BX) mode frequencies for X=Cl, Br, and I. Also, the spectra of the fluorides are less complex than those of the other three halides. It has therefore been proposed that in the fluoride derivatives, there is delocalization of the π-electrons into the empty boron $p\pi$ orbital. This delocalization would increase the B-C and B-F bond orders and give a structure of C_{2v} symmetry in which the phenyl ring(s) is (are) in the skeletal plane of the molecule (515,516). The infrared spec-

TABLE 3.26
The q-Mode and Boron-Halide Stretching and Deformation Mode Assignments for Phenyl-Boron Halides

Compound	q-Mode	ν(BX)	Deform.	Refs.
$(C_6H_5)_2BF$	1375 1338	1348	696 582	441,515
$(C_6H_5)_2BCl$	1335 1222	910	580 565	441,515
$(C_6H_5)_2BBr$	1261 1167	837	580 550	441,515
$(C_6H_5)_2BI$	1260 1166	811	438	515
$C_6H_5BF_2$	728	1338 1078	570	441,515
$C_6H_5BCl_2$	1230	910 383	633 230	441,515
$C_6H_5BBr_2$	1225	808 278	603 206	441,515
$C_6H_5BI_2$	1208	717 691	500	515
C_6H_5BFCl	1180	1325[a] 460[b]		441
C_6H_5BFBr	1180	428[c]		441
C_6H_5BClBr		898[b] 350[c]		441

[a]ν(BF) mode.
[b]ν(BCl) mode
[c]ν(BBr) mode.

trum of a mixture of $C_6H_5BCl_2$ and $C_6H_5BBr_2$ exhibited bands consistent with the presence of C_6H_5BClBr (518). Vibrational data for various mixed halides are included in Table 3.26.

The infrared spectra have been assigned for C_6H_5M-X_2 (M=Al,Ga;X=Cl,Br), which have a dimeric, halogen-bridged structure in benzene (519). Infrared, Raman, ^{115}In nqr, mass spectral, and conductivity data have been used to characterize solid $(C_6H_5)_nInX_{3-n}$ (n=1,2; X=Cl,Br,I) (520). A polymeric structure with multiply-associated halogen atoms linking nearly linear $(C_6H_5)_2In$ units was found for $(C_6H_5)_2InX$ (X=Cl,Br,I). While C_6H_5-InX_2 (X=Cl,Br) are also polymeric, the data for $C_6H_5InI_2$ are more consistent with the ionic formulation $[(C_6H_5)_2$-$In]^+InI_4^-$.

The infrared spectra of solid $C_6H_5TlX_2$ (X=F,Cl (521),Br) have been interpreted in terms of a monomeric structure (522), with the ν(TlX) modes assigned in the 525 to 480 cm^{-1} and 342 to 335 cm^{-1} ranges for X=F and Cl, respectively. Although the infrared spectra of solid $(C_6H_5)_2TlX$ (X=Cl (483),Br (484,485)) are reported to be consistent with the presence of ionic Tl-X interactions, polymeric structures with weakly bridging halide groups cannot be ruled out. Table 3.24 includes the X-sensitive mode assignments for $(C_6H_5)_2Tl^+$

Group IVb. The derivatives $(C_6H_5)_nMX_{4-n}$ are monomeric with approximately tetrahedral skeletons for M=Si, Ge, and (with the exception of $(C_6H_5)_3SnF$) Sn. Solid $(C_6H_5)_3SnF$ is polymeric (466,523). The assignments for the phenyl halides of silicon, germanium and tin are summarized in Table 3.27.

A polymeric structure with bridging chloride ligands and six-coordinate lead atoms (1.31) has been found in a single crystal X-ray study of $(C_6H_5)_2PbCl_2$ (526). Analogous polymeric structures have been proposed for solid $(C_6H_5)_2PbX_2$ (X=Br,I) (454). In solution, however, $(C_6H_5)_2PbX_2$ (X=Cl,Br,I) become monomeric (454). Similarly, it has been proposed that $(C_6H_5)_3PbX$ (X=F,Cl,Br,I) have polymeric solid-state structures with bridging halide ligands and five-coordinate lead atoms, and monomeric structures in benzene (454) or N,N'-dimethylformamide (527) solutions. A complete analysis has been reported of the infrared (4000 to 70 cm^{-1}) and Raman (500 to 70 cm^{-1}) spectra of $(C_6H_5)_2PbX_2$ (X=Cl,Br,I) and $(C_6H_5)_3PbX$ (X=F,Cl,Br,I) in the solid state and of the infrared and Raman spectra (500 to 70 cm^{-1}) of $(C_6H_5)_2PbI_2$ and $(C_6H_5)_3PbX$ (X=Cl,Br,I) in benzene (454). These spectral data are consistent with the proposed monomeric solution structures and polymeric solid-state structures. As observed in Table 3.28, a major difference between the solid-state and benzene solution spectra of the phenyl-lead halides is present in the ν(Pb-halide) mode region with the terminal ν(Pb-halide) modes appearing at higher frequencies than the bridging ν(Pb-halide) modes.

The solid-state infrared (1300 to 100 cm^{-1}) and Raman (400 to 100 cm^{-1}) spectra have been tabulated and assigned for the tetraalkylammonium salts of $(C_6H_5)_3SnCl_2^-$ (ν(SnCl)=240 cm^{-1}), $(C_6H_5)_2SnCl_3^-$ (ν(SnCl)=330 and 250 cm^{-1}) and $(C_6H_5)_2SnCl_4^{2-}$ (ν(SnCl)=267, 254, and 237 cm^{-1}) (528). No attempt has been made to discuss the structures of these anions.

Group Vb. A new interpretation (511) has been offered for the infrared and Raman data originally thought (457) to have been for $C_6H_5PF_2$. This compound has now been shown (529,530) to have been $C_6H_5PF_4$. A more recent vibrational study has been reported for $C_6H_5PF_2$ (530). Although the phenyl halides of phosphorus(III) and arsenic(III) have monomeric structures, the structures of the corresponding bismuth(III) halides are uncertain. It has been suggested that in the solid state, $C_6H_5BiX_2$ (X=Cl,Br,I) have a polymeric structure with

TABLE 3.27
X-Sensitive and Metal-Halide Mode Assignments (cm^{-1}) for Phenyl Halides of Silicon, Germanium, and Tin

Compd.[a]	q-Mode	r-Mode	y-Mode	t-Mode	u-Mode	x-Mode	ν(MX)	δ(MX)	Refs.
$RSiF_3$			472		269	238	970		445
							862		
R_2SiF_2	1140		515	396	256	228	918		445
	1128						681		
R_3SiF	1128		515	355		231	852		445,524
$RSiCl_3$	1122	719	460	346	284	244	619		443,445,449
							596		
							518		
R_2SiCl_2	1120	717	481	444	283	211	576	166	443,445
	1108	697		318	245		540	136	
								132	
R_3SiCl	1117	716	504	434	250	212	552	120	443,445
	1104	699		306	240	170			
R_3GeF	1098	681	464	336	246	216	653		438,448,452,524
		668	420	287	237	169			
					228				
$RGeCl_3$	1085	677	454	289	255	216	427	159	449,451,472,473
							407	142	
$RGeCl_3$[b]	1037	651		293	249	207	426	159	451
							412	143	
R_2GeCl_2	1095	684	459	338	238	199	403	145	438,508
	1088	675		278				120	
R_3GeCl	1092	678	469	329	258	205		89	440,448,473
		668	460	237	249	171			
					232				
$RGeBr_3$	1083	678	451	328	200	59	314	115	473
							229	97	

$RGeBr_3$[b]	1031	647	405	323	187	58	311	114	451
							229	94	
R_2GeBr_2	1089	680	462	329	238	190	224	94	438,473,5-8
		672	453	314				58	
R_3GeBr	1090	678	463	329	249	190	221		438,448,473
		666	455	308	235	170			
					230				
R_3GeI	1089						283		473
R_3SnF	1079	616	453	283		174	555[c]	206-	466,476,524,525
				214			371[d]	190	
$RSnCl_3$	1067	660	440	253	218	185	380	132	449,472,476,525
							370	114	
R_2SnCl_2	1069	614	442	273	213	180	364	115	472,475,476,524,
				235			356	95	525
R_3SnCl	1072	614	452	268	217	182	344	87	450,472,475,476
			448	233	210	176		77	
R_3SnBr	1074	616	449	276	232	173	251		466,476,524,525
			444	269	222				
					209				
$RSnI_3$	1055		442	259			228		476,525
R_2SnI_2	1065	614	443	273		166	208		476,524,525
				248			183		
R_3SnI	1075	617	445	270	216		170		466,476,525
				240					

[a] $R=C_6H_5$ unless otherwise indicated.
[b] $R=C_6D_5$.
[c] For vapor-state sample.
[d] For solid-state sample.

TABLE 3.28
Lead-Halide Stretching Mode Assignments (cm^{-1}) for Phenyl-Lead Halides in the Solid State and Benzene Solution[a]

Compound	Phase[b]	ν(Pb-halide) Mode Assignments Infrared		Raman	
$(C_6H_5)_3PbF$	Solid	343s,br	299m,br		
$(C_6H_5)_3PbCl$	Solid	180s			
	Solution	290s	284sh	301s	
$(C_6H_5)_3PbBr$	Solid	121vs,br			
	Solution	191sh		191vs,p	
$(C_6H_5)_3PbI$	Solid	122vs		121vvs	
	Solution	149s		151 s,p	
$(C_6H_5)_2PbBr_2$	Solid	122s			
$(C_6H_5)_2PbI_2$	Solid	136s	121s	140m	118s
	Solution	160s	144s	162sh	152vs,p

[a]Reference 454.
[b]Infrared spectra recorded as Nujol or hexachlorobutadiene mulls or in benzene solution; Raman spectra recorded as powders or in benzene solution.

bridging halogen ligands (531). This is offered as a possible explanation for the fact that the ν(Bi-halide) modes show an increase in frequency of approximately 20 to 30 cm^{-1} when these compounds form complexes with 1,10-phenanthroline or 2,2'-bipyridine, rather than a frequency decrease, which might be expected if the uncomplexed compounds were monomeric. The ν(M-halide) mode assignments for trivalent phenyl-group Vb halides are summarized in Table 3.29.

As was true for the corresponding alkyl derivatives, there is an increasing tendency toward covalence in the pentavalent phenyl-group Vb halides as the mass of the central atom increases for a given series or as the mass of the halogen decreases. Therefore, although the vibrational spectra of the known tetraphenylphosphorus(V) halides (486,487) and tetraphenylarsenic(V) halides (438,487,534) show them to have four-coordinate, ionic structures, the solid tetraphenylantimony(V) halides have covalent structures with trigonal bipyramidal skeletons (447,487,535). Also, while $(C_6H_5)_4SbF$ retains a covalent structure in methanol solution, vibrational data indicate that the other tetraphenylantimony(V) halides dissociate to give tetrahedral $(C_6H_5)_4Sb^+$ cations (487). In contrast, a recent pulsed

TABLE 3.29
Metal-Halide Stretching Mode Assignments (cm^{-1}) for Trivalent and Pentavalent Phenyl-Group Vb Halides

Compound	ν(MX)	Refs.
$C_6H_5PF_2$	817 793	530
$(C_6H_5)_2PCl$	500	455,511,532
$C_6H_5PCl_2$	500 495	455,511,532
$C_6H_5PBr_2$	403 374	457,511
$(C_6H_5)_2AsCl$	372	533
$C_6H_5AsCl_2$	395 369	455,533
$(C_6H_5)_2AsBr$	315[a] 291[a]	533
$C_6H_5AsBr_2$	312[a] 290[a] 276[a]	533
$C_6H_5BiBr_2$	120	531
$C_6H_5BiI_2$	90	531
$C_6H_5AsCl_4$	420[b] 398	533
$(C_6H_5)_2AsCl_3$	423[b]	438
$(C_6H_5)_3AsCl_2$[c]	270	438
$(C_6H_5)_3SbF_2$	509 485	538,540
$(C_6H_5)_4SbF$	400	487,535
$(C_6H_5)_3SbCl_2$	275	538
$(C_6H_5)_4SbCl$	353	487
$(C_6H_5)_3SbBr_2$	188 161	538
$(C_6H_5)_4SbBr$	263[d]	487
$(C_6H_5)_3BiF_2$	434 410	537,541
$(C_6H_5)_3BiCl_2$	255 230	537,541,542
$(C_6H_5)_3BiBr_2$	153	537,542
$(C_6H_5)_2SbF_4^-$[e]	515 496	545
$(C_6H_5)_2SbCl_4^-$[e]	273 267	545
$(C_6H_5)_2SbBr_4^-$[e]	193	545
$(C_6H_5)_2SbCl_3Br^-$[e]	267[f]	545
$(C_6H_5)_2SbCl_3N_3^-$[e]	266[f]	545
$(C_6H_5)_2SbCl_3NCS^-$[e]	270[f]	545

[a]The t- and ν(AsBr) modes overlap.
[b]Axial mode.
[c]In a more recent study (533), the ν(AsCl) mode was not assigned because of mixing between this mode and the t-modes which appear in the same frequency region.
[d]Assigned to the same band as the u-mode.
[e]$(C_6H_5)_4As^+$ salt.
[f]ν(SbCl) mode.

nmr study of the nuclear relaxation of the ^{121}Sb nucleus of $(C_6H_5)_4SbCl$ dissolved in methanol indicates that the coordination number of antimony is not four (536). This nmr study does, however, support the conclusion of the vibrational study (487) that in an aqueous solution the tetraphenylantimony(V) halides form five-coordinate $(C_6H_5)_4SbOH_2^+$ cations. Also, as was noted for the corresponding alkyl derivatives, covalence increases as phenyl groups are replaced with halide ligands. In another recent study, it has been concluded that the infrared and Raman spectra of $(C_6H_5)_3BiX_2$ (X=F,Cl,Br) in the solid state and benzene solution are consistent with the presence of trigonal bipyramidal skeletons for all three compounds (537). Therefore, all of the pentavalent triphenylgroup Vb dihalides have trigonal bipyramidal skeletons with the halides in the axial positions (438,533,537-544). The skeletal mode vibrational assignments for $M[(C_6H_5)_2SbX_4]$ ($M=(C_6H_5)_4As$, $X_4=F_4,Cl_4,Br_4,Cl_3Br,Cl_3N_3,Cl_3NCS$; $M=(CH_3)_4N$, $X_4=Cl_4,Cl_3Br,Cl_3N_3$) in the solid state have been reported to be consistent with the *trans*-$(C_6H_5)_2SbX_4^-$ species (545). The ν(M-halide) mode assignments for neutral and anionic, pentavalent phenylgroup Vb halides are summarized in Table 3.29.

Group VIb. The phenyl modes have been assigned for $(C_6H_5)_nSeCl_{4-n}$ (n=1 to 3) (482). A monomeric structure has been found in a single crystal X-ray study of $(C_6H_5)_2TeBr_2$ (3.66) (546). The solid-state infrared and

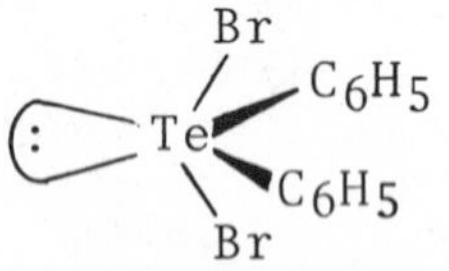

(3.66)

Raman spectra below 400 cm^{-1} are consistent with this structure for $(C_6H_5)_2TeX_2$ (X=Cl,Br,I) (504). The assignment of these spectra is complicated by the overlap of the ν(TeX) and X-sensitive modes. Although the ν_s(TeCl) mode has been assigned at a higher frequency than the ν_a(TeCl) mode for $(C_6H_5)_2TeCl_2$, the ν_a(TeX) mode has been assigned at a higher frequency than the ν_s(TeX) mode for $(C_6H_5)_2TeX_2$ (X=Br,I). The assignment of the infrared and Raman spectra below 400 cm^{-1} for $(C_6H_5)_2TeX_2$ (X=Cl,Br) has been aided by the preparation of these complexes using the stable isotopes ^{126}Te and

^{130}Te (547). It has been concluded in this study that while the ν_s(TeCl) and ν_s(TeC) modes of $(C_6H_5)_2TeCl_2$ are coupled, the ν_s(TeBr) mode of $(C_6H_5)_2TeBr_2$ is a relatively pure vibration.

The solid-state infrared and Raman spectra below 400 cm^{-1}, as well as other data, suggest that $C_6H_5TeX_3$ have a dimeric structure (3.67) for X=Cl and I, and a more highly associated structure for X=Br (548). More complex structures, however, are also possible for these compounds

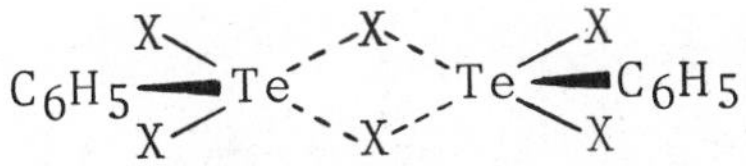

(3.67)

The low-frequency infrared (solid state) and Raman (solid state and CH_3CN solution) spectra of the triphenylmethylphosphate salts of $C_6H_5TeX_4^-$ (X=Cl,Br) are consistent with monomeric structures and square pyramidal skeletons (549). In Table 3.30, the ν(Te-halide) mode assignments are summarized for neutral and anionic phenyl-tellurium halides.

c. Pseudohalides

The nature of the vibrational modes expected for pseudohalide ligands have been discussed in Sec. E.3 of Chapter 1.

Complete infrared and Raman data for C_6H_5HgSCN have been interpreted in terms of a dimeric structure and a Hg_2S_2 ring of D_{2h} symmetry (440).

The low frequency of the $\nu_a(N_3)$ mode in $(C_6H_5)_2TlN_3$ (2000 cm^{-1}) has been interpreted as indicating the presence of a highly ionic Tl-N interaction, although the unusually high thermal stability of this compound has been attributed to nitrogen to thallium dative π-bonding (550). From the group frequencies of the NCS ligand in the infrared spectra and the integrated intensity of the ν(CN) mode in acetone solution, it has been concluded that a nitrogen-bonded isothiocyanate structure with a highly ionic Tl-N interaction is present for $(C_6H_5)_2Tl$-NCS (ν(CN)=2050 (vs) cm^{-1}) while a sulfur-bonded thiocyanato group is found for $C_6H_5Tl(SCN)_2$ (ν(CN)=2130 (m,br) cm^{-1}) (551). The appearance of a sharp band at 2090 cm^{-1} for $(C_6H_5)_2TlSeCN$ has been given as evidence that the SeCN ligand is selenium-bonded to thallium (552,553).

TABLE 3.30
Tellurium-Halide Stretching Mode Assignments (cm^{-1}) for Neutral and Anionic Phenyl-Tellurium Halides

	ν(Te-halide) Mode Assignments[a]				
	Terminal		Bridging[c]		
Compound[b]	Infrared	Raman	Infrared	Raman	Refs.
$R_2{}^{126}TeCl_2$	286.5	289			504,547
	265	267.5			
$R_2{}^{130}TeCl_2$	284.5	287			504,547
	263.5	267			
$R_2{}^{126}TeBr_2$	186[d]	184[d]			504,547
	160	158.5			
$R_2{}^{130}TeBr_2$	185[d]	d			504,547
	160	158.5			
R_2TeI_2	129	116			504
$RTeCl_3$	337s	342s	178s	174w	548
	317sh	318m	153sh	143m	
	306s	303m			
$RTeBr_3$	220sh		134m	132w	548
	213s	210sh	123m	120vw	
	198m	198vs	113sh		
$RTeI_3$	169s	168s	97m	94m	548
	158s	153s	86m		
$RTeCl_4^-$	282vw,sh	282s			549
		273w,sh			
	256s	250s			
$RTeBr_4^-$		181w-m			549
		167s			
		148m			

[a]All data from solid samples.
[b]$R=C_6H_5$.
[c]Bridging ν(TeX) and δ(TeX_2) modes are coupled.
[d]Overlaps with the phenyl x-mode at 184 cm^{-1}.

The frequency trend of the $\nu_a(N_3)$ mode for $(C_6H_5)_3$-MN_3 (M=Si (2149 cm^{-1}),Ge (2100 cm^{-1}),Sn (2093 cm^{-1},Pb (2046 cm^{-1})) has been attributed to an increasing degree of ionic character in the metal-azide bond as the mass of the metal atom increases (554). Nitrogen-bonded, isothiocyanate structures have been proposed for $(C_6H_5)_3$-GeNCS (555) and $(C_6H_5)_3SnNCS$ (556) with the aid of infrared data. A recent single crystal X-ray study of $(C_6H_5)_3SnNCS$, however, has shown it to have a polymeric solid-state structure consisting of planar C_3Sn units and zigzag Sn-NCS···Sn-NCS chains bent only at the sulfur (557). The strong intensity infrared ν_a(NCS) band appearing at 2100 cm^{-1} in solid $(C_6H_5)_2Sn(NCS)_2$ as compared to a range of 2040 to 2020 cm^{-1} for the solid, monomeric, six-coordinate complexes $(C_6H_5)_2Sn(NCS)_2X$ (X=Bipyr,Phen) indicates that $(C_6H_5)_2Sn(NCS)_2$ has a polymeric solid-state structure with bridging NCS ligands (558). Similarly, on the basis of solid-state infrared data, isothiocyanate structures have been proposed for $(C_6H_5)_nPb(NCS)_{4-n}$ (n=2,3) with some indication that these compounds are polymeric with bridging NCS ligands (559). Using Mossbauer, Raman, and infrared data, trigonal bipyramidal skeletons with apical pseudohalide ligands have been proposed for the tetraphenylarsonium and tetramethylammonium salts of $(C_6H_5)_3Sn(N_3)_2^-$ and $(C_6H_5)_3Sn(N_3)NCS^-$ while a *trans* octahedral skeleton has been proposed for the tetraphenylarsonium salt of $(C_6H_5)_2Sn(N_3)_2(NCS)_2^{2-}$ (560). Furthermore, the NCS ligands are nitrogen-bonded for these monomeric derivatives. Monomeric structures have also been proposed for $(C_6H_5)_2Pb(NCS)_4^{2-}$ and $(C_6H_5)_3Pb(NCS)_2^-$ (559). The broad infrared band appearing at 2108 cm^{-1} for solid $(C_6H_5)_3SnNCSe$ is resolved into a weak band at 2144 cm^{-1} and a strong, broad band at 2055 cm^{-1} (552). From these data, it has been concluded that $(C_6H_5)_3SnNCSe$ has a nitrogen-bonded isoselenocyanate structure (552). Both $(C_6H_5)_3PbSeCN$ and $(C_6H_5)_2Pb(SeCN)_2$, however, show a sharp, single infrared band of moderate intensity at approximately 2100 cm^{-1}, indicating the presence of selenium-bonded selenocyanate ligands (552).

On standing, liquid $C_6H_5P(NCO)_2$ has been reported to polymerize to a viscous, yellow-brown glass. The change was accompanied by a decrease in the intensity of the ν_a(NCO) mode (2240 cm^{-1}) of the liquid and the coincidental appearance of a ν(C=O) mode (1696 cm^{-1}) (561). A trimeric structure (3.68) has been proposed for polymeric $C_6H_5P(NCO)_2$ (561). Vibrational data have also been re-

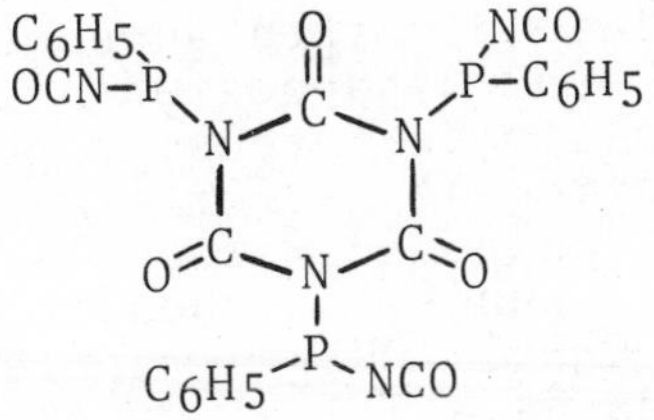

(3.68)

ported for the trivalent derivatives $(C_6H_5)_2AsN_3$ and several other covalent phenyl-arsenic(III) azides (562), and $(C_6H_5)_2BiSCN$ (563). More data are available for pentavalent phenyl-group Vb pseudohalides. It has been proposed that tetraphenylarsenic(V) tellurocyanate (564) and cyanate dihydrate (534) have ionic structures with $(C_6H_5)_4As^+$ cations, while $(C_6H_5)_4SbX$ (X=N_3 (460,565), NCO (460,565),NCS (538,565)), $[(C_6H_5)_3SbX]_2O$ (X=N_3,NCO, NCS) (565), $(C_6H_5)_3BiX_2$ (X=N_3 (566),NCO (537,541),CN (566)), and $[(C_6H_5)_3BiNCO]_2O$ (567) have covalent structures and axial pseudohalide ligand(s). A single crystal X-ray study has confirmed this structure for $[(C_6H_5)_3SbN_3]_2O$ (568). The skeletal mode vibrational assignments for $M[(C_6H_5)_2SbX_4]$ (M=$(C_6H_5)_4As$,X_4=$(N_3)_4$, $(NCS)_4$,Cl_3N_3,Cl_3NCS;M=$(CH_3)_4N$,X_4=Cl_3N_3) in the solid state have been reported as consistent with the *trans*-$(C_6H_5)_2SbX_4^-$ species (545).

Infrared assignments have been reported for three polymeric forms of a compound with the empirical formula $(C_6H_5)_2Se(NCO)_2$ (482). The assignments made for the modes arising from the pseudohalide ligand for several phenyl pseudohalide derivatives are summarized in Table 3.31.

d. *Hydroxides, Oxides, and Peroxides*

The structure of a compound formulated as C_6H_5HgOH has yet to be fully characterized. In addition to a strong intensity infrared band that is due to the ν(OH) mode (3385 cm^{-1}), this compound exhibits three medium intensity bands at 620, 600, and 516 cm^{-1} that cannot be attributed to phenyl modes and are more than the number expected for the ν(HgO) mode of C_6H_5HgOH (81). The two alternative structures that have therefore been suggested (81) as more consistent with these infrared data are $[(C_6H_5Hg)_3O]OH$, which has a distorted, pyramidal skeleton and three ν(HgO) modes, and $[(C_6H_5Hg)_2OH]OH$. Heating this compound at 250°C for 45 minutes resulted in the almost complete disappearance of the ν(OH) band. Also, the 516 cm^{-1} band disappeared and the 620 and 600 cm^{-1} bands became a strong intensity, broad band at 618 cm^{-1} with an inflection at 592 cm^{-1}. It has been suggested that the new compound formed is $(C_6H_5Hg)_2O$ with the 618 and 592 cm^{-1} bands assigned to ν(HgO) modes (81).

The fact that $C_6H_5Tl(OH)_2$ is appreciably soluble in methanol and behaves as a 1:1 electrolyte in this solvent indicates that it has a partially ionic structure (522). The ν(OH) mode of solid $C_6H_5Tl(OH)_2$ is easily

TABLE 3.31
Vibrational Assignments (cm^{-1}) for the Pseudohalide Modes of Phenyl-Metal Pseudohalides[a]

Compound[b]	ν_a(XYZ)	ν_s(XYZ)	δ(XYZ)	ν(M-Pseud.)	Refs.
$RZnN_3$	2108	1276	639		569
			590		
$RHgN_3$	2076	1278	652		570
R_2BN_3	2120				571
R_2TlN_3	2000	1325	648		550
$R_2Si(N_3)_2$	2136	1324	679		570
R_3SiN_3	2149	1308	660		544
R_3GeN_3	2100	1280	660		544
R_3SnN_3	2093	1265	658		544
$R_3Sn(N_3)_2^-$ [c]	2070				560
	2050				
$R_3Sn(N_3)NCS^-$ [c]	2070[d]		478	305[d]	560
	2060[d]		472[d]		
	2020[e]				
$R_2Sn(N_3)_2$-$(NCS)_2^{2-}$ [f]	2065[e]				560
	2050[d]				
R_3PbN_3	2046	1261	655		554
$R_2Pb(N_3)_2$	2040	1280	650		572
R_2AsN_3	2084	1250		441	562
$R_3Sb(N_3)_2$	2072	1268	648	355	460,464
R_4SbN_3	2045	1328	636		535
$R_2Sb(N_3)_4^-$ [f]	2070	1275	650		545
$R_2SbCl_3N_3^-$ [c]	2085	1285		346	545
$(R_3SbN_3)_2O$	2071	1284	658	319	565
$R_3Bi(N_3)_2$	2050	1265		328[g]	566
				303	
$RB(NCO)_2$	2257	1220	607		573
R_2BNCO	2283	1220	625		573
R_3GeNCO	2242	1430[h]	624	450	438,555
			610		
R_3SnNCO			610	380	574
$RP(NCO)_2$ [i]	2240				561
$R_3Sb(NCO)_2$	2208	1370[g]	633	335	460,565
R_4SbNCO	2180		640		535
$(R_3SbNCO)_2O$	2209		642	332	565
			622		
$R_3Bi(NCO)_2$				310	537
				274	
R_4BiNCO	2160	1250	636		489
			626		
$(R_3BiNCO)_2O$	2190				567

TABLE 3.31 (continued)

Compound[b]	ν_a(XYZ)	ν_s(XYZ)	δ(XYZ)	ν(M-Pseud.)	Refs.
$RB(NCS)_2$	2024	849	633		573
R_2BNCS	2075	891	660		573
RHgSCN	2182	729	446-	230[g]	440
	2125[g]	668[g]	223	223	
R_2TlNCS	2050	755	480		551
$RTl(SCN)_2$	2130		455	280	551
			420		
R_3GeNCS	2060	904	478	368	438,555
$R_2Sn(NCS)_2$	2100				558
R_3SnNCS			465	326	556,574
R_3PbNCS	2092[j]		464[j]		559
	2050[j]		458[j]		
	2039[k]				
$R_2Pb(NCS)_2$	2098[j]	832[j]	462[j]		559
	2050[j]	765[j]	440[j]		
$R_3Pb(NCS)_2^-$[c]	2090[j]		462[j]		559
	2045[j]		452[j]		
	2040[k]				
$R_2Pb(NCS)_4^-$[c]	2054[j]	832[j]	490[j]		559
	2039[k]	790[j]	472[j]		
$R_3As(NCS)_2$	2000	861	479		563
$R_3Sb(NCS)_2$	2022	865	495	265	538,563,
			481		565
R_4SbNCS	2040	795	480		535
$R_2Sb(NCS)_4^-$[f]	1990	870		302	545
$R_2SbCl_3NCS^-$[f]	1985				545
$(R_3SbNCS)_2O$		854		382	565
R_2BiSCN	2080	764	444		563
			438		
R_4BiNCS	2045	762	470		489
RHgSeCN	2129	542	389	246[l]	575
			375		
$R_2TlSeCN$	2090	571	402	348	552,553
$R_3SnNCSe$	2108[j]	615			552
	2144[m]				
	2055[m]				
$R_3PbSeCN$	2102	564			552
$R_2Pb(SeCN)_2$	2100	552			552
RHgCNO	2194	1200	484	314	485,576
R_2TlCNO	2042	1072	488	260	485
R_3SiCNO	2200	1302	526	562	577
R_3GeCNO	2164	1276			577
R_3SnNCO	2156	1165	484	499	577
R_3PbCNO	2123	1149	483	492	577

TABLE 3.31 (continued)

Compound[b]	ν_a(XYZ)	ν_s(XYZ)	δ(XYZ)	ν(M-Pseud.)	Refs.
R_3GeCN	2135				555
$R_3Bi(CN)_2$	2135			289[g]	566
				275	
RSeCN	2154		390	523	575
			349		
$RPd(CN)_3^{2-f}$	2120				578
	2101				

[a]All data are from infrared spectra unless otherwise indicated.
[b]$R=C_6H_5$.
[c]$(CH_3)_4N^+$ salt.
[d]NCS mode.
[e]N_3 mode.
[f]$(C_6H_5)_4As^+$ salt.
[g]Data taken from the Raman spectrum.
[h]This band was also assigned to the n-mode.
[i]Data for the monomer as a liquid.
[j]Data recorded for solid-state samples as mulls.
[k]Data recorded for acetone solutions of the sample.
[l]This band was also assigned to the t-mode.
[m]Data recorded for $CHCl_3$ solutions of the sample.

assigned to a medium intensity infrared band at 3575 cm^{-1}, while bands at 698 and 550 cm^{-1} have been assigned to possibly the δ(TlOH) modes (522). No assignment has been made of the ν(TlO) modes. Some phenyl and X-sensitive mode assignments have been reported for $[(C_6H_5)_2Tl]_2O$ (483).

Complete vibrational data have been reported for the silanols $(C_6H_5)_3SiOH$ (ν(OH)=3664 cm^{-1}) (445,579) and $(C_6H_5)_2Si(OH)_2$ (ν(OH)=3600 cm^{-1}) (445), and the oxides $[(C_6H_5)_3Si]_nO$ (n=2 to 4) (445). The ν_a(SiO) modes of these oxides have been assigned to a strong intensity infrared band at 1075, 1018, and 1085 cm^{-1} for n=2, 3, and 4, respectively (445). A strong intensity infrared band at 858 cm^{-1} has been assigned to the ν_a(GeO) mode of $[(C_6H_5)_3Ge]_2O$ (438,473). The infrared spectra have been assigned for $[(C_6H_5)_3GeOGaR_2]_2$ ($R=CH_3,C_6H_5$) (580). In a complete infrared and Raman study of $(C_6H_5)_3SnOH$, a medium intensity infrared band at 3620 cm^{-1}, which was not observed in the infrared spectrum of $[(C_6H_5)_3Sn]_2O$, was assigned to the ν(OH) mode (466). In another infrared study of both $(C_6H_5)_3SnOH$ and $[(C_6H_5)_3Sn]_2O$, however, it was reported that a band at 3646 cm^{-1} appears in the spectra of both compounds and cannot be explained as due simply to the ν(OH) mode of $(C_6H_5)_3Sn$-OH (581). More recently, however, it has been reported that a 3620 cm^{-1} band of $(C_6H_5)_3SnOH$ is indeed absent from the infrared spectrum of $[(C_6H_5)_3Sn]_2O$ and that it shifts to 2669 cm^{-1} on deuteration of the OH group of $(C_6H_5)_3SnOH$ (582). Furthermore, deuteration of $(C_6H_5)_3$-SnOH shifted the δ(SnOH) mode, which had been assigned to a strong intensity doublet at 910/890 cm^{-1}, to a single band at 637 cm^{-1} (582). The ν(SnO) mode of $(C_6H_5)_3SnOH$ has been assigned at 456 cm^{-1} using infrared data (466). The ν_a(SnO) mode of $[(C_6H_5)_3Sn]_2O$ is easily assigned to a strong intensity infrared band at 774 cm^{-1} (466,525,581,582) while the ν_s(SnO) mode has been assigned to a Raman band at 240 cm^{-1} (466). Although solid $(C_6H_5)_2SnO$ is probably polymeric (523), two different structures have been proposed as possible. In the first, it is assumed that linear Sn-O chains are present and that there are weak interactions between these chains (583). The second proposed structure consists of Sn_2O_2 rings (523,584). Infrared bands at 571 and 553 cm^{-1} have been assigned to the ν_a(SnO) modes of $(C_6H_5)_2SnO$ (525,582), while strong intensity infrared bands at 326 and 320 cm^{-1} have been assigned to ν_s(SnO) modes (523). A polymeric structure has also been proposed for C_6H_5SnO with an infrared band at 573 cm^{-1} assigned to an ν(SnO) mode (582).

Limited vibrational assignments have been made for the peroxide derivatives $[(C_6H_5)_3Si]_2OO$ (ν(OO)=895 cm^{-1}, ν(SiO)=800 cm^{-1}) and $(C_6H_5)_3SiOOGe(C_6H_5)_3$ (ν(OO)=905 cm^{-1}, ν(SiO)=780 cm^{-1}, ν(GeO)=640 cm^{-1}) (585). For R_3Si-$OOM(C_6H_5)_3$ (M=Si, R=CH_3, C_2H_5, *n*-C_3H_7, *n*-C_4H_9; M=Ge, R=CH_3, C_2H_5, *n*-C_4H_9) a characteristic band in the 935 to 890 cm^{-1} region has been assigned to the ν(OO) mode while bands in the 797 to 770 cm^{-1} and 638 to 635 cm^{-1} regions have been assigned to the ν(SiO) and ν(GeO) modes, respectively (585).

Complete vibrational assignments have been reported for $(C_6H_5)_3PO$ (586) and $(C_6H_5)_3AsO$ (586,587) with the ν(MO) mode assigned at 1193 and 880 cm^{-1}, respectively. The assignment of the ν(AsO) mode has been confirmed using ^{18}O-labeled $(C_6H_5)_3AsO$ (587). In an infrared study (500 to 200 cm^{-1}) of $(C_6H_5)_3MO$ (M=P,As), the frequencies of the X-sensitive phenyl modes have been correlated with the reduced mass of M-C_6H_5 (M=P,As) and the electronegativity sums of the substituents (588). Although $(C_6H_5)_3MO$ (M=P,As) are both monomeric in the solid state and solution, this is not true for the analogous antimony and bismuth compounds. Monomeric $(C_6H_5)_3SbO$ had been reported as a product of heating $(C_6H_5)_4SbOH$ in xylene (589). Another group, however, reports that heating $(C_6H_5)_4SbOH$ in *p*-xylene gives a product that does not analyze for $(C_6H_5)_3SbO$, and that the products prepared by other methods that do analyze for $(C_6H_5)_3SbO$ are not monomeric but polymeric (590). Although the $\nu_a(SbO)$ and $\nu_s(SbO)$ modes of polymeric $(C_6H_5)_3SbO$ had been assigned at 744 and 669 cm^{-1}, respectively (590), it has more recently been concluded (537) that the assignment of the $\nu_s(SbO)$ mode is incorrect. A polymeric structure with five-coordinate bismuth atoms and bent Bi-O-Bi bonds has been proposed for solid $(C_6H_5)_3BiO$ (537,567); the $\nu_a(BiO)$ and $\nu_s(BiO)$ modes have been assigned at 620 (537,567) and 336 cm^{-1} (537), respectively.

Infrared assignments have been reported for $(C_6H_5)_3POSn(C_6H_5)_3$ and $(C_6H_5)_2P(=O)OSn(C_6H_5)_3$ (591). For the former compound, the $\nu(PO)$ and $\nu(SnO)$ modes have been assigned at 555 and 535 cm^{-1}, and 394 cm^{-1}, respectively. Infrared data have also been reported for $[(C_6H_5)_3MX]_2O$ (M=Sb,X=Cl (592),NO_3 (592),N_3 (565),NCO (565),NCS (565);M=Bi,X=Cl (537),Br (537),NCO (567)).

e. Sulfides

An infrared assignment of $(C_6H_5Hg)_2S$ did not include data for the $\nu(HgS)$ modes (81). Some phenyl and X-sensitive mode assignments have been reported for $[(C_6H_5)_2Tl]_2S$ (483).

Assignments made for the metal-sulfur modes of $(C_6H_5)_3MSM'(C_6H_5)_3$ (M and M'=Ge,Sn,Pb) are summarized in Table 3.32. Diphenyltin sulfide is known to be trimeric with infrared and Raman bands at 377 and 345 cm^{-1}, respectively, assigned to $\nu_a(SnS)$ modes (523,594,595) and a band at approximately 321 cm^{-1} in both the infrared (523,595) and Raman (594) spectra assigned to a $\nu_s(SnS)$ mode. Some Raman assignments have also been reported for $(C_6H_5)_4Sn_4S_6$ ($\nu_a(SnS)$=340,333 cm^{-1},$\nu_s(SnS)$=319 cm^{-1} (594).

Vibrational assignments have been reported for $(C_6H_5)_3MS$ (M=P,As (596),Sb) (481) assuming monomeric structures. The $\nu(AsS)$ mode assignment for C_6H_5AsS (465 cm^{-1}), however, together with a molecular weight determination in bromoethane, indicates that it is a tetramer (596). A complete infrared and Raman study and a normal coordinate analysis have been reported for diphenyldiarsine trisulfide (3.69) ($\nu(SS)$=492 cm^{-1})

TABLE 3.32
Metal-Sulfur Mode Assignments (cm^{-1}) for Symmetrical and Unsymmetrical Hexaphenyl-Di(Group IVb) Sulfides[a]

Compound	ν_a(MSM') IR	ν_a(MSM') Raman	ν_s(MSM') IR	ν_s(MSM') Raman	δ(MSM') IR	δ(MSM') Raman
$R_3GeSGeR_3$	417s		385m		125s	
$R_3GeSSnR_3$	404s	410	355m	359	101m	
$R_3GeSPbR_3$	400s		303m		94m	
$R_3SnSSnR_3$	376s	373m	330m	327vs	73m	82
$R_3SnSPbR_3$	365s		305m		72m	
$R_3PbSPbR_3$	336s		278s			

[a]Data for all compounds from Ref. 593; data for $[(C_6H_5)_3Sn]_2S$ also found in Refs. 523, 574, and 594.
[b]$R=C_6H_5$.

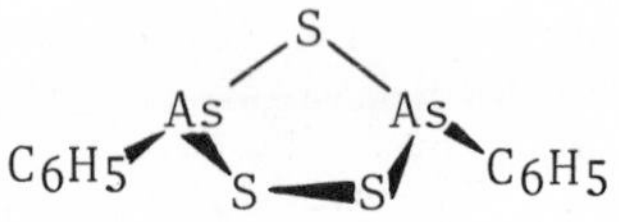

(3.69)

(459). Infrared bands at 648, 528, and 340 cm^{-1} have been assigned to the ν(P=S), ν(PS), and ν(SnS) modes, respectively, for $(C_6H_5)_2P(=S)SSn(C_6H_5)_3$ (591).

f. *Alkoxides, Organoperoxides, Ethers, Alkyl- and Phenylsulfides, and Diorganosulfides*

The infrared and Raman spectra of $(C_6H_5)_nInCl_{3-n}$·Dioxane (n=0 to 3) are consistent with the presence of monomeric structures (482). Several phenyl and X-sensitive modes have been assigned for $(C_6H_5)_2TlOC_6H_5$ (483).

Tentative assignments have been made of the infrared and Raman spectra below 600 cm^{-1} for $(C_6H_5)_{4-n}Si(OCH_3)_n$ (n=1 to 3) (472). In $(C_6H_5)_4SbOR$ ($R=CH_3,C_2H_5$, n-C_3H_7,i-C_3H_7), the alkoxide ligand occupies an axial position with the ν(SbO) mode assigned in the 335 to 320 cm^{-1} region (597). Detailed infrared assignments have been reported for $C_6H_5Se(OCH_3)_3$ (598), $(C_6H_5)_2Se(OCH_3)_2$ (599), $C_6H_5SeX_2(OCH_3)$ (X=Cl,Br), $C_6H_5SeX(OCH_3)_2$ (X=Cl,Br,NO_3), and $(C_6H_5)_2SeX(OCH_3)$ (X=Cl,Br) (600). A correlation has been found between the number and nature of the X substituents, the position of the ν(SeO) and ν(CO) modes, and the nature of the decomposition processes for these phenyl-selenium methoxides (600). Thermal decay of $[(C_6H_5)_2Zr{\cdot}O(C_2H_5)_2]_2$ has been reported to produce $C_6H_5ZrOC_2H_5$, which shows characteristic phenyl and ethoxy bands at 780 and 1280 cm^{-1}, respectively (601). The infrared spectra of both $[(C_6H_5)_2Zr{\cdot}O(C_2H_5)_2]_2$ (1700 to 550 cm^{-1}) and C_6H_5Zr-OC_2H_5 (1900 to 700 cm^{-1}) were illustrated in this study.

Vibrational data for alkyl- and phenylperoxides have been limited to compounds of silicon. The ν(OO) and ν(SiO) modes have been assigned in the 920 to 895 cm^{-1} and 810 to 800 cm^{-1} regions, respectively, in the infrared spectra of $(C_6H_5)_3SiOOC(CH_3)_n(C_6H_5)_{3-n}$ (n=0 to 3) (585,602).

The infrared data reported for $C_6H_5B(SCH_3)_2$ (ν_a(BS)=936/908 (vs) cm^{-1},ν_s(BS)=632 (m) cm^{-1}) and $(C_6H_5)_2BSCH_3$ (ν(BS)=899 (s) cm^{-1}) have been interpreted as indicating the absence of π-bonding between the boron and sulfur atoms (603). The vibrational spectrum of 2-phenyl-1,3,2-dithioborolane (3.70) has also been

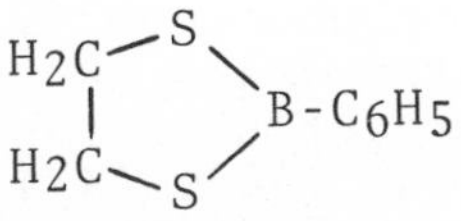

(3.70)

been assigned (604). The infrared spectra below 600 cm^{-1} have been assigned for $(C_6H_5)_3SnSC_6H_5$ (603). Infrared bands at 502 and 495 cm^{-1} have been assigned to the accidentally degenerate t- and ν(PtC) modes of the *trans* isomer and the *cis* isomer, respectively, of $(C_6H_5)_2Pt[S(C_2H_5)_2]_2$ (605).

g. *Carboxylates and Thiocarboxylates*

The infrared spectra of $C_6H_5Tl(O_2CR)_2$ (R=CH_3 (606), $(CH_3)_2CH$,CF_3) (522) indicate the presence of two different types of carboxylate groups in each compound. The large frequency separation of one pair of ν(CO) bands (ν(C=O)=1595 cm^{-1} and ν(CO)=1375 cm^{-1} for C_6H_5Tl-$(O_2CCH_3)_2$) indicates that one of the carboxylate groups is monodentate in each compound. The nature of the second carboxylate ligand is not as easily resolved. The smaller frequency separation of a second pair of ν(CO) bands (1520 and 1405 cm^{-1} for $C_6H_5Tl(O_2CCH_3)_2$) has been interpreted as indicating the presence of a weakly bridging carboxylate, which would produce a dimeric or polymeric structure, or an ionic carboxylate, which would give the structure $[C_6H_5TlO_2CR]^+O_2CR^-$ (522, 606). Several of the phenyl and X-sensitive modes have been assigned for $(C_6H_5)_2TlO_2CCH_3$ (483).

The frequency separation of the ν(CO) modes in the solid-state infrared spectra of 13 triphenyltin carboxylates in which there is no branching at the carboxylate α-carbon atom indicates that these compounds have a polymeric structure with bridging carboxylate ligands and five-coordinate tin atoms (607). In a CCl_4 solution, the larger frequency separation of the ν(CO) modes indicates that these triphenyltin carboxylates become monomeric. For example, the ν(CO) modes of solid $(C_6H_5)_3SnO_2CCH_3$ appear at 1548 and 1420 cm^{-1}, while in CCl_4 solution, they appear at 1640 and 1370 cm^{-1} (607). For triphenyltin carboxylates in which the carboxylate is branched at the α-carbon atom, the positions of the ν(CO) modes are the same in both the solid state and CCl_4 solution and are in the range expected for monodentate carboxylate ligands (607). Therefore, it has been proposed that these branched carboxylate derivatives are monomeric in the solid state and solution. Infrared data below 600 cm^{-1} have been reported for $(C_6H_5)_3SnO_2CR$ ($R=CH_3$ (574), C_6H_5 (523,574)). The position of the ν(CO) modes for $(C_6H_5)_2Sn(O_2CC_6H_5)_2$ in the solid state (1567 and 1392 cm^{-1}) and $CHCl_3$ solution (1567 and 1361 cm^{-1}) has been interpreted as indicating structural rearrangement from bridging benzoate ligands in the solid state to chelating benzoate ligands in solution (608). For $(C_6H_5)_2SnCl(O_2CC_6H_5)$, the ν_a(CO) mode frequency remains constant while the ν_s(CO) mode frequency increases by 11 cm^{-1} on dissolving the solid in $CHCl_3$. This suggests that $(C_6H_5)_2SnCl(O_2CC_6H_5)$ has a different structure from $(C_6H_5)_2Sn(O_2CC_6H_5)_2$ in solution (608). Infrared data for $(C_6H_5)_3SnO_2CR$ (*p*-$CH_3C_6H_4$; 2,4,6-$(CH_3)_3C_6H_2$; *p*-OHC_6H_4; *p*-$CH_3OC_6H_4$; *p*-$H_2NC_6H_4$; 3,5$(O_2N)_2C_6H_3$) are consistent with monomeric structures in both the solid state and solution (CCl_4), thus suggesting that steric hindrance between the substituted benzoate ligands and the phenyl groups bonded to the tin is sufficient to prevent polymer formation (609). The positions of the ν(CO) modes for $(C_6H_5)_4Sn_2(O_2CR)_2$ ($R=CH_3, CH_2Cl, CHCl_2, CCl_3, CF_3$) are approximately the same in $CHCl_3$ solutions, in which monomeric structures have been found, and the solid state and seem consistent with the presence of bridging carboxylate ligands (610). A polymeric structure (3.71) has therefore been proposed as consistent with the solid-state data, while in solution, a structure

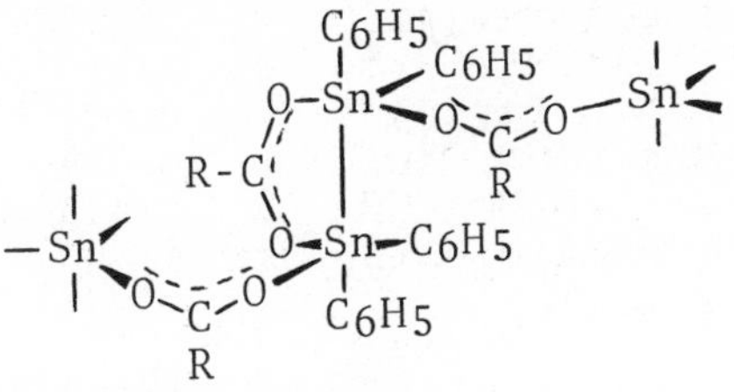

(3.71)

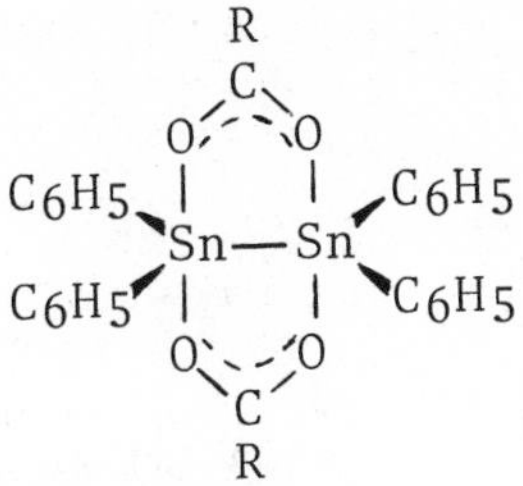

(3.72)

has been proposed in which both carboxylate ligands bridge the same two tin atoms (3.72).

The frequency separation of the ν_a(CO) and ν_s(CO) modes for $(C_6H_5)_4SbO_2CR$ (R=CCl_3,CF_3), $(C_6H_5)_3SbX_2$ (538, 540,611), and $(C_6H_5)_3BiX_2$ (489,537,541,612) (X=acetate, haloacetate, cyanoacetate) indicates that the carboxylate ligands are monodentate. However, while $(C_6H_5)_4$-SbO_2CCH_3 has a five-coordinate structure with a monodentate acetate ligand in $CHCl_3$ and $CHBr_3$ solutions (ν(C=O)=1625 cm^{-1},ν(CO)=1370 cm^{-1}), in the solid state, the acetate ligand becomes chelating to give a six-coordinate structure as indicated by the smaller separation of the ν(CO) mode frequencies (1555 and 1395 cm^{-1}) (535).

A relatively complete discussion has been presented of the infrared spectra of $(C_6H_5)_2SeO_2CCH_3$, $(C_6H_5)_2Se$-$(O_2CCH_3)_2$ (ν(SeO)=348,326 cm^{-1}), and $(C_6H_5)_2Se(O_2CCF_3)_2$ (482). The infrared spectra of $(C_6H_5)_2Te(O_2CCH_3)_2$ (ν(C=O)=1644 cm^{-1},ν(CO)=1288 cm^{-1}) and $(C_6H_5)_2Te$-$(O_2CC_6H_5)_2$ in both the solid state and CCl_4 solution are consistent with monomeric structures and monodentate carboxylate ligands (613). A band at approximately 280 cm^{-1} in both phases has been assigned to a ν(TeO) mode of $(C_6H_5)_2Te(O_2CCH_3)_2$ (613). The frequency separation of the ν_a(CO) and ν_s(CO) modes indicate the presence of monodentate carboxylate ligands and that all four of the carboxylate ligands are in similar environments in the dimeric derivatives $(C_6H_5)_2TeL$ (L=*o*-phthalate, tetrabromo *o*-phthalate) (614).

h. Oxo- and Halo-Acid Salts

The infrared spectrum of solid $C_6H_5HgNO_3$ is consistent with the presence of an ionic nitrate group (81). The infrared spectrum of a compound with the composition corresponding to the formulation C_6H_5Hg-$NO_3 \cdot C_6H_5HgOH$ suggests that the nitrate group is coordinated to the mercury atom with a strong intensity band at 562 cm^{-1} assigned to the ν(Hg-ONO_2) mode (81). The infrared spectrum of solid $(C_6H_5Hg)_2CO_3$ is consistent with the presence of a bidentate carbonate ligand that bridges the mercury atoms (81).

Infrared data have been reported for a compound originally formulated as the benzenesulphinato complex $(C_6H_5)_2In(O_2SC_6H_5)$ (615). It has more recently been concluded (616), however, that the infrared data are only consistent with a structure containing a coordinated SO_2 molecule, $(C_6H_5)_3InOSO$, with the large frequency separation of the $\nu(SO_2)$ modes in the complex ($\nu_a(SO_2)$=1053 cm^{-1},$\nu_s(SO_2)$=854 cm^{-1}) relative to the values found for free SO_2 ($\nu_a(SO_2)$=1340 cm^{-1},$\nu_s(SO_2)$= 1150 cm^{-1}) (617) indicating that the SO_2 molecule is coordinated to the indium atom through one of the oxygen atoms. Structures with ionic nitrate groups have been proposed for solid $[(C_6H_5)_2Tl]NO_3$ (618) and $(C_6H_5TlOH)NO_3$ (522). Although the sulfate bands have been reported from infrared data for $C_6H_5TlSO_4$, it has been noted that the compound is hydroscopic and therefore the spectrum may not be for the pure compound (522).

The infrared spectra of $(C_6H_5)_nSn(NO_3)_{4-n}$ (n=2,3) and $[(C_6H_5)_2SnNO_3]_2O$ are consistent with the presence of covalent nitrate groups (619). Infrared bands at 570 and 515 cm^{-1} may be due to the ν(SnO) modes of $(C_6H_5)_3SnNO_3$ and $[(C_6H_5)_3Sn]_2SeO_3$, respectively (574). Infrared assignments below 650 cm^{-1} have also been reported for $[(C_6H_5)_3Sn]_2XO_4$ (X=S,Se) (574). Both infrared and nmr data have been used in characterizing the structures of $(C_6H_5)_2SnSO_3$ and $[(C_6H_5)_3Sn]_2SO_3$ (620). The infrared and Raman spectra of $[(C_6H_5)_2Sn]_2O(CO_3)$ are consistent withaa polymeric structure with bridging carbonate groups and Sn_2O_2 rings (621).

Vibrational data have been reported for the covalent derivatives $(C_6H_5)_3M(NO_3)_2$ (M=As (622),Sb (540), Bi (541)) and $(C_6H_5)_4MNO_3$ (M=Sb (535),Bi (489,537,612)) in which the unidentate nitrate group(s) is (are) in the axial position(s). Although the covalent compound $[(C_6H_5)_3SbNO_3]_2O$ has been characterized, attempts to prepare the corresponding phosphorus and arsenic complexes have resulted in the formation of the nitric acid adducts $(C_6H_5)_3M \cdot HNO_3$ (M=P,As); infrared assignments have been reported for all three compounds (622). The infrared data originally reported to have been for $[(C_6H_5)_3Sb]_2O(ClO_4)_2$ (540) have recently been interpreted (623) as having been for the hydrated species. An investigation of the splitting of the perchlorate bands has led to the conclusion that this compound has a nonionic structure (623). This is supported by the fact that the perchlorate mode frequencies are in the same region as those of anhydrous $[(C_6H_5)_3Bi]_2O(ClO_4)_2$ (567), which a single crystal X-ray study has shown (568) to have a slightly distorted trigonal bipyramidal skeletal structure with coordinated perchlorate groups. The splitting of the $\nu_a(ClO_4)$ band in $[(C_6H_5)_3Sb]_2O(ClO_4)_2$ (140 cm^{-1}), $[(C_6H_5)_3Bi]_2O(ClO_4)_2$ (100 cm^{-1}), and $[(CH_3)_3Sb]_2O(ClO_4)_2$ (13 cm^{-1}) has been interpreted (623) as implying that the $M-OClO_3$ bond is strongest in $[(C_6H_5)_3Sb]_2O(ClO_4)_2$ and weakest in $[(CH_3)_3Sb]_2O(ClO_4)_2$. The low-frequency Raman spectrum of solid $[(C_6H_5)_4Sb]ClO_4$ (487), and the infrared (622,623) and low-frequency Raman spectra (489) of solid $[(C_6H_5)_4Bi]ClO_4$ indicate that both compounds have structures with ionic perchlorate groups. The aqueous solution Raman spectrum (350 to 125 cm^{-1}) of $[(C_6H_5)_4As]_2SO_4$ proved difficult to interprete in structural terms, although the compound is known to dissociate in this phase (620). The infrared results for $(C_6H_5)_3SbX$ and $[(C_6H_5)_3SbX]_2O$ ($X=SO_4$ (540), SeO_4 (624),CrO_4 (624)) and $(C_6H_5)_3BiX$ ($X=CO_3,SO_4,SeO_4,CrO_4$) (625) in the solid state indicate the presence of nonionic, polymeric structures with bridging anions and five-coordinate antimony or bismuth atoms. Solid-state infrared spectra, molecular weight measurements in benzene, and conductivity measurements in nitromethane have been reported for $[(C_6H_5)_4Sb]_2X$ ($X=SO_4,SeO_4,CrO_4$) (535). These data are consistent with covalent structures in both the solid state and benzene and nitromethane solutions. Conductivity measurements and solid-state vibrational data show $(C_6H_5)_4BiX$ ($X=BF_4,PF_6$) to have ionic structures (489).

The infrared spectra have been assigned for $(C_6H_5)_nSe(NO_3)_{4-n}$ (n=2,3), $(C_6H_5)_2SeSO_4$, and $(C_6H_5)_2Se(CH_3SO_3)_2$ (482). Although the infrared data reported for $(C_6H_5)_2Te(NO_3)_2$ indicate that the nitrate groups are not ionic, it was not possible to definitely interprete these data in structural terms (613). It has been suggested, however, that the structure of this compound is monomeric with unidentate or very unsymmetrically bidentate nitrate groups (613).

i. Lewis-Base Adducts and Miscellaneous Chelates

Partial infrared data have been reported for the pyridine derivatives $C_6H_5ZnN_3Py$ (569) and $C_6H_5AuCl_2Py$ (514). Complete infrared and Raman assignments have been reported for $(C_6H_5)_2BN(CH_3)_2$ ($\nu(BN)$=1518 cm^{-1}), $(C_6H_5)_2BN(C_6H_5)CH_3$ ($\nu(BN)$=1388 cm^{-1}), and $CH_3(C_6H_5)B$-$N(C_6H_5)CH_3$ ($\nu(BN)$=1387 cm^{-1}) (445). It has been noted in this study that substitution of a nitrogen-bonded methyl group with a phenyl group causes a characteristic lowering of the $\nu(BN)$ mode frequency. A similar substitution at the boron atom, however, produces an insignificant shift in the $\nu(BN)$ mode frequency. The infrared spectra of $C_6H_5TlCl_2L$ (L=$(C_6H_5)_3PO$,Bipyr,Phen) and $C_6H_5Tl(O_2CCF_3)$Phen show bands typical of coordinated ligands (522).

The $\nu(SnCl)$ and t-mode assignments have been reported for several amine complexes of phenyl-tin chlorides (626-628). On adduct formation, the coordination number of the tin atom increases from four to six. Similar monomeric structures with six-coordinate tin atoms are indicated by infrared data for $(C_6H_5)_2Sn$-$(NO_3)_2$Phen (619) and $(C_6H_5)_2Sn(NCS)_2L$ (L=Bipyr,Phen) (558), while infrared data for $(C_6H_5)_2Sn(NCS)$(oxinate) in the solid state and benzene solution indicate a monomeric structure with five-coordinate tin atoms (558). Infrared assignments for $(C_6H_5)_nSn_{4-n}L$ (n=2,3) have been made in the $\nu(NO)$, $\nu(PO)$, and $\nu(AsO)$ mode regions for L=pyridine N-oxide, $(C_6H_5)_3PO$, and $(C_6H_5)_3AsO$, respectively (629). The infrared spectra have been obtained for the 1:1 adducts obtained on reacting N,N'-ethylenebis(salicylideneiminato)Ni(II) with $(C_6H_5)_nSn$-Cl_{4-n} (n=1,2) (630). The structures of $(C_6H_5)_2SnL$ (L= tridentate ONO and SNO ligands) have been investigated by vibrational, pmr, and electronic spectroscopy (631). Several phenyl-tin(IV) complexes with quadri- and tetradentate anionic Schiff base ligands have been investigated in the solid state using ^{119}Sn Mossbauer and infrared spectroscopy (632). Infrared data, including $\nu(MS)$ mode assignments, have been reported for $(C_6H_5)_2M$-(Mnt) and the tetraalkylammonium and tetraphenylarsonium salts of $(C_6H_5)_2M(Mnt)Cl^-$ and $(C_6H_5)_2M(Mnt)_2^{2-}$ (M=Sn,Pb; Mnt=dicyanoethylene-1,2-dithiolate (633). Similar data and assignments have been reported for $(C_6H_5)_3MS(O)$-COC_2H_5 (M=Ge,Sn,Pb) and $(C_6H_5)_3MS(S)COC_2H_5$ (M=Sn,Pb) in which the ligands are monodentate and sulfur-bonded (634).

Infrared assignments have been reported for C_6H_5As-$(X)NR_2$ (X=Cl,N_3;R=CH_3,C_2H_5) (562) and $[(C_6H_5)_3SbCl]_2NH$

(592). While the complexes $(C_6H_5)_3MC_2O_4$ (M=Sb (625), Bi (626)) and $[(C_6H_5)_3SbC_2O_4]_2O$ (625) are polymeric with five-coordinate central atoms and bridging oxalate ligands, six-coordinate antimony atoms with a symmetrically bichelating oxalate ligand has been proposed for $[(C_6H_5)_4Sb]_2C_2O_4$ (535). This is indicated by the fact that while $(C_6H_5)_3SbC_2O_4$ shows two strong intensity infrared bands in the $\nu(CO_2)$ mode region (1740 and 1655 cm^{-1}), $[(C_6H_5)_4Sb]_2C_2O_4$ gives one band in this region (1625 cm^{-1}). Solid-state infrared data for $(C_6H_5)_nSbCl_{4-n}Acac$ (n=1 to 4) are consistent with six-coordinate structures and chelating acetylacetonate ligands (635). A single crystal X-ray study of $(C_6H_5)_2SbCl_2Acac$ has shown that in the solid state, the Cl ligands are mutually *cis* and coplanar with the chelating acetylacetonate ligand, while the phenyl groups are mutually *trans* (636). It had originally been proposed (637) that in solution $(C_6H_5)_2SbCl_2Acac$ consists of an equilibrium mixture of the *trans* dichloro structure with a chelating acetylacetonate ligand, and a structure with a nonchelating acetylacetonate ligand. This equilibrium has been questioned (638), and it has now been proposed that in solution not only $(C_6H_5)_2SbCl_2Acac$ (635,638) but also $(C_6H_5)_2SbX_2Acac$ (X=Br,NCS) (638) consist of an equilibrium mixture between molecules with mutually *trans* phenyl groups and molecules with mutually *trans* X groups. Spectroscopic data (infrared, uv, and nmr) indicate that $(C_6H_5)_nSbCl_{4-n}Ox$ (n=1 to 4) consist of six-coordinate structures with chelating oxinate (Ox) ligands in benzene solution (639). In polar solvents, however, complete rupture of the Sb-N bond can occur (639). In the solid state, spectral data indicate $(C_6H_5)_3Bi(X)Ox$ (X=Cl,Br) to have six-coordinate structures with chelating oxinate ligands (542). The Bi-N bond is broken on dissolving $(C_6H_5)_3Bi(X)Ox$ (X=Cl,Br) in CH_2Cl_2 or benzene (542). Infrared data have been used to help characterize $(C_6H_5)_3BiX_2L$ (X=Cl,Br,I;L=Phen,Bipyr). Infrared data and conductance measurements have also been used to characterize the structures of the following complexes which contain neutral, oxygen-donor Lewis bases: $(C_6H_5)_3SbCl_3L$ (L=DMSO,HMPA) (640), $[(C_6H_5)_3SbL]_2O(ClO_4)_2$ (L=DMA,DMSO,PyO,$(C_6H_5)_3PO$,$(C_6H_5)_3AsO$) (623), $[(C_6H_5)_3BiL_2]X_2$, and $[((C_6H_5)_3BiL)_2O]X_2$ (L=DMSO,PyO, $(C_6H_5)_3PO$,$(C_6H_5)_3AsO$;X=ClO_4,BF_4,PF_6) (641).

j. Metal-Metal Bonds and Complexes

Infrared data reported for $(C_6H_5)_6Ge_2$ did not include the assignment of the ν(GeGe) mode (438). A ν(GeGe) mode has been assigned from infrared data reported for $[(C_6H_5)_3Ge]_3GeH$ (228 cm^{-1}) (642). Infrared bands originally assigned (643) to the ν(SnSn) modes of $(C_6H_5)_6Sn_2$ (208 cm^{-1}) and $[(C_6H_5)_3Sn]_4Sn$ (207 cm^{-1}) have more recently been reassigned to an internal phenyl mode on the basis of a Raman study of both compounds and a partial normal coordinate analysis of $(C_6H_5)_6Sn_2$ (644). The revised ν(SnSn) mode assignments have been made at 140 and 136 cm^{-1} for $(C_6H_5)_6Sn_2$ (the splitting was attributed to crystal effects) and 159 (T_2) and 103 (A_1) cm^{-1} for $[(C_6H_5)_3Sn]_4Sn$. Detailed infrared (4000 to 70 cm^{-1}) (644) and Raman (below 450 cm^{-1}) (454,644) assignments have been reported for $(C_6H_5)_6Pb_2$ (ν(PbPb)= 109 cm^{-1}).

Vibrational data have been reported for $(C_6H_5)_4P_2$ (511,645) and $(C_6H_5P)_4$ (646). The latter compound contains a puckered P_4 ring and has been reported to exist in the solid state in two different stereoisomeric forms (647). The Raman spectra of both forms have been assigned (646). Although the ν_a(PP) modes have been assigned at 488 and 501 cm^{-1} for these two isomers, the ν_s(PP) modes could not be identified because of mixing of these and other low-frequency modes.

Through a comparison of infrared and Raman coincidences and of the relative intensities of ν_a(TeC) and ν_s(TeC) modes in the infrared and Raman spectra, it has been concluded that $(C_6H_5)_2Te_2$ (ν(TeTe)=167 cm^{-1}) has a structure intermediate between C_{2v} and C_{2h} symmetry (648).

Vibrational data have been reported for several complexes formed when $(C_6H_5)_3M$ (M=Si,Ge,Sn,P,As,Sb) molecules coordinate to a metal atom in complexes of the main group or transition elements. The ν(MM') mode assignments for several of these complexes are summarized in Table 3.33. The failure to observe the ν(PM) and ν(PC) modes in the infrared spectra of triphenylphosphine-iron CO and -molybdenum CO complexes has been attributed to coupling between the ν(PM) and ν(MC) modes of these derivatives (663). An analysis of the infrared and Raman spectra of the *cis* square-planar complexes $[(CH_3)_2AuL_2]ClO_4$ and $(CH_3)_2Au(X)L$ (L=$(C_6H_5)_3P$, $(C_6H_5)_3As$;X=Cl,SCN) also did not include assignments for the ν(AuP) and ν(AsAs) modes (664). The infrared bands have been listed for $(R_2N)_3TiGe(C_6H_5)_3$ (R=CH_3, C_2H_5) (665).

TABLE 3.33
The ν(MM') Mode Assignments (cm^{-1}) for Main Group and Transition Element Complexes of $(C_6H_5)_3M$ (M=Si,Ge,Sn,P, As)

Compound[a]	ν(MM') Mode	Frequency	Refs.
$R_3PAuSiR_3$	Si-Au	305,IR	649
$(R_3Ge)_3P$	Ge-P	366 (E),R 311 (A_1),R	650
trans-$(R_3Ge)_2Pt[P(C_2H_5)_3]_2$	Ge-Pt	242,IR	651
trans-$(R_3Ge)_2Pt[P(n\text{-}C_3H_7)_3]_2$	Ge-Pt	244,IR	651
cis-$(R_3Ge)_2Pt(PR_3)_2$	Ge-Pt	224,IR 222,IR	651
$R_3SnMn(CO)_3$	Sn-Mn	174,IR	652,653
$R_3SnFe(CO)_2C_5H_5$	Sn-Fe	174,IR	652,653
$R_3SnMo(CO)_3C_5H_5$	Sn-Mo	169,IR	652,653
$(R_3P)_2ZnCl_2$	P-Zn	166,IR 140,IR	654,655
$(C_2H_5)_4N[(R_3P)_3ZnCl_3]$	P-Zn	146,IR	655
$(R_3P)_2ZnBr_2$	P-Zn	156,IR 138,IR 128,IR	655
$(C_2H_5)_4N[(R_3P)_3ZnBr_3]$	P-Zn	141,IR	655
$R_4N[(R_3P)_3ZnBr_3]$	P-Zn	134,IR	655
$(R_3P)_2ZnI_2$	P-Zn	153,IR	654
$(R_3P)_2CdCl_2$	P-Cd	136,IR	654
$(R_3P)_2CdBr_2$	P-Cd	134,IR	654
$(R_3P)_2CdI_2$	P-Cd	133,IR	654
$(R_3P)_2HgCl_2$	P-Hg	137,IR 108,IR	654
$(R_3P)_2HgBr_2$	P-Hg	132,IR 104,IR	654
$(R_3P)_2HgI_2$	P-Hg	133,IR 99,IR	654
R_3PGaBr_3	P-Ga	350,IR	656
$(R_3P)_2SnCl_4$	P-Sn	513,IR	657
R_3PSbCl_4	P-Sn	529,IR	657
$(R_3P)_2CoCl_2$	P-Co	187,IR 151,IR	655
$(C_2H_5)_4N(R_3PCoCl_3)$	P-Co	167,IR	655
$(R_3P)_2CoBr_2$	P-Co	187,IR 142,IR	655
$(R_3P)_2{}^{58}NiCl_2$	P-Ni	189.6,IR 164.0,IR	655,658
$(R_3P)_2{}^{58}NiBr_2$	P-Ni	196.8,IR 189.5,IR	659

TABLE 3.33 (continued)

Compound[a]	ν(MM') Mode	Frequency	Refs.
$(C_2H_5)_4N(R_3PNiCl_3)$	P-Ni	177,IR	655
$R_3PNi(CO)_3$	P-Ni	197,R	660,661
trans-$(R_3P)_2PdCl_2$	P-Pd	191,IR 155,R	658
trans-$(R_3P)_2PdBr_2$	P-Pd	185,IR 150,R	658
$(R_3P)_2Os(CO)_2Cl$	P-Os	159,R,IR	662
$(R_3P)_2Os(CO)_2Br$	P-Os	158,R	662
$(R_3P)_2Os(CO)_2I$	P-Os	156,R,IR	662
R_3PAuCl	P-Au	182,R,IR	661
R_3PAuBr	P-Au	173,R	661
$(R_3AsHgCl_2)_2$	As-Hg	138,IR	586
$(R_3AsHgBr_2)_2$	As-Hg	112,IR	586
$(R_3AsHgI_2)_2$	As-Hg	70,IR	586
$(R_3As)_2SnCl_4$	As-Sn	330,IR	657
$(R_3As)_2SnBr_4$	As-Sn	334,IR	657
$R_3AsNi(CO)_3$	As-Ni	182,R	661
trans-$(R_3As)_2PdCl_2$	As-Pd	180,IR 132,R	658
trans-$(R_3As)_2PdBr_2$	As-Pd	180,IR 130,R	658
$R_3AsAuCl$	As-Au	171,R	661
$R_3AsAuBr$	As-Au	163,R	661

[a]$R=C_6H_5$.

5. Ring-Substituted Phenyl Complexes

From the pattern of the infrared spectra of the tetraethylammonium salts of $XC_6H_4AuCl_3^-$ (X=Cl,Br,NO_2) in the overtone and combination band region (2000 to 1600 cm^{-1}), it has been concluded that the X-substituent is *para* to the gold atom (513). This study also includes an assignment of the infrared and Raman spectra below 400 cm^{-1} for these square-planar complexes. Partial infrared data have been repoıted for *p*-$XC_6H_4AuCl_2$ (X=CH_3,C_2H_5,*i*-C_3H_7,*t*-C_4H_9,C_6H_5) which are presumed to have a dimeric, chloride-bridged structure (514).

The infrared spectra (3550 to 375 cm^{-1}) have been assigned for *p*-XC_6H_4HgCl (X=CH_3,CH_3O) (81). A fairly good correlation is found between the assignments for these compounds and those of the corresponding *para*-substituted iodobenzene molecules (81).

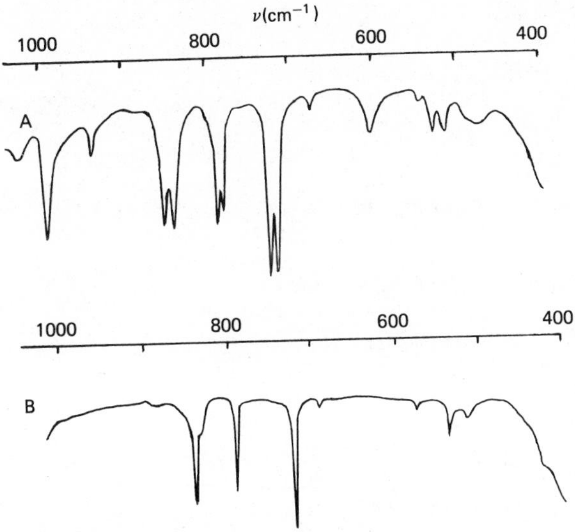

Figure 3.31. Infrared spectra of (A) mesitylTl(O_3CCF_3)$_2$ and (B) (mesityl)$_2$TlO_2CCF_3 (667).

An attempt has been made to study the mass and electronic effects of the X substituents on the infrared spectra (650 to 200 cm^{-1}) of K[(*p*-XC_6H_4)$_4$B] (X=F,Cl,Br, CH_3,CF_3,OCH_3) (666). The position of the y-mode (522 to 487 cm^{-1}) in these derivatives is dependent to a small extent on the electronegativity of the *para*-substituents. The t-mode position, however (565 to 506 cm^{-1}), which exhibits some ν(BC) character, is chiefly mass dependent with the electronic effects of the substituents having only a secondary influence. Therefore, infrared data cannot distinguish very small electronic effects that arise from the inductive effects of the *para*-substituents and influence the B-C bond in tetraarylborate anions (669). The pseudohalide stretching and bending modes have been assigned for the covalent derivatives R_2TlX (R= *o*-tolyl,*m*-tolyl,*p*-tolyl;X=N_3 (550),SeCN (553)). The infrared spectra of (aryl)$_n$Tl(O_2CCF_3)$_{3-n}$ (n=1,2) show three sets of bands at approximately 835, 800, and 720 cm^{-1}, which have been assigned to the ν(CC), ν_s(CF_3), and δ(C-CO_2) modes, respectively. These bands appear as sharp singlets for (aryl)$_2$TlO_2CCF_3 and as doublets for arylTl(O_2CCF_3)$_2$ as illustrated by the infrared spectra of (mesityl)$_n$Tl(O_2CCF_3)$_{3-n}$ (n=1,2) in Figure 3.31 (667). It has been suggested that the doublets in the infrared spectra of aryl$Tl(O_2CCF_3)_2$ are consistent with the formulation $[(aryl)Tl(O_2CCF_3)]^+O_2CCF_3^-$ (667). As noted in the discussion of the infrared spectra of C_6H_5Tl(carboxylate)$_2$ derivatives, however, it is difficult to distinguish between ionic and weakly bridging carboxylate ligands using infrared data. Therefore, the spectra of aryl$Tl(O_2CCF_3)_2$ derivatives might also be interpreted in terms of the presence of a monodentate and a weakly bridging trifluoroacetate ligand, which would give a dimeric or polymeric structure.

The infrared and Raman spectra have been assigned for $XC_6H_4(CH_3)_2SiH$ (X=*p*-F,*p*-Cl,*p*-Br,*p*-CH_3,*p*-CH_3O,*m*-Cl, *m*-Br,*m*-CH_3) (506). Furthermore, the influence of the substituents in terms of their Hammett substituent constants and also of the solvent (CCl_4, hexamethylenesiloxane, and hexane) on the frequency, intensity, and half-band width of the ν(SiH) mode have been discussed for these derivatives (507). Characteristic infrared bands below 459 cm^{-1} have been listed for $(XC_6H_4)_mGe_n$ (X=*o*-CH_3,*m*-CH_3,*p*-CH_3;n=6,m=2;X=*m*-CH_3,*p*-CH_3,n=4,m=1), $(XC_6H_4)_3GeH$ (X=*o*-CH_3,*p*-CH_3), and (*o*-$CH_3C_6H_4)_3GeBr$ (473), while pseudohalide mode assignments have been reported for (*p*-$XC_6H_4)_3GeX$ (X=NCS,NCO,CN) (555).

Infrared (2000 to 100 cm^{-1}) and Raman (1600 to 100 cm^{-1}) data have been listed for (*p*-$FC_6H_4)_3P$ (668). This study also included the assignment of these data below 450 cm^{-1} for (*p*-$FC_6H_4)_3P$ and (*p*-$FC_6H_4)_3PAuX$ (X=Cl,Br); the position of the ν(AuP) mode for X=Cl (184 cm^{-1}) and Br (175 cm^{-1}) is essentially the same as that found for the corresponding mode in the analogous phenyl compounds (661) and therefore appears to be little affected by *para* substitution of the phenyl ring. Detailed infrared and Raman assignments (4000 to 100 cm^{-1}) have been reported for $(XC_6H_4)_3M$ (X=*m*-F,*m*-Cl,*p*-F,*p*-Cl;M=P,As (669),Sb,Bi) (670). The infrared and Raman spectra of $(XC_6H_4)_{3-n}(C_6H_5)_nPY$ (X=*m*-F,*m*-Cl,*p*-F,*p*-Cl;n=0 to 2) and (*m*-$FC_6H_4)_n$-(*p*-$FC_6H_4)_{3-n}PY$ (n=1,2;Y=O,S,Se) have been studied and assignments proposed for the ν(P=O), ν(P=S), and ν(P=Se) modes (671). It has been concluded that the frequency shifts caused by the substituents bonded to phosphorus are more important in the P=S and P=Se bonds than in the P=O bond. The infrared and Raman spectra have also been assigned for $(XC_6H_4)_3AsO$ (X=*m*-F,*m*-Cl,*p*-F,*p*-Cl) (669). The infrared intensities for the ring stretching mode near 1600 cm^{-1} of several *meta*- and *para*-substituted aryldialkylphosphines, -dichlorophosphines, and -dialkylphosphines (570), and triarylphosphines, -phosphine sulfides (571) have been used to determine the σ_R^0 terms for the substituents.

The infrared and Raman spectra of ^{126}Te and ^{130}Te substituted $(p\text{-}CH_3C_6H_4)_2TeBr_2$ have been assigned below 400 cm^{-1} (547). The low-frequency infrared and Raman assignments have been reported for XC_6H_4TeY (X=p-CH_3, p-CH_3O,p-C_2H_5O,p-C_6H_5O;Y=Cl,Br,I) (548) and $[(C_6H_5)_3(CH_3)P](p\text{-}C_6H_5OC_6H_4TeX_4)$ (X=Cl,Br) (538); it has been proposed that these aryl derivatives are isostructural with the corresponding phenyl complexes. Although the position of the ν(TeTe) mode (187 to 169 cm^{-1}) for R_2Te_2 (R=p-$CH_3C_6H_4$,p-$CH_3OC_6H_4$,p-$C_2H_5OC_6H_4$,p-$C_6H_5OC_6H_4$, naphthyl) is sensitive to the nature of R, it has been concluded that this can be attributed as easily to crystal effects as to electronic effects (648). The frequency separation of the ν_a(CO) and ν_s(CO) modes indicates the presence of monodentate carboxylate ligands and that all four carboxylate ligands are in similar environments in the dimeric o-phthalate and tetrabromo o-phthalate derivatives $(p\text{-}XC_6H_4)_2Te(C_4Y_4O_4)$ (X=CH_3O C_2H_5O,Y=H;X=CH_3O,Y=Br) (614).

Infrared data have been listed for the uranium complex $(\eta^5\text{-}C_5H_5)_3U(p\text{-}C_6H_4)U(\eta^5\text{-}C_5H_5)_3$ (82).

6. Perhalogenated Phenyl Complexes

Vibrational data in various phases have been reported and assigned for C_6X_6 (X=F,Cl,Br,I) (674-677). The vibrational data indicate that C_6F_6 has a planar structure of D_{6h} symmetry (674,675). Vibrational as well as other data reported for C_6H_6 (X=Cl,Br,I) in various phases, however, have been interpreted in terms of either a planar structure or of a nonplanar, puckered geometry of D_{3d} symmetry (674).

Although some pentabromophenyl organometallic derivatives have been characterized, these studies (678, 679) did not include vibrational data. Infrared data have been reported for several pentachlorophenyl derivatives of both the main group (680-683) and transition (683,687) elements. Infrared bands at 1515, 1330, and 1305 cm^{-1} have been assigned to ν(CC) modes in pentachlorophenyl-mercury derivatives, while characteristic bands at approximately 853 and 684 cm^{-1} have been assigned to X-sensitive and ν(CCl) modes, respectively, in these same derivatives (680,681). Included in discussions of infrared data reported for $C_6Cl_5MX[P(C_6H_5)_3]_2$ (M=Ni,X=NCS,NCO,N_3;M=Pd,X=Cl,Br,I,NCS,NCO,N_3) was the assignment of a medium intensity band in the 590 to 575 cm^{-1} region to the ν(MC) mode (686).

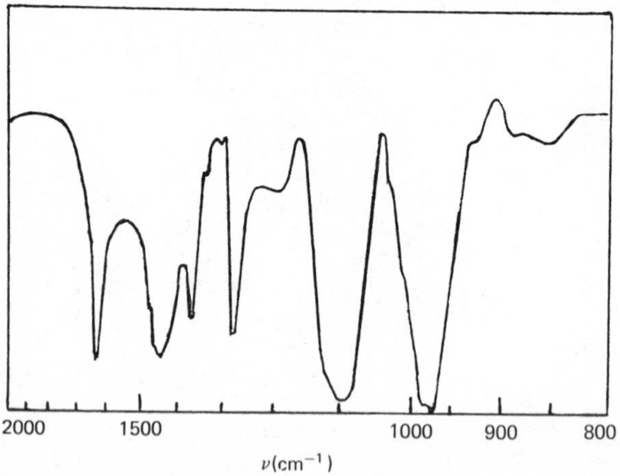

Figure 3.32. Infrared spectrum of ($C_6F_5)_3B$ in chloroform (692).

A few detailed vibrational studies have been reported (688-691) for pentafluorophenyl organometallic derivatives. The infrared spectrum (2000 to 800 cm^{-1}) of $(C_6F_5)_3B$ is illustrated in Figure 3.32 (692). Although vibrational assignments for pentafluorophenyl organometallic derivatives are relatively complex, fundamentals involving ν(CC) and ν(CF) and an X-sensitive mode that has mainly ν(MC) character have been assigned in the infrared region above approximately 750 cm^{-1}. The infrared assignments are summarized in Table 3.34 for several $(C_6F_5)_nM$ derivatives. The Raman spectrum has been assigned for $(C_6F_5)_3Sb$ (691). The number of infrared bands observed for the X-sensitive mode at 800 cm^{-1} for several complexes containing the $(C_6F_5)_2Tl^+$ unit have been correlated with the symmetry of the C-Tl-C skeleton (689). The presence of two bands in this region is characteristic of a nonlinear C-Tl-C skeleton, while only one band is found for those derivatives for which a linear C-Tl-C skeleton has been proposed.

In Table 3.35, the ν(M-ligand) mode assignments are summarized for pentafluorophenyl organometallic hydrides, halides, and oxides. A dimeric structure with bridging hydride ligands has been proposed for $M[(C_6F_5)_2ZnH]$ (M=Li,Na) and the analogous deuterides

TABLE 3.34
Selected Infrared Assignments (cm^{-1}) for $(C_6F_5)_nM$ Derivatives of the Main Group Elements

Compound	ν(CC) Mode	ν(CF) Mode	X-Sensitive Mode	Refs.
$(C_6F_5)_2Hg$	1639 1504 1481	1370 1078s,sh 1066 970	806	693
$(C_6F_5)_3B$	1642w 1475m,sh 1451s	1374s 1316s 1292m 1151m 1139m 1110m 1093w,sh 1012 973s		692
$(C_6F_5)_3In$	1644s 1511vs 1470vs	1371s 1277sh 1136w 1080vs 1010m 956vs	793sh 788	688
$(C_6F_5)_4Si$	1641m 1516s 1466s	1379s 1292s 1140w,sh 1098s 1023w 970s		186
$(C_6F_5)_4Ge$	1671m 1539s 1479s	1411s 1313s 1140m 1106m,sh 1087s 1015w 970s	818	186,694

TABLE 3.34 (continued)

Compound	ν(CC) Mode	ν(CF) Mode		X-Sensitive Mode	Refs.
$(C_6F_5)_4Sn$	1640m	1378s		803	694,695
	1509s	1281m			
	1479s	1137m			
		1087s	1077s,sh		
		1015m			
		964s			
$(C_6F_5)_4Pb$	1632m	1375s			696
	1509s	1275m			
	1469s	1134m			
		1078s	1075m,sh	782	
		963s			
$(C_6F_5)_3P$	1650s	1392s		849m	690
	1522vs	1296s			
	1485vs	1145m			
		1098vs			
		980vs			
$(C_6F_5)_3As$	1645s	1390s			694
	1520vs	1290s			
	1486vs	1085vs			
		975vs			
$(C_6F_5)_3Sb$	1645s	1386s			691,697
	1520vs	1285s			
	1480vs	1087vs			
		968vs			
$(C_6F_5)_3Bi$	1633	1370		770	698
	1509	1273			
	1473	1070	1049		
		1003			
		965			

TABLE 3.35
The ν(M-Ligand) Mode Assignments (cm^{-1}) for C_6F_5-Metal Hydrides, Halides, and Oxides[a]

Compound[b]	ν(M-ligand)	Refs.
$Na[R_2ZnH]$[c]	1700-1300	505
$Na[R_2ZnD]$[c]	1200-1000	505
$RGeH_3$	2120	699
R_2GeH_2	2142	699
R_3GeH	2224	699
$R_2Ge(C_2H_5)H$	2120[d]	699
$R_2Ge(Br)H$	2150[d]	699
$(R_2GeH)_2$	2170[d]	699
RPH_2	2350 2325	690
R_2TlF[c]	320 165	689,700
R_2TlCl[c]	215 130	689,700
R_2TlBr[c]	151 74	689,700
R_3SnF	328	701
$RPCl_2$	504[e] 387	690
$RPBr_2$	416 351	690
R_3SbCl_2	327 294[f]	691
R_3SbBr_2	240 188[f]	691
R_2TeCl_2	269 264	504
R_2TeBr_2	184 168[f]	504
R_3PO	1242	690

[a]Data from infrared spectra unless otherwise indicated.
[b]$R=C_6F_5$.
[c]Data for a dimeric structure.
[d]Data for the ν(GeH) mode.
[e]Overlaps with the ν(PC) mode.
[f]Data from the Raman spectrum.

(505). A dimeric structure with bridging halide ligands has also been proposed for $(C_6F_5)_2TlX$ (X=F,Cl,Br) (618, 700) although in acetone the bromide is monomeric because of complex formation (618). The compounds $(C_6F_5)_2Tl[OP(C_6H_5)_3]X$ are also monomers with the ν(TlX) mode assigned at 243 and 169 cm^{-1} for X=Cl and Br, respectively (700). Similarly, while $(C_6F_5)_2TlO_2CR$ ($R=CH_3,C_6H_5,C_6F_5$) are dimeric with bridging carboxylate ligands, the addition of 1,10-phenanthroline to a benzene or acetone solution of these carboxylate derivatives produces monomeric complexes (702). The frequency separation of the ν_a(CO) and ν_s(CO) modes for these carboxylate derivatives has been used to distinguish between bridging and monodentate carboxylate ligands (702).

The X-sensitive mode at approximately 800 cm^{-1} has been observed at slightly lower frequencies in several complexes of $(C_6F_5)_2Hg$ (703) and $(C_6F_5)_3In$ (704) than in the uncomplexed derivatives. This frequency change is to be expected because of the increased coordination number of the metal atom in the complexes; the frequency change is small, however, since this band is not due to a pure ν(MC) mode (704). In addition to the specific derivatives discussed above, infrared data have also been reported for pentafluorophenyl derivatives of mercury (492,693,703), boron (692), indium (688,704,705), thallium (618,689,700,702,706), silicon (707,708), germanium (694,708,709), tin (695), phosphorus (690), titanium (185,186), zirconium (186), hafnium (710), vanadium (711), tungsten (712), manganese (713), iron (713), nickel (684,714), palladium (685,715,716), platinum (685,715), gold (717), and ytterbium (718). Infrared data have been listed for μ-(*o*-tetrafluorophenylene)-diiron ocatcarbonyl (719,720). A single crystal X-ray study of this compound has shown the presence of a four-membered, cyclic structure (3.73)

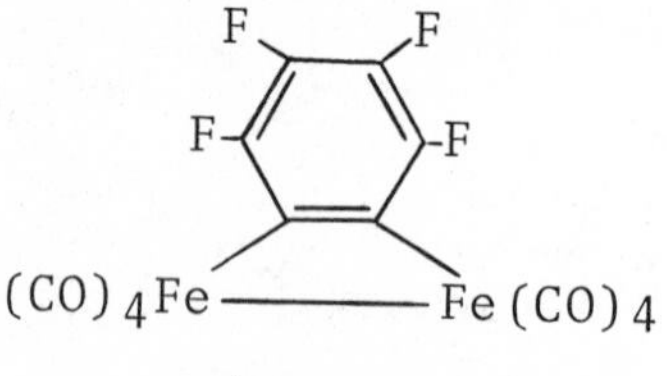

(3.73)

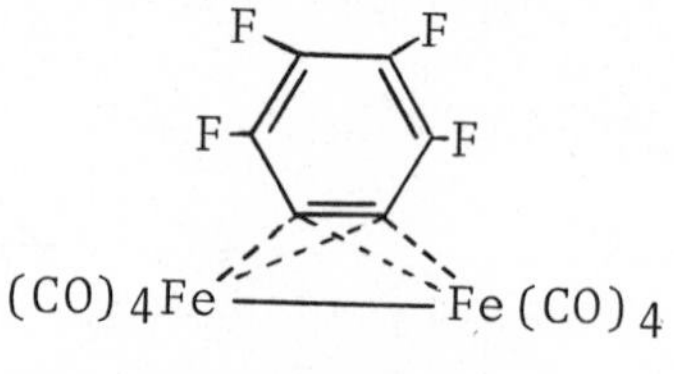

(3.74)

(720), as had originally been proposed (719), rather than a structure in which tetrafluorobenzene is bonded to the two iron atoms (3.74) in a manner similar to the dicobalt-acetylene interaction in $HC{\equiv}CHCo_2(CO)_6$.

7. Benzene Complexes

Several vibrational studies have been reported for $(\eta^6\text{-}C_6H_6)_2Cr$, the infrared spectrum of which is illustrated in Figure 3.33 (721). In the earliest of these studies (721-724), which also included data for $(\eta^6\text{-}C_6D_6)_2Cr$, different conclusions were reached as to whether the C-C bond lengths in the complexed benzene ring remain equal to give $(\eta^6\text{-}C_6H_6)_2Cr$ a point symmetry of D_{6h}, or become alternately longer and shorter to give a point symmetry of D_{3h}. A detailed vapor-state infrared study (725) and solid-state Raman

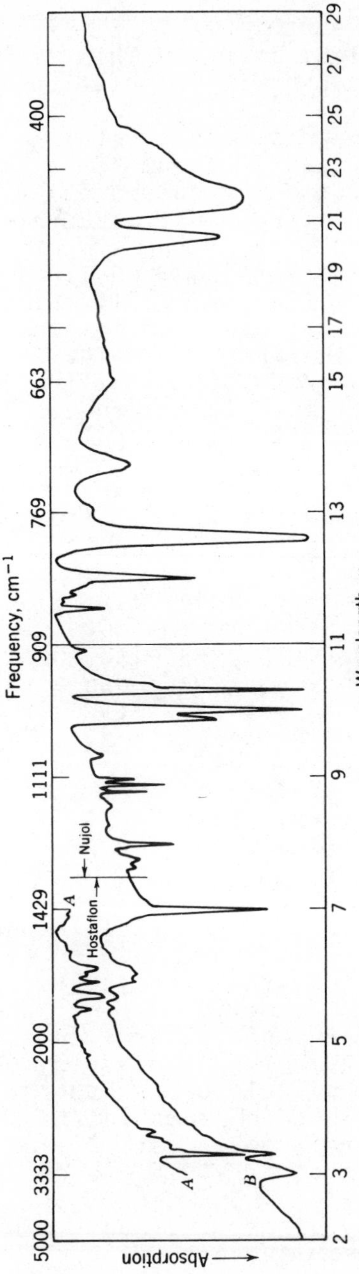

Figure 3.33. *Infrared spectrum of crystalline* (η^6-C_6H_6)$_2$*Cr as a (A) KBr pellet, (B) Hostaflon-oil suspension (2–7.5 μ) and as a Nujol mull (721).*

studies (726,727) support the conclusion that the symmetry of (η^6-C_6H_6)$_2$Cr is D_{6h}. This structure is also supported by thermodynamic calculations that were based on the vibrational data (728) as well as by electron-diffraction (729) and single crystal X-ray (730) studies. The infrared spectrum has also been reported for (η^6-C_6H_6)$_2$Cr formed by codepositing benzene and chromium atoms in an argon matrix after annealing at 45°K (731). A structure of D_{6h} symmetry has consistently been proposed for [(η^6-C_6H_6)$_2$Cr]I on the basis of vibrational data (721,723,732). Similar structures have been proposed for all other dibenzene metal complexes. Although (η^6-C_6H_6)$_2$V is found in either a cubic or monoclinic crystalline modification (733), the infrared spectra of both forms are nearly identical (724,733). The vibrational assignments for the benzene modes of several dibenzene metal complexes are compared in Table 3.36 together with analogous assignments for the free benzene molecule (734). Limited infrared data have also been listed for (η^6-C_6H_6)$_2$Ti and [(η^6-C_6H_6)$_2$-Fe](PF_6)$_2$ (737).

Dibenzene metal complexes give rise to metal-ring skeletal modes analogous to those illustrated in Figure 3.7 for (η^5-C_5H_5)$_2$M complexes with parallel rings. Although the ν_a(M-ring) mode had been assigned at a lower frequency than the asymmetric metal-ring tilting mode in several of the earlier infrared studies (48,721,722, 725), in more recent vibrational studies, these assignments have been reversed with the tilting mode assigned at the lower frequency. In Table 3.37, the metal-ring skeletal mode assignments are compared for dibenzene metal complexes.

Total normal coordinate analyses have been reported for several dibenzene metal complexes (726,727, 738,739). These studies have been reported to show that the frequency changes observed for the benzene molecule on complex formation are not due to significant differences between the force constants found for free benzene and the corresponding force constants of the complexed benzene molecule; rather they arise from kinematic coupling between the benzene and metal-ring skeletal modes in the complexes. It has also been reported (738) that frequency differences between the infrared active skeletal modes of the various complexes are due less to the electronic configurations of the central metal atom in these complexes than to the mass of the metal atom. These results contrast with those reached in a normal coordinate analysis reported for

TABLE 3.36
Vibrational Assignments (cm^{-1}) for Benzene and the Benzene Modes of $(\eta^6\text{-}C_6H_6)_2M$

Mode			Compound[a]									
Sym.[b]	Desc.[c]	No.[d]	C_6H_6[e]	R_2Cr h_{12}[f]	R_2Cr d_{12}[g]	R_2Cr^+[h]	R_2Mo[i]	R_2W[i]	R_2V[j] cubic	R_2V[j] monocl.	R_2Tc^+[k]	R_2Re^+[k]
A_{1g}	ν_s(CH)	2s	3073	3053	2726							
	ν_s(CC)	1s	993	970	920							
	π_s(CH)	11s	673	794	566							
A_{2g}	δ(CH)	3s	1350		1021							
A_{1u}	δ(CH)	3a	1350									
A_{2u}	ν_s(CH)	2a	3073	3053	2267	3095	3030	3012	3058	3062	3096	3086
	ν_s(CC)	1a	993	971	931	972	966	963	959	957	919	923
	π_s(CH)	11a	673	794	664	795	773	798	739	742		
B_{1g}	ν(CC)	14s	1309									
	δ(CH)	15s	1146									
B_{2g}	ν(CH)	13s	3057	2855	2125							
	δ(CCC)	12s	1010		955							
	π(CH)	5s	990		835							
	π(CCC)	4s	707									
B_{1u}	ν(CH)	13a	3057									
	δ(CCC)	12a	673									
	π(CH)	5a	990									
	π(CCC)	4a	707									
	ν(CC)	14a	1309	1308	1282							
	δ(CH)	15a	1146	1142	826							
E_{1g}	ν(CH)	20s	3064	2904	2212							
	ν(CC)	19s	1482	1430	1271							
	δ(CH)	18s	1037	999	802							
	π(CH)	10s	846	811	699							
E_{2g}	ν(CH)	7s	3056	2955	2212	3074						
	ν(CC)	8s	1599	1631	1562	1620						

TABLE 3.36 (continued)

Sym.[b]	Desc.[c]	No.[d]	C_6H_6[e]	R_2Cr h_{12}[f]	R_2Cr d_{12}[g]	R_2Cr^+[h]	R_2Mo[i]	R_2W[i]	R_2V[j] cubic	R_2V[j] monocl.	R_2Tc^+[k]	R_2Re^+[k]
	δ(CH)	9s	1178	1143	868	1148						
	δ(CCC)	6s	606	604	566	616						
	π(CCC)	17s	967	910	735							
	π(CCC)	16s	404	409								
E_{1u}	ν(CH)	20a	3064	2904	2212	3040	2916	2898		2928	2933	2941
	ν(CC)	19a	1482	1426	1271	1430	1425	1412	1416	1415	1443	1437
	δ(CH)	18a	1037	999	802	1000	995	985	985	988	978	979
	π(CH)	10a	846	860	669	857	811	882	818	818		
E_{2u}	ν(CH)	7a	3056									
	ν(CC)	8a	1599									
	δ(CH)	9a	1178									
	δ(CCC)	6a	606									
	π(CCC)	17a	967									
	π(CCC)	16a	404									

[a] $R=\eta^6-C_6H_6$.
[b] Symmetry species for $(\eta^6-C_6H_6)_2M$ complex with eclipsed rings and D_{6h} symmetry.
[c] References 734 and 735.
[d] Reference 735; numbering of the modes corresponds to that found for free benzene with the s and a subscripts used to distinguish between modes that involve the in-phase and out-of-phase combinations, respectively, of the two rings.
[e] Reference 734.
[f] References 721, 722, 726, and 727.
[g] References 721 and 722.
[h] References 721 and 726; data for the iodide salt.
[i] References 721 and 726.
[j] References 721, 726, and 733.

TABLE 3.37
Metal-Ring Skeletal Mode Assignments (cm^{-1}) for $(\eta^6\text{-}C_6H_6)_2M$ Complexes

Compound[a]	ν_a(MR) (A_{2u})	ν_s(MR) (A_{1g})	Asym. Tilt (E_{1u})	Symm. Tilt (E_{1g})	δ(RMR) (E_{1u})	Torsion (A_{1u})	Refs.
R_2Ti	452[b]		411[b]				736
R_2V: Monoclinic	478		439		138.5	121	724
			428		134		
Cubic	478		429		138	118	724
R_2Cr[c]	423	277[d]	479	301[d]	158[d]	139[d]	727
R_2Cr	490	277	459	335	171	152	724,726,727
R_2Mo	424		362				721,738
R_2W	386		331				721,738
R_2Cr^+	466	279	415				721,738
R_2Mo^+	410		333				721,738
R_2W^+	378		306				721,738
R_2Tc^+	431		358				48,738
R_2Re^+	396		336				48,738

[a] $R=\eta^6\text{-}C_6H_6$ unless otherwise indicated.
[b] Assignment made using data listed in Ref. 736.
[c] $R=\eta^6\text{-}C_6D_6$.
[d] Frequency calculated in a complete normal coordinate analysis of the entire molecule.

η^6-$C_6H_6Cr(CO)_3$ (740), in which it was concluded that changes in the force constants of the benzene ring are largely responsible for the frequency shifts observed in the benzene molecule on complex formation.

A limited amount of infrared data have been used to help characterize the structures of $(C_6H_6RuX)_2$ and $C_6H_6RuX_2PR_3$ (X=Cl,Br,I,SCN) (741). The benzene solution vibrational spectra of various metal halides show differences from the corresponding spectra of the isolated metal halides and pure benzene. These spectral changes have been interpreted as arising from a weak π-complex formed between the metal halide and benzene. The number and position of the ν(FeCl) modes in the infrared spectrum of iron(III) chloride in benzene (423 (vvs) and 363 (w) cm^{-1}) suggest that monomeric $FeCl_3$ is present with a nonplanar structure of C_{3v} symmetry (742). This solution also exhibited bands at 2932±7 and 2864±7 cm^{-1} that are absent in the infrared spectrum of pure benzene. These bands are in the region expected for the ν(CH) modes in dibenzene metal complexes (see Table 3.36). The above data have been interpreted as consistent with the adduct $C_6H_6FeCl_3$ in solution. The structures of gallium dichloride and gallium dibromide have been established as $Ga^+(GaX_4)^-$ (X=Cl,Br) (743). The infrared spectra (500 to 200 cm^{-1}) of these two halides in benzene have led to the proposal that the $Ga^+(GaX_4)^-$ ion pairs form a weak η^6-benzene complex (3.75) (743). No bands were observed

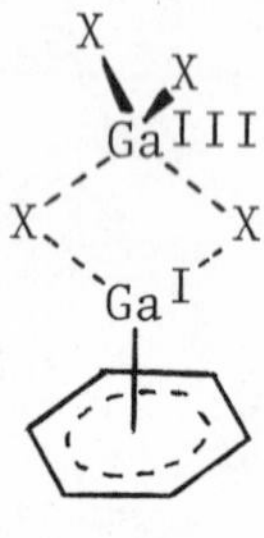

(3.75)

that could be attributed to Ga-C_6H_6 skeletal modes since the interaction is relatively weak and these modes are therefore expected below 200 cm^{-1}. Infrared (744) and Raman (745-747) studies have also been reported for the benzene complexes formed with $SbCl_3$ (744-747) or $SbBr_3$ (747) in both solution (744,745) and the solid state (744,746,747). A single crystal

X-ray study of $C_6H_6AgClO_4$ (748) shows the presence of $-C_6H_6-Ag^+-C_6H_6-Ag^+-$ chains and perchlorate ions in the solid state. The two silver ions associated with each benzene ring lie above and below the ring and over bonds one and four of the ring rather than being equidistant from each carbon atom of the ring. The benzene rings are therefore distorted with the two C-C bond lengths of each ring nearest to the silver ions equal to 1.35 A and the other four equal to 1.43 A. The infrared spectrum of solid $C_6H_6AgClO_4$ (5000 to 650 cm^{-1}) is relatively simple even though the complex has a relatively low symmetry (744). The Raman spectrum of solid $C_6H_6AgClO_4$ has also been reported (749) as well as the infrared spectrum of $AgClO_4$ in benzene (744). Single crystal X-ray studies of $C_6H_6MAlCl_4$ (M=Ag (750), Cu (751)) show the M atom to be bonded to only one C-C bond of each benzene ring. The solid-state infrared spectra (4000 to 33 cm^{-1}) have been assigned for $C_6H_6MAlCl_4$ (M=Ag,Cu) and their deuterated analogs (752). The complexity of these spectra is illustrated by comparing the infrared spectrum of $C_6H_6AgAlCl_4$ (Figure 3.34) (752) with that of $(\eta^6\text{-}C_6H_6)_2Cr$ (Figure 3.33) (721). The complexity of the spectra for $C_6H_6MAlCl_4$ (M=Ag,Cu) is consistent with the lowering of the benzene ring symmetry on complex formation. Very weak intensity bands at approximately 100 cm^{-1} have been assigned to the ν(MC) mode (M=Ag,Cu) (757). A single crystal X-ray study of $C_6H_6U(AlCl_4)_3$ shows the uranium atom to be coordinated to two Cl atoms of each of the three $AlCl_4^-$ tetrahedra through U-Cl-Al bridges and bonded to a η^6-benzene ring (753). The simplicity of the infrared spectrum of this complex, with benzene bands at 3075, 1020, 740, 682, and 667 cm^{-1}, is also consistent with the presence of a η^6-benzene ring (753). The bridging ν(AlCl) modes appear at 550 and 480 cm^{-1}.

Several infrared and/or Raman studies of varying degrees of completeness have been reported for $\eta^6\text{-}C_6H_6Cr(CO)_3$ (726,740,754-763) and $\eta^6\text{-}C_6D_6Cr(CO)_3$ (740,754, 755,762). Some of the original infrared assignments for these two complexes (754-756) have been revised on the basis of Raman data (740,759,762) and normal coordinate analyses for the entire molecules (726,740,764, 765). Conflicting conclusions have been reached in the normal coordinate studies. One group has concluded (726,764,765) that the calculations show that frequency differences between the modes of free benzene and complexed benzene are due mainly to kinematic coupling of

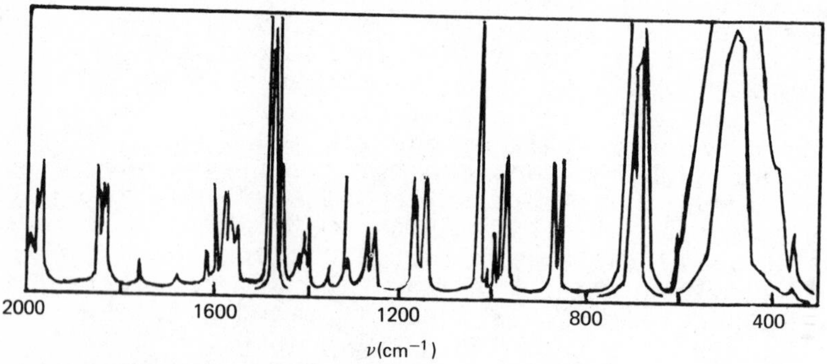

Figure 3.34. Infrared spectrum of $C_6H_6AgAlCl_4$ at -180°C (752).

the benzene modes and metal-ring skeletal modes rather than to force-constant changes in the benzene molecule brought about by complex formation. Another group, however, has concluded (740) that changes in the force constants associated with the benzene ring are largely responsible for the observed coordination shifts. Although the assignment of the benzene modes requires the assumption of a C_{3v} local symmetry, this does not imply that the benzene ring is distorted, as had been indicated in earlier vibrational studies (755,756). A single crystal X-ray study has, in fact, shown that the complexed benzene ring is not distorted but retains a symmetry of D_{6h} (766). Rather, it illustrates the need to take the C_{3v} local symmetry of the entire molecule into account for a satisfactory analysis of the vibrational data (759,760). Three infrared and three Raman bands have been observed in the ν(CO) mode region of solid η^6-$C_6H_6Cr(CO)_3$ rather than the two expected on the basis of the C_{3v} local symmetry of the isolated molecule. This has been attributed to strong intermolecular interactions for η^6-$C_6H_6Cr(CO)_3$, which are as important as the intramolecular interactions. A factor-group analysis of solid η^6-$C_6H_6Cr(CO)_3$ is therefore required for an understanding of the ν(CO) mode region (757,758). In solution, only two ν(CO) bands have been observed in both the infrared (760) and Raman (758,760) spectra of η^6-$C_6H_6Cr(CO)_3$. The benzene modes have been assigned using infrared data for η^6-$C_6H_6Mo(CO)_3$ and its deuterated analog (756). Some revisions have been made in the assignments for these two complexes on the basis of normal coordinate calculations (236). Infrared data have been listed (236) and some assignments made (48, 236) for η^6-$C_6H_6W(CO)_3$. The nonbenzene modes have been assigned for η^6-$C_6H_6Cr(CO)_2X$ (X=$(C_4H_9)_3P$,$(C_6H_5)_3P$, $(C_6H_5)_3Sb$) (313) while infrared data have been listed for η^6-$C_6H_6Cr(PF_6)_3$ (767).

8. Ring-Substituted Benzene Complexes

Several vibrational studies have been reported for η^6-arene complexes in which one or more substituents are present on the benzene ring.

Limited infrared data have been listed for R_2Ti (R=$CH_3C_6H_5$;1,3,5-$(CH_3)_3C_6H_3$) (736). Complete solution and solid-state infrared and Raman studies have been reported for $RCr(CO)_3$ (R=$CH_3C_6H_5$ (760),1,3,5-$(CH_3)_3C_6H_3$ (759,768)). The study of $CH_3C_6H_5Cr(CO)_3$ also includes solution-state infrared and Raman data in the ν(CO) mode region for $RCr(CO)_3$ (R=FC_6H_5,$CH_3C_6H_5$,1,2-$(CH_3)_2C_6H_4$, 1,2,3-$(CH_3)_3C_6H_3$) (760). These compounds all show two bands (A_1 + E) in the ν(CO) mode region using CH_2Cl_2 as the solvent as expected for a local symmetry of C_{3v}. Since these two bands are relatively broad because of solvent-solute interactions, data were also obtained using dilute cyclohexane solutions. This resulted in a splitting of the lower-frequency ν(CO) band of E symmetry for all of the compounds except for $C_6H_6Cr(CO)_3$, because these compounds have an effective symmetry lower than the C_{3v} local symmetry. It has been suggested, therefore, that it is of limited value to use the local symmetry in analyzing the ν(CO) modes of these substituted benzene derivatives. The infrared and Raman active modes of solid $RCr(CO)_3$ (R=$CH_3C_6H_5$, 1,2-$(CH_3)_2C_6H_4$,$(CH_3)_5C_6H$,$(CH_3)_6C_6$) have been discussed in terms of a vibrational factor group that is different from the crystallographic factor group (769). To a first approximation, however, the spectra can be considered as arising in band sets from the ν(CO) modes of the molecule in dilute solution. As has been found for $C_6H_6Cr(CO)_3$. a factor-group analysis of the solid-state infrared and Raman active ν(CO) modes of $RCr(CO)_3$ (R= 1,3-$(CH_3)_2C_6H_4$,1,4-$(CH_3)_2C_6H_4$) shows that intermolecular interactions are comparable in importance to intramolecular coupling (770). Single crystal Raman studies of $RCr(CO)_3$ (R=$(CH_3)_6C_6$,$(CH_3)_5C_6H$) show that while vibrational factor-group analysis offers the simplest explanation of the 2000 cm^{-1} region, some features remain unexplained (771). Complete infrared and Raman studies have also been reported for 1,3,5-$(CH_3)_3C_6H_3M(CO)_3$ (M=

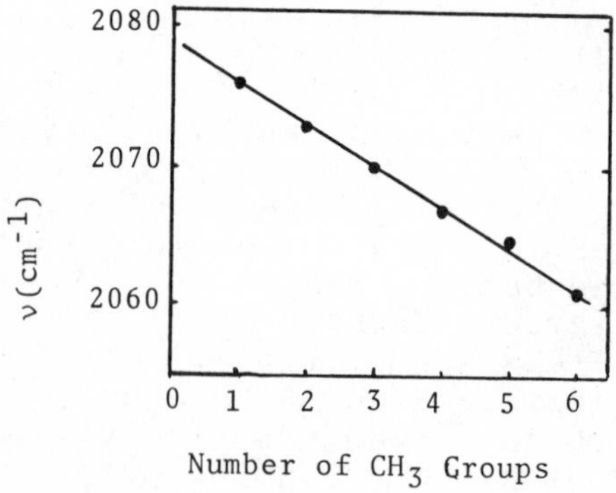

Figure 3.35. Correlation between the frequencies of the A_1 (CO) mode for $[(CH_3)_nC_6H_{6-n}Mn(CO)_3]I_3$ (n=0-6) and the number of methyl groups on the benzene ring (772).

Mo (759,768),W (236,759)) and $RCr(CO)_3$ (R=anisole, methylbenzoate) (759), while infrared data have been listed for $RW(CO)_3$ ($R=CH_3C_6H_5$,1,4-$(CH_3)_2C_6H_4$) (236). In Figure 3.35, a correlation is shown between the frequencies of the A_1 ν(CO) mode for $[(CH_3)_nC_6H_{6-n}Mn(CO)_3]I_3$ (n=0 to 6) and the number of methyl groups on the benzene ring (772). The same type of correlation has been found with analogous chromium(0) complexes (773). Similarly, the ν(CO) mode frequencies observed in CH_2Cl_2 solutions decrease in the order $FC_6H_5Cr(CO)_3$ > $C_6H_6Cr(CO)_3$ > $CH_3C_6H_5Cr(CO)_3$ > 1,2-$(CH_3)_2C_6H_4Cr(CO)_3$ > 1,2,3-$(CH_3)_3C_6H_3Cr(CO)_3$, as expected because of differences in the electron-donating abilities of the substituents (760). The infrared assignments in the 2100 to 1800 cm^{-1} and 700 to 225 cm^{-1} regions have been discussed for $ClC_6H_5Cr(CO)_2L$ ($L=(C_6H_5)_3M$;M=P,As,Sb) (313). A valence force field has been determined for methylbenzoate$Cr(CO)_3$ (774).

Infrared data have been listed for η^6-hexamethylbenzene derivatives of titanium (775), zirconium (775), niobium (775), tantalum (775), chromium (776,777), iron (776,777), cobalt (776,778), and rhodium (778). The hexakis(trifluoromethyl)benzene ring in $(CF_3)_6C_6Rh(\eta^5-C_5H_5)$ is bonded to the rhodium atom by two σ-bonds and one π-bond with the ν(C=C) mode of the remaining uncoordinated C=C bond assigned to an infrared band at 1623 cm^{-1} (26). The bonding is therefore similar to that found in the corresponding tetrakis(trifluoromethyl)cyclopentadienone complex (3.3) with a $CF_3C{=}CCF_3$ group replacing the ketonic carbonyl and bent out of the plane of the other four carbon atoms of the ring (26). Infrared data have been listed for $C_6F_6Cr(CO)_3$ (767).

F. SEVEN-CARBON RINGS

Using infrared data, it has been reported that the $\nu(C{=}C)$ mode frequency of cycloheptene, C_7H_{12}, (1651 cm^{-1}) decreases to 1526 cm^{-1} in $C_7H_{12}AuCl$ (779). Cycloheptadiene, C_7H_{10}, and cycloheptatriene, C_7H_8, both act as four-electron donors in $RFe(CO)_3$ ($R{=}C_7H_{10},C_7H_8$) (780). The former compound shows no bands in the $\nu(C{=}C)$ region while the latter gives an infrared band at 1660 cm^{-1} due to the uncomplexed C=C bond. Infrared data have also been listed for $C_7H_{10}Cr(\eta^5\text{-}C_5H_5)$, $(C_7H_8)_2V$ (782), C_7H_8-$W(CO)_3$ (236), and C_7H_8RuR (R=norbornadiene, 1,5-cyclooctadiene) (783).

Cycloheptatriene can give rise to three species, namely, $C_7H_7^+$, $C_7H_7\cdot$, and $C_7H_7^-$. The tropylium ion, $C_7H_7^+$, which is the most stable of these three species because of its aromatic character, has a planar structure of D_{7h} symmetry. No detailed vibrational studies have been reported for organometallic complexes of the planar η^7-cycloheptatrienyl cation, although infrared data have been reported for the following such complexes: $(C_7H_7)_2$-V^{2+} (782), $C_7H_7M(CO)_3^+$ (M=Cr (48),Mo (48,784),W (236)), $C_5H_5VC_7H_7^+$ (785), and $C_5H_5MC_7H_7$ (M=Ti (786,787),Zr (787), V (786,787),Nb (787),Cr (788,789),Mo (788),W (788)).

It has been suggested that the complexity of the infrared spectrum of $C_7H_7V(CO)_3$ is not consistent with a planar cycloheptatrienyl ligand of C_{7v} symmetry (790). An unsymmetrical structure in which the metal atom is bonded to the neutral η^3-cycloheptatrienyl ring (3.76)

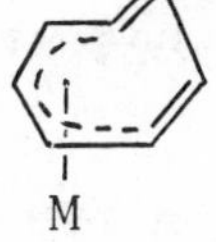

(3.76)

has been proposed for $C_7H_7M(CO)_2X$ (M=Mo (789),W (236), $X{=}C_5H_5$;M=W,X=I (236);M=Co,X=CO (789)); infrared data

have been listed for all of these complexes. The complex nature of the infrared data listed for $C_7H_7CeCl_2$
(791), $(C_7H_7)_2MCl_2$ (M=Ti,Zr) (792), and $C_7H_7M(=O)Cl_2$
(M=Mo,W) (792) has led to the proposal that these complexes formally contain the $C_7H_7^-$ group; two bands in
the 1675 to 1630 cm^{-1} region of each compound have been
assigned to the $\nu(C=C)$ mode.

G. EIGHT-CARBON RINGS

The infrared $\nu(C=C)$ mode frequency of *cis*-cyclooctene,
cis-C_8H_{14} (1648 cm^{-1}) has been reported at 1532 cm^{-1}
in *cis*-$C_8H_{14}AuCl$ (779). Several complexes of *cis,trans*-
1,3-cyclooctadiene, *cis-trans*-1,3-C_8H_{12}, have been characterized. The presence of a single infrared band at
1505 cm^{-1} for *cis-trans*-1,3-$C_8H_{12}Cu_2Cl_2$ suggests that both
C=C bonds of the diene are coordinated. These data,
together with the appeareance of a terminal $\nu(CuCl)$ mode
(375 cm^{-1}), suggest that the olefin is bidentate and
bridges the two copper atoms (793). Although one infrared band, arising from a coordinated C=C bond (1580
cm^{-1}), is observed for *cis,trans*-1,3-$C_8H_{12}AgNO_3$, it has
been suggested that both C=C bonds can interact with the
silver atom to different degrees and that the bands
arising from the more strongly coordinated C=C bond
might overlap a nitrate band at 1490 cm^{-1} (793). Since
the unstable complex *cis-trans*-1,3-$C_8H_{12}AuCl_3$ shows infrared bands at 1625 and 1530 cm^{-1} corresponding to free
and coordinated C=C bonds, respectively, it has tentatively been suggested that the olefin behaves as a monodentate ligand (793). The KBr pellet infrared spectrum
of *cis,trans*-1,3-$C_8H_{12}PtCl_2$ also shows bands corresponding
to free and coordinated C=C bonds at 1640 and 1510 cm^{-1},
respectively, although the KBr pellet infrared spectrum
of *cis,trans*-1,3-$C_8H_{12}PdCl_2$ shows only one band (1475 cm^{-1}),
which has been assigned to a coordinated C=C bond (794).
In CHCl$_3$ solution, however, infrared data and a molecular weight of 603 (as compared to a calculated value of
570 for $(C_8H_{12}PdCl_2)_2$) suggest the presence of an equilibrium (3.77 and 3.78) in which both C=C bonds of the

$$2 C_8H_{12}PdCl_2 \quad \rightleftharpoons \quad \begin{array}{c} C_8H_{12} \diagdown \quad \diagup Cl \diagdown \quad \diagup C_8H_{12} \\ \qquad Pd \qquad \qquad Pd \\ C_8H_{12} \diagup \quad \diagdown Cl \quad \diagdown C_8H_{12} \end{array}$$

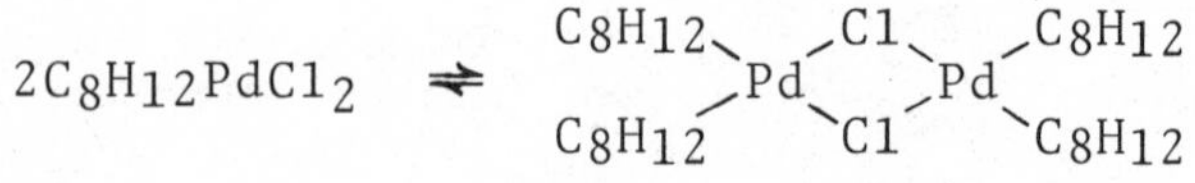

(3.77)   (3.78)

olefin are coordinated in the monomeric complex (3.77) while one C=C bond of each olefin is coordinated in the dimeric complex (3.78) (794).

Infrared and Raman studies have been reported for the 1,5-cyclooctadiene (1,5-C_8H_{12}) complexes formed with the d^8 metals rhodium(I), platinumII), and palladium(II) (795), and the d^{10} metals copper(I), silver(I), and gold(I) (796) to assess the relative strength of the metal-olefin interaction. The infrared and Raman data discussed in Sec. B of Chapter 2 for Zeise's salt and other noncyclic monoolefin complexes show strong coupling of the ν(C=C) and δ(CH) modes, which made it improper to assign the ν(C=C) mode to only one band. Rather, two olefin bands shift significantly to lower frequencies on complex formation. These bands both have ν(C=C) and δ(CH) mode character; the higher-frequency band is called Band I and the lower-frequency band is called Band II. A similar treatment has been presented for the 1,5-cyclooctadiene complexes of d^8 and d^{10} metal ions (795,796). Although both Band I and Band II are easily identified in the Raman spectra of these complexes, in the infrared spectra, Band I is of weak to moderate intensity while Band II is not observed. This is illustrated by the infrared spectra of 1,5-C_8H_{12}, (1,5-C_8H_{12}-Ag)BF_4, and (1,5-C_8H_{12}CuCl)$_2$ in Figure 3.36 and the Raman spectra of the same compounds in Figure 3.37 (795). In Table 3.38, the assignments for Bands I and II are compared for 1,5-cyclooctadiene and several d^8 and d^{10} metal complexes of this ligand, together with the percentage by which the frequency of these modes is lowered on complex formation. The larger decrease in the frequency of Band I relative to that of Band II indicates that the former band has a greater percentage of ν(C=C) mode character. The order of the frequency decrease, namely, Rh(I) > Pt(II) > Pd(II) for the d^8 ions and Au(I) > Cu(I) > Ag(I) for the d^{10} ions, is that expected for a corresponding decrease in the metal-olefin bond strength. With the exception of 1,5-$C_8H_{12}Au_2Cl_2$, all of the complexes in Table 3.38 have been proposed to have a structure in which the 1,5-C_8H_{12} ligand has a tub configuration (3.79). For such a structure, two metal-olefin tilting

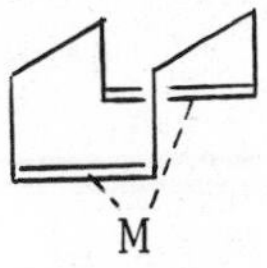

(3.79)

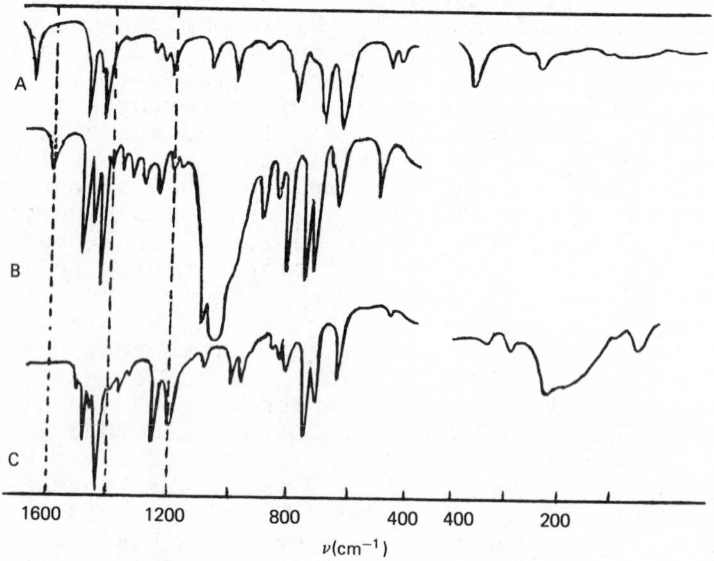

Figure 3.36. Infrared spectra of (A) 1,5-C_8H_{12} (liquid), (B) (1,5-C_8H_{12}Ag)BF_4, and (C) (1,5-C_8H_{12}CuCl)$_2$ (795).

and two metal-olefin stretching modes are expected. These are illustrated in Figure 3.38 while the assignments proposed for these modes in 1,5-$C_8H_{12}MCl_2$ (M=Pd, Pt) and (1,5-$C_8H_{12}RhCl)_2$ are summarized in Table 3.39. Three types of 1,5-cyclooctadiene-gold complexes have been characterized using infrared data (796). For the first, (1,5-$C_8H_{12}AuX_2$)X (X=Cl,Br), it has been proposed that the olefin retains the tub configuration on coordination to the gold(III) ion. The simplicity of the infrared spectra of 1,5-$C_8H_{12}Au_2X_2$ (X=Cl,Br), together with the appearance of the ν(AuX) modes (ν(AuCl)=254 and 238 cm^{-1};ν(AuBr)=194 and 183 cm^{-1}) in the region expected for bridging halogens, suggests a polymeric structure in which the olefin with a chair configuration bridges gold(I) atoms. Although the infrared spectrum of the third gold complex, 1,5-$C_8H_{12}AuCl_2$, is different from those of the other two types of complexes, its structure could not be characte ized. The replacement of the 1658 cm^{-1} infrared band for free 1,5-C_8H_{12} with a new band at 1450 cm^{-1} in 1,5-$C_8H_{12}[Fe(CO)_4]_2$ is con-

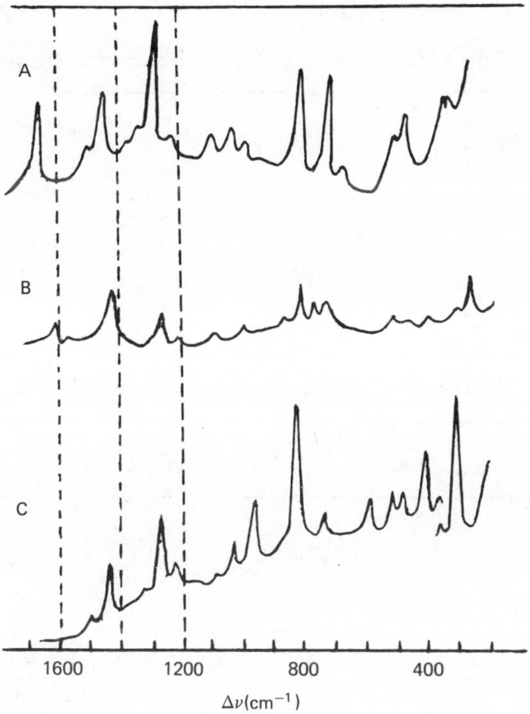

Figure 3.37. Raman spectra of (A) 1,5-C_8H_{12} (liquid), (B) (1,5-$C_8H_{12}Ag$)BF_4, and (C) (1,5-$C_8H_{12}CuCl$)$_2$ (795).

sistent with a structure in which each C=C bond of the olefin is coordinated to an iron atom of a $Fe(CO)_4$ unit (797). The absence of infrared bands characteristic of the free olefin indicates that both olefin C=C bonds are coordinated in 1,5-$C_8H_{12}NiX$ (X=Br,I) (798). The reaction of amines with (1,5-$C_8H_{12}RhCl$)$_2$ gives complexes whose infrared and nmr spectra are consistent with the structure 1,5-$C_8H_{12}Rh$(amine)Cl (799). Infrared data and assignments have been reported for several 1,5-C_8H_{12} complexes of rhodium(I) (800) and iridium (801), while infrared data have been listed for (1,5-C_8H_{12})$_2$Pt (802), 1,5-$C_8H_{12}W(CO)_4$ (236), and 1,5-$C_8H_{12}RuC_7H_8$ (783).

TABLE 3.38
Olefin Frequencies (cm^{-1}) for 1,5-Cyclooctadiene and its Metal Complexes[a]

	Frequency		% Lowering		
Compound[b]	Band I	Band II	Band I	Band II	Total
1,5-C_8H_{12}	1658s,IR				
	1644,R	1280s,R			
$RPdCl_2$	1534s,IR				
	1511mw,IR				
	1522vs,R	1271vs,R	8.5	0.5	9.0
$RPtCl_2$	1496m,IR				
	1500vs,R	1267vs,R	9.5	1.0	10.5
$(RRhCl)_2$	1475m,IR				
	1476vs,R	1241s,R	11.5	3.0	14.5
$(RCuCl)_2$	1490w,IR				
	1490m,R	1265s,R	10.3	1.2	11.5
$(RAg)BF_4$	1602ms,IR				
	1605s,R	1276s,R	3.8	0.3	4.1
RAu_2Cl_2	1488w,IR		10.5		

[a]References 795 and 796.
[b]R=1,5-cyclooctadiene (1,5-C_8H_{12}).

Although infrared data have been listed, no detailed assignments have been reported for the following 1,3,5-cyclooctatriene (1,3,5-C_8H_{10}) complexes: 1,3,5-$C_8H_{10}CrC_5H_5$ (803), 1,3,5-$C_8H_{10}M(CO)_3$ (M=Mo (804), W (236)), and 1,3,5-$C_8H_{10}MC_8H_{10}$ (M=Fe,Ru) (783).

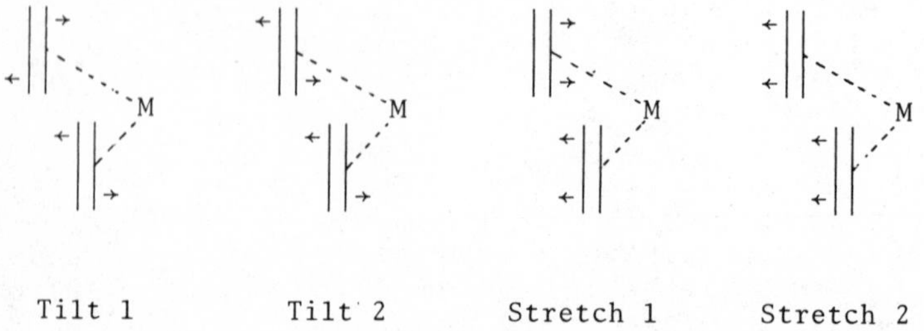

Figure 3.38. Metal-olefin skeletal modes for 1,5-cyclooctadiene complexes

TABLE 3.39
Skeletal Mode Assignments (cm^{-1}) for 1,5-Cyclooctadiene Complexes[a]

	Compound[c]		
Skeletal Mode[b]	$RPdCl_2$	$RPtCl_2$	$(RRhCl)_2$
Tilt 1: IR	570m	588m	583w
R	569mw	587m	586mw
Tilt 2: IR	464vs	480s	490vs
R	464m	482m	480vs
Stretch 1: IR	415w	461m	476vs
R	413vs	461vs	480vs
Stretch 2: IR	350w	378w	388ms
R	352m	385m	393s

[a]Reference 795.
[b]Modes illustrated in Figure 3.38.
[c]R=1,5-cyclooctadiene.

Cyclooctatetraene (C_8H_8) is found with various configurations in metal complexes. A single crystal X-ray study of $C_8H_8AgNO_3$ shows each silver ion to interact with two nonadjacent C=C bonds of the tub form of each cyclooctatetraene molecule (3.80) (805). Although the infrared spectrum of solid $C_8H_8AgNO_3$ exhibits a band at 1449 cm^{-1} that might be attributed to a complexed C=C bond (806), this frequency is relatively low in light of the weak interaction generally found in silver(I)-olefin complexes. The infrared spectrum of

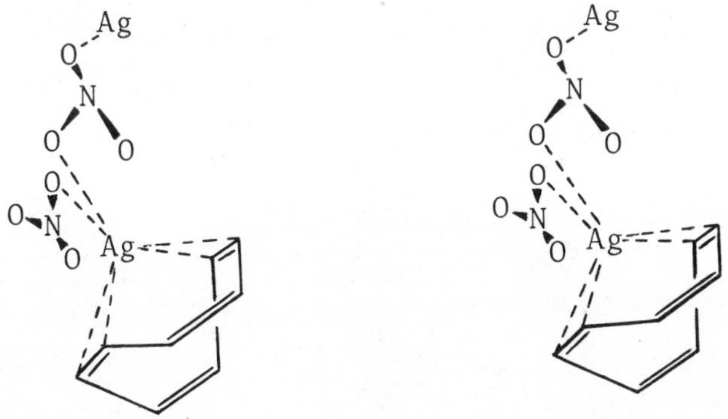

(3.80)

$(C_8H_8Tl)Cl$ shows a band at 1610 cm^{-1}, indicating the presence of a free C=C bond, and suggests the presence of the tub configuration for the cyclooctatetraene ring (807). The infrared spectrum of $K(C_8H_8Tl)$ shows bands similar to those of $(C_8H_8Tl)Cl$ (807). Since infrared bands characteristic of both free (1630 cm^{-1}) and complexed (1410 cm^{-1}) C=C bonds have been reported for $(C_8H_8RhCl)_2$, it has been proposed that the olefin retains a tub configuration with only two nonadjacent C=C bonds involved in bonding (808). The absence of a band at approximately 1600 cm^{-1} for $(C_8H_8Rh_2Cl_2)_n$ has been interpreted as showing that all of the C=C bonds of the olefin in the tub configuration are involved in bonding (3.81) (808). For the same reason, a similar structure

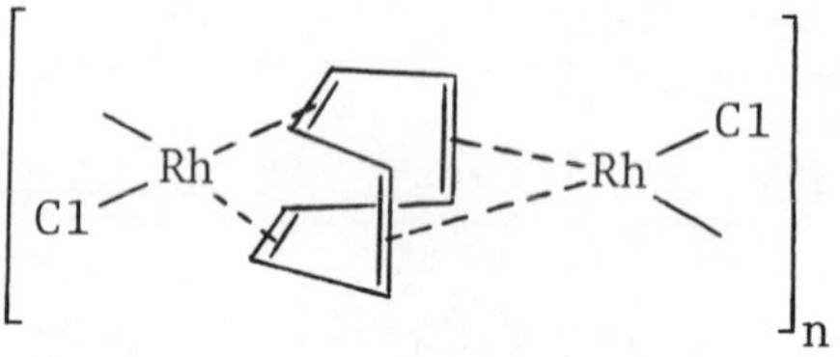

(3.81)

has been proposed for the monomeric compounds C_8H_8-$(PtR_2)_2$ ($R=CH_3,C_6H_5$) (809).

A single crystal X-ray study of $C_8H_8Fe(CO)_3$ (810) has shown that the iron atom of the $Fe(CO)_3$ unit is bonded to two adjacent C=C bonds of the olefin (3.82).

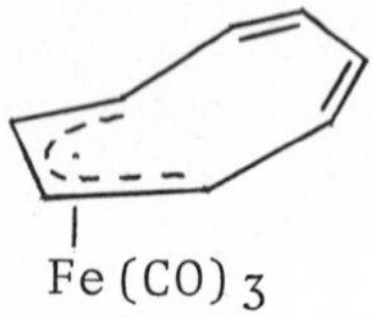

(3.82)

The assignments made in a detailed infrared (4000 to 110 cm^{-1}) and Raman (1650 to 100 cm^{-1}) study of C_8H_8-$Fe(CO)_3$ are summarized in Table 3.40. The similarity of the solid-state and solution (CS_2,CCl_4 and C_2Cl_4) infrared spectra of $C_8H_8Fe(CO)_3$ has been offered as strong evidence for identical structures in both phases. Infrared data have been listed for $C_8H_8[Fe(CO)_3]_2$ (806) in which the iron atom of each $Fe(CO)_3$ unit is bonded to a pair of adjacent C=C bonds and the olefin has a chair configuration (810).

TABLE 3.40
Fundamental Mode Assignments (cm^{-1}) for $C_8H_8Fe(CO)_3$[a]

Mode Description	Assignment Infrared	Raman
ν(CH)	3075w 3040vs 3022s,sh	
ν(CO)	2061vs 1993vs 1976vs	
ν_s(C=C)	1562m	1563m,p
ν(C=C)	1490m	
ν(C=C)	1460m	1460s
Ring deformation	1431w	1431m,p
Ring deformation	1419s	
δ(CH)	1400- 1100m-w	1400- 1100m-w
δ(FeCO)	750- 500vs-s	750- 500s-w
ν(Fe-CO)	475- 350vs-w	475- 350vs-w
Fe-ring tilt	404w	
ν(Fe-ring)	330m	330vs,br,p
δ(CFeC)		137m
δ(CFeC)		100vs,br

[a]Data and assignments from Ref. 811; infrared data obtained using CS_2, CCl_4, and/or C_2Cl_4 solutions; Raman data obtained usind C_6H_6 and/or CS_2 solutions.

Solid $C_8H_8Mo(CO)_3$ has a structure in which only three of the cyclooctatetraene C=C bonds interact with the molybdenum atom (3.83), although the molybdenum

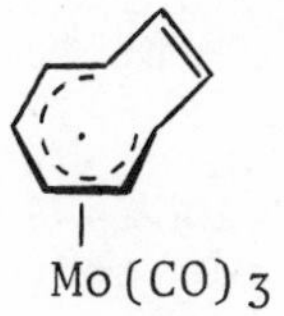

(3.83)

atom is not equidistant from the carbon atoms of these three C=C bonds (812). An analogous structure is sup-

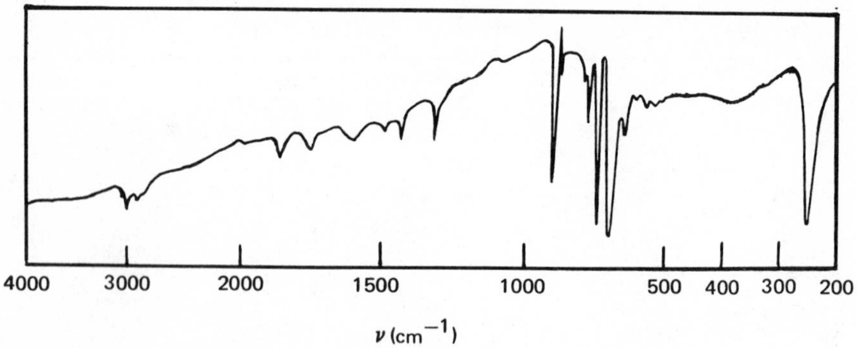

Figure 3.39. Infrared spectrum of (η^8-C_8H_8)$_2$Th (815).

ported by infrared data reported for $C_8H_8W(CO)_3$ with a medium intensity band at 1670 cm^{-1} assigned to the uncomplexed olefin C=C bond (236).

The cyclooctatetraene dianion ($C_8H_8^{2-}$) has 10 π-electrons and a delocalized aromatic structure of D_{8h} symmetry. Because of its high symmetry, the dianion has only four infrared active (A_{2u} + $3E_{1u}$) and 12 Raman active ($2A_{1g}$ + A_{2g} + E_{1g} + $4E_{2g}$ + $4E_{3g}$) modes. Infrared data have been listed for the potassium salt of $C_8H_8^{2-}$ with the infrared active ν(CH) (E_{1u}), ν(CC) (E_{1u}), δ(CH) (E_{1u}), and π(CH) (A_{2u}) modes assigned at 2994, 1431, 880, and 684 cm^{-1}, respectively (48,503). The synthesis and characterization of "uranocene," $(C_8H_8)_2U$, (813) has produced a great deal of interest in organometallic derivatives of the rare earth elements. Single crystal X-ray studies of $(C_8H_8)_2M$ (M=U,Th) show both to have a sandwich-type structure with eclipsed, *octahapto* rings (814). The infrared (815) and Raman (816) spectra of $(C_8H_8)_2Th$ are illustrated in Figures 3.39 and 3.40, respectively, while the descriptions of the infrared and Raman active modes of (η^8-C_8H_8)$_2$M complexes of D_{8h} symmetry are summarized in Table 3.41. The infrared (815-820) and Raman (816,820,821) assignments reported for dicyclooctatetraene actinide complexes have been qualitative in nature and far from complete. In Table 3.42, the infrared and Raman data and some approximate assignments are summarized for dicyclooctatetraene actinide complexes. Infrared data have also been listed for $(RC_8H_7)_2U$ (R=C_2H_5,*n*-C_3H_7,C_6H_5,CH_2=CH,*c*-C_3H_5) and [1,3,5,7-$(CH_3)_4C_8H_4$]$_2$U (818).

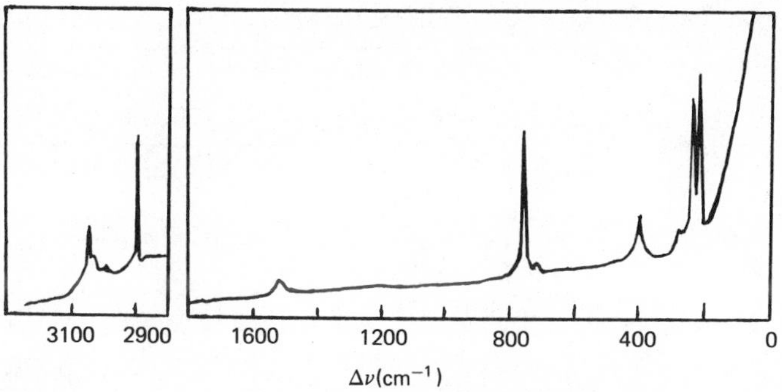

Figure 3.40. Raman spectrum of $(\eta^8\text{-}C_8H_8)_2Th$ *(816).*

The Nujol mull infrared spectra (1200 to 600 cm^{-1}) of the lanthanide complexes $K[(C_8H_8)_2M]$ (M=Ce,Pr,Nd,Sm, Tb) have been reported to be identical to that of $(C_8H_8)_2U$, suggesting that these lanthanide complexes also have an *octahapto* structure of D_{8h} symmetry (821, 822). This has been confirmed in a single crystal X-ray study of $K[(C_8H_8)_2Ce]$ (823). Raman bands at 370 and 200 cm^{-1} for $K[(C_8H_8)_2Ce]$ have been assigned to the asymmetric cerium-ring tilting and stretching modes, respectively (821). A single crystal X-ray study of $(C_8H_8CeCl\cdot 2THF)_2$ has shown the presence of planar cyclooctatetraene rings with aromatic C-C bond lengths (824). The similarity of the infrared spectra of $(C_8H_8MCl\cdot 2THF)_2$ (M=Pr,Nd,Sm) to that of the cerium(IV) analog suggests that these compounds are isostructural (821). A structure of D_{8h} symmetry has also been suggested on the basis of the infrared spectrum of $(C_8H_8)_2Ce$ (825).

A single crystal X-ray study of $(C_8H_8)_2Ti$ has shown that while one of the rings is planar and bonded in an *octahapto* manner to the titanium atom, the other has a boat configuration with only four carbon atoms bonded to the titanium atom (826). The similarity of the infrared spectra (4000 to 200 cm^{-1}) of $(C_8H_8)_2Ti$ and $(C_8H_8)_2V$ indicates that both compounds have similar structures (815). The asymmetric metal-ring tilting and stretching modes have been assigned at 688 and 455 cm^{-1}, respectively, for $(C_8H_8)_2V$ (815).

TABLE 3.41
Description of the Infrared and Raman Active Modes for $(\eta^8\text{-}C_8H_8)_2M$ Complexes of D_{8h} symmetry[a]

Irreducible Representation	Approximate Description	Activity
A_{2u}	ν(CH)	Infrared
	ν(CC)	Infrared
	π(CH)	Infrared
	ν_a(M-ring)[b]	Infrared
E_{1u}	ν(CH)	Infrared
	ν(CC)	Infrared
	δ(CH)	Infrared
	π(CH) or π(CCC)[c]	Infrared
	Asymmetric M-Ring Tilt[b]	Infrared
	δ(Ring-M-Ring)[b]	Infrared
A_{1g}	ν(CH)	Raman
	ν(CC)	Raman
	π(CH)	Raman
	ν_s(M-Ring)[b]	Raman
E_{1g}	ν(CH)	Raman
	ν(CC)	Raman
	δ(CH)	Raman
	π(CH) or π(CCC)[c]	Raman
	Symmetric M-Ring Tilt[b]	Raman
E_{2g}	ν(CH)	Raman
	ν(CC)	Raman
	δ(CH)	Raman
	δ(CH)	Raman
	π(CH)	Raman
	π(CCC)	Raman

[a]Reference 815.
[b]The M-ring skeletal modes are analogous to those illustrated in Figure 3.7 for $(\eta^5\text{-}C_5H_5)_2M$ complexes.
[c]The possibility of coupling makes it impossible to distinguish between these modes without a normal coordinate analysis.

H. NINE-CARBON RINGS

The diene *cis,cis*-1,5-cyclononadiene (*cis,cis*-1,5-C_9H_{14}) forms a 1:1 $AgNO_3$-olefin complex, a 2:1 CuCl-olefin complex, and a 1:1 RhCl-olefin complex (827). Infrared bands at 1646 and 1633 cm^{-1} of the uncomplexed olefin appear at 1602 and 1592 cm^{-1} in the silver nitrate complex while the copper(I) chloride complex shows a new

TABLE 3.42
Vibrational Data and Approximate Assignments (cm^{-1}) for $(\eta^8\text{-}C_8H_8)_2M$ (M=Actinide Element) Complexes

	R_2Th[b]		R_2Pa[c]	R_2U[d]		R_2Np[e]	R_2Pu[e]
Assignment[a]	Infrared	Raman	Infrared	Infrared	Raman	Infrared	Infrared
		3045m				3020sh	3040w
		3022m					
ν(CH)	3005m			3000w		3005vs	2970vs
	2920m	2928vw				2920s	2920s
	2880w	2855vw				2840sh	2860m
	2830w						
	1965w			1960w			
	1865w		1850w	1870w			
	1765w,br		1750w	1765w,br			
		1505w					
	1450w					1470vw	
ν(CC)	1430m						
	1315m	1315vw	1310m	1320m			
						1180m	1210w
						1120s	1120s
						960m	
δ(CH)	895s	901vw	895s	900s		890w	905w
		860vw					
	848w			845w		830w	
	810w						
π(CH)	790w	790vw	795m	792w			
π(CH)	775s	775s	775m	777m			
		770vw					
Ring breathing	742s		745s	746s	750m		

TABLE 3.42 (continued)

	R_2Th[b]		R_2Pa[c]	R_2U[d]		R_2Np[e]	R_2Pu[e]
Assignment[a]	Infrared	Raman	Infrared	Infrared	Raman	Infrared	Infrared
		726w					
						680sh	690sh
Asym. M-R tilt	695vs		695vs	69 vs		670s	660m
	642w						
	608w						
				595s			
	565w						
	525w						
Sym. M-R tilt	375m,br	391m		385m,br	380w		
		270w					
ν_a(MR)	250vs	242s		242/237s			
ν_s(MR)		228/225s					
		200vw					
		185vw					

[a] $R = \eta^8\text{-}C_8H_8$.
[b] Raman data from Ref. 816; infrared data from Refs. 815-817.
[c] Reference 817.
[d] Raman data from Ref. 821; infrared data from Refs. 815-819.
[e] Reference 820.

infrared band at 1615 cm^{-1}; no infrared bands have been observed in the 1675 to 1600 cm^{-1} region for the rhodium(I) chloride complex, indicating a strong rhodium-olefin interaction (827). Infrared data have also been listed for (*cis,cis*-1,5-$C_9H_{14})_2MCl_2$ (M=Pd,Pt) (828).

While free 1,2,6-cyclononatriene shows infrared bands at 1950 and 1650 cm^{-1}, which have been assigned to the ν_a(C=C=C) and ν(C=C) modes, respectively, the 1:1 $AgNO_3$-olefin complex shows infrared bands at 1900 and 1740 cm^{-1}, which have been assigned to the free and complexed allenic C=C bonds, respectively, and 1595 cm^{-1}, which has been assigned to the isolated, complexed C=C bond (829). A monomeric structure in which the silver ion is coordinated to the isolated C=C bond and one of the allenic C=C bonds of the triene (3.84) is con-

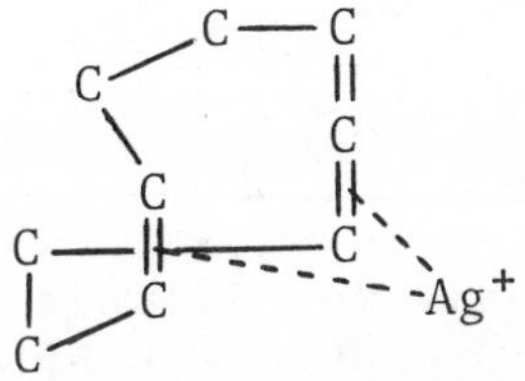

(3.84)

sistent with the infrared data. The copper(I) chloride complex with 1,2,6-cyclononatriene exhibits infrared bands at 1845 and 1660 cm^{-1}, which have been assigned to the free and complexed allenic C=C bonds, respectively, of the triene: a polymeric structure with the copper(I) ions bridged by the Cl ligands seems probable for this compound (830). Both nmr and infrared data (831) and a single crystal X-ray study (832) of the *cis,cis,cis*-1,4,7-cyclononatriene complex *cis,cis,cis*-1,4,7-$C_9H_{12}\cdot 3AgNO_3$ have shown the olefin to have a crown configuration with each silver ion associated with one of the C=C bonds (3.85). Infrared data have also been

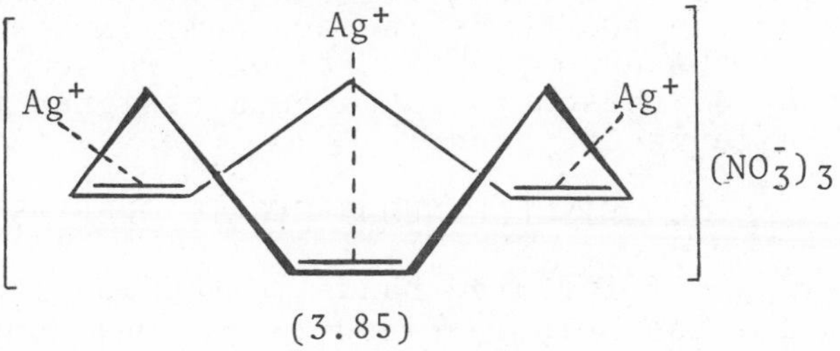

(3.85)

listed for *cis,cis,cis*-1,4,7-$C_9H_{12}Mo(CO)_3$ for which a structure in which the molybdenum atom is coordinated to all three C=C bonds of the olefin with the crown configuration has been proposed (831).

I. TEN-CARBON RINGS

The ν(C=C) mode frequency of *trans*-cyclodecene (1657 cm^{-1}) changes to 1532 cm^{-1} in *trans*-$C_{10}H_{18}AuCl$ (779). Infrared data have been discussed for several silver nitrate and copper(I) chloride and bromide complexes of both *cis,trans*-1,5-cyclodecadiene and *cis,cis*-1,6-cyclodecadiene in terms of their probable structures (833). Infrared data have been listed for *cis,cis*-1,6-$C_{10}H_{16}$-$PdCl_2$ (834).

The olefin 1,2,6-cyclodecatriene (1,2,6-$C_{10}H_{14}$) forms a 1:1 $AgNO_3$-olefin complex and a 2:1 CuCl-olefin complex (830). The infrared spectrum of the silver nitrate complex shows bands at 1905, 1745, and 1600 cm^{-1}, similar to the spectrum of the analogous 1,2,6-cyclononatriene complex (3.84), which have been assigned to the uncomplexed allenic C=C bond, complexed C=C bond, and the isolated C=C bond, respectively, of the olefin. The silver complex has either a monomeric or polymeric structure. The CuCl complex shows infrared bands at 1840 and 1660 cm^{-1}, which have been assigned to the free and complexed allenic C=C bonds, respectively, of the olefin, this again being similar to what was found for the analogous 1,2,6-cyclononatriene complex. A polymeric structure with bridging Cl ligands is probable for 1,2,6-$C_{10}H_{14}(CuCl)_2$.

The structures of the 1,2,6,7-cyclodecatetraene (1,2,6,7-$C_{10}H_{12}$) complexes 1,2,6,7-$C_{10}H_{12}\cdot 2AgNO_3$ and 1,2,6,7-$C_{10}H_{12}\cdot 2CuCl$, have been characterized using infrared data (830). The silver nitrate complex shows infrared bands at 1890 and 1750 cm^{-1} and is either polymeric or monomeric, while the CuCl complex shows analogous bands at 1840 and 1650 cm^{-1} and appears to be polymeric with bridging Cl ligands. The higher frequency of the two bands in each complex has been assigned to the uncomplexed C=C bond, while the lower frequency of these bands has been assigned to the complexed allenic C=C bond.

J. MISCELLANEOUS CYCLIC OLEFINS

The absence of infrared bands in the 1650 to 1550 cm^{-1} region for the norbornadiene (*nor*-C_7H_8) complex *nor*-$C_7H_8\cdot 2AgNO_3$, indicates that both of the olefinic C=C bonds are involved in complex formation (835). A single crystal X-ray study of *nor*-$C_7H_8 \cdot 2AgNO_3$ (836) has confirmed this and has also shown the silver ions to be linked in chains by the nitrate groups (3.86). Simi-

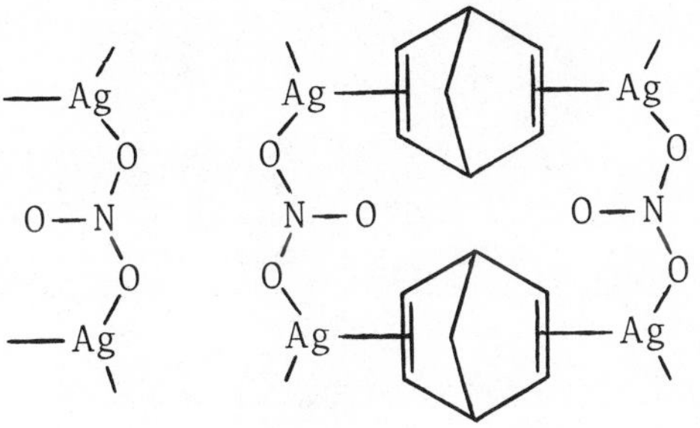

(3.86)

larly, an infrared band at 1470 cm^{-1} indicates that both olefinic C=C bonds are coordinated in *nor*-$C_7H_8(CuCl)_2$ (837). This conclusion is also supported by the absence of bands that are found at 730 and 660 cm^{-1} for free norbornadiene. Infrared bands at 1555 and 1470 cm^{-1} have been given as evidence that *nor*-$C_7H_8(CuBr)_2$ has both coordinated and free C=C bonds and that its structure is different from that of *nor*-$C_7H_8(CuCl)_2$ (837); this conclusion differs from one reached earlier (838). A monomeric structure in which both C=C bonds of the norbornadiene ligand are coordinated to the same metal atom (3.87) is found for several complexes. The major

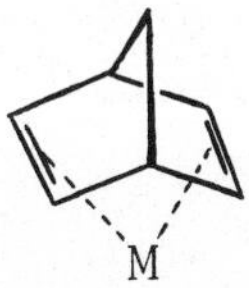

(3.87)

changes noted in the vibrational spectrum of norbornadiene after formation of complexes such as 3.87 include a decrease in the ν_a(C=C) and ν_s(C=C) mode frequencies by over 170 cm^{-1} to the region expected for the $\delta_s(CH_2)$ mode. Table 3.43 compares the ν_a(C=C), ν_s(C=C), and $\delta_s(CH_2)$ mode assignments for free norbornadiene and several metal complexes of this ligand. These data have been interpreted as indicating that the metal-olefin bond strength increases in the order Fe > Cr > Mo

TABLE 3.43

Selected Vibrational Assignments for Norbornadiene and Several Norbornadiene-Metal Complexes[a]

Compound[b]	ν_s(C=C)	ν_a(C=C)	$\delta_s(CH_2)$	Tilt 2[c]	Stretch 1[c]	Stretch 2[c]	ν(M-halide)
nor-C_7H_8	1579	1547	1455				
$RCr(CO)_4$	1457	1427	1432				
$RMo(CO)_4$	1456	1423	1431				
$RFe(CO)_3$	1400	1371	1438	250		304	
$(RRhCl)_2$	1395	1380	1430	296	373	358	247 244
$RPtCl_2$	1414	1394	1440	265	342	319	295
$RPtBr_2$	1412	1395	1440	260	335	298	225 222
$RPtI_2$	1407	1389	1434	258	322	288	180 176
$RPdCl_2$	1429	1409	1452	235	333	302	268
$RPdBr_2$	1422	1409	1444	296	324	302	225 208

[a]Reference 839.

[b]R=*nor*-C_7H_8

[c]Illustrated in Figure 3.38.

for the CO complexes and Rh > Pt > Pd for the halide complexes (839). For a given metal, the strength of this interaction also increases in the order I > Br > Cl. Another change in the spectrum of norbornadiene upon coordination is a decrease in the infrared intensity of the A_1 and B_1 ρ(=CH) modes at 730 and 660 cm^{-1} respectively, and their simultaneous increase in frequency (839). Metal-olefin skeletal modes analogous to those illustrated in Figure 3.38 for 1,5-cyclooctadiene complexes are expected for norbornadiene complexes. Various metal-ligand skeletal mode assignments reported for norbornadiene-metal complexes are summarized in Table 3.43. Limited infrared data have been reported for *nor*-$C_7H_8W(CO)_4$ (840), *nor*-$C_7H_8RuC_7H_8$ (783), (*nor*-$C_7H_8RuX_2)_n$ (X=Cl,Br) (838), and (*nor*-$C_7H_2Cl_6$-$MCl)_2$ (M=Rh,Ir) (841).

The hexamethyl Dewar benzene ($(CH_3)_6C_6$) complexes $[(CH_3)_6C_6]M(CO)_4$ (3.86) have structures similar to those

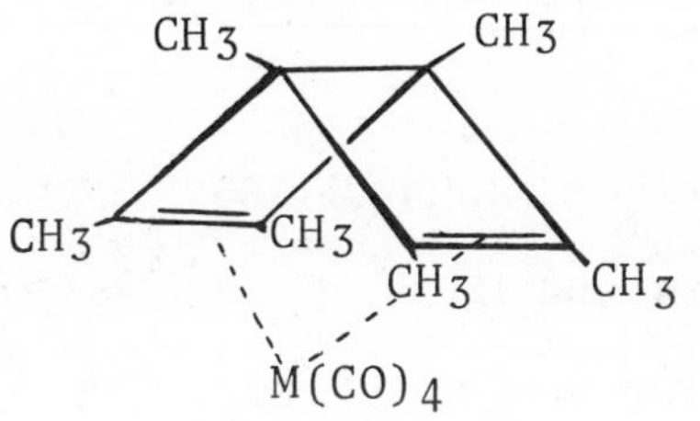

(3.88)

of the analogous norbornadiene complexes with an infrared band at 1694 cm^{-1} in free hexamethyl Dewar benzene shifted to 1543, 1544, and 1530 cm^{-1} for M=Cr, Mo, and W, respectively (842,843). Infrared data have also been reported for $[((CH_3)_6C_6)RhCl]_2$ (844) and $[(CH_3)_6C_6]$-PdX_2 (X=Cl,Br) (845).

As is true for norbornadiene and hexamethyl Dewar benzene, both C=C bonds in dicyclopentadiene ($C_{10}H_{12}$) coordinate to a metal atom. Infrared data have been listed for $C_{10}H_{12}PtCl_2$ (846,847) and $C_{10}H_{12}W(CO)_4$ (236).

Infrared data have been listed for complexes in which there is both a η^5-cyclopentadienyl ring and either a η^4-cyclobutadienyl (848), η^6-benzene (220,849), η^7-cycloheptatrienyl (785-788), or η^8-cyclooctatetraenyl (405,850) ring, as well as for $[(C_6H_5)_4C_4]TiC_8H_8$ (851) and several mixed-ring complexes of iron and ruthenium (783). The structures of both (η^7-azulenium)Cr(0)-(η^5-azuleniate) in which the chromium atom is bonded to an anionic five-membered ring and a cationic seven-membered ring (3.89), and (azulene)$_2$Fe(0) in which the iron atom is bonded to an uncharged five-membered ring

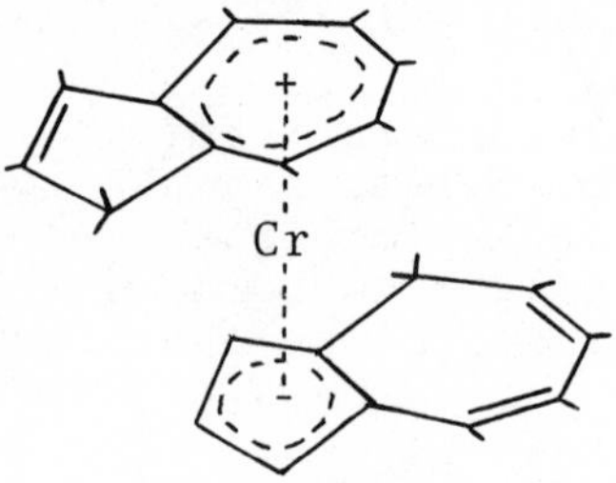

(3.89)

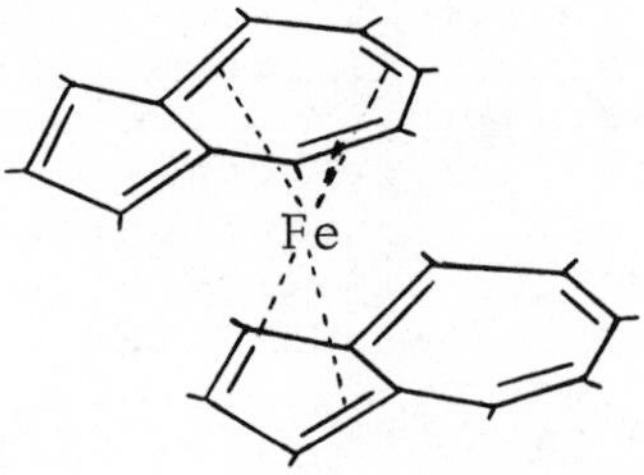

(3.90)

portion and an uncharged seven-membered ring portion of the azulene ligands (3.90) have been proposed on the basis of infrared and nmr data.

REFERENCES

1. D. C. Harris and H. B. Gray, *Inorg. Chem.*, *13*, 2250 (1974).
2. W. K. Olander and T. L. Brown, *J. Amer. Chem. Soc.*, *94*, 2139 (1972).
3. E. W. Gowling and S. F. A. Kettle, *Inorg. Chem.*, *3*, 604 (1964).
4. P. S. Welcker and L. J. Todd, *Inorg. Chem.*, *9*, 286 (1970).
5. E. Osawa, K. Kitamura, and Z. Yoshida, *J. Amer. Chem. Soc.*, *89*, 3814 (1967).
6. A. Krebs and B. Schrader, *Z. Naturforsch.*, *21b*, 194 (1966).
7. C. W. Bird and E. M. Briggs, *J. Chem. Soc. A*, 1004 (1967).
8. W. L. Fichteman, P. Schmidt, and M. Orchin, *J. Organometal. Chem.*, *12*, 249 (1968).
9. P. W. Jolly, M. I. Bruce, and F. G. A. Stone, *J. Chem. Soc.*, 5830 (1965).
10. R. E. Bichler, M. R. Booth, and H. C. Clark, *Inorg. Nucl. Chem. Lett.*, *3*, 71 (1967).
11. D. C. Andrews and G. Davidson, *J. Organometal. Chem.*, *36*, 349 (1972).
12. D. C. Andrews and G. Davidson, *J. Organometal. Chem.*, *76*, 373 (1974).

13. H. P. Fritz, J. F. W. McOmie, and N. Sheppard, *Tetrahedron Lett.*, *26*, 35 (1960).
14. R. G. Amiet, P. C. Reeves, and R. Pettit, *Chem. Commun.*, 1208 (1967).
15. H. P. Fritz, *Z. Naturforsch*, *16b*, 415 (1961).
16. R. Bruce and P. M. Maitlis, *Can. J. Chem.*, *45*, 2017 (1967).
17. H. A. Brune, W. Eberius, and H. P. Wolff, *J. Organometal. Chem.*, *12*, 485 (1968).
18. H. A. Brune, G. Horlbeck, and H. P. Wolff, *Z. Naturforsch*, *25b*, 326 (1970).
19. A. T. Blomquist and P. M. Maitlis, *J. Amer. Chem. Soc.*, *84*, 2329 (1962).
20. P. M. Maitlis and M. L. Games, *Can. J. Chem.*, *42*, 182 (1964).
21. F. Canziani, P. Chini, A. Quarta, and A. Di Martino, *J. Organometal. Chem.*, *26*, 285 (1971).
22. R. B. King and A. Efraty, *J. Organometal. Chem.*, *24*, 241 (1970).
23. D. W. Wertz, D. F. Boclan, and M. J. Hazouri, *Spectrochim. Acta*, *29A*, 1439 (1973).
24. E. O. Fischer and K. Ulm, *Chem. Ber.*, *94*, 2413 (1961).
25. E. O. Fischer and H. Werner, *Chem. Ber.*, *95*, 695 (1962).
26. R. S. Dickson and G. Wilkinson, *J. Chem. Soc.*, 2699 (1964).
27. K. E. Blick, J. W. DeHaan, and K. Niedenzu, *Spectrochim. Acta*, *26A*, 2319 (1970).
28. E. Gallinella, B. Fortunato, and P. Mirone, *J. Mol. Spectrosc.*, *24*, 345 (1967).
29. B. Fortunato, E. Gallinella, and P. Mirone, *Gazz. Chim. Ital.*, *101*, 543 (1971).
30. E. Castellucci, P. Manzelli, B. Fortunato, E. Gallinella, and P. Mirone, *Spectrochim. Acta*, *31A*, 451 (1975).
31. R. K. Kochhar and R. Pettit, *J. Organometal. Chem.*, *6*, 272 (1966).
32. M. L. H. Green, L. Pratt, and G. Wilkinson, *J. Chem. Soc.*, 3753 (1959).
33. P. M. Treichel and R. L. Shubkin, *Inorg. Chem.*, *6*, 1328 (1967).
34. A. Davidson, M. L. H. Green, and G. Wilkinson, *J. Chem. Soc.*, 3172 (1961).
35. M. L. H. Green, L. Pratt, and G. Wilkinson, *J. Chem. Soc.*, 989 (1960).
36. H. P. Fritz, *Chem. Ber.*, *92*, 780 (1959).
37. E. O. Fischer and U. Zahn, *Chem. Ber.*, *92*, 1624 (1959).
38. E. O. Fischer and H. Werner, *Chem. Ber.*, *92*, 1423 (1959).

39. E. O. Fischer and H. Werner, *Chem. Ber.*, *93*, 2075 (1960)
40. E. O. Fischer and B. J. Weimann, *Z. Naturforsch*, *21b*, 84 (1966).
41. E. O. Fischer and B. J. Weimann, *J. Organometal. Chem.*, *8*, 535 (1967).
42. R. S. Dickson and H. P. Kirsch, *Aust. J. Chem.*, *26*, 1911 (1973).
43. R. Markby, H. W. Sternberg, and I. Wender, *Chem. Ind.*, 1381 (1959).
44. M. I. Bruce, M. Cooke, and M. Green, *J. Organometal. Chem.*, *13*, 227 (1968).
45. M. I. Bruce and J. R. Knight, *J. Organometal. Chem.*, *12*, 411 (1968).
46. M. I. Bruce, M. Cooke, M. Green, and D. J. Westlake, *J. Chem. Soc. A*, 987 (1969).
47. H. P. Fritz and L. Schäfer, *Chem. Ber.*, *97*, 1829 (1964).
48. H. P. Fritz, *Adv. Organometal. Chem.*, *1*, 239 (1964).
49. W. T. Ford, *J. Organometal. Chem.*, *32*, 27 (1971).
50. E. O. Fischer and G. Stölzle, *Chem. Ber.*, *94*, 2187 (1961).
51. K. A. Allan, B. G. Gowenlock, and W. E. Lindsell, *J. Organometal. Chem.*, *55*, 229 (1973).
52. R. Zerger and G. Stucky, *J. Organometal. Chem.*, *80*, 7 (1974).
53. F. A. Cotton and L. T. Reynolds, *J. Amer. Chem. Soc.*, *80*, 269 (1958).
54. E. R. Lippincott, J. Xavier, and D. Steele, *J. Amer. Chem. Soc.*, *83*, 2262 (1961).
55. R. T. Bailey and A. H. Curran, *J. Mol. Struct.*, *6*, 391 (1970).
56. G. Wilkinson, F. A. Cotton, and J. M. Birmingham, *J. Inorg. Nucl. Chem.*, *2*, 95 (1956).
57. E. O. Fischer and H. Fischer, *Angew. Chem., Inter. Ed. Engl.*, *3*, 132 (1964).
58. E. O. Fischer and H. Fischer, *J. Organometal. Chem.*, *3*, 181 (1965).
59. E. O. Fischer and H. Fischer, *Angew. Chem., Inter. Ed. Engl.*, *4*, 246 (1965).
60. E. O. Fischer and H. Fischer, *J. Organometal. Chem.*, *6*, 141 (1966).
61. R. A. Forder and K. Prout, *Acta Crystallogr.*, *B30*, 491 (1974).
62. J. L. Calderon, F. A. Cotton, and P. Legzdins, *J. Amer. Chem. Soc.*, *91*, 2528 (1969).
63. A. E. Smith, *Inorg. Chem.*, *11*, 165 (1972).
64. R. B. King, *Inorg. Chem.*, *7*, 90 (1968).
65. D. A. Drew and A. Haaland, *Acta Chem. Scand.*, *27*, 3735 (1973).

66. W. R. Kroll and W. Naegele, *Chem. Commun.*, 246 (1969).
67. A. Haaland and J. Weidlein, *J. Organometal. Chem.*, *40*, 29 (1972).
68. E. Maslowsky, Jr. and K. Nakamoto, *Inorg. Chem.*, *8*, 1108 (1969).
69. F. A. Cotton and T. J. Marks, *J. Amer. Chem. Soc.*, *91*, 7281 (1969).
70. E. R. Lippincott and R. D. Nelson, *Spectrochim. Acta*, *10*, 307 (1958).
71. D. Hartley and M. J. Ware, *J. Chem. Soc. A*, 138 (1969).
72. J. S. Poland and D. G. Tuck, *J. Organometal. Chem.*, *42*, 307 (1972).
73. F. W. B. Einstein, M. M. Gilbert, and D. G. Tuck, *Inorg. Chem.*, *11*, 2832 (1972).
74. J. L. Atwood and K. D. Smith, *J. Amer. Chem. Soc.*, *95*, 1488 (1973).
75. L. J. Guggenberger, *Inorg. Chem.*, *12*, 294 (1973).
76. R. Hoxmeier, B. Deubzer, and H. D. Kaesz, *J. Amer. Chem. Soc.*, *93*, 536 (1971).
77. F. N. Tebbe and L. J. Guggenberger, *Chem. Commun.*, 227 (1973).
78. E. Maslowsky, Jr., *J. Mol. Struct.*, *21*, 464 (1974).
79. J. N. Willis, Jr., M. T. Ryan, F. L. Hedberg, and H. Rosenberg, *Spectrochim. Acta*, *24A*, 1561 (1968).
80. D. H. Whiffen, *J. Chem. Soc.*, 1350 (1956).
81. J. H. S. Green, *Spectrochim. Acta*, *24A*, 863 (1968).
82. M. Tsutsui, N. Ely, and A. Gebala, *Inorg. Chem.*, *14*, 78 (1975).
83. E. Frasson, F. Menegus, and C. Panattoni, *Nature*, *199*, 1087 (1963).
84. G. Bombieri and C. Panattoni, *Acta Crystallogr.*, *20* 595 (1966).
85. C. Panattoni, G. Bombieri, and U. Croatto, *Acta Crystallogr.*, *21*, 823 (1966).
86. E. G. Lindstrom and M. R. Barusch, Abst. 77, 131st ACS Meeting, Miami, Fla., 1957; cited by F. A. Cotton, in *Modern Coordination Chemistry*, J. Lewis and R. G. Wilkins, Eds., Interscience, New York, 1961, p. 301.
87. E. O. Fischer and H. P. Hofmann, *Chem. Ber.*, *92*, 482, (1959).
88. E. O. Fischer and S. Schreiner, *Chem. Ber.*, *92*, 938, (1959).
89. A. Almenninger, O. Bastiansen, and A. Haaland, *J. Chem. Phys.*, *40*, 3434 (1964).
90. A. Haaland, *Acta Chem. Scand.*, *22*, 3030 (1968).
91. H. P. Fritz and D. Sellman, *J. Organometal. Chem.*, *5*, 501 (1966).

93. C. Wong, T. Y. Lee, T. J. Lee, T. W. Chang, and C. S. Liu, *Inorg. Nucl. Chem. Lett.*, *9*, 667 (1973).
94. D. A. Drew and A. Haaland, *Acta Crystallogr.*, *B28*, 3671 (1972).
95. G. B. McVicker and G. L. Morgan, *Spectrochim. Acta*, *26A*, 23 (1970).
96. D. A. Coe, J. W. Nibler, T. H. Cook, D. Drew, and L. Morgan, *J. Chem. Phys.*, *63*, 4842 (1975).
97. G. Wilkinson and T. S. Piper, *J. Inorg. Nucl. Chem.*, *2*, 32 (1956).
98. G. M. Whitesides and J. S. Fleming, *J. Amer. Chem. Soc.*, *89*, 2855 (1967).
99. F. A. Cotton and J. Takats, *J. Amer. Chem. Soc.*, *92*, 2353 (1970).
100. L. T. J. Delbaere, D. W. McBride, and R. B. Ferguson, *Acta Crystallogr.*, *B26*, 515 (1970).
101. J. L. Calderon, F. A. Cotton, and J. Takats, *J. Amer. Chem. Soc.*, *93*, 3587 (1971).
102. A. G. Lee and G. M. Sheldrick, *Chem. Commun.*, 441 (1969).
103. G. Ortaggi, *J. Organometal. Chem.*, *80*, 275 (1974).
104. S. W. Krauhs, G. C. Stocco, and R. S. Tobias, *Inorg. Chem.*, *10*, 1365 (1971).
105. A. N. Nesmeyanov, G. G. Dvoryantseva, N. S. Kochetkova, R. B. Materikova, and Y. N. Sheinker, *Dokl. Akad. Nauk SSSR*, *159*, 847 (1964).
106. J. Stadelhofer, J. Weidlein, and A. Haaland, *J. Organometal. Chem.*, *84*, C1 (1975).
107. S. Mehra and R. K. Multani, *J. Inorg. Nucl. Chem.*, *37*, 2315 (1975).
108. S. Shibata, L. S. Bartell, and R. M. Gavin, Jr., *J. Chem. Phys.*, *41*, 717 (1964).
109. J. G. Contreras and D. G. Tuck, *Inorg. Chem.*, *12*, 2596 (1973).
110. J. G. Contreras and D. G. Tuck, *J. Organometal. Chem.*, *66*, 405 (1974).
111. P. Krommes and J. Lorberth, *J. Organometal. Chem.*, *88*, 329 (1975).
112. J. K. Tyler, A. P. Cox, and J. Sheridan, *Nature*, *183*, 1182 (1959).
113. N. Kumar, B. L. Kalsotra, and R. K. Multani, *J. Inorg. Nucl. Chem.*, *35*, 3019 (1973).
114. N. Kumar and R. K. Sharma, *J. Inorg. Nucl. Chem.*, *36*, 2625 (1974).
115. N. Kumar, B. L. Kalsotra, and R. K. Multani, *J. Inorg. Nucl. Chem.*, *35*, 4295 (1973).
116. N. Kumar, B. L. Kalsotra, and R. K. Multani, *J. Inorg. Nucl. Chem.*, *36*, 1157 (1974).

117. T. Abe and R. Okawara, *J. Organometal. Chem.*, *35*, 27 (1972).
118. A. Almenningen, A. Haaland, and T. Motzfeldt, *J. Organometal. Chem.*, *7*, 97 (1967).
119. L. D. Dave, D. F. Evans, and G. Wilkinson, *J. Chem. Soc.*, 3684 (1959).
120. H. P. Fritz, *Chem. Ber.*, *90*, 780 (1959).
121. H. P. Fritz and E. O. Fischer, *J. Chem. Soc.*, 547 (1961).
122. E. Weiss, *Z. Anorg. Allg. Chem.*, *287*, 236 (1956).
123. J. V. Scibelli and M. D. Curtis, *J. Amer. Chem. Soc.*, *95*, 924 (1973).
124. P. G. Harrison and J. J. Zuckerman, *J. Amer. Chem. Soc.*, *91*, 6885 (1969).
125. P. G. Harrison and M. A. Healy, *J. Organometal. Chem.*, *51*, 153 (1973).
126. P. G. Harrison and J. J. Zuckerman, *J. Amer. Chem. Soc.*, *92*, 2577 (1970).
127. A. Davidson and P. E. Rakita, *Inorg. Chem.*, *9*, 289 (1970).
128. I. M. Shologon, M. K. Romantsevich, and S. V. Kul'kova, *Zh. Obshch. Khim.*, *37*, 2315 (1967).
129. P. C. Angus and S. R. Stobart, *J. Chem. Soc., Dalton Trans.*, 2374 (1973).
130. H. P. Fritz and K.-E. Schwarzhans, *Chem. Ber.*, *97*, 1390 (1964).
131. B. Deubzer, M. Elian, E. O. Fischer, and H. P. Fritz, *Chem. Ber.*, *103*, 799 (1970).
132. P. Krommes and J. Lorberth, *J. Organometal. Chem.*, *88*, 329 (1975).
133. R. S. P. Coutts and P. C. Wailes, *Chem. Ber.*, *108*, 2439 (1975).
134. R. S. P. Coutts and P. C. Wailes, *J. Organometal. Chem.*, *25*, 117 (1970).
135. G. W. Watt, L. J. Baye, and F. O. Drummond, Jr., *J. Amer. Chem. Soc.*, *88*, 1138 (1966).
136. G. W. Watt and F. O. Drummond, Jr., *J. Amer. Chem. Soc.*, *88*, 5926 (1966).
137. G. W. Watt and F. O. Drummond, Jr., *J. Amer. Chem. Soc.*, *92*, 826 (1970).
138. J.-J. Salzmann and P. Mosimann, *Helv. Chim. Acta*, *50*, 1831 (1967).
139. H. H. Brintzinger and J. E. Bercaw, *J. Amer. Chem. Soc.*, *92*, 6182 (1970).
140. L. J. Guggenberger and F. N. Tebbe, *J. Amer. Chem. Soc.*, *95*, 7870 (1973).
141. A. Davidson and S. D. Wreford, *J. Amer. Chem. Soc.*, *96*, 3017 (1974).

142. E. O. Fischer and A. Löckner, *Z. Naturforsch, 15b,* 266 (1960).
143. F. W. Siegert and H. J. de Liefde Meijer, *J. Organometal. Chem., 20,* 141 (1969).
144. J. L. Calderon, F. A. Cotton, B. G. DeBoer, and J. Takats, *J. Amer. Chem. Soc., 93,* 3592 (1971).
145. V. I. Kulishov, E. M. Brainina, N. G. Bokiy, and Yu. T. Struchkov, *Chem. Commun.,* 475 (1970).
146. V. I. Kulishov, N. G. Bokiy, and Yu. T. Struchkov, *Zh. Strukt. Khim., 11,* 700 (1970).
147. V. I. Kulishov, E. M. Brainina, N. G. Bokiy, and Yu. T. Struchkov, *J. Organometal. Chem., 36,* 333 (1972).
148. B. V. Lokshin and E. M. Brainina, *Zh. Strukt. Khim., 12,* 1001 (1971).
149. H. Brintzinger, *J. Amer. Chem. Soc., 88,* 4305 (1966).
150. B. D. James, R. K. Nanda, and M. G. H. Wallbridge, *Inorg. Chem., 6,* 1979 (1967).
151. P. C. Wailes and H. Weigold, *J. Organometal. Chem., 24,* 405 (1970).
152. T. J. Marks, W. J. Kennelly, J. R. Kolb, and L. A. Shimp, *Inorg. Chem., 11,* 2540 (1972).
153. H. Nöth and R. Hartwimmer, *Chem. Ber., 93,* 2238 (1960).
154. F. Klanberg, E. L. Muetterties, and L. J. Guggenberger, *Inorg. Chem., 7,* 2272 (1968).
155. N. Davies, B. D. James, and M. G. H. Wallbridge, *J. Chem. Soc. A,* 2601 (1969).
156. G. Natta, G. Dall'Asta, G. Mazzanti, U. Giannini, and S. Cesca, *Angew. Chem., 71,* 205 (1959).
157. P. C. Wailes and H. Weigold, *J. Organometal. Chem., 28,* 91 (1971).
158. E. Samuel and R. Setton, *C. R. Acad. Sci., Paris, 256,* 443 (1963).
159. E. M. Brainina, R. Kh. Freidlina, and A. N. Nesmeyanov, *Dokl. Akad. Nauk SSSR, 154,* 1113 (1964).
160. A. F. Reid, J. S. Shannon, J. M. Swan, and P. C. Wailes, *Aust. J. Chem., 18,* 173 (1965).
161. S. A. Giddings, *Inorg. Chem., 3,* 684 (1964).
162. R. S. P. Coutts, R. L. Martin, and P. C. Wailes, *Aust. J. Chem., 24,* 2533 (1971).
163. P. M. Druce, B. M. Kingston, M. F. Lappert, T. R. Spalding, and R. C. Srivastava, *J. Chem. Soc. A,* 2106 (1969).
164. P. M. Druce, B. M. Kingston, M. F. Lappert, R. C. Srivastava, M. J. Frazer, and W. E. Newton, *J. Chem. Soc. A,* 2814 (1969).
165. E. Maslowsky, Jr. and K. Nakamoto, *Appl. Spectrosc., 25,* 187 (1971).

166. E. Samuel, R. Ferner, and M. Bigorgne, *Inorg. Chem.*, *12*, 881 (1973).
167. A. F. Reid and P. C. Wailes, *J. Organometal. Chem.*, *2*, 329 (1964).
168. I. A. Ronova and N. V. Alekseev, *Dokl. Akad. Nauk SSSR*, *174*, 614 (1967).
169. A. Clearfield, D. K. Warner, C. H. Saldarriaga-Molina, R. Ropal, and I. Bernal, *Can. J. Chem.*, *53*, 1622 (1975).
170. M. A. Bush and G. A. Sion, *J. Chem. Soc. A*, 2225 (1971).
171. H. A. Martin, M. van Gorkom, and R. O. de Jongh, *J. Organometal. Chem.*, *36*, 93 (1972).
172. C. L. Sloan and W. A. Barber, *J. Amer. Chem. Soc.*, *81*, 1364 (1959).
173. J. A. Waters and G. A. Mortimer, *J. Organometal. Chem.*, *22*, 417 (1970).
174. J. L. Burmeister, E. A. Deardorff, A. Jensen, and V. H. Christiansen, *Inorg. Chem.*, *9*, 58 (1970).
175. R. S. P. Coutts, R. L. Martin, and P. C. Wailes, *Aust. J. Chem.*, *26*, 941 (1973).
176. R. S. P. Coutts and P. C. Wailes, *Aust. J. Chem.*, *20*, 1579 (1967).
177. G. G. Dvoryantseva, Yu. N. Sheinker, A. N. Nesmeyanov, O. V. Nogina, N. A. Lazareva, and V. A. Dubovitskii, *Dokl. Akad. Nauk SSSR*, *161*, 603 (1965).
178. N. N. Vyshinskii, T. I. Ermolaeva, V. N. Latyaeva, A. N. Lineva, and N. E. Lukhton, *Dokl. Akad. Nauk SSSR*, *198*, 1081 (1971).
179. R. S. P. Coutts and P. C. Wailes, *Aust. J. Chem.*, *21*, 373 (1968).
180. R. S. P. Coutts and P. C. Wailes, *Aust. J. Chem.*, *22*, 1547 (1969).
181. G. Doyle and R. S. Tobias, *Inorg. Chem.*, *6*, 1111 (1967).
182. R. S. P. Coutts and P. C. Wailes, *J. Organometal. Chem.*, *84*, 47 (1975).
183. R. S. P. Coutts, R. L. Martin, and P. C. Wailes, *Aust. J. Chem.*, *25*, 1401 (1972).
184. J. H. Teuben and H. J. de Liefde Meijer, *J. Organometal. Chem.*, *46*, 313 (1972).
185. M. A. Chaudhari, P. M. Treichel, and F. G. A. Stone, *J. Organometal. Chem.*, *2*, 206 (1964).
186. C. Tamborski, E. J. Soloski, and S. M. Dec, *J. Organometal. Chem.*, *4*, 446 (1965).
187. H. Köpf and M. Schmidt, *J. Organometal. Chem.*, *10*, 383 (1967).
188. H. Köpf and M. Schmidt, *Z. Anorg. Allg. Chem.*, *340*, 340 (1965).

189. H. Köpf, B. Block, and M. Schmidt, *Z. Naturforsch.*, *22b*, 1077 (1967).
190. H. Köpf, B. Block, and M. Schmidt, *Chem. Ber.*, *101*, 272 (1968).
191. F. Calderazzo, J. J. Salzmann, and P. Mosimann, *Inorg. Chim. Acta*, *1*, 65 (1967).
192. J. L. Thomas and K. T. Brown, *J. Organometal. Chem.*, *111*, 297 (1976).
193. G. Fachinetti, G. Fachi, and C. Floriani, *Chem. Commun.*, 230 (1976).
194. B. Demerseman, G. Bouquet, and M. Bigorgne, *J. Organometal. Chem.*, *107*, C19 (1976).
195. H. A. Martin, P. J. Lemaire, and F. Jellinek, *J. Organometal. Chem.*, *14*, 149 (1968).
196. P. C. Wailes, H. Weigold, and A. P. Bell, *J. Organo-Metal. Chem.*, *34*, 155 (1972).
197. H. Köpf, *J. Organometal. Chem.*, *14*, 353 (1968).
198. H. P. Fritz and R. Schneider, *Chem. Ber.*, *93*, 1171 (1960).
199. F. N. Tebbe and G. W. Parshall, *J. Amer. Chem. Soc.*, *93*, 3793 (1971).
200. L. J. Guggenberger and F. N. Tebbe, *J. Amer. Chem. Soc.*, *93*, 5924 (1971).
201. F. W. Siegert and H. J. de Liefde Meijer, *J. Organometal. Chem.*, *15*, 131 (1968).
202. F. W. Siegert and H. J. de Liefde Meijer, *J. Organometal. Chem.*, *23*, 177 (1970).
203. E. O. Fischer and A. Treiber, *Chem. Ber.*, *94*, 2193 (1961).
204. F. W. Siegert and H. J. de Liefde Meijer, *Rec. Trav. Chim. Pays-Bas*, *87*, 1445 (1968).
205. M. L. H. Green, J. A. McCleverty, L. Pratt, and G. Wilkinson, *J. Chem. Soc.*, 4854 (1961).
206. G. Wilkinson and J. M. Birmingham, *J. Amer. Chem. Soc.*, *76*, 4281 (1954).
207. P. M. Treichel and G. P. Werber, *J. Organometal. Chem.*, *12*, 479 (1968).
208. R. B. King, *Inorg. Chem.*, *5*, 2231 (1966).
209. G. M. Larkin, V. T. Kalinnikov, G. G. Aleksandrov, Yu. T. Struchkov, A. A. Pasnskii, and N. E. Kolobova, *J. Organometal. Chem.*, *27*, 53 (1971).
210. A. T. Casey and J. R. Thackeray, *Aust. J. Chem.*, *25*, 2085 (1972).
211. A. T. Casey and J. R. Thackeray, *Aust. J. Chem.*, *28*, 471 (1975).
212. J. R. Durig, R. B. King, L. W. Houk, and A. L. Marston, *J. Organometal. Chem.*, *16*, 425 (1969).

213. B. V. Lokshin, A. A. Pasinsky, N. E. Kolobova, K. N. Anisimov, and Yu. V. Makarov, *J. Organometal. Chem.*, *55*, 315 (1969).
214. F. Calderazzo and J. Bacciarelli, *Inorg. Chem.*, *2*, 721 (1963).
215. E. O. Fischer, R. J. J. Schneider, and J. Müller, *J. Organometal. Chem.*, *14*, P4 (1968).
216. G. C. Faber and R. J. Angelici, *Inorg. Chem.*, *9*, 1586 (1970).
217. R. B. King, *Z. Naturforsch.*, *18b*, 157 (1963).
218. S. Kapur, B. L. Kalsotra, and R. K. Multani, *Indian J. Chem.*, *10*, 947 (1972).
219. G. W. A. Fowles, L. S. Pu, and D. A. Rice, *J. Organometal. Chem.*, *54*, C17 (1973).
220. H. P. Fritz and J. Manchot, *J. Organometal. Chem.*, *2*, 8 (1964).
221. E. O. Fischer and Y. Hristidu, *Chem. Ber.*, *95*, 253 (1962).
222. H. P. Fritz, Y. Hristidu, H. Hummel, and R. Schneider, *Z. Naturforsch.*, *15b*, 419 (1960).
223. M. J. D'Aniello, Jr. and E. K. Barefield, *J. Organometal. Chem.*, *76*, C50 (1974).
224. R. L. Cooper, M. L. H. Green, and J. T. Moelwyn-Hughes, *J. Organometal. Chem.*, *3*, 261 (1963).
225. A. Davison, J. A. McCleverty, and G. Wilkinson, *J. Chem. Soc.*, 1133 (1963).
226. M. Gerloch and R. Mason, *J. Chem. Soc.*, 296 (1965).
227. E. O. Fischer, K. Ulm, and P. Kuzel, *Z. Anorg. Allg. Chem.*, *319*, 253 (1963).
228. B. Müller and J. Krausse, *J. Organometal. Chem.*, *44*, 141 (1972).
229. R. L. Cooper and M. L. H. Green, *J. Chem. Soc. A*, 1155 (1967).
230. S. P. Anand, *J. Inorg. Nucl. Chem.*, *36*, 925 (1974).
231. K. L. T. Wong and H. H. Brintzinger, *J. Amer. Chem. Soc.*, *97*, 5143 (1975).
232. R. D. Fischer and K. Noack, *J. Organometal. Chem.*, *16*, 125 (1969).
233. D. J. Parker, *J. Chem. Soc., Dalton Trans.*, 155 (1974).
234. D. J. Parker, *J. Chem. Soc. A*, 1382 (1970).
235. A. Davison, W. McFarlane, L. Pratt, and G. Wilkinson, *J. Chem. Soc.*, 3653 (1962).
236. R. B. King and A. Fronzaglia, *Inorg. Chem.*, *5*, 1837 (1966).
237. E. O. Fischer and E. Moser, *J. Organometal. Chem.*, *2*, 230 (1964).
238. H. Brunner, *J. Organometal. Chem.*, *16*, 119 (1969).

239. M. Graziani, J. P. Bibler, R. M. Montesano, and A. Wojcicki, *J. Organometal. Chem.*, *16*, 507 (1969).
240. D. J. Parker and M. H. B. Stiddard, *J. Chem. Soc. A*, 480 (1970).
241. F. C. Wilson and D. P. Showmaker, *J. Chem. Phys.*, *27*, 809 (1957).
242. R. D. Adams and F. A. Cotton, *Inorg. Chim. Acta*, *7*, 153 (1973).
243. A. Alich, N. J. Nelson, D. Strope, and D. F. Shriver, *Inorg. Chem.*, *11*, 2976 (1972).
244. G. Davidson and E. M. Riley, *J. Organometal. Chem.*, *51*, 297 (1973).
245. D. J. Cardin, S. A. Keppie, M. F. Lappert, M. R. Litzow, and T. R. Spalding, *J. Chem. Soc. A*, 2262 (1971).
246. N. Flitcroft and D. Sutton, *Chem. Ind.*, 201 (1969).
247. E. O. Fischer and M. W. Schmidt, *Angew. Chem., Inter. Ed. Engl.*, *6*, 93 (1967).
248. I. J. Hyams, R. T. Bailey, and E. R. Lippincott, *Spectrochim. Acta*, *23A*, 273 (1967).
249. D. M. Adams and A. Squire, *J. Organometal. Chem.*, *63*, 381 (1973).
250. B. V. Lokshin, Z. S. Klemenkova, and Yu. V. Makarov, *Spectrochim. Acta*, *28A*, 2209 (1972).
251. P. S. Braterman and J. D. Black, *J. Organometal. Chem.*, *39*, C3 (1972).
252. A. E. Fenster and I. S. Butler, *Inorg. Chem.*, *13*, 915 (1974).
253. I. S. Butler and D. A. Johansson, *Inorg. Chem.*, *14*, 701 (1975).
254. G. G. Barna, I. S. Butler, and K. R. Plowman, *Can. J. Chem.*, *54*, 110 (1976).
255. D. Sellmann, *J. Organometal. Chem.*, *36*, C27 (1972).
256. W. K. Winter, B. Curnutte, and S. E. Whitcomb, *Spectrochim. Acta*, *15*, 1085 (1959).
257. T. V. Long, Jr. and F. R. Huege, *Chem. Commun.*, 1239 (1968).
258. J. Bodenheimer, E. Lowenthal, and W. Low, *Chem. Phys. Lett.*, *3*, 715 (1969).
259. M. L. H. Green and J. S. Ogden, unpublished results cited by A. J. Downs, in *Spectroscopic Methods in Organometallic Chemistry*, W. O. George, Ed., Butterworth, London, 1970, p. 1.
260. K. Nakamoto, C, Udovich, J. R. Ferraro, and A. Quattrochi, *Appl. Spectrosc.*, *24*, 606 (1970).
261. R. T. Bailey, *Spectrochim. Acta*, *27A*, 199 (1971).
262. I. J. Hyams, *Chem. Phys. Lett.*, *18*, 399 (1973).
263. I. J. Hyams, *Spectrochim. Acta*, *29A*, 839 (1973).

264. F. Rocquet, L. Berreby, and J. P. Marsault, *Spectrochim. Acta*, *29A*, 1101 (1973).
265. J. S. Bodenheimer and W. Low, *Spectrochim. Acta*, *29A*, 1733 (1973).
266. B. V. Lokshin, V. T. Aleksanian, and E. B. Rusach, *J. Organometal. Chem.*, *86*, 253 (1975).
267. J. Bodenheimer, *Chem. Phys. Lett.*, *6*, 519 (1970).
268. D. M. Adams and W. S. Fernando, *J. Chem. Soc., Dalton Trans.*, 2507 (1972).
269. E. O. Fischer and H. Schuster-Woldan, *Chem. Ber.*, *100*, 705 (1967).
270. J. Brunvoll, S. J. Cyvin, and L. Schäfer, *J. Organometal. Chem.*, *27*, 107 (1971).
271. I. J. Hyams, *Chem. Phys. Lett.*, *15*, 88 (1972).
272. L. Schäfer, J. Brunvoll, and S. J. Cyvin, *J. Mol. Struct.*, *11*, 459 (1972).
273. J. Brunvoll, S. J. Cyvin, and L. Schäfer, *Chem. Phys. Lett.*, *13*, 286 (1972).
274. H. P. Fritz, and L. Schäfer, *Z. Naturforsch.*, *19b*, 169 (1964).
275. I. Pavlík and J. Klikorka, *Collect. Czech. Chem. Commun.*, *30*, 664 (1965).
276. P. M. Maitlis and J. D. Brown, *Z. Naturforsch*, *20b*, 597 (1965).
277. I. J. Spilners, *J. Organometal. Chem.*, *11*, 381 (1968).
278. P. Sohár and J. Kuszmann, *J. Mol. Struct.*, 359 (1969).
279. I. Pavlík and V. Plechácek, *Collect. Czech. Chem. Commun.*, *31*, 2083 (1966).
280. H. Werner, G. Mattmann, A. Salzer, and T. Winkler, *J. Organometal. Chem.*, *25*, 461 (1970).
281. T. L. Court and H. Werner, *J. Organometal. Chem.*, *65*, 245 (1974).
282. Y. S. Sohn, A. W. Schlueter, D. N. Hendrickson, and H. B. Gray, *Inorg. Chem.*, *13*, 301 (1974).
283. H. Werner and A. Salzer, *Synthesis Inorg. Metal-Org. Chem.*, *2*, 239 (1972).
284. E. Dubler, M. Textor, H.-R. Oswald, and A. Salzer, *Angew. Chem., Inter. Ed. Engl.*, *12*, 135 (1974).
285. J. Müller and H. Dorner, *Angew. Chem., Inter. Ed. Engl.*, *12*, 843 (1973).
286. J. Müller, H. Dorner, G. Huttner, and H. Lorenz, *Angew. Chem., Inter. Ed. Engl.*, *12*, 1005 (1973).
287. O. S. Mills, *Acta Crystallogr.*, *11*, 620 (1958).
288. R. F. Bryan and P. T. Greene, *J. Chem. Soc. A*, 3064 (1970).
289. R. F. Bryan, P. T. Greene, D. S. Field, and M. J. Newlands, *Chem. Commun.*, 1477 (1969).

290. R. F. Bryan, P. T. Greene, M. J. Newlands, and D. S. Field, *J. Chem. Soc. A*, 3068 (1970).
291. O. S. Mills and J. P. Nice, *J. Organometal. Chem.*, *9*, 339 (1967).
292. R. D. Fischer, A. Vogler, and K. Noack, *J. Organometal. Chem.*, *7*, 135 (1967).
293. F. A. Cotton and G. Yagupsky, *Inorg. Chem.*, *6*, 15 (1967).
294. A. R. Manning, *J. Chem. Soc. A*, 1319 (1968).
295. P. McArdle and A. R. Manning, *J. Chem. Soc. A*, 2128 (1970).
296. J. G. Bullitt, F. A. Cotton, and T. J. Marks, *Inorg. Chem.*, *11*, 671 (1972).
297. S. Onaka and D. F. Shriver, *Inorg. Chem.*, *15*, 915 (1976).
298. M. A. Neuman, Trinh-Toan, and L. F. Dahl, *J. Amer. Chem. Soc.*, *94*, 3383 (1972).
299. A. Terzis and T. G. Spiro, *Chem. Commun.*, 1160 (1970).
300. F. A. Cotton and J. D. Jamerson, *J. Amer. Chem. Soc.*, *98*, 1273 (1976).
301. E. F. Paulus, E. O. Fischer, H. P. Fritz, and H. Schuster-Woldan, *J. Organometal. Chem.*, *10*, P3 (1967).
302. P. McArdle and A. R. Manning, *J. Chem. Soc. A*, 717 (1971).
303. A. R. Manning, *J. Chem. Soc. A*, 2321 (1971).
304. T. S. Piper, F. A. Cotton, and G. Wilkinson, *J. Inorg. Nucl. Chem.*, *1*, 165 (1955).
305. E. O. Fischer and K. Bittler, *Z. Naturforsch.*, *16b*, 225 (1961).
306. E. O. Fischer and K. S. Brenner, *Z. Naturforsch.*, *17b*, 774 (1962).
307. E. O. Fischer and K. Bittler, *Z. Naturforsch.*, *16b*, 835 (1961).
308. E. O. Fischer, H. Schuster-Woldan, and K. Bittler, *Z. Naturforsch.*, *18b*, 429 (1963).
309. P. McArdle and A. R. Manning, *J. Chem. Soc. A*, 2133 (1970).
310. N. E. Kim, N. J. Nelson, and D. F. Shriver, *Inorg. Chim. Acta*, *7*, 393 (1973).
311. J. S. Kristoff and D. F. Shriver, *Inorg. Chem.*, *13*, 499 (1974).
312. J. S. Kristoff and D. F. Shriver, *Inorg. Chem.*, *12*, 1788 (1973).
313. A. R. Manning, *J. Chem. Soc. A*, 106 (1971).
314. M. Ahmad, R. Bruce, and G. R. Knox, *J. Organometal. Chem.*, *6*, 1 (1966).
315. R. B. King and G. R. Knox, *J. Organometal. Chem.*, *6*, 67 (1966).

316. W. R. Cullen and R. G. Hayter, *J. Amer. Chem. Soc.*, *86*, 1030 (1964).
317. J. P. Bibler and A. Wojcicki, *J. Amer. Chem. Soc.*, *88*, 4862 (1966).
318. D. S. Field and M. J. Newlands, *J. Organometal. Chem.*, *27*, 221 (1971).
319. T. Blackmore, M. I. Bruce, and F. G. A. Stone, *J. Chem. Soc. A*, 2376 (1971).
320. W. H. Morrison, Jr. and D. N. Hendrickson, *Inorg. Chem.*, *11*, 2912 (1972).
321. K. R. Mann, W. H. Morrison, Jr., and D. N. Hendrickson, *Inorg. Chem.*, *13*, 1180 (1974).
322. G. Paliani, R. Cataliotti, A. Poletti, and A. Foffani, *J. Chem. Soc., Dalton Trans.*, 1741 (1972).
323. J. R. Hall and B. E. Smith, *Aust. J. Chem.*, *24*, 911 (1971).
324. A. Poletti, R. Cataliotti, and G. Paliani, *Spectrochim. Acta*, *29A*, 277 (1973).
325. R. J. Cross and R. Wardle, *J. Chem. Soc. A*, 2000 (1971).
326. B. L. Kalsotra, S. P. Anand, R. K. Multani, and B. D. Jain, *J. Organometal. Chem.*, *28*, 87 (1971).
327. S. Kapur, B. L. Kalsotra, and R. K. Multani, *J. Inorg. Nucl. Chem.*, *36*, 932 (1974).
328. B. L. Kalsotra, R. K. Multani, and B. D. Jain, *J. Inorg. Nucl. Chem.*, *34*, 2265 (1972).
329. S. Kapur and R. K. Multani, *J. Organometal. Chem.*, *63*, 301 (1973).
330. B. L. Kalsotra, R. K. Multani, and B. D. Jain, *J. Chinese Chem. Soc.*, *18*, 189 (1971).
331. S. Kapur, B. L. Kalsotra, and R. K. Multani, *J. Inorg. Nucl. Chem.*, *35*, 3966 (1973).
332. B. Kanellakopulos, E. O. Fischer, E. Dornberger, and F. Baumgärtner, *J. Organometal. Chem.*, *24*, 507 (1970).
333. B. Kanellakopulos, E. Dornberger, and F. Baumgärtner, *Inorg. Nucl. Chem. Lett.*, *10*, 155 (1974).
334. E. O. Fischer and A. Treiber, *Z. Naturforsch.*, *17b*, 276 (1962).
335. F. Baumgärtner, E. O. Fischer, B. Kanellakopulos, and P. Laubereau, *Angew. Chem., Inter. Ed. Engl.*, *8*, 202 (1969).
336. F. Baumgärtner, E. O. Fischer, B. Kanellakopulos, and P. Laubereau, *Angew. Chem., Inter. Ed. Engl.*, *7*, 634 (1968).
337. E. O. Fischer and Y. Hristidu, *Z. Naturforsch.*, *17b*, 275 (1962).

338. M. L. Anderson and L. R. Crisler, *J. Organometal. Chem.*, *17*, 345 (1969).
339. J. H. Burns, *J. Organometal. Chem.*, *69*, 225 (1974).
340. R. D. Fischer, R. von Ammon, and B. Kanellakopulos, *J. Organometal. Chem.*, *25*, 123 (1970).
341. B. Kanellakopulos, E. Dornberger, R. von Ammon, and R. D. Fischer, *Angew. Chem., Inter. Ed. Engl.*, *9*, 957 (1970).
342. J. Müller, *Chem. Ber.*, *102*, 152 (1969).
343. R. R. Ryan, R. A. Penneman, and B. Kanellakopulos, *J. Amer. Chem. Soc.*, *97*, 4258 (1975).
344. C. H. Wong, T. U. Yen, and T. Y. Lee, *Acta Crystallogr.*, *18*, 340 (1965).
345. L. T. Reynolds and G. Wilkinson, *J. Inorg. Nucl. Chem.*, *2*, 246 (1956).
346. G. L. Ter Haar and M. Dubeck, *Inorg. Chem.*, *3*, 1648 (1964).
347. E. O. Fischer, P. Laubereau, F. Baumgärtner, and B. Kanellakopulos, *J. Organometal. Chem.*, *5*, 583 (1966).
348. T. J. Marks and G. W. Grynkewich, *Inorg. Chem.*, *15*, 1302 (1976).
349. R. von Ammon, B. Kanellakopulos, R. D. Fischer, and P. Laubereau, *Inorg. Nucl. Chem. Lett.*, *5*, 219 (1969).
350. T. J. Marks and J. R. Kolb, *J. Amer. Chem. Soc.*, *97*, 27 (1975).
351. M. Tsutsui and N. M. Ely, *J. Amer. Chem. Soc.*, *97*, 1280 (1975).
352. M. Tsutsui and N. M. Ely, *J. Amer. Chem. Soc.*, *97*, 3551 (1975).
353. N. M. Ely and M. Tsutsui, *Inorg. Chem.*, *14*, 2681 (1975).
354. T. J. Marks and A. Wachter, *J. Amer. Chem. Soc.*, *98*, 703 (1976).
355. G. Brandi, M. Brunelli, G. Lugli, and A. Mazzei, *Inorg. Chim. Acta*, *7*, 319 (1973).
356. T. J. Marks, A. M. Seyam, and J. R. Kolb, *J. Amer. Chem. Soc.*, *95*, 5529 (1973).
357. A. E. Gebala and M. Tsutsui, *J. Amer. Chem. Soc.*, *95*, 91 (1973).
358. G. W. Halstead, E. C. Baker, and K. N. Raymond, *J. Amer. Chem. Soc.*, *97*, 3049 (1975).
359. R. T. Bailey and E. R. Lippincott, *Spectrochim. Acta*, *21*, 389 (1965).
360. M. Rosenblum, *Chem. Ind.*, 953 (1958).
361. K. L. Rinehart, Jr., K. L. Motz, and S. Moon, *J. Amer. Chem. Soc.*, *79*, 2749 (1957).

362. R. A. Benkeser, J. Hooz, and Y. Nagai, *Bull. Chem. Soc. Japan, 37,* 53 (1964).
363. M. Rosenblum and R. B. Woodward, *J. Amer. Chem. Soc., 80,* 5443 (1958).
364. R. B. King and M. B. Bisnette, *J. Organometal. Chem., 8,* 287 (1967).
365. A. N. Nesmeyanov, L. A. Kazitsyna, B. V. Lokshin, and I. I. Kritskaya, *Dokl. Akad. Nauk SSSR, 117,* 433, (1957).
366. A. N. Nesmeyanov, L. A. Kazitsyna, B. V. Lokshin, and V. D. Vil'chevskaya, *Dokl. Akad. Nauk SSSR, 125,* 1037 (1959).
367. P. L. Pauson, *J. Amer. Chem. Soc., 76,* 2187 (1954).
368. M. Rosenblum, *J. Amer. Chem. Soc., 81,* 4530 (1959).
369. M. Rosenblum and W. G. Howells, *J. Amer. Chem. Soc., 84,* 1167 (1962).
370. M. Rosenblum, W. G. Howells, A. K. Banerjee, and C. Bennett, *J. Amer. Chem. Soc., 84,* 2726 (1962).
371. G. D. Broadhead, J. M. Osgerby, and P. L. Pauson, *J. Chem. Soc.,* 650 (1958).
372. M. D. Rausch, E. O. Fischer, and H. Grubert, *J. Amer. Chem. Soc., 82,* 78 (1960).
373. G. G. Dvoryantseva and Yu. N. Sheinker, *Izvest. Akad. Nauk SSSR, Otd. Khim. Nauk,* 924 (1963).
374. K. L. Rinehart, Jr., R. J. Curby, Jr., and P. E. Sokol, *J. Amer. Chem. Soc., 79,* 3420 (1957).
375. K. L. Rienhart, Jr. and R. J. Curby, Jr., *J. Amer. Chem. Soc., 79,* 3290 (1957).
376. K. Schlögl and H. Seiler, *Tetrahedron Lett., 7,* 4 (1960).
377. K. Schlögl and M. Peterlik, *Tetrahedron Lett., 13,* 573 (1962).
378. K. Schlögl, M. Peterlik, and H. Seiler, *Monatsh. Chem., 93,* 1309 (1962).
379. P. Carty and M. F. A. Dove, *J. Organometal. Chem., 28,* 125 (1971).
380. M. Rosenblum, *Chemistry of the Iron Group Metallocenes,* Interscience, New York, 1965.
381. M. D. Rausch, E. O. Fischer, and H. Grubert, *Chem. Ind.,* 756 (1958).
382. M. F. Sullivan and W. F. Little, *J. Organometal. Chem., 8,* 277 (1967).
383. D. J. Parker, *Spectrochim. Acta, 31A,* 1789 (1975).
384. R. T. Bailey, *Spectrochim. Acta, 25,* 1127 (1969).
385. J. Doe, S. Borkett, and P. G. Harrison, *J. Organometal. Chem., 52,* 343 (1973).
386. L. M. C. Shen, G. G. Long, and C. G. Moreland, *J. Organometal. Chem., 5,* 362 (1966).

387. E. O. Fischer and W. Fellmann, *J. Organometal. Chem.*, *1*, 191 (1963).
388. F. Kaufman and D. O. Cowan, *J. Amer. Chem. Soc.*, *92*, 6198 (1970).
389. F. L. Hedberg and H. Rosenberg, *J. Amer. Chem. Soc.*, *91*, 1258 (1969).
390. M. F. Semmelhack, J. S. Foos, and S. Katz, *J. Amer. Chem. Soc.*, *95*, 7325 (1973).
391. T. J. Katz and N. Acton, *J. Amer. Chem. Soc.*, *94*, 3281 (1972).
392. A. Salzer and H. Werner, *Angew. Chem., Inter. Ed. Engl.*, *11*, 930 (1972).
393. R. E. Christopher and L. M. Venanzi, *Inorg. Chim. Acta*, *7*, 489 (1973).
394. R. B. King, A. Efraty, and W. M. Douglas, *J. Organometal. Chem.*, *56*, 345 (1973).
395. R. B. King and A. Efraty, *J. Amer. Chem. Soc.*, *94*, 3773 (1972).
396. C. White, A. J. Oliver, and P. M. Maitlis, *J. Chem. Soc., Dalton Trans.*, 1901 (1973).
397. J. E. Bercaw, *J. Amer. Chem. Soc.*, *96*, 5087 (1974).
398. J. M. Manriquez and J. E. Bercaw, *J. Amer. Chem. Soc.*, *96*, 6229 (1974).
399. E. Samuel and M. Bigorgne, *J. Organometal. Chem.*, *19*, 9 (1969).
400. E. Samuel and M. Bigorgne, *J. Organometal. Chem.*, *30*, 235 (1971).
401. E. Samuel and R. Setton, *J. Organometal. Chem.*, *4*, 156 (1965).
402. E. Samuel, *J. Organometal. Chem.*, *19*, 87 (1969).
403. J. Trotter, *Acta Crystallogr.*, *11*, 355 (1958).
404. J. Goffart, J. Fuger, B. Gilbert, and G. Duyckaerts, *Inorg. Nucl. Chem. Lett.*, *11*, 569 (1975).
405. M. E. E. Veldman and H. O. van Oven, *J. Organometal. Chem.*, *84*, 247 (1975).
406. J. Goffart and G. Duyckaerts, *J. Organometal. Chem.*, *94*, 29 (1975).
407. J. H. Osiecki and C. J. Hoffman, *J. Organometal. Chem.*, *3*, 107 (1965).
408. G. Wulfsberg and R. West, *J. Amer. Chem. Soc.*, *93*, 4085 (1971).
409. G. Wulfsberg, R. West, and V. N. M. Rao, *J. Amer. Chem. Soc.*, *95*, 8658 (1973).
410. G. M. Brown, F. L. Hedberg, and H. Rosenberg, *Chem. Commun.*, 5 (1972).
411. F. L. Hedberg and H. Rosenberg, *J. Amer. Chem. Soc.*, *95*, 870 (1973).

412. F. L. Hedberg and H. Rosenberg, *J. Amer. Chem. Soc.*, *92*, 3239 (1970).
413. G. Wulfsberg and R. West, *J. Amer. Chem. Soc.*, *94*, 6069 (1972).
414. K. J. Reimer and A. Shaver, *Inorg. Chem.*, *14*, 2702 (1975).
415. D. C. Beer, V. R. Miller, L. G. Sneddon, R. N. Grimes, M. Mathew, and G. J. Palenik, *J. Amer. Chem. Soc.*, *95*, 3046 (1973).
416. R. N. Grimes, *J. Amer. Chem. Soc.*, *93*, 201 (1971).
417. J. P. Brennan, R. N. Grimes, R. Schaeffer, and L. G. Sneddon, *Inorg. Chem.*, *12*, 2266 (1973).
418. R. N. Grimes, W. J. Rademaker, M. L. Denniston, R. F. Bryan, and P. T. Greene, *J. Amer. Chem. Soc.*, *94*, 1865 (1972).
419. R. N. Grimes and W. J. Rademaker, *J. Amer. Chem. Soc.*, *91*, 6498 (1969).
420. J. W. Howard and R. N. Grimes, *J. Amer. Chem. Soc.*, *91*, 6499 (1969).
421. R. Cataliotti, A. Foffani, and S. Pignataro, *Inorg. Chem.*, *9*, 2594 (1970).
422. B. V. Lokshin, E. B. Rusach, V. N. Setkina, and N. I. Pyshnograeva, *J. Organometal. Chem.*, *77*, 69 (1974).
423. K. Öfele and E. Dotzauer, *J. Organometal. Chem.*, *30*, 211 (1971).
424. D. C. Andrews and G. Davidson, *J. Organometal. Chem.*, *74*, 441 (1974).
425. B. V. Lokshin, V. T. Aleksanyan, and Z. S. Klemenkova, *J. Organometal. Chem.*, *70*, 437 (1974).
426. B. V. Lokshin, V. T. Aleksanyan, Z. S. Klemankova, L. V. Rybin, and N. T. Gubenko, *J. Organometal. Chem.*, *74*, 97 (1974).
427. H. Hosoya and S. Nagakura, *Bull. Chem. Soc. Japan*, *37*, 249 (1964).
428. G. Winkhaus, L. Pratt, and G. Wilkinson, *J. Chem. Soc.*, 3807 (1961).
429. P. H. Bird and M. R. Churchill, *Chem. Commun.*, 777 (1967).
430. F. Calderazzo, *Inorg. Chem.*, *5*, 429 (1966).
431. J. F. Helling and D. M. Braitsch, *J. Amer. Chem. Soc.*, *92*, 7207 (1970).
432. J. F. Helling and D. M. Braitsch, *J. Amer. Chem. Soc.*, *92*, 7209 (1970).
433. D. Jones, L. Pratt, and G. Wilkinson, *J. Chem. Soc.*, 4458 (1962).
434. E. O. Fischer and M. W. Schmidt, *Chem. Ber.*, *102*, 1954 (1969).

435. G. E. Herberich and J. Schwarzer, *Angew. Chem., Inter. Ed. Engl.*, *8*, 143 (1969).
436. J. C. Lockhart, *J. Chem. Soc. A*, 1552 (1966).
437. H. S. Kimmel, *J. Mol. Struct.*, *12*, 373 (1972).
438. K. M. Mackay, D. B. Sowerby, and W. C. Young, *Spectrochim. Acta*, *24A*, 611 (1968).
439. G. J. Kubas and D. F. Shriver, *J. Amer. Chem. Soc.*, *92*, 1949 (1970).
440. K. Dehnicke, *J. Organometal. Chem.*, *9*, 11 (1967).
441. W. Haubold and J. Weidlein, *Z. Anorg. Allg. Chem.*, *420*, 251 (1976).
442. H. T. Baechle and H. J. Becher, *Spectrochim. Acta*, *21*, 579 (1965).
443. A. L. Smith, *Spectrochim. Acta*, *23A*, 1075 (1967).
444. J. R. Durig, K. L. Hellams, and J. H. Mulligan, *Spectrochim. Acta*, *28*, 1039 (1972).
445. H. Kriegsmann and K.-H. Schowatka, *Z. Phys. Chem. (Leipzig)*, *209*, 261 (1958).
446. J. R. Durig, C. W. Sink, and S. F. Bush, *J. Chem. Phys.*, *45*, 66 (1966).
447. J. R. Durig, C. W. Sink, and J. B. Turner, *Spectrochim. Acta*, *26A*, 557 (1970).
448. J. R. Durig, C. W. Sink, and J. B. Turner, *Spectrochim. Acta*, *25A*, 629 (1969).
449. J. R. Durig, C. W. Sink, and J. B. Turner, *J. Chem. Phys.*, *49*, 3422 (1968).
450. J. R. Durig, B. M. Gibson, and C. W. Sink, *J. Mol. Struct.*, *2*, 1 (1968).
451. J. R. Durig and C. W. Sink, *Spectrochim. Acta*, *24A*, 575 (1968).
452. K. Licht and P. Koehler, *Z. Anorg. Allg. Chem.*, *383*, 174 (1971).
453. H. Kriegsmann and K. Ulbricht, *Z. Anorg. Allg. Chem.*, *328*, 90 (1964).
454. R. J. H. Clark, A. G. Davies, and R. J. Puddephatt, *Inorg. Chem.*, *8*, 457 (1969).
455. H. Schindlbauer and H. Stenzenberger, *Spectrochim. Acta*, *26A*, 1707 (1970).
456. H. Stenzenberger and H. Schindlbauer, *Spectrochim. Acta*, *26A*, 1713 (1970).
457. J. Goubeau and D. Langhardt, *Z. Anorg. Allg. Chem.*, *338*, 163 (1965).
458. E. Steger and K. Stopperka, *Chem. Ber.*, *94*, 3023 (1961).
459. R. L. Amster and N. B. Colthup, *Spectrochim. Acta*, *19*, 1849 (1963).
460. K. Volka, P. Adamek, H. Schulze, and H. J. Barber, *J. Mol. Struct.*, *21*, 457 (1974).

461. G. L. Kok, *Spectrochim. Acta, 30A,* 961 (1974).
462. R. G. Goel and D. R. Ridley, *Inorg. Chem., 13,* 1252 (1974).
463. J. Mink, G. Végh, and Yu. A. Pentin, *J. Organometal. Chem., 35,* 225 (1972).
464. N. I. Ruch'eva, A. N. Rodion v, I. M. Viktorova, G. V. Zenina, D. N. Shigorin, and N. I. Sheverdina, *Zh. Prikl. Spectrosk., 13,* 322 (1970).
465. A. N. Rodionov, I. E. Paleeva, K. L. Rogozhin, D. N. Shigorin, N. I. Sheverdina, and K. A. Kocheshkov, *Zh. Prikl. Spectrosk., 10,* 797 (1969).
466. H. Kriegsmann and H. Geissler, *Z. Anorg. Allg. Chem., 323,* 170 (1963).
467. G. Costa, A. Camus, N. Marsich, and L. Gatti, *J. Organometal. Chem., 8,* 339 (1967).
468. A. N. Rodionov, N. I. Ruch'eva, I. M. Viktorova, D. N. Shigorin, N. I. Sheverdina, and K. A. Kocheshkov, *Izvest. Akad. Nauk SSSR, Ser. Khim.,*1047 (1969).
469. H. F. Shurvell, *Spectrochim. Acta, 23A,* 2925 (1967).
470. L. A. Harrah, M. T. Ryan, and C. Tamborski, *Spectrochim. Acta, 18,* 21 (1962).
471. M. C. Henry and J. G. Noltes, *J. Amer. Chem. Soc., 82,* 555 (1960).
472. A. L. Smith, *Spectrochim. Acta, 24A,* 695 (1968).
473. R. J. Cross and F. Glockling, *J. Organometal. Chem., 3,* 146 (1965).
474. D. H. Brown, A. Mohammed, and D. W. A. Sharp, *Spectrochim. Acta, 21,* 659 (1965).
475. V. S. Griffiths and G. A. W. Derwish, *J. Mol. Spectrosc., 5,* 148 (1960).
476. P. C. Poller, *Spectrochim. Acta, 22,* 935 (1966).
477. K. Shobatake, C. Postmus, J. R. Ferraro, and K. Nakamoto, *Appl. Spectrosc., 23,* 12 (1969).
478. G. B. Deacon and J. H. S. Green, *Spectrochim. Acta, 24A,* 845 (1968).
479. J. H. S. Green, W. Kynaston, and G. A. Rodley, *Spectrochim. Acta, 24A,* 853 (1968).
480. A. H. Norbury, *Spectrochim. Acta, 26A,* 1635 (1970).
481. K. A. Jensen and P. H. Nielsen, *Acta Chem. Scand., 17,* 1875 (1963).
482. V. Horn and R. Paetzold, *Spectrochim. Acta, 30A,* 1489 (1974).
483. T. N. Srivastava, S. K. Tandon, and K. K. Bajpai, *Spectrochim. Acta, 28A,* 455 (1972).
484. G. B. Deacon and J. H. S. Green, *Spectrochim. Acta, 24A,* 885 (1968).
485. W. Beck and E. Shuierer, *J. Organometal. Chem., 3,* 55 (1965).

486. G. B. Deacon, R. A. Jones, and P. E. Rogasch, *Aust. J. Chem., 16,* 360 (1963).
487. J. B. Orenberg, M. D. Morris, and T. V. Long, II, *Inorg. Chem., 10,* 933 (1971).
488. G. O. Doak, G. G. Long, S. K. Kakar, and L. D. Freedman, *J. Amer. Chem. Soc., 88,* 2342 (1966).
489. R. E. Beaumont and R. G. Goel, *J. Chem. Soc., Dalton Trans.,* 1394 (1973).
490. R. West and W. Glaze, *J. Amer. Chem. Soc., 83,* 3580 (1961).
491. G. Costa, A. Camus, L. Gatti, and N. Marsich, *J. Organometal. Chem., 5,* 568 (1966).
492. D. Seybold and K. Dehnicke, *J. Organometal. Chem., 11,* 1 (1968).
493. F. Zettler, H. D. Hausen, and H. Hess, *J. Organometal. Chem., 72,* 157 (1974).
494. J. F. Malone and W. S. McDonald, *Chem. Commun.,* 444 (1967).
495. J. F. Malone and W. S. McDonald, *Chem. Commun.,* 591 (1969).
496. F. A. Hart, A. G. Massey, and N. S. Saran, *J. Organometal. Chem., 21,* 147 (1970).
497. P. J. Wheatley, *J. Chem. Soc.,* 2206 (1964).
498. P. J. Wheatley, *J. Chem. Soc.,* 3718 (1964).
499. A. L. Beaucha p, M. J. Bennett, and F. A. Cotton, *J. Amer. Chem. Soc., 90,* 6675 (1968).
500. I. R. Beattie, K. M. S. Livingston, G. A. Ozin, and R. Sabine, *J. Chem. Soc., Dalton Trans.,* 784 (1972).
501. A. J. Downs, R. Schmutzler, and I. A. Steer, *Chem. Commun.,* 221 (1966).
502. E. L. Muetterties, W. Mahler, and R. Schmutzler, *Inorg. Chem., 2,* 613 (1963).
503. H. P. Fritz and H. Keller, *Chem. Ber., 94,* 1524 (1961).
504. W. R. McWhinnie and M. G. Patel, *J. Chem. Soc., Dalton Trans.,* 199 (1972).
505. G. J. Kubas and D. F. Shriver, *Inorg. Chem., 9,* 1951 (1970).
506. P. Reich and H. Kriegsmann, *Z. Anorg. Allg. Chem., 334,* 272 (1965).
507. P. Reich and H. Kriegsmann, *Z. Anorg. Allg. Chem., 334,* 283 (1965).
508. J. R. Durig, J. B. Turner, B. M. Gibson, and C. W. Sink, *J. Mol. Struct., 4,* 79 (1969).
509. W. P. Neumann and H. Niermann, *Justus Liebigs Ann. Chem., 653,* 164 (1962).
510. W. P. Neumann and R. Sommer, *Angew. Chem., 75,* 788 (1963).

511. J. H. S. Green and W. Kynaston, *Spectrochim. Acta, 25A,* 1677 (1969).
512. A. E. Borisov, N. V. Novikova, N. A. Chumaevskii, and E. B. Shkirtil, *Ukr. Fiz. Zh., 13,* 75 (1968).
513. P. Braunstein and R. J. H. Clark, *Inorg. Chem., 13,* 2224 (1974).
514. P. W. J. de Graaf, J. Boersma, and G. J. M. Van der Kerk, *J. Organometal. Chem., 105,* 399 (1976).
515. F. C. Nahm, E. F. Rothergy, and K. Niedenzu, *J. Organometal. Chem., 35,* 9 (1972).
516. N. N. Greenwood and J. C. Wright, *J. Chem. Soc.,* 448 (1965).
517. A. N. Rodionov, N. I. Ruch'eva, G. V. Zenina, I. M. Viktorova, V. A. Gribova, N. I. Sheverdina, D. N. Shigorin, and K. A. Kocheshkov, *Zh. Prikl. Spectrosk., 16,* 740 (1972).
518. J. C. Lockhart, *Spectrochim. Acta, 24A,* 1205 (1968).
519. P. G. Perkins and M. E. Twentyman, *J. Chem. Soc.,* 1038 (1965).
520. S. B. Miller, B. L. Jelus, and T. B. Brill, *J. Organometal. Chem., 96,* 1 (1975).
521. H. B. Stegmann, K. B. Ulmschneider, and K. Scheffler, *J. Organometal. Chem., 72,* 41 (1974).
522. A. G. Lee, *J. Organometal. Chem., 22,* 537 (1970).
523. J. R. Mays, W. R. McWhinnie, and R. C. Poller, *Spectrochim. Acta, 27A,* 969 (1971).
524. K. Licht, P. Koehler, and H. Kriegsmann, *Z. Anorg. Allg. Chem., 415,* 31 (1975).
525. R. C. Poller, *J. Inorg. Nucl. Chem., 24,* 593 (1962).
526. M. Mammi, V. Busetti, and A. Del Pra, *Inorg. Chim. Acta, 1,* 419 (1967).
527. A. B. Thomas and E. G. Rochow, *J. Amer. Chem. Soc., 79,* 1843 (1957).
528. I. Wharf, J. Z. Lobos, and M. Onyszchuk, *Can. J. Chem., 48,* 2787 (1970).
529. R. Schmutzler, *Inorg. Chem., 3,* 410 (1964).
530. J. H. S. Green, D. J. Harrison, and H. A. Lauwers, *Bull. Soc. Chim. Belges, 79,* 567 (1970).
531. S. Faleschini, P. Zanella, L. Doretti, and G. Faraglia, *J. Organometal. Chem., 44,* 317 (1972).
532. J. Goubeau, R. Baumgärtner, W. Koch, and U. Müller, *Z. Anorg. Allg. Chem., 337,* 174 (1965).
533. D. M. Revitt and D. B. Sowerby, *Spectrochim. Acta, 26A,* 1581 (1970).
534. O. H. Ellestad, P. Klaeboe, E. E. Tucker, and J. Songstad, *Acta Chem. Scand., 26,* 1721 (1972).
535. R. G. Goel, *Can. J. Chem., 47,* 4607 (1969).

536. G. L. Kok, M. D. Morris, and R. R. Sharp, *Inorg. Chem., 12,* 1709 (1973).
537. R. G. Goel and H. S. Prasad, *Spectrochim. Acta, 32A,* 569 (1976).
538. R. G. Goel, E. Maslowsky, Jr., and C. V. Senoff, *Inorg. Chem., 10,* 2572 (1971).
539. W. C. Smith, *J. Amer. Chem. Soc., 82,* 6176 (1969).
540. G. O. Doak, G. G. Long, and L. D. Freedman, *J. Organometal. Chem., 4,* 82 (1965).
541. R. G. Goel and H. S. Prasad, *Can. J. Chem., 48,* 2488 (1970).
542. G. Faraglia, E. Rivarola, and F. Di Bianca, *J. Organometal. Chem., 38,* 91 (1972).
543. T. N. Polynova and M. A. Porai-Koshits, *Zh. Strukt. Khim., 7,* 742 (1966).
544. D. M. Hawley and G. Ferguson, *J. Chem. Soc. A,* 2539 (1968).
545. N. Bertazzi, *J. Organometal. Chem., 110,* 175 (1976).
546. G. D. Christofferson and J. D. McCullough, *Acta Crystallogr., 11,* 249 (1958).
547. N. S. Dance and W. R. McWhinnie, *J. Chem. Soc., Dalton Trans.,* 43 (1975).
548. W. R. McWhinnie and P. Thavornyutikarn, *J. Chem. Soc., Dalton Trans.,* 551 (1972).
549. N. Petragnani, L. Torres Castellanos, K. J. Wynne, and W. Maxwell, *J. Organometal. Chem., 55,* 295 (1973).
550. T. N. Srivastava and K. K. Bajpai, *J. Inorg. Nucl. Chem., 34,* 1458 (1972).
551. N. Bertazzi, G. C. Stocco, L. Pellerito, and A. Silvestri, *J. Organometal. Chem., 81,* 27 (1974).
552. E. E. Aynsley, N. N. Greenwood, G. Hunter, and M. J. Sprague, *J. Chem. Soc. A,* 1344 (1966).
553. T. N. Srivastava and K. K. Bajpai, *J. Organometal. Chem., 31,* 1 (1971).
554. J. S. Thayer, *Inorg. Chem., 3,* 406 (1964).
555. T. N. Srivastava and S. K. Tandon, *J. Inorg. Nucl. Chem., 30,* 1399 (1968).
556. T. N. Srivastava and S. N. Bhattacharya, *J. Inorg. Nucl. Chem., 28,* 2445 (1966).
557. A. M. Domingos and G. M. Sheldrick, *J. Organometal. Chem., 67,* 257 (1974).
558. M. A. Mullins and C. Curran, *Inorg. Chem., 7,* 2584 (1968).
559. N. Bertazzi, G. Alonzo, A. Silvestri, and G. Consiglio, *J. Organometal. Chem., 37,* 281 (1972).
560. R. Barbieri, N. Bertazzi, C. Tomarchio, and R. H. Herber, *J. Organometal. Chem., 84,* 39 (1975).

561. J. J. Pitts, M. A. Robinson, and S. I. Trotz, *Inorg. Nucl. Chem. Lett.*, *4*, 483 (1968).
562. D. M. Revitt and D. B. Sowerby, *J. Chem. Soc., Dalton Trans.*, 847 (1972).
563. T. Wizemann, N. Müller, D. Seybold, and K. Dehnicke, *J. Organometal. Chem.*, *20*, 211 (1969).
564. O. H. Ellestad, P. Klaeboe, and J. Songstad, *Acta Chem. Scand.*, *26*, 1724 (1972).
565. R. G. Goel and D. R. Ridley, *Inorg. Nucl. Chem. Lett.*, *7*, 21 (1971).
566. R. G. Goel and H. S. Prasad, *J. Organometal. Chem.*, *50*, 129 (1973).
567. R. G. Goel and H. S. Prasad, *J. Organometal. Chem.*, *36*, 323 (1972).
568. G. Ferguson, R. G. Goel, F. C. March, D. R. Ridley, and H. S. Prasad, *J. Chem. Soc. D*, 1547 (1971).
569. H. Müller and K. Dehnicke, *J. Organometal. Chem.*, *10*, P1 (1967).
570. J. H. Thayer, *Organometal. Chem. Rev.*, *1*, 157 (1966).
571. P. I. Paetzold, *Z. Anorg. Allg. Chem.*, *326*, 53 (1963).
572. E. Lieber, C. N. R. Rao, and F. M. Keane, *J. Inorg. Nucl. Chem.*, *25*, 631 (1963).
573. M. F. Lappert and H. Pyszora, *J. Chem. Soc. A*, 854 (1967).
574. T. N. Srivastava and S. K. Tandon, *Spectrochim. Acta*, *27A*, 593 (1971).
575. E. E. Aynsley, N. N. Greenwood, and M. J. Sprague, *J. Chem. Soc.*, 2395 (1965).
576. M. V. Kashutina and O. Y. Okhlobystin, *J. Organometal. Chem.*, *9*, 5 (1967).
577. W. Beck and E. Schuierer, *Chem. Ber.*, *97*, 3517 (1964).
578. R. Nast, J. Bülck, and R. Kramolowsky, *Chem. Ber.*, *108*, 3461 (1975).
579. K. Licht and H. Kriegsmann, *Z. Anorg. Allg. Chem.*, *323*, 190 (1963).
580. B. Armer and H. Schmidbaur, *Chem. Ber.*, *100*, 1521 (1967).
581. B. Kushlefsky, I. Simmons, and A. Ross, *Inorg. Chem.*, *2*, 187 (1963).
582. R. A. Cummins, *Aust. J. Chem.*, *18*, 98 (1965).
583. D. H. Lohmann, *J. Organometal. Chem.*, *4*, 382 (1965).
584. V. I. Gol'danskii, E. F. Makarov, R. A. Stukan, V. A. Trukhtanov, and V. V. Khrapov, *Dokl. Akad. Nauk SSSR*, *151*, 357 (1963).
585. V. A. Yablokov, S. Ya. Khorshev, A. P. Tarabarina, and A. N. Sunis, *Zh. Obshch. Khim.*, *43*, 607 (1973).

586. G. B. Deacon and J. H. S. Green, *Spectrochim. Acta, 25A,* 355 (1969).
587. J. Bernstein, M. Halmann, S. Pinchas, and D. Samuel, *J. Chem. Soc.*, 821 (1964).
588. S. Miličev, *Spectrochim. Acta, 30A,* 255 (1974).
589. D. N. Shuly and W. E. McEwen, Ph. D. Thesis, University of Massachusetts, Amherst, Massachusetts, February, 1971.
590. D. L. Venezky, C. W. Sink, B. A. Nevett, and W. F. Fortescue, *J. Organometal. Chem., 35,* 131 (1972).
591. H. Schumann, P. Jutzi, A. Roth, P. Schwabe, and E. Schauer, *J. Organometal. Chem., 10,* 71 (1967).
592. R. L. McKenney and H. H. Sisler, *Inorg. Chem., 6,* 1178 (1967).
593. H. Schumann and P. Reich, *Z. Anorg. Allg. Chem., 377,* 63 (1970).
594. P. G. Harrison and S. R. Stobart, *J. Organometal. Chem., 47,* 89 (1973).
595. H. Schumann, *Z. Anorg. Allg. Chem., 354,* 192 (1967).
596. R. A. Zingaro, R. E. McGlothlin, and R. M. Hedges, *Trans. Faraday Soc., 59,* 798 (1963).
597. G. O. Doak, G. G. Long, and L. D. Freedman, *J. Organometal. Chem., 12,* 443 (1968).
598. V. Horn and R. Paetzold, *Z. Anorg. Allg. Chem., 398,* 173 (1973).
599. R. Paetzold and U. Lindner, *Z. Anorg. Allg. Chem., 350,* 295 (1967).
600. V. Horn and R. Paetzold, *Z. Anorg. Allg. Chem., 398,* 179 (1973).
601. G. A. Razuvaev, V. N. Latyaeva, L. I. Vishinskaya, and A. M. Rabinovitch, *J. Organometal. Chem., 49,* 441 (1973).
602. Y. L. Fan and R. G. Shaw, *J. Org. Chem., 38,* 2410 (1973).
603. H. Vahrenkamp, *J. Organometal. Chem., 28,* 181 (1971).
604. A. Finch, J. Pearn, and D. Steele, *Trans. Faraday Soc., 62,* 1688 (1966).
605. S. Sergi, V. Marsala, R. Pietropaolo, and F. Faraone, *J. Organometal. Chem., 23,* 281 (1970).
606. H. Kurosawa and R. Okawara, *J. Organometal. Chem., 19,* 253 (1969).
607. B. F. E. Ford, B. V. Liengme, and J. R. Sams, *J. Organometal. Chem., 19,* 53 (1969).
608. A. D. Cohen and C. R. Dillard, *J. Organometal. Chem., 25,* 421 (1970).
609. M. A. Mesubi, *Spectrochim. Acta, 32A,* 1327 (1976).
610. G. Plazzogna, V. Peruzzo, and G. Tagliavini, *J. Organometal. Chem., 24,* 667 (1970).

611. H. Schmidbaur and H. Stuhler, *Angew. Chem., Inter. Ed. Engl., 11,* 145 (1972).
612. R. E. Beaumont and R. G. Goel, *Inorg. Nucl. Chem. Lett., 8,* 989 (1972).
613. B. C. Pant, W. R. McWhinnie, and N. S. Dance, *J. Organometal. Chem., 63,* 305 (1973).
614. N. Dance and W. R. McWhinnie, *J. Organometal. Chem., 104,* 317 (1976).
615. A. T. T. Hsieh, *J. Organometal. Chem., 27,* 293 (1971).
616. A. T. T. Hsieh and G. B. Deacon, *J. Organometal. Chem., 70,* 39 (1974).
617. E. R. Lippincott and F. E. Welsh, *Spectrochim. Acta, 17,* 123 (1961).
618. G. B. Deacon, J. H. S. Green, and R. S. Nyholm, *J. Chem. Soc.,* 3411 (1965).
619. A. N. Fenster and E. I. Becker, *J. Organometal. Chem., 11,* 549 (1968).
620. U. Kunze and H. P. Völker, *Chem. Ber., 107,* 3818 (1974).
621. R. G. Goel, H. S. Prasad, G. M. Bancroft, and T. K. Sham, *Can. J. Chem., 54,* 711 (1976).
622. G. C. Tranter, C. C. Addison, and D. B. Sowerby, *J. Organometal. Chem , 12,* 369 (1968).
623. R. G. Goel and H. S. Prasad, *Inorg. Chem., 11,* 2141 (1972).
624. R. G. Goel, P. N. Joshi, D. R. Ridley, and R. E. Beaumont, *Can. J. Chem., 47,* 1423 (1969).
625. R. G. Goel and H. S. Prasad, *Can. J. Chem., 49,* 2529 (1971).
626. R. C. Poller, J. N. R. Ruddick, M. Thevarasa, and W. R. McWhinnie, *J. Chem. Soc. A,* 2327 (1969).
627. R. C. Poller and D. L. B. Toley, *J. Chem. Soc. A,* 1578 (1967).
628. K. L. Jaura and K. R. Sharma, *J. Indian Chem. Soc., 48,* 965 (1971).
629. V. G. K. Das and W. Kitching, *J. Organometal. Chem., 13,* 523 (1968).
630. L. Pellerito, R. Cefalù, A. Gianguzza, and R. Barbieri, *J. Organometal. Chem., 70,* 303 (1974).
631. R. Cefalù, R. Bosco, F. Bonati, F. Maggio, and R. Barbieri, *Z. Anorg. Allg. Chem., 376,* 180 (1970).
632. J. N. R. Ruddick and J. R. Sams, *J. Organometal. Chem., 60,* 233 (1973).
633. E. S. Bretschneider and C. W. Allen, *Inorg. Chem., 12,* 623 (1973).
634. M. Schmidt, H. Schumann, F. Gliniecki, and J. F. Jaggard, *J. Organometal. Chem., 17,* 277 (1969).
635. H. A. Meinema, A. Mackor, and J. G. Noltes, *J. Organometal. Chem., 37,* 285 (1972).

636. J. Kroon, J. B. Hulscher, and A. F. Peerdeman, *J. Organometal. Chem.*, *37*, 297 (1972).
637. H. A. Meinema and J. G. Noltes, *J. Organometal. Chem.*, *16*, 257 (1969).
638. N. Nishii, Y. Matsumura, and R. Okawara, *Inorg. Nucl. Chem. Lett.*, *5*, 703 (1969).
639. H. A. Meinema, E. Rivarola, and J. G. Noltes, *J. Organometal. Chem.*, *17*, 71 (1969).
640. N. Nishi, Y. Matsumura, and R. Okawara, *J. Organometal. Chem.*, *30*, 59 (1971).
641. R. E. Beaumont, R. G. Goel, and H. S. Prasad, *Inorg. Chem.*, *12*, 944 (1973).
642. F. Glockling and K. A. Hooton, *J. Chem. Soc.*, 1849 (1963).
643. H. M. Gager, J. Lewis, and M. J. Ware, *Chem. Commun.*, 616 (1966).
644. P. A. Bulliner, C. O. Q icksall, and T. G. Spiro, *Inorg. Chem.*, *10*, 13 (1971).
645. A. H. Cowley and W. D. White, *Spectrochim. Acta*, *22*, 1431 (1966).
646. R. L. Amster, W. A. Henderson, and W. B. Colthup, *Can. J. Chem.*, *42*, 2577 (1964).
647. W. A. Henderson, Jr., M. Epstein, and F. S. Seichter, *J. Amer. Chem. Soc.*, *85*, 2462 (1963).
648. W. R. McWhinnie and P. Thavornyutikarn, *J. Organometal. Chem.*, *35*, 149 (1972).
649. M. C. Baird, *J. Inorg. Nucl. Chem.*, *29*, 367 (1968).
650. G. Engelhardt, P. Reich, and H. Schumann, *Z. Naturforsch.*, *22b*, 352 (1967).
651. U. Belluco, G. Deganello, R. Pietropaolo, and P. Uguagliati, *Inorg. Chim. Acta*, *4*, 7 (1970).
652. N. A. D. Carey and H. C. Clark, *Chem. Commun.*, 292 (1967).
653. N. A. D. Carey and H. C. Clark, *Inorg. Chem.*, *7*, 94 (1968).
654. G. B. Deacon and J. H. S. Green, *Chem. Commun.*, 629 (1966).
655. J. Bradbury, K. P. Forest, R. H. Nuttall, and D. W. A. Sharp, *Spectrochim. Acta*, *23A*, 2701 (1967).
656. A. Balls, N. N. Greenwood, and B. P. Straughan, *J. Chem. Soc. A*, 753 (1968).
657. R. Rivest, S. Singh, and C. Abraham, *Can. J. Chem.*, 3138 (1967).
658. K. Shobatake and K. Nakamoto, *J. Amer. Chem. Soc.*, *92*, 3332 (1970).
659. J. T. Wang, C. Udovich, K. Nakamoto, A. Quattrochi, and J. R. Ferraro, *Inorg. Chem.*, *9*, 2675 (1970).

660. W. F. Edgell and M. P. Dunkle, *Inorg. Chem.*, *4*, 1629 (1965).
661. A. G. Jones and D. P. Powell, *Spectrochim. Acta*, *30A*, 563 (1974).
662. C. W. Bradford, W. van Bronswijk, R. J. H. Clark, and R. S. Nyholm, *J. Chem. Soc. A*, 2889 (1970).
663. S. Singh, P. P. Singh, and R. Rivest, *Inorg. Chem.*, *7*, 1236 (1968).
664. G. C. Stocco and R. S. Tobias, *J. Amer. Chem. Soc.*, *93*, 5057 (1971).
665. H. Burger and H.-J. Neese, *J. Organometal. Chem.*, *32*, 223 (1971).
666. J. T. Vandeberg, C. E. Moore, and F. P. Cassaretto, *Spectrochim. Acta*, *27A*, 501 (1971).
667. A. McKillop, J. D. Hunt, and E. C. Taylor, *J. Organometal. Chem.*, *24*, 77 (1970).
668. A. G. Jones and D. B. Powell, *Spectrochim. Acta*, *30A*, 1001 (1974).
669. F. T. Delbeke, R. De Ketelaere, and G. P. Van der Kelen, *J. Organometal. Chem.*, *28*, 225 (1971).
670. R. F. De Ketelaere and G. P. Van der Kelen, *J. Organometal. Chem.*, *73*, 251 (1974).
671. R. F. De Ketelaere and G. P. Van der Kelen, *J. Mol. Struct.*, *23*, 233 (1974).
672. M. I. Kabachnik, I. G. Malakhov, E. N. Tsvetrov, K. F. Johnson, A. R. Katritzky, A. J. Sparrow, and R. D. Topsom, *Aust. J. Chem.*, *28*, 755 (1975).
673. T. B. Grindley, K. F. Johnson, H. Kaack, A. R. Katritzky, G. P. Schiemenz, and R. D. Topsom, *Aust. J. Chem.*, *28*, 327 (1975).
674. S. Abramowitz and I. W. Levin, *Spectrochim. Acta*, *26A*, 2261 (1970).
675. P. Delorme, F. Denisselle, and V. Lorenzelli, *J. Chim. Phys.*, *64*, 591 (1967).
676. D. Steele and D. H. Whiffen, *Trans. Faraday Soc.*, *55*, 369 (1959).
677. J. P. Marsault and F. Marsault-Herail, *C. R. Acad. Sci. Paris, Ser. B*, *723*, 723 (1971).
678. C. F. Smith, G. J. Moore, and C. Tamborski, *J. Organometal. Chem.*, *33*, C21 (1971).
679. C. F. Smith, G. J. Moore, and C. Tamborski, *J. Organometal. Chem.*, *42*, 257 (1972).
680. G. B. Deason and P. W. Felder, *Aust. J. Chem.*, *20*, 1587 (1967).
681. G. B. Deacon and P. W. Felder, *J. Chem. Soc. C*, 2313 (1967).
682. K. Shiina, T. Brennan, and D. Gilman, *J. Organometal. Chem.*, *11*, 471 (1968).

683. M. D. Rausch, F. E. Tibbetts, and H. B. Gordon, *J. Organometal. Chem.*, *5*, 493 (1966).
684. M. D. Rausch and F. E. Tibbetts, *Inorg. Chem.*, *9*, 512 (1970).
685. M. D. Rausch and F. E. Tibbetts, *J. Organometal. Chem.*, *21*, 487 (1970).
686. J. Casabo, J. M. Coronas, and J. Sales, *Inorg. Chim. Acta*, *11*, 5 (1974).
687. M. Wada, *Inorg. Chem.*, *14*, 1415 (1975).
688. G. B. Deacon and J. C. Parrott, *Aust. J. Chem.*, *24*, 1771 (1971).
689. G. B. Deacon and J. H. S. Green, *Spectrochim. Acta*, *24A*, 1125 (1968).
690. M. Fild, I. Hollenberg, and O. Glemser, *Z. Naturforsch.*, *22b*, 248 (1967).
691. B. A. Nevett and A. Perry, *Spectrochim. Acta*, *31A*, 101 (1975).
692. A. G. Massey and A. J. Park, *J. Organometal. Chem.*, *2*, 245 (1964).
693. R. D. Chambers, G. E. Coates, J. G. Livingston, and W. K. R. Musgrave, *J. Chem. Soc. C*, 4367 (1962).
694. D. E. Fenton, A. G. Massey, and D. S. Urch, *J. Organometal. Chem.*, *6*, 352 (1966).
695. R. D. Chambers and T. Chivers, *J. Chem. Soc.*, 4782 (1964).
696. D. E. Fenton and A. G. Massey, *J. Inorg. Nucl. Chem.*, *27*, 329 (1965).
697. M. Fild, O. Glemser, and G. Christoph, *Angew. Chem., Inter. Ed. Engl.*, *4*, 801 (1964).
698. G. B. Deacon and I. K. Johnston, *Inorg. Nucl. Chem. Lett.*, *8*, 271 (1972).
699. M. N. Bochkarev, L. P. Maiorova, S. P. Korneva, L. N. Bochkarev, and N. S. Vyazankin, *J. Organometal. Chem.*, *73*, 229 (1974).
700. G. B. Deacon, J. H. S. Green, and W. Kynaston, *J. Chem. Soc. A*, 158 (1967).
701. D. H. Brown, A. Mohammed, and D. W. A. Sharp, *Spectrochim. Acta*, *21*, 1013 (1965).
702. G. B. Deacon, *Aust. J. Chem.*, *20*, 459 (1967).
703. A. J. Canty and G. B. Deacon, *Aust. J. Chem.*, *24*, 489 (1971).
704. G. B. Deacon and J. C. Parrott, *Aust. J. Chem.*, *25*, 1169 (1972).
705. G. B. Deacon and J. C. Parrott, *Aust. J. Chem.*, *27*, 2547 (1974).
706. G. B. Deacon and R. S. Nyholm, *J. Chem. Soc.*, 6107 (1965).

707. A. J. Oliver and W. A. G. Graham, *J. Organometal. Chem.*, *19*, 17 (1969).
708. M. Weidenbruch and N. Wessal, *Chem. Ber.*, *105*, 173 (1972).
709. M. N. Bochkarev, L. P. Maiorova, N. S. Vyazankin, and G. A. Razuvaev, *J. Organometal. Chem.*, *82*, 65 (1974).
710. M. D. Rausch, H. B. Gordon, and E. Samuel, *Coord. Chem.*, *1*, 141 (1971).
711. G. A. Razuvaev, V. N. Latyaeva, A. N. Lineva, and N. N. Spiridonova, *J. Organometal. Chem.*, *46*, C13 (1972).
712. E. Kinsella, V. B. Smith, and A. G. Massey, *J. Organometal. Chem.*, *34*, 181 (1972).
713. P. M. Treichel, M. A. Chaudhari, and F. G. A. Stone, *J. Organometal. Chem.*, *1*, 98 (1963).
714. J. R. Phillips, D. T. Rosevear, and F. G. A. Stone, *J. Organometal. Chem.*, *2*, 455 (1964).
715. R. Usón, J. Fornies, J. Gimeno, P. Espinet, and R. Navarro, *J. Organometal. Chem.*, *81*, 115 (1974).
716. R. Usón, J. Fornies, and S. Gonzalo, *J. Organometal. Chem.*, *104*, 253 (1976).
717. R. Usón, P. Royo, and A. Laguna, *Rev. Acad. Ciencias Zaragosa*, *27*, 19 (1972).
718. G. B. Deacon and D. G. Vince, *J. Organometal. Chem.*, *112*, C1 (1976).
719. D. M. Roe and A. G. Massey, *J. Organometal. Chem.*, *23*, 547 (1970).
720. M. J. Bennett, W. A. G. Graham, R. P. Stewart, Jr., and R. M. Tuggle, *Inorg. Chem.*, *12*, 2944 (1973).
721. H. P. Fritz, W. Lüttke, H. Stammreich, and R. Forneris, *J. Chem. Soc. A*, 1068 (1961).
722. R. G. Snyder, *Spectrochim. Acta*, *15*, 807 (1959).
723. H. P. Fritz, W. Lüttke, H. Stammreich, and R. Forneris, *Chem. Ber.*, *92*, 3246 (1959).
724. H. P. Fritz and E. O. Fischer, *J. Organometal. Chem.*, *7*, 121 (1967).
725. L. H. Ngai, F. E. Stafford, and L. Schäfer, *J. Amer. Chem. Soc.*, *91*, 48 (1969).
726. L. Schäfer, J. F. Southern, and S. J. Cyvin, *Spectrochim. Acta*, *27A*, 1083 (1971).
727. S. J. Cyvin, J. Brunvoll, and L. Schäfer, *J. Chem. Phys.*, *54*, 1517 (1971).
728. J. T. S. Andrews, E. F. Westrum, Jr., and N. Bjerrum, *J. Organometal. Chem.*, *17*, 293 (1969).
729. L. Schäfer, J. F. Southern, S. J. Cyvin, and J. Brunvoll, *J. Organometal. Chem.*, *24*, C13 (1970).
730. E. Keulen and F. Jellinek, *J. Organometal. Chem.*, *5*, 490 (1966).

731. J. W. Boyd, J. M. Lavoie, and D. M. Gruen, *J. Chem. Phys., 60,* 4088 (1974).
732. H. Saito, Y. Kakiuti, and M. Tsutsui, *Spectrochim. Acta, 23A,* 3013 (1967).
733. E. O. Fischer, H. P. Fritz, J. Manchot, E. Priebe, and R. Schneider, *Chem. Ber., 96,* 1418 (1963).
734. G. Varsányi with a contribution from S. Szöke, *Vibrational Spectra of Benzene Derivatives,* Academic Press, New York, 1969.
735. E. B. Wilson, Jr., *Phys. Rev., 45,* 706 (1934).
736. M. T. Anthony, M. L. H. Green, and D. Young, *J. Chem. Soc., Dalton Trans.,* 1420 (1975).
737. J. F. Helling, S. L. Rice, D. M. Braitsch, and T. Mayer, *Chem. Commun.,* 930 (1971).
738. S. J. Cyvin, B. N. Cyvin, J. Brunvoll, and L. Schäfer, *Acta Chem. Scand., 24,* 3420 (1970).
739. J. Brunvoll, S. J. Cyvin, and L. Schäfer, *J. Organometal. Chem., 27,* 69 (1971).
740. D. M. Adams, R. E. Christopher, and D. C. Stevens, *Inorg. Chem., 14,* 1562 (1975).
741. R. A. Zelonka and M. C. Baird, *Can. J. Chem., 50,* 3063 (1972).
742. R. A. Work, III and R. L. McDonald, *Inorg. Chem., 12,* 1936 (1973).
743. E. Kinsella, J. Chadwick, and J. Coward, *J. Chem. Soc. A,* 969 (1968).
744. L. W. Daasch, *Spectrochim. Acta, 15,* 726 (1959).
745. Sh. Sh. Raskin, *Dokl. Akad. Nauk SSSR, 15,* 485 (1955).
746. Sh. Sh. Raskin, *Opt. Spectrosk., 1,* 516 (1956).
747. A. T. Kozulin, *Opt. Spectrosk., 18,* 337 (1965).
748. H. G. Smith and R. E. Rundle, *J. Amer. Chem. Soc., 80,* 5075 (1958).
749. Sh. Sh. Raskin, *Dokl. Akad. Nauk SSSR, 141,* 900 (1961).
750. R. W. Turner and E. L. Amma, *J. Amer. Chem. Soc., 88,* 3243 (1966).
751. R. W. Turner and E. L. Amma, *J. Amer. Chem. Soc., 88,* 1877 (1966).
752. C. Sourisseau and B. Pasquier, *Spectrochim. Acta, 26A,* 1279 (1970).
753. M. Cesari, U. Pedretti, A. Zazzatta, G. Lugli, and W. Marconi, *Inorg. Chim. Acta, 5,* 439 (1971).
754. R. E. Humphrey, *Spectrochim. Acta, 17,* 93 (1961).
755. H. P. Fritz and J. Manchot, *Spectrochim. Acta, 18,* 171 (1962).
756. H. P. Fritz and J. Manchot, *Z. Naturforsch., 17b,* 711 (1962).
757. H. J. Buttery, G. Keeling, S. F. A. Kettle, I. Paul, and P. J. Stamper, *Discuss. Faraday Soc., 47,* 48 (1969).

758. H. J. Buttery, G. Keeling, S. F. A. Keeling, I. Paul, and P. J. Stamper, *J. Chem. Soc. A*, 2077 (1969).
759. D. M. Adams and A. Squire, *J. Chem. Soc. A*, 814 (1970).
760. G. Davidson and E. M. Riley, *Spectrochim. Acta*, *27A*, 1649 (1971).
761. L. Schäfer, G. M. Begun, and S. J. Cyvin, *Spectrochim. Acta*, *28A*, 803 (1972).
762. I. J. Hyams and E. R. Lippincott, *Spectrochim. Acta*, *28A*, 1741 (1972).
763. M. Bigorgne, O. Kahn, M. F. Koenig, and A. Loutellier, *Spectrochim. Acta*, *31A*, 741 (1975).
764. S. Kjelstrup, S. J. Cyvin, J. Brunvoll, and L. Schäfer, *J. Organometal. Chem.*, *36*, 137 (1972).
765. J. Brunvoll, S. J. Cyvin, and L. Schäfer, *J. Organometal. Chem.*, *36*, 143 (1972).
766. M. F. Bailey and L. F. Dahl, *Inorg. Chem.*, *4*, 1314 (1965).
767. R. Middleton, J. R. Hull, S. R. Simpson, C. H. Tomlinson, and P. L. Timms, *J. Chem. Soc., Dalton Trans.*, 120 (1973).
768. G. Davidson and E. M. Riley, *J. Organometal. Chem.*, *19*, 101 (1969).
769. H. J. Buttery, G. Keeling, S. F. A. Kettle, I. Paul, and P. J. Stamper, *J. Chem. Soc. A*, 2224 (1969).
770. H. J. Buttery, G. Keeling, S. F. A. Kettle, I. Paul, and P. J. Stamper, *J. Chem. Soc. A*, 471 (1970).
771. H. J. Buttery, S. F. A. Kettle, and I. Paul, *J. Chem. Soc., Dalton Trans.*, 969 (1975).
772. T. B. Brill and A. J. Kotlar, *Inorg. Chem.*, *13*, 470 (1974).
773. R. D. Fischer, *Chem. Ber.*, *93*, 165 (1960).
774. P. Caillet, *J. Organometal. Chem.*, *102*, 481 (1975).
775. E. O. Fischer and F. Röhrscheid, *J. Organometal. Chem.*, *6*, 53 (1966).
776. E. O. Fischer and H. H. Lindner, *J. Organometal. Chem.*, *2*, 222 (1964).
777. E. O. Fischer and C. G. Kreiter, *J. Organometal. Chem.*, *14*, P25 (1968).
778. E. O. Fischer and H. H. Lindner, *J. Organometal. Chem.*, *1*, 307 (1964).
779. R. Hüttel, H. Reinheimer, and H. Dietl, *Chem. Ber.*, *99*, 462 (1966).
780. R. Burton, L. Pratt, and G. Wilkinson, *J. Chem. Soc.*, 594 (1961).

781. E. O. Fischer, A. Reckziegel, J. Müller, and P. Goser, *J. Organometal. Chem.*, *11*, P13 (1968).
782. J. Müller and B. Mertschenk, *J. Organometal. Chem.*, *34*, C41 (1972).
783. J. Müller and E. O. Fischer, *J. Organometal. Chem.*, *5*, 275 (1966).
784. H. J. Dauben, Jr. and L. R. Honnen, *J. Amer. Chem. Soc.*, *80*, 5570 (1958).
785. J. Müller, P. Göser, and P. Laubereau, *J. Organometal. Chem.*, *14*, P7 (1968).
786. H. O. van Oven and H. J. de Liefde Meijer, *J. Organlmetal. Chem.*, *23*, 159 (1970).
787. H. O. van Oven, C. J. Groenboom, and H. J. de Liefde Meijer, *J. Organometal. Chem.*, *81*, 379 (1974).
788. H. W. Wehner, E. O. Fischer, and J. Müller, *Chem. Ber.*, *103*, 2258 (1970).
789. R. B. King and M. B. Bisnette, *Inorg. Chem.*, *3*, 785 (1964).
790. H. P. Fritz and C. G. Kreiter, *Chem. Ber.*, *97*, 1398 (1964).
791. B. L. Kalsotra, R. K. Multani, and B. D. Jain, *J. Organometal. Chem.*, *31*, 67 (1971).
792. K. M. Sharma, S. K. Anand, R. K. Multani, and B. D. Jain, *J. Organometal. Chem.*, *23*, 173 (1970).
793. H. A. Tayim and A. Vassilian, *Inorg. Nucl. Chem. Lett.*, *8*, 215 (1972).
794. H. A. Tayim and A. Vassilian, *Inorg. Nucl. Chem. Lett.*, *8*, 659 (1972).
795. D. B. Powell and T. J. Leedham, *Spectrochim. Acta*, *28A*, 337 (1972).
796. T. J. Leedham, D. B. Powell, and J. G. V. Scott, *Spectrochim. Acta*, *29A*, 559 (1973).
797. C. Krüger, *J. Organometal. Chem.*, *22*, 697 (1970).
798. L. Porri, G. Vitulli, and M. C. Gallazzi, *Angew. Chem., Inter. Ed. Engl.*, *6*, 452 (1967).
799. P. Fou eroux, B. Denise, R. Bonnaire, and G. Pannetier, *J. Organometal. Chem.*, *60*, 375 (1973).
800. D. Brodzki and G. Pannetier, *J. Organometal. Chem.*, *63*, 431 (1973).
801. G. Pannetier, R. Bonnaire, and P. Fougeroux, *J. Organometal. Chem.*, *30*, 411 (1971).
802. J. Müller and P. Göser, *Angew. Chem., Inter. Ed. Engl.*, *6*, 364 (1967).
803. E. O. Fischer and J. Müller, *Z. Naturforsch.*, *18b*, 1137 (1963).
804. C. Palm and H. P. Fritz, *Chem. Ber.*, *92*, 2645 (1959).

805. F. S. Mathews and W. N. Lipscomb, *J. Phys. Chem.*, *63*, 845 (1959).
806. H. P. Fritz and H. Keller, *Chem. Ber.*, *95*, 158 (1962).
807. N. Kumar and R. K. Multani, *J. Organometal. Chem.*, *63*, 47 (1973).
808. M. A. Bennett and J. D. Saxby, *Inorg. Chem.*, *7*, 321 (1968).
809. J. R. Doyle, J. H. Hutchinson, N. C. Baenziger, and L. W. Tresselt, *J. Amer. Chem. Soc.*, *83*, 2768 (1961).
810. B. Dickens and W. N. Lipscomb, *J. Chem. Phys.*, *37*, 2084 (1962).
811. R. T. Bailey, E. R. Lippincott, and D. Steele, *J. Amer. Chem. Soc.*, *87*, 5346 (1965).
812. J. D. McKechnie and I. C. Paul, *J. Amer. Chem. Soc.*, *88*, 5927 (1966).
813. A. Streitwieser, Jr. and U. Müller-Westerhoff, *J. Amer. Chem. Soc.*, *90*, 7364 (1968).
814. A. Avdeef, K. N. Raymond, K. O. Hodgson, and A. Zalkin, *Inorg. Chem.*, *11*, 1083 (1972).
815. L. Hocks, J. Goffart, G. Duyckaerts, and P. Teyssié, *Spectrochim. Acta*, *30A*, 907 (1974).
816. J. Goffart, J. Fuger, B. Gilbert, B. Kanellakopulos, and G. Duyckaerts, *Inorg. Nucl. Chem. Lett.*, *8*, 403 (1972).
817. J. Goffart, J. Fuger, D. Brown, and G. Duychaerts, *Inorg. Nucl. Chem. Lett.*, *10*, 413 (1974).
818. A. Streitwieser, Jr. and C. A. Harmon, *Inorg. Chem.*, *12*, 1102 (1973).
819. A. Streitwieser, Jr., U. Muller-Westerhoff, G. Sonnichsen, F. Mares, D. G. Morrell, K. O. Hodgson, and C. A. Harmon, *J. Amer. Chem. Soc.*, *95*, 8644 (1973).
820. D. G. Karraker, J. A. Stone, E. R. Jones, Jr., and N. Edelstein, *J. Amer. Chem. Soc.*, *92*, 4841 (1973).
821. K. O. Hodgson, F. Mares, D. F. Starks, and A. Streitwieser, Jr., *J. Amer. Chem. Soc.*, *95*, 8650 (1973).
822. F. Mares, K. Hodgson, and A. Streitwieser, Jr, *J. Organometal. Chem.*, *24*, C68 (1970).
823. K. O. Hodgson and K. N. Raymond, *Inorg. Chem.*, *11*, 3030 (1972).
824. K. O. Hodgson and K. N. Raymond, *Inorg. Chem.*, *11*, 171 (1972).
825. B. L. Kalsotra, R. K. Multani, and B. D. Jain, *Chem. Ind (London)*, 339 (1972).
826. H. Dietrich and M. Solwisch, *Angew. Chem., Inter. Ed. Engl.*, *8*, 765 (1969).
827. G. Nagendrappa and D. Devaprabhakara, *J. Organometal. Chem.*, *15*, 225 (1968).

828. G. Nagendrappa and D. Devaprabhakara, *J. Organometal. Chem.*, *17*, 182 (1969).
829. G. Nagendrappa and D. Devaprabhakara, *J. Organometal. Chem.*, *16*, P73 (1969).
830. G. Nagendrappa, G. C. Joshi, and D. Devaprabhakara, *J. Organometal. Chem.*, *27*, 421 (1971).
831. K. G. Untch and D. J. Martin, *J. Org. Chem.*, *29*, 1903 (1964).
832. R. B. Jackson and W. E. Streib, *J. Amer. Chem. Soc.*, *89*, 2539 (1967).
833. J. C. Trebellas, J. R. Olechowski, and H. B. Jonassen, *Inorg. Chem.*, *4*, 1818 (1965).
834. J. C. Trebellas, J. R. Olechowski, and H. B. Jonassen, *J. Organometal. Chem.*, *6*, 412 (1966).
835. J. G. Traynham, *J. Org. Chem.*, *26*, 4694 (1961).
836. N. C. Baenziger, H. L. Haight, R. Alexander, and J. R. Doyle, *Inorg. Chem.*, *5*, 1399 (1966).
837. B. W. Cook, R. G. J. Miller, and P. F. Todd, *J. Organometal. Chem.*, *19*, 421 (1969).
838. E. W. Abel, M. A. Bennett, and G. Wilkinson, *J. Chem. Soc.*, 3178 (1959).
839. I. A. Zakharova, Ya. V. Salyn, I. A. Garbouzova, V. A. Aleksanyan, and M. A. Prianichnicova, *J. Organometal. Chem.*, *102*, 227 (1975).
840. D. F. Hunt, C. P. Lillya, and M. D. Rausch, *Inorg. Chem.*, *8*, 446 (1968).
841. G. Winkhaus, M. Kricke, and H. Singer, *Z. Naturforsch.*, *22b*, 893 (1967).
842. E. O. Fischer, C. G. Kreiter, and W. Berngruber, *Angew. Chem., Inter. Ed. Engl.*, *6*, 634 (1967).
843. E. O. Fischer, W. Berngruber, and C. G. Kreiter, *Chem. Ber.*, *101*, 824 (1968).
844. B. L. Shaw and G. Shaw, *J. Chem. Soc. A*, 1560 (1969).
845. B. L. Shaw and G. Shaw, *J. Chem. Soc. A*, 602 (1969).
846. J. R. Doyle and H. B. Jonassen, *J. Amer. Chem. Soc.*, *78*, 3965 (1956).
847. J. Chatt, L. M. Vallarino, and L. M. Venanzi, *J. Chem. Soc.*, 2496 (1957).
848. M. Rosenblum and B. North, *J. Amer. Chem. Soc.*, *90*, 1060 (1968).
849. E. O. Fischer and R. D. Fischer, *Z. Naturforsch.*, *16b*, 556 (1961).
850. H. O. van Oven and H. J. de Liefde Meijer, *J. Organometal. Chem.*, *19*, 373 (1969).
851. H. W. Wehner, E. O. Fischer, and J. Müller, *Chem. Ber.*, *103*, 2258 (1970).
852. E. O. Fischer and J. Müller, *J. Organometal. Chem.*, *1*, 464 (1964).

Index

Numbers in italic indicate figures.